DIVERSITY OF ECOSYSTEMS

Edited by **Mahamane Ali**

INTECHOPEN.COM

Diversity of Ecosystems
Edited by Mahamane Ali

CBS, Edition **2016**

Published by InTech
Janeza Trdine 9, 51000 Rijeka, Croatia

Notice

Publishing Process Manager Jana Sertic

Technical Editor Teodora Smiljanic

Cover Designer InTech Design Team

Additional hard copies can be obtained from orders@intechopen.com

Diversity of Ecosystems, Edited by Mahamane Ali
p. cm.
ISBN 978-953-51-0572-5

INTECH
open science | open minds

Contents

Preface

If we unanimously agree that the survival of each living organism depends on the nature and ecological services rendered by ecosystems, we must equally agree unanimously that these ecosystems do not always benefit the required attention. Hence they are subjected to many treats. Indeed, the ecosystems are subjected to many pressing usages at unknown tolerable levels. It generally results to equilibrium breaks leading ineluctably to their degradation. This tendency has engendered many international initiatives to prevent ecosystems degradation. For instance, the Biodiversity Convention, United Nations Convention to Combact Desertification (UNCCD), etc.

Regarding to the magnitude and modification consequences of ecosystems, the United Nations have commended the study on Millennium Ecosystems Assessment (MA). In order to better conserve the ecosystems and their services, it will be better to understand these ecosystems in all their complexity. It is to this aim that this book suggests some case studies undertaking all continents.

Indeed we try to fill the gap on the knowledge on ecosystems diversity and functioning. Since we have started the book project, we were invaded by many chapters on current environmental issues from reputable international research teams and laboratories. This shows that this book has aroused many interests from international scientific community given the important number of chapters submitted form the beginning. Consequently, we were subjected to make selection. We use this opportunity to thank the Publisher for publishing this book to the benefit of the international scientific community.

This book is educational and useful to students, researchers and all those are interested in environmental issues.

This volume offers a compilation of 20 chapters on the sampling methods of terrestrial and aquatic ecosystems, the algorithms meant for phytoplankton evaluation using satellites data, the biodiversity of arid regions ecosystems, the dynamics of grazed ecosystems in arid regions, the primary production of oceanic and terrestrial ecosystems, the development of new techniques on ecosystems analysis, the dynamics of carbon on forestry ecosystems, the prey-predator relationship within ecosystems, the dynamics of animal population based on many environmental gradients, etc.

We are inviting students, researchers, teachers and people interested in environmental issues to read this book which is educational considering the different methods which are presented.

Dr. Mahamane Ali
Deputy Vice Chancellor and Dean of Faculty of Sciences and Technics (FST),
University of Maradi, Maradi
Niger

Macrofaunistic Diversity in *Vallisneria americana* Michx. in a Tropical Wetland, Southern Gulf of Mexico

Alberto J. Sánchez[1], Rosa Florido[1],
Miguel Ángel Salcedo[1], Violeta Ruiz-Carrera[1],
Hugo Montalvo-Urgel[2] and Andrea Raz-Guzman[3]
*[1]Diagnóstico y Manejo de Humedales Tropicales, CICART,
División Académica de Ciencias Biológicas,
Universidad Juárez Autónoma de Tabasco, Tabasco
[2]Posgrado en Ciencias Ambientales, División Académica de Ciencias Biológicas, UJAT
[3]Instituto de Ciencias del Mar y Limnología, UNAM
México*

1. Introduction

The variety of macrofauna in limnetic and estuarine ecosystems is related to the spatial arrangement of habitats with different quantitative and qualitative complexities (Heck & Crowder, 1991; Taniguchi et al., 2003; Genkai-Kato, 2007; Gullström et al., 2008). Among these habitats, submerged aquatic vegetation (SAV), harbour a high diversity of molluscs, macrocrustaceans and fish, by favouring a greater survival and growth of the associated populations (Minello & Zimmerman, 1991; Pelicice & Agostinho, 2006; Rozas & Minello, 2006; Cetra & Petrere, 2007; Genkai-Kato, 2007; Hansen et al., 2011). In structured habitats, as SAV, the faunistic diversity tends to be greater and mortality rate have a tendency to be lower than in non-structured habitats (Taniguchi et al., 2003; Gullström et al., 2008), although with exceptions (Bogut et al., 2007; Florido & Sánchez, 2010; Schultz & Kruschel, 2010). In particular, SAV in coastal ecosystems provides structured habitats that shelter a greater abundance and diversity of invertebrates and fish, where species use the habitat to obtain protection against predators, and as feeding and reproduction areas (Minello & Zimmerman, 1991; Rozas & Minello, 2006; Genkai-Kato, 2007; Florido &Sánchez, 2010; Hansen et al., 2011).

SAV has been considered a key component in maintaining the functions of shallow aquatic ecosystems with bottom-up type trophic dynamics, as it affects the physical, chemical and biological processes of coastal ecosystems worldwide. At present, its vulnerability in face of the eutrophication of coastal aquatic ecosystems and the declination or disappearance of populations with the resulting loss of biodiversity are a matter of concern (Wigand et al., 2000; Ni, 2001; Bayley et al., 2007; Duarte et al., 2008; Orth et al., 2010). SAV populations, including those of the American Wildcelery *Vallisneria americana* Michx., have however decreased drastically or disappeared in coastal ecosystems throughout the world (Short et al., 2006; Best et al., 2008).

The American Wildcelery populations in the Biosphere Reserve of Pantanos de Centla (BRPC) have recorded wide fluctuations in space and time with respect to density, biomass, patch size and distribution (Sánchez et al., 2007). The high variability of this grass and other macrophytes has been associated both with an increase in total suspended solids (TSS), nutrients and physical disturbances caused by human activities (Touchette & Burkholder, 2000; Ni, 2001; Best et al., 2008), and with the low persistence (below 50%) of the patches on a local scale (Capers, 2003). In the BRPC, the marked variations in the *V. americana* patches have been analysed only with respect to the enrichment in N, with no symptoms of lethal stress or direct toxicity recorded experimentally in young plants as a result of enrichment in N by NH_4, NO_3 and $NO_3:NH_4$ up to 2000 µg L^{-1}, though variations in growth at the sublethal level were recorded (Ruiz-Carrera & Sánchez, 2012).

Aquatic macrofauna, that present distribution patterns associated with particular habitats, is more vulnerable in face of anthropogenic threats. This is reflected in the high number of species that are considered at risk and in the conservation status of the macrofauna itself (Revenga et al., 2005; Dudgeon et al., 2006). The high diversity values that have been recorded for aquatic invertebrates and fish in structured habitats are threatened by the drastic reduction in the surface area of limnetic ecosystems, as has been documented for several areas of the USA (Revenga et al., 2005; Dudgeon et al., 2006). The records of species that are threatened or in danger of extinction should thus be complemented in the short term with an analysis of their distribution patterns in habitats with a high biodiversity such SAV (for examples; Rozas & Minello, 2006; Genkai-Kato, 2007; Hansen et al., 2011), as these are drastically decreasing or disappearing in the wetlands of southeastern Mexico and, in general, on a global scale (Sheridan et al., 2003; Sánchez et al., 2007; Schloesser & Manny, 2007; Best et al., 2008). Considering the above, it has become necessary to carry out short term studies focused on understanding the dynamics, reproduction and production of macrophytes (Ruiz-Carrera & Sánchez, 2008; Liu et al., 2009), and to prepare inventories of species associated with SAV, together with their distribution, particularly in ecosystems and regions where information is still limited (Lévêque et al., 2005) and social and economic situations prevent conservation programmes from being successful (Fisher & Christopher, 2007; Kiwango & Wolanski, 2008). This is the case of tropical fluvial wetlands located in Mesoamerica.

Freshwater ecosystems are rich in species diversity and endemisms, but only a small proportion of species have been assessed in freshwater ecosystems of tropical areas (Lévêque et al., 2005; Dudgeon et al., 2006), including Mesoamerica. This lack of biodiversity data for tropical areas becomes critical considering the high rates of extinction that have been recorded (Revenga et al., 2005). The BRPC is a freshwater Protected Area that, together with the freshwater ecosystem of Pom-Atasta and the estuarine ecosystem of Laguna de Términos in Campeche, forms one the most extensive tropical wetlands in Mesoamerica. In spite of the increase in records of freshwater fauna for the BRPC over the last years, it is believed it is still underestimated (Reséndez & Salvadores 2000; Mendoza-Carranza et al., 2010; Montalvo-Urgel et al., 2010; Maccosay et al., 2011; Sánchez et al., 2012).

Vallisneria americana shelters a high diversity of macrofauna in the BRPC and other ecosystems (Rozas & Minello, 2006; Sánchez et al., 2012). The spatial and temporal variations of the fauna associated with *V. americana* may be explained by quantitative changes in habitat complexity (Rozas & Minello, 2006) and by the effects of flood pulses on the physicochemical properties of the water column (Thomaz et al., 2007; Souza-Filho, 2009). However, the reasons behind the drastic decrease or disappearance of *V. americana* patches in the BRPC have not been well documented, in spite of the SAV patches occupying less than 1% of the total area of the aquatic substrate (Sánchez et al., 2007). For this reason, *in vitro* micropropagation of *V. americana* has been carried out (Ruiz-Carrera & Sánchez, 2008) considering the possibility of: 1) testing hypotheses through experimental designs focused on phytodiagnosis, and 2) generating a germoplasm bank for future repopulation programmes (Ruiz-Carrera & Sánchez, 2012).

Notwithstanding that the BRPC and most American tropical wetlands with *V. americana* shelter a high faunal diversity, these ecosystems still present a scarcity of information regarding the associated fauna and the spatial-temporal variations in the ecological condition. This lays emphasis on the need to update data bases, as well as ecological scenarios for freshwater ecosystems with few studies that, at the same time, receive anthropological pressures from agriculture, cattle ranching and oil industry activities, together with the construction of dams and hydrological structures to control floods and produce electricity. Moreover, the BRPC is located in the ichthyofaunistic Usumacinta Province where the greatest diversity has been recorded for Mexico (Miller et al., 2005), and in the only hydrological area with a high availability of water resources in the country (Sánchez et al., 2008). This chapter includes a 10 year checklist of macrofauna species associated with *V. americana*, together with an analysis of whether lagoons with a great number of species and high density (org/m^2) of fauna, maximum values of quantitative habitat complexity of SAV and a minimum degree of perturbation, present the most favourable ecological condition in the BRPC. This point is considered in three steps through the analysis of the spatial and temporal variations of 1) the environmental quality of the water column, 2) the quantitative habitat complexity of *V. americana*, and 3) the abundance and diversity of molluscs, crustaceans and fish.

2. Materials and methods

2.1 Study area and habitat

The BRPC is a tropical fluvial wetland that covers an area of 302,000 ha, with approximately 110 lentic ecosystems and 2,934.1 km^2 of areas prone to flooding, where the volume of water increases by about 50% during the flood seasons (Sánchez et al., 2007). It is located in the low basin of the rivers Grijalva and Usumacinta (17°57'53"-18°39'03" N, 92°06'39"-92°45'58" W) and receives discharges from these two rivers and another four, all of which define the water changes in volume by flooding cycles (Salcedo et al., 2012). The volume of water discharged from the rivers Grijalva and Usumacinta is the third in importance in the Gulf of Mexico after those of the Mississippi and Atchafalaya (Velázquez-Villegas, 1994; Collier & Halliday, 2000).

The American Wildcelery, *V. americana*, is the dominant submerged aquatic vegetation species in the BRPC (Sanchez et al., 2007). It is widely distributed from Nova Scotia to La Libertad, Petén and Lago Petén, Itza in Guatemala (Korschgen & Green, 1988). It was selected for this study as it is the habitat with the greatest diversity of associated fauna in comparison with two other structured habitats present in the BRPC (Sánchez et al., 2012). A situation similar to this has been recorded in other wetlands (Pelicice & Agostinho, 2006; Rozas & Minello, 2006; Cetra & Petrere, 2007; Genkai-Kato, 2007).

2.2 Sampling and laboratory analyses and procedures

2.2.1 Macrofauna

Faunal specimens were collected from 1999 to 2010 in six lagoons and Polo Bank, a low-energy shallow area along the Usumacinta river. Sites with *V. americana* patches were selected for sampling. Sampling varied among the seasons and the years following the distribution and persistence of the patches. For example, Polo Bank and Tronconada lagoon present a spatially and temporally low persistence and were thus sampled only in 2005 (Table 1). Sampling took place during daylight hours in the minimum (April to March) and maximum (October to November) flood seasons.

Aquatic invertebrates and fish associated with submerged aquatic vegetation were collected with a drop net and a Renfro beam net. The drop net covers an area of 0.36 m², five random repetitions were carried out, and the macrofauna caught in the net was collected with a small dip net. The Renfro beam net has a 1.8 m mouth and a 0.8 mm mesh size. Two 25 m long transects were sampled at each site, each covering an area of 45 m².

2.2.2 Ecological condition

The spatial and temporal variations in the ecological condition were analysed considering the fauna associated with *V. americana*, the quantitative habitat complexity of *V. americana*, the degree of perturbation and the trophic state, in six lagoons and a low-energy shallow bank (Table 1) during two flood seasons (minimum and maximum) in the year 2005. The physicochemical and biological variables of the water column, the quantitative habitat complexity of *V. americana*, and the faunal samples were recorded simultaneously.

Sampling sites	UTM	1999 - 2010	2005 [2]
Laguna El Viento	536096 – 2015690	X	X
Laguna San Pedrito	542550 – 2030632	X	X
Laguna Chichicastle	559375 – 2014741	X	X
Polo Bank [1]	536869 - 2046013		X
Laguna El Guanal	558711 – 2022995	X	X
Laguna El Sauzo	567364 – 2013952	X	X
Laguna Tronconada	539661 - 2011309		X

Table 1. Sampling sites in six lagoons and a low-energy shallow bank along the Usumacinta river [1]. Geographical positions are Universal Transverse Mercator (UTM) units. [2] = sampling sites for the ecological condition analysis.

Sixteen physicochemical and biological variables were recorded at each sampling site. Five variables were quantified *in situ*: water temperature with a conventional thermometer (0-50°C), visibility with a Secchi disc (VSD), depth with a dead weight, pH with a pH meter with an accuracy of ± 0.05 (Hanna model HI98128), and electric conductivity (EC) with a conductivity meter (Yellow Springs Instruments [YSI] model 30). Water was collected with a van Dorn bottle at mid depth and stored at less than 4 °C. The 12 variables analysed in the laboratory were dissolved oxygen saturation (DOS), total suspended solids (TSS), ammonium (NH_4), nitrites (NO_2), total phosphorus (TP), orthophosphates (PO_4), biochemical oxygen demand (BOD_5), chemical oxygen demand

(COD), fats and oils (FO), chlorophyll a (chl *a*) and fecal coliforms (FC). All samples were preserved and analysed following the techniques established by Scientific Committee on Oceanic Research-United Nations Educational, Scientific and Cultural Organisation [SCOR-UNESCO] (1966), Wedepohl et al., (1990) and American Public Health Association [APHA] (1998).

Vallisneria americana stems were collected with a 0.0625 m^2 quadrant. Three replicas were taken per sampling site and the leaves and roots were frozen to measure leaf area (cm^2), plant density (stems/m^2) and biomass as ash free dry weight (g_{AFDW}/m^2), as metrics of quantitative habitat complexity. The animal specimens obtained for this analysis were collected with a drop net as indicated in section 2.3.1. Faunal metrics were species richness (S´) and density (org/m^2).

2.2.3 Species identification and determination of trophic groups

Mollusc species were identified mainly based on the taxonomic characters proposed by García-Cubas (1981), Hershler & Thompson (1992) and Taylor (2003) for gastropods and bivalves. Macrocrustaceans were identified following the taxonomic characters published by Bousfield (1973), Lincoln (1979), Villalobos-Figueroa (1983), Williams (1984), Nates & Villalobos-Hiriart (1990), and Pérez-Farfante & Kensley (1997). Fish species were identified considering the criteria established by Castro-Aguirre et al., (1999), Smith & Thacker (2000), Miller et al., (2005) and Marceniuk & Bentacur (2008). The trophic groups of all the species associated with *V. americana* were defined based on the information published by Hershler & Thompson (1992), Schmitter-Soto (1998), Miller et al., (2005), Rocha-Ramírez et al., (2007) and Froese & Pauly (2011).

2.3 Data analyses

2.3.1 Estimation of the Perturbation Degree Index (PDI) and Trophic State Index (TSI)

The variations in the 16 physicochemical and biological parameters quantified in the water column made it possible to select those that explain the variability through the estimation of the PDI, and to group the lagoons into perturbation categories for each flood season (Salcedo et al., 2012). A first selection of the physicochemical and biological metrics was based on their chemical and statistical effects on the temporal and spatial variations in the sampling sites. This selection eliminated eight parameters. A second selection was carried out with a principal components analysis (PCA) using the JMP vs 9.0 programme (Statistical Analysis System Institute [SAS Institute], 2010), with metrics values previously transformed into natural and standardised logarithms (Legendre & Legendre, 1998). The metrics were selected with values above 40 for the more important weights (Weilhoefer et al., 2008), and considered the correlations among the metrics with the greater weights (Liou et al., 2004). The five metrics selected were EC, DOS, TSS, NH_4 and PO_4.

Reference values for these five metrics were obtained through the analysis of their spatial and temporal variations in the seven sampling sites, and their averages and standard errors (average ± standard error), using a data matrix where the seven sampling sites were placed on files and the five metrics in columns. The calculation of the averages and standard errors was based on 70 data (5 metrics x 7 sampling sites x 2 flood seasons). The reference values were calculated with the average plus the standard error in the case of the parameters of which the maximum values represented conditions of environmental alteration (EC, TSS, NH_4 and PO_4). In contrast, the reference value was estimated with the average minus the standard error in the case of the metric (DOS) of which the minimum value defined conditions of environmental alteration.

The value of each metric per sampling site (5 metrics x 7 sites x 2 seasons = 70) was compared with its reference value to establish its effect. The negative effect of DOS was determined when its value was below or equal to the reference value. In contrast, the negative effect of EC, TSS, NH_4 and PO_4 was defined when its value was greater or equal to the reference value. Thus, positive effects were established in an opposite way for both groups of metrics. The effects of each metric were substituted by a value of cero when positive and a value of one when negative. The resulting binary matrix (n = 70) was divided into two independent matrices per season (5 metrics x 7 sites = 35). The independent evaluation of each flood season was carried out considering that the water changes in volume affect the temporal and spatial variations of the physicochemical parameters of the water, and the biota, in BRPC (Salcedo et al., 2012). Each binary or pondered values matrix included 35 data (5 metrics x 7 sites) with sites on files and metrics in columns.

The pondered value of each of the five metrics was averaged per sampling site (5 metrics x 7 sites = 35), for each of the two binary matrices. The seven averages of the pondered values were analysed through percentiles, and each sampling site was placed in a category of degree of perturbation. The perturbation categories were minimum (< 25%), medium-low (≥ 25 – < 50%), medium-high (≥ 50 – < 75%) and maximum (> 75%). In this study, the PDI was calculated only per sampling site.

The sum of the pondered values of each site was calculated for each metric ($\sum$ 7 sites and 5 metrics = 5 sums). The sum per metric was divided by the number of sampling sites (7) to estimate the persistence as a percentage. The persistence reflects the negative effect of the metrics, as the positive effect was substituted by cero and the negative effect by one in the binary matrix with pondered metrics.

TSI is a parametric index with multimetric applications (Carlson, 1977), and is not referential. It defines trophic states with four trophic categories in a scale of 0 to 100 as: oligotrophic, mesotrophic, eutrophic and hypereutrophic (Carlson, 1977). Only the TSI for phosphorus (TSI_{TP}) was calculated in this study for each season, as its interpretation is comparative and complementary to the PDI (Salcedo et al., 2012).

2.3.2 Estimation of quantitative habitat complexity and faunal metrics

The physical complexity of the habitat has been determined qualitatively and quantitatively, and it has been related to increases in survival and growth of its associated populations (Stoner & Lewis, 1985; Heck & Crowder, 1991; Minello & Zimmerman, 1991; Rozas & Minello, 2006; Genkai-Kato, 2007). This study determined the quantitative habitat complexity through the SAV density (stems/m^2), SAV leaf area (cm^2) and SAV biomass (g_{AFDW}/m^2), per sampling site for the two flood seasons (minimum and maximum), and excluded the architecture or qualitative complexity (Stoner & Lewis, 1985). These three habitat complexity variables were transformed into logarithms (Debels et al., 2005) and analysed with a PCA. The metrics were selected as is mentioned in section 2.3.1, and the same programme was used. The SAV leaf area and biomass were selected in this process.

Density (org/m²), species richness (S´) and the invasive/native species rate were the metrics quantified for the macrofauna. The invasive/native rate was determined based on the density values (org/m²) of the macrofauna collected. The invasive species were those recorded by the Global Invasive Species Database (Invasive Species Specialist Group [ISSG], 2011). All biological metrics were estimated per sampling site for the two flood seasons.

2.3.3 Estimation of ecological condition

The ecological condition was estimated for each of the seven sampling sites, and each of the flood seasons, following the "Marco de Evaluación de Sistemas de Manejo de Recursos Naturales" (MESMIS) procedure (López-Ridaura et al., 2002). The MESMIS procedure was applied considering the PDI, the TSI_{TP}, two habitat complexity indices and the three faunal metrics mentioned in the previous section. The maximum reference value for each of the seven metrics was calculated based on 1) the value defined by the scale of each metric, as in the case of the PDI, the TSI_{TP} and the invasive/native rate, or 2) the value recorded in the study area for macrofauna density (org/m²) and species richness (S´), and SAV leaf area (cm²) and biomass (g_{AFDW}/m²). From these maximum reference values per metric, the indicator values of the MESMIS were calculated and expressed on a scale of averages (0-100%), with 100% corresponding to the greatest reference value per metric. The ecological condition was obtained 1) per sampling site by averaging the seven indicator values of the MESMIS, and 2) per season through the pondered average of the seven sampling sites. The inter-seasonal variation of each of the seven metrics included in the MESMIS, and of the estimated values for the ecological condition, was analysed independently with Kruskal-Wallis tests using the JMP vs 9.0 programme (SAS Institute, 2010), as the data did not satisfy the conditions of homocedasticity and normality (Underwood, 1997). The spatial distribution of the environmental condition of the seven sampling sites was grouped using average-linkage hierarchical clustering (Legendre & Legendre, 1998) and the JMP vs 9.0 programme (SAS Institute, 2010).

3. Results

3.1 Checklist

A total of 53 species of molluscs, macrocrustaceans and fish were collected. Fish dominated with 30 species, followed by macrocrustaceans with 14 species and lastly molluscs with 9 species. Two invasive species were recorded, the gastropod red rimmed Melania *Thiara tuberculata* and the amazon sailfin catfish *Pterygoplichthys pardalis*. Of the 53 species distributed in the SAV, the omnivores represented 36% and included 10 species of fish and nine of crustaceans. The carnivores included 13 fish and one crustacean species, representing 28% of the total. The detritivores made up 15% with two mollusk, three crustacean, and two fish species, including the two invasive species, *T. tuberculata* and *P. pardalis*. The herbivores represented 15% with eight species, and the planctivores and benthic filter feeders represented 4% each (Table 2).

species	species
molluscs	*Astyanax aeneus* [a, b, 2] (Günther, 1860)
Neritina reclivata [a, b, 3] (Say, 1822)	*Hyphessobrycon compressus* [a, 5] (Meek, 1904)
Cochliopina francesae [a, b, 4] (Goodrich & Van der Schalie 1937)	*Pterygoplichthys pardalis* [a, c, 3] (Castelnau, 1855)
Pyrgophorus coronatus [a, 4] (Pfeiffer, 1840)	*Rhamdia quelen* [1] (Quoy & Gaimard, 1824)
Aroapyrgus clenchi [a, b, 4] (Goodrich & Van der Schalie 1937)	*Opsanus beta* [1] (Goode & Bean, 1880)
Pomacea flagellata [a, 4] (Say, 1827)	*Atherinella alvarezi* [1] (Díaz-Pardo, 1972)
Thiara tuberculata [a, b, c, 3] (Müller, 1774)	*Carlhubbsia kidderi* [a, 1] (Hubbs, 1936)
Mexinauta impluviatus [a, 4] (Morelet, 1849)	*Gambusia yucatana* [2] Regan, 1914
Rangia cuneata [a, 6] (Sowerby I, 1831)	*Gambusia sexradiata* [a, 1] Hubbs, 1936
Cyrtonaias tampicoensis [a, 6] (Lea, 1838)	*Heterophallus (aff) rachovii* [1]
Crustaceans	*Poecilia mexicana* [3] Steindachner, 1863
Discapseudes sp. [a, b, 3]	*Ophisternon aenigmaticum* [1] Rosen & Greenwood, 1976
Hyalella azteca [a, b, 2] Saussure, 1857	*Amphilophus robertsoni* [a, 2] (Regan, 1905)
Litopenaeus setiferus [2] (Linnaeus, 1767)	*Rocio octofasciata* [a, 2] (Regan, 1903)
Potimirim mexicana [3] (Saussure, 1857)	*Cichlasoma pearsei* [a, 4] (Hubbs, 1936)
Macrobrachium acanthurus [a, b, 2] (Wiegmann, 1836)	*"Cichlasoma" salvini* [a, 2] (Günther, 1862)
Macrobrachium hobbsi [a, 2] Nates and Villalobos, 1990	*"Cichlasoma" urophthalmum* [a, 1] (Günther, 1862)
Macrobrachium olfersii [2] (Wiegmann, 1836)	*Parachromis friedrichsthalii* [a, 1] (Heckel, 1840)
Procambarus (Austrocambarus) llamasi [4] Villalobos, 1954	*Paraneetroplus synspilus* [a, 4] (Hubbs, 1935)
Callinectes sapidus [a, 2] Rathbun, 1896	*Petenia splendida* [1] Günther, 1862
Callinectes rathbunae [a, 2] Contreras, 1930	*Theraps heterospilus* [a, 2] (Hubbs, 1936)
Rhithropanopeus harrisii [a, 1] (Gould, 1841)	*Thorichthys helleri* [a, 1] (Steindachner, 1864)
Platychirograpsus spectabilis [a, 3] de Man, 1896	*Thorichthys meeki* [a, 2] Brind, 1918
Armases cinereum [2] (Bosc, 1802)	*Thorichthys pasionis* [a, 1] (Rivas, 1962)
Goniopsis cruentata [2] (Latreille, 1802)	*Dormitator maculatus* [a, 2] (Bloch, 1792)
Fish	*Eleotris amblyopsis* [a, 2] (Cope, 1871)
Anchoa parva [5] (Meek & Hildebrand, 1923)	*Gobionellus oceanicus* [1] (Pallas, 1770)
Dorosoma petenense [2] (Günther, 1867)	*Microdesmus longipinnis* [a, d, 3] (Weymouth, 1910)

Table 2. Species list of macrofauna associated with *Vallisneria americana* in Pantanos de Centla. [a] = species included in the ecological condition analysis (2005); [b] = dominant species; [c] = invasive species; [d] = first record in the study area; [1] = carnivores; [2] = omnivores; [3] = detritivores; [4] = herbivores; [5] = planctivores; [6] = benthic filter feeders.

3.2 Ecological condition

3.2.1 Water quality

In general, the seven sampling sites were placed in the category of minimum perturbation during the minimum flood season, whereas during the maximum flood season the perturbation was medium-low. However, the spatial distribution of the perturbation categories was more homogeneous during minimum flooding, when only Polo Bank recorded a medium perturbation and the six lagoons presented a minimum perturbation. In contrast, during maximum flooding two lagoons and Polo Bank were recorded with minimum perturbation, three lagoons with medium-low perturbation and one lagoon with medium-high perturbation (Table 3). The variations in the season of minimum floods were mainly due to the effect of the EC and the TSS, whereas during maximum flooding the metrics that increased and explained the variability were DOS, NH_4 and PO_4.

Prevailing conditions were eutrophic (four lagoons) in the minimum flood season and hypereutrophic (five sites) in the maximum flood season. This difference was observed in the TSI_{TP} values which were lower in the minimum (55 to 73) than in the maximum (65 to 73) flood season, and in the tendency of TP to increase in most lagoons to hypereutrophic or to remain in this category (Table 3). The increase in TP was notable in Polo Bank and was reflected in a 20-unit increase in the TSI_{TP} (Table 3).

The PDI and TSI_{TP} values increased with a directly proportional tendency in the two flood seasons. During minimum flooding, four lagoons with a minimum perturbation coincided with a eutrophic state, and Polo Bank with a medium-low perturbation presented a hypereutrophic state. During maximum flooding, two lagoons with a minimum perturbation were eutrophic, and two lagoons and Polo Bank with a medium-high perturbation were hypereutrophic.

3.2.2 Habitat complexity and macrofauna

The greatest values of habitat complexity were recorded for San Pedrito lagoon and Polo Bank in the two seasons. However, during the minimum flood season the San Pedrito values of plant density (stems/m²) and leaf area occupied second place after Polo Bank and the lagoon Tronconada, respectively (Table 3). Polo Bank also occupied second place in biomass during minimum flooding, and in stem density during maximum flooding (Table 3). El Guanal lagoon stood out for its lack of SAV during minimum flooding and for its minimum values of structural complexity during maximum flooding (Table 3).

Species richness was greater in the minimum (S' = 28) than in the maximum (S' = 23) flood seasons. The greatest number of species in the minimum flood season was recorded for San Pedrito (11 species), followed by three lagoons with 10 species. No species were collected in El Guanal as there was no SAV. In contrast, the greatest number of species in the maximum flood season was recorded for El Guanal and the lowest number for Chichicastle (Table 3). The species collected with the greatest frequencies differed between the seasons. These were *Macrobrachium acanthurus*, *Dormitator maculatus*, *Hyalella azteca* and *Neritina reclivata* in the minimum flood season, and *N. reclivata*, *Astyanax aeneus* and *"Cichlasoma" salvini* in the maximum flood season.

With respect to macrofauna density, four molluscs, three crustaceans and one fish species were dominant (Table 2). On a spatial scale, the greatest value was recorded for El Sauzo in the two flood seasons (Table 3). The molluscs *Thiara tuberculata* and *Aroapyrgus clenchi* and the fish *Astyanax aeneus* and *Carlhubbsia kidderi* represented 87 and 11% of the density in the minimum flood season, whereas the gastropods *T. tuberculata*, *Cochliopina francesae* and *A. clenchi* contributed 87% of the density in the maximum flood season. In this locality, El

Sauzo, crustaceans were totally absent during the maximum flood season, and only *Macrbrachuim acanthurus* was collected during the minimum flood season.

The two sampling sites with the greatest quantitative habitat complexity, San Pedrito and Polo Bank, recorded high densities during both seasons, although with lower values than those of El Sauzo (Table 3). The gastropod *Neritina reclivata* was dominant in density in these two localities, and was followed by the crustaceans *Discapseudes* sp, *Hyalella azteca* and *Macrobrachium acanthurus* in the minimum flood season. However, in the maximum flood season, the densities of *Discapseudes* sp and *H. azteca* were greater than that of *Neritina reclivata* in San Pedrito and Polo Bank.

Sample site	PDI	TSI_{TP}	SAV biomass	SAV leaf area	S'	org/m^2	inv/nat
Minimum flood season							
San Pedrito	20	75	174.2	32.3	11	24.9	0
Polo Bank	40	73.8	158.7	39.7	6	25.4	0
El Guanal	0	55	0	0	0	0	0
Chichicastle	20	65	56.6	27.2	10	12.7	0
El Sauzo	20	58	117.3	24.9	8	55.3	77
Tronconada	0	73	129.4	40.0	10	10.6	2
El Viento	20	68	133.2	28.3	10	8.4	0
Maximum flood season							
San Pedrito	60	83	217.7	54.6	8	17.5	0
Polo Bank	40	93	151.5	39.2	7	11.7	0
El Guanal	20	73	57.9	37.6	11	11.9	63
Chichicastle	40	71	130.1	40.1	3	8.3	0
El Sauzo	20	65	100	44.6	7	20.6	40
Tronconada	20	65	130.3	37.4	9	8.1	4
El Viento	40	84	181.3	49.3	7	4.7	8

Table 3. Values of water quality indices (PDI, TSI_{TP}), quantitative habitat complexity of *Vallisneria americana* as SAV biomass (g_{AFDW}/m^2) and SAV leaf area (cm^2), macrofauna species richness (S') and density (org/m^2), and invasive/native species rate (inv/nat), in the Biosphere Reserve of Pantanos de Centla.

The lagoon of El Guanal had no SAV and no fauna during the minimum flood season, however it recorded 11.9 org/m^2, and occupied third place in density during the maximum flood season (Table 3). The invasive snail *Thiara tuberculata* represented 61% of the total density in this sampling site, where the invasive fish *Pterygoplichthys pardalis* was also collected. The fish presented high densities (13 - 13.8 org/m^2), whereas the crustaceans were scantily (2.7 - 4.6 org/m^2).

With respect to the invasive species, the gastropod *Thiara tuberculata* was collected in El Sauzo during both seasons, in Tronconada during minimum flooding and in El Guanal during maximum flooding. The amazon sailfin catfish, *Pterygoplichthys pardalis*, was collected only during maximum flooding in these same three sites, El Sauzo, Tronconada and El Guanal. The greatest value of the invasive/native rate was calculated for El Sauzo for the minimum flood season (Table 3), in response to the high density of this mollusc (77% of the total density) in this sampling site in this season. The second greatest value of the invasive/native rate was calculated for El Guanal (Table 3), as the two invasive species represented 63% of the total density in this site in this season.

3.2.3 Spatial and temporal variation of the ecological condition

The PDI and the invasive/native species rate acted positively on the ecological condition with values above 60% in both seasons, although the PDI increased 17% during the minimum flood season. In contrast, the TSI_{TP}, S´ and macrofauna density values were below 35%. The percentages calculated for SAV leaf area and SAV biomass were intermediate (38 – 60%) and, inversely to the PDI and TSI_{TP}, decreased by 22 and 11% during the minimum flood season (Figure 1). The SAV decrease was mainly associated with the effect of the absence of SAV in El Guanal. Notwithstanding the temporal variation of the seven metrics described in this paragraph, only the SAV leaf area and PDI values were significantly different in the two seasons (Kruskal-Wallis; p = 0.0181 and p = 0.0527, respectively), as the other five metrics were statistically similar.

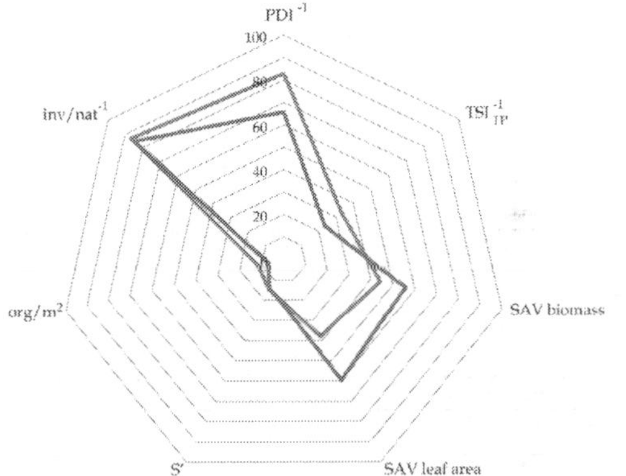

Fig. 1. Temporal variation of the ecological condition for macrofauna associated with *Vallisneria americana* in the Biosphere Reserve Pantanos de Centla (minimum flood season = blue line; maximum flood season = red line) (PDI⁻¹ = Perturbation Degree Index; TSI_{TP}^{-1} = Trophic State Index for total phosphorus; S´ = species richness of macrofauna; inv/ nat⁻¹ = Invasive/native species rate; org/m² = macrofauna density; SAV leaf area = *Vallisneria americana* leaf area; SAV biomass = *Vallisneria americana* biomass).

The ecological condition estimated for the minimum (45%) and the maximum (44%) flood seasons was statistically similar (Kruskal-Wallis; p = 0.9491). This similarity is explained by the fact that the effect of the temporal variation of the two environmental indices of the water (PDI and TSI$_{TP}$) was neutralised by similar opposite values recorded for the two habitat complexity metrics, SAV leaf area and SAV biomass (Figure 1).

The increase in the quantitative complexity of the habitat during the maximum flood season is reflected in the increase in the number of species and density of the associated macrofauna on a local scale, as occurred in El Guanal lagoon (11 species). However, this was not the case in general, as the S′ (23 species) was lower during this season than during the minimum flood season (28 species) (Figure 1). In contrast, the decrease in the environmental condition of the water (PDI and TSI$_{TP}$) was associated with the decrease in the macrofauna collected during maximum flooding, particularly in San Pedrito lagoon and Polo Bank where the only medium-high perturbation (PDI) and the highest value of hypereutrophia (TSI$_{TP}$), respectively, were recorded (Table 3).

This temporal variability in the metrics explained the spatial distribution of the lagoons and Polo Bank in three different groups for the two seasons. During minimum flooding (Figure 2), a group with the greatest ecological condition values (> 45%) was formed by three lagoons and Polo Bank, characterised by the greatest values of habitat complexity. A second group with intermediate values was formed by El Sauzo with high habitat complexity values and the greatest density of the invasive mollusc *Thiara tuberculata*, and Chichicastle with low values of habitat complexity. Lastly, the lagoon of El Guanal remained separated due to the disappearance of its SAV patch and the resulting absence of macrofauna (Table 3).

During the season of maximum floods, three groups were formed (Figure 3) where, in contrast with the other season, Polo Bank moved to the group with intermediate values of ecological condition, in response to the increase in TP that was reflected in the extreme value of hypereutrophic condition detected by the TSI$_{TP}$. Although El Guanal recorded 11 species of macrofauna in SAV, this locality again remained separated, with the smallest value of ecological condition as a result of 1) the density of two invasive species, *T. tuberculata* and *Pterygoplichthys pardalis*, and 2) the change from an eutrophic to a hypereutrophic condition (Table 3).

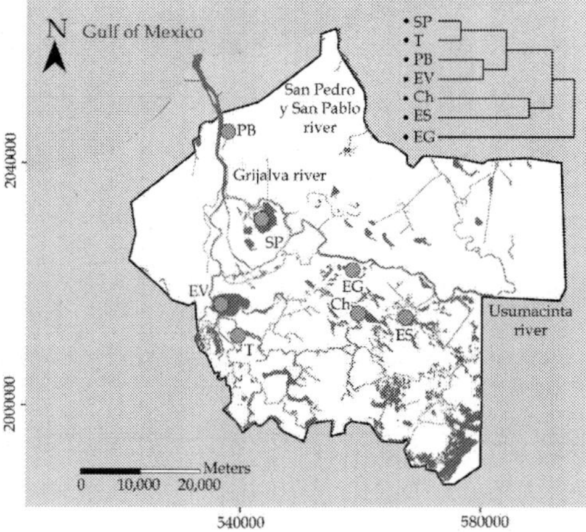

Fig. 2. Spatial variation of the ecological condition for macrofauna associated with *Vallisneria americana* in the minimum flood season in the Biosphere Reserve Pantanos de Centla.

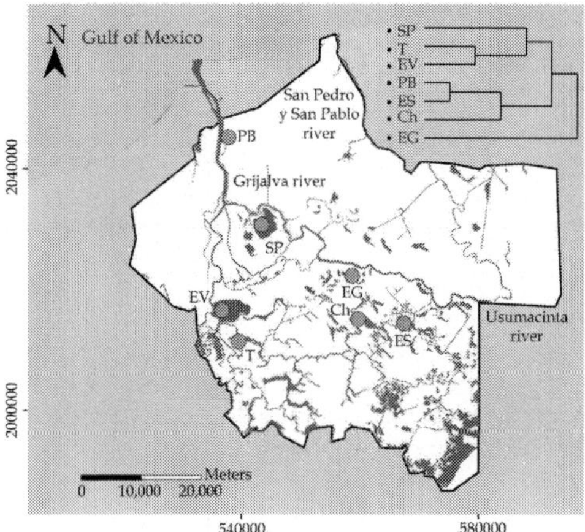

Fig. 3. Spatial variation of the ecological condition for macrofauna associated with *Vallisneria americana* in the maximum flood season in the Biosphere Reserve Pantanos de Centla.

4. Discussion

The water quality indicators (PDI and TSI_{TP}) had greater values in the minimum flood season than in the maximum flood season, although the degree of perturbation (PDI) in the SAV sampling sites was lower than that recorded for SBS in 2000-2001 (Salcedo et al., 2012). In contrast, the quantitative complexity of *V. americana* increased during maximum flooding, although this was not reflected in a significant increase in species richness and density of associated macrofauna. This is contrary to the general situation where an increase in habitat complexity, understood as a greater availability of microhabitats, favours a greater abundance and diversity of associated fauna (Heck & Wilson, 1987; York et al., 2006; Gullström et al., 2008). However, quantitative habitat complexity not always increase the relative habitat value of SAV (Seitz et al., 2006; Bogut et al., 2007; Florido & Sánchez et al., 2010; Schultz & Kruschel, 2010).

The greater abundance and diversity of fauna associated with SAV is explained considering the part the vegetation plays as feeding and nursery areas, and as physical refuges against predators (Moskness & Heck, 2006; Pelicice & Agostinho, 2006; Rozas & Minello, 2006; Genkai-Kato, 2007; Hansen et al., 2011). SAV is also a key component in maintaining ecosystem function, particularly in shallow systems with bottom-up trophic dynamics (Ni, 2001; Jaschinski & Sommer, 2008). The record of 53 species of macrofauna associated with SAV in this study was greater than those published for other similar ecosystems with structured habitats (Vega-Cendejas, 2004; Bogut et al., 2007; Quiroz et al., 2007). One species, the omnivorous fish *Microdesmus longipinnis*, was not recorded previously by Reséndez and Salvadores (2000), Mendoza-Carranza et al. (2010), Macossay et al. (2011) or Sánchez et al. (2012).

The high number of species associated with *V. americana* that was recorded in this study and other studies carried out in the BRPC (Mendoza-Carranza et al., 2010; Macossay et al., 2011;

Sánchez et al., 2012) emphasizes the importance of the *V. americana* patches, that occupy less than 1% of the substrate of the permanent aquatic ecosystems (Sánchez et al., 2007). Among the molluscs that were collected, most of the species of the Hydrobiidae family are recorded as herbivorous (Hershler & Thompson, 1992) and *Neritina reclivata* is a microphage with a distribution associated with SAV (as examples, García-Cubas et al., 1990; Lane, 1991). The invasive detritivorous snail *Thiara tuberculata* is widely distributed, numerically dominant and was collected both in SAV and SBS (Contreras-Arquieta et al., 1995; Sánchez et al., 2012).

The distribution of macrocrustaceans in fringe wetlands (Brinson, 1993) has been frequently associated with macroalgae and aquatic angiosperms (Minello et al., 1990; Sheridan, 1992; Sánchez et al., 2012), although their dependence on SAV and their feeding habits vary according to the stage of ontogenetic development, behaviour, foraging strategies and predator-prey interactions, as is the case of the swimming crabs (Kuhlmann & Hines, 2005; Florido & Sánchez, 2010). Their use of the habitat also varies, as for example penaeid shrimp during the various phases of epibenthic postlarvae and juveniles in estuarine ecosystems (McTigue & Zimmerman, 1991; Sánchez, 1997; Pérez-Castañeda & Defeo, 2001; Beseres & Feller, 2007). The brachyurans *Platychirograpsus spectabilis*, *Armases cinereum* and *Goniopsis cruentata* collected in SAV were small in size, as the adults are semi-terrestrial (Schubart et al., 2002; Álvarez et al., 2005).

Most of the 12 cichlids collected in SAV in this study were not associated with any particular habitat (Miller et al., 2005; Macossay et al., 2011). This may be because only *Paraneetroplus synspilus* and *Cichlasoma pearsei* are herbivores, while the other species are carnivores or omnivores (Schmitter-Soto, 1998; Miller et al., 2005; Froese & Pauly, 2011). The smaller number of species of the family Poeciliidae collected in this study, in comparison with Macossay et al. (2011), is related to most of the species inhabiting MV (Sánchez et al., 2012) and feeding mainly on insect larvae (Schmitter-Soto, 1998; Miller et al., 2005), which is not the case of the omnivore *Poecilia mexicana*. The batracoid *Opsanus beta* was the only species associated exclusively with SAV in Pantanos de Centla (Sánchez et al., 2012) and seagrasses in other ecosystems (Schofield, 2003; Vega-Cendejas, 2004). The preference of the carnivore *O. beta* for epibenthic prey distributed in SAV and its stalking strategy are common in fish that are associated with this habitat (Schultz & Kruschel, 2010).

Electric conductivity (EC) and TSS increased markedly during the minimum flood season. The increase in EC was related to the effect of the tidal currents, as the water volume decreases by 40% (18,722 million m^3) during this season (Salcedo et al., 2012). This increase in EC, during the smaller phase of the flood pulse, coincides with reports for other fringe wetlands that consider water chemistry and ecology at different scales (see Thomaz et al., 2007; Fernandes et al., 2009; Souza-Filho, 2009). The resuspension of TSS in response to the high energy in tidal-current dominated environments (Brinson, 1993) and the decrease in DOS (Varona-Cordero & Gutiérrez, 2003) have been related to an increase in EC.

Ecologically, the spatial and temporal variation of the estuarine condition has helped explain the distribution of species in SAV in coastal wetlands, the temporal immigration of marine fauna and the establishment of estuarine fauna (Pérez-Castañeda & Defeo, 2001; Barba et al., 2005; Sosa-López et al., 2007). In the case of the BRPC, the effect of sea water by the tidal currents influence is restricted during the high tide in the minimum flood season with EC long term values (2001-2010) below 10,880 µScm^{-1} (Salcedo, unpublish data). This restricted estuarine effect may be observed through the dominance of freshwater aquatic angiosperms and the minimum amount of estuarine fauna in the BRPC. For example, the presence of *V. americana* is indicative of frequent freshwater or oligohaline conditions, as it is common in 1.3 to 5 PSS and has been found in 12 PSS under natural conditions (Korschgen & Green, 1988; Boustany et al., 2010). Thus, the absence of estuarine molluscs and the occasional immigration of estuarine or marine crustaceans are to be noted, though expected, in this study, with only the blue crab *Callinectes sapidus* collected once in Polo Bank, in the minimum flood season, in 0.5 PSS. This euryhaline swimming crab is numerically dominant in many estuarine ecosystems of the Western Atlantic, and shows no preference for a particular type of structured or unstructured habitat (Kuhlmann & Hines, 2005; Florido & Sánchez, 2010). Only the cirripedian *Balanus improvisus* Darwin, 1854, a stenohaline macrocrustacean distributed in polyhaline and marine environments and was reported with a distribution restricted to the area with marine influence (Montalvo-Urgel et al., 2010), where Polo Bank is located. Moreover, there were no estuarine species and only two marine fish species (*Anchoa parva* and *Gobionellus oceanicus*) among the 30 collected in SAV in this study, as well as among the 12 estuarine and marine species previously recorded for the BRPC (Macossay et al., 2011). Also, the macrofauna collected from *V. americana* in the area that receives the effect of the tides differs markedly from that recorded for oligo-mesohaline environments in other coastal wetlands (Zimmerman et al., 1990; Domínguez et al., 2003).

The increase in TSS during minimum flooding is explained by the resuspension of sediments generated by the dominant northerly and southeasterly winds. Sediment resuspension by the wind has been reported for other ecosystems (Schallenberg & Burns, 2004; Thomaz et al., 2007). The resuspended TSS affect the presence, diffusion and assimilation of nutrients that, in turn, affect the transparency of the water and the primary production in shallow aquatic ecosystems (Ni, 2001; Schallenberg & Burns, 2004; Ahearn et al., 2006). The obstacle that the TSS present to light penetration has been considered one of the main hypotheses to explain the drastic decrease and disappearance of SAV in coastal wetlands (Boustany et al., 2010). However, *V. americana* has a competitive advantage over other aquatic angiosperms in conditions of limited growth caused by the availability of light, as its rate of photosynthesis is more efficient and it rapidly acclimatises to light variations (Barko et al., 1982; Catling et al., 1994; Harley & Findlay, 1994). Under experimental conditions of low light intensity, *V. americana* responds with a decrease in the below and above sediment biomass and in the length and width of the leaves (Kurtz et al., 2003). However, light in wetlands is irregular in intensity, and rooted macrophytes recuperate by the translocation of resources through interconnected ramets, that allows nutrient and metabolic products to move among the patches that receive light (Catling et al., 1994).

The increase in TSS deteriorates the ecological condition by favouring eutrophic-hypereutrophic conditions (Salcedo et al., 2012). The eutrophic state obtained with the TSI$_{TP}$ may be related to the homogeneous distribution of the TSS, TP and PO$_4$ that was established by the flooding cycle (Salcedo et al., 2012). The deterioration of the ecological condition due to the increase in TSS, and the eutrophic state, have also been related to the loss of the indirect ecological services generated by the SAV, such as the control of the bottom-up dynamics in aquatic ecosystems (Ni, 2001; Bayley et al., 2007; Genkai-Kato, 2007; Heck & Valentine, 2007). In contrast, the perturbation categories (PDI) resulted lower than those estimated in 2000-2001 for sampling sites with SBS in the BRPC (Salcedo et al., 2012). The local disappearance of the *V. americana* patch in El Guanal during the minimum flood season, as well as the marked variation in SAV biomass and SAV leaf area, negatively affected the value of the ecological condition estimated for the minimum flood season. This disappearance coincided with the low persistence on a local scale recorded in the freshwater wetland of Whalebone Cove, that is connected to Connecticut river in the USA (Capers, 2003). This point is relevant as the value of EC (8000 μScm^{-1}) quantified in SAV sites in the BRPC is not reported as critical for the growth and survival of *V. americana* (Boustany et al., 2010). In contrast, the maximum value of TSS (42 mg L^{-1}) is near the critical value for the survival of SAV (Catling et al., 1994).

During the maximum flood season, the decrease in DOS and the increase in NH_4 and PO_4 recorded for the six sampling sites with the lower degree of perturbation (PDI) occurred in response to the homogenisation generated by river transportation at the time of the maximum level of the flood pulse, as has been reported in this and other coastal wetlands (Contreras-Espinoza & Wagner, 2004; Debels et al., 2005; Thomaz et al., 2007; Salcedo et al., 2012). In the case of the decrease in DOS, this has been related to the greater dispersion of the algal community generated by the increase in the water volume of the rivers throughout the flood plain (Ahearn et al., 2006; Weilhoefer et al., 2008), as has occurred in other neotropical and tropical wetlands (Thomaz et al., 2007).

The decrease in DOS has also been related to the remineralisation of organic matter that was observed through the increase in NH_4 (Carvalho et al., 2003; Ahearn et al., 2006). The greater amounts of PO_4 during maximum flooding in comparison with minimum flooding have been related to fluvial and agricultural runoff, and input of waste water from villages along the river margins (Contreras-Espinoza & Wagner, 2004; Debels et al., 2005; Salcedo et al., 2012). This pattern of spatial distribution has been reported for other fringe wetlands (Contreras-Espinoza & Wagner, 2004; Debels et al., 2005). The increase in TP, detected through the dominance of the hypereutrophic condition of TSI_{TP} during maximum flooding, is related to the effect of the greatest flood pulse level. It has also been associated with the remineralisation of organic matter from macrophytes established in areas of temporal flooding, from where it is freed into the water column (Carvalho et al., 2003; Kansiime et al., 2007).

The ecological integrity of coastal rooted macrophytes is directly threatened on a worldwide scale by the decrease in DOS and the overload of N, a chronic stress situation that has recently been recognised (Burkholder et al., 2007; Waycott et al., 2009). In contrast, the impact of the deficiency of P has been more significant with respect to the growth and survival of SAV (Touchette & Burkholder, 2000; Leoni et al., 2008). American Wildcelery has been proposed as a bioindicator of eutrophication considering its $\delta^{15}N$ values (Benson et al., 2007). However, no symptoms of lethal stress or direct toxicity were experimentally recorded in the leaves and roots of young plants of the population established in the BRPC, as a result of the enrichment of N with NH_4, NO_3 and the $NO_3:NH_4$ ratio of 2000 μg L^{-1}, although variations in growth were detected at the sublethal level (Ruiz-Carrera & Sánchez, 2012). In the case of *Vallisneria spinulosa* Yan, the interaction between N and P produces differences in the stress response in a wide range of N concentrations that explain the absence of toxicity (Li et al., 2008), implying that the N/P synergism may be associated with the vitality of the plants. In this season, the increase in SAV quantitative habitat complexity may be explained by the minimum values of DOS (18%) and maximum values of NH_4 (321 μgL^{-1}), TP (460 μgL^{-1}) and PO_4 (100 μgL^{-1}) that have not yet generated lethal effects for the *V. americana* in the BRPC, as well as by the resilience mechanism that is active in the aquatic ecosystems with *V. americana* populations that have been affected by increases in nutrient concentrations (Morris et al., 2003a; 2003b).

Also to be considered is that, at the start of the maximum flood season in the BRPC, *V. americana* presents a direct relationship between the growth in length of the leaves and the depth of the water column, which implies growth occurs in a short time, as depth increases from 0.2-0.4 to 2.5-5 m. The hypotheses to explain this growth of more than 1 m in a short time have not been solved, although it is believed that phototropism and gas accumulation in the system may play a part. The variation in the length of the grass leaves has been related to the increase in the abundance and diversity of the associated fauna (Bell & Westoby, 1986). This however has not been experimentally tested for the BRPC. The effect of the increase in quantitative habitat complexity provided by the greater on the density of the fauna is expected to be minimised and remain similar, especially in the case of fish, as mobile fauna colonises areas with emergent rooted vegetation that are prone to flooding, in coastal plain wetlands that are regulated by flood pulses (Fernandes et al., 2009; da Silva et al., 2010).

5. Conclusions

The ecological condition of the sampling sites with *V. americana* was favourable for the growth and survival of SAV and the associated macrofauna, in both flood seasons. In spite of the spatial differences in the ecological condition, as well as in the eutrophic and hypereutrophic conditions estimated in the SAV patches, the perturbation degree (PDI) was lower than that recorded for sites with SBS in the BRPC. Similarly, the macrofauna species number in the SAV sites was greater than in the SBS sites. The effect of two invasive species, the snail *Thiara tuberculata* and the fish *Pterygoplichthys pardalis*, requires a special analysis considering the high density of the snail. The similarity in the ecological condition estimated for both flood seasons is explained by the fact that the effect of the temporal variation of the two water environmental indices (PDI and TSI_{TP}) was neutralised by similar opposite values recorded for the two quantitative habitat complexity metrics (SAV leaf area and SAV biomass). This means that the water changes in volume by flooding cycles has a regulation effect on the physical structure of the SAV and an indirect effect on the faunal metrics, as the variations in the macrofauna are regulated by the physical structure of the SAV. The importance of the SAV patches and the flood pulses in the ecological regulation in the BRPC emphasize the necessity to review: 1) the programmes for the construction of flood control structures that alter the natural flood cycles, and 2) the hypotheses concerning the wide spatial and temporal variations and the local disappearance of the SAV patches that shelter the greatest biodiversity and occupy less than 1% of the substrate of the aquatic ecosystem in the tropical wetland of the BRPC.

6. References

Ahearn, D.S., Viers, J.H., Mount, J.F., & Dahlgren, R.A. (2006). Priming the productivity pump: flood pulse driven trend in suspended algal biomass distribution across a restored floodplain. *Freshwater Biology*, Vol. 51, No. 8, (August 2006), pp. 1417-1433, ISSN 1365-2427

Álvarez, F., Villalobos-Hiriart, J.L., & Robles, R. (2005). Crustáceos, In: *Biodiversidad del estado de Tabasco*, J. Bueno, F. Álvarez & S. Santiago (Eds.), 177-194, Instituto de Biología, Universidad Nacional Autónoma de México–Comisión Nacional para el Conocimiento y Uso de la Biodiversidad, ISBN 970-9000-26-8, D.F.

APHA. (1998). *Standard methods for the examination of water and wastewater*. 20th edn. American Public Health Association, ISSN 0875532357 9780875532356, Washington, DC

Barba, E., Raz-Guzman, A., & Sánchez, A.J. (2005). Distribution patterns of caridean shrimps in the Southwestern Gulf of Mexico. *Crustaceana*, Vol. 78, No. 6, (Jun 2005), pp. 709-726, ISSN 1568-5403

Barko, W.J., Hardin, G.D., & Mathews, S.M. (1982). Growth and morphology of submersed freshwater macrophytes in relation to light and temperature. *Canadian Journal of Botany*, Vol. 60, No. 6, (1982), pp. 877-887, ISSN 1480-3305

Bayley, C.E., Creed, I.F., Sass G.Z., & Wong, A.S. (2007). Frequent regime shifts in trophic states in shallow lakes on the Boreal Plain: Alternative "unstable" states? *Limnology and Oceanography*, Vol. 52, No. 5, (September 2007), pp. 2002-2012, ISSN 1541-5856

Bell, J.D., & Westoby, M. (1986). Variation in seagrass height and density over a wide spatial scale: effects on fish and decapods. *Journal of Experimental Marine Biology and Ecology*, Vol. 104, No. 1-3, (December 1986), pp. 275-295, ISSN 0022-0981

Benson, R.E., O'Neil, M.J., & Dennison, C.W. (2007). Using the aquatic macrophyte *Vallisneria americana* (wild celery) as a nutrient bioindicator. *Hydrobiologia*, Vol. 596, No. 1, (July 2007), pp. 187-196, ISSN 1573-5117

Beseres, J.J., & Feller, R.J. (2007). Importance of predation by white shrimp *Litopenaeus setiferus* on estuarine subtidal macrobenthos. *Journal of Experimental Marine and Biology and Ecology* , Vol. 344, No. 2, (June 2007), pp. 193-205, ISSN 0022-0981

Best, E.P.H., Teeter, A.M., Landwehr, K.J., James, W.F., & Nair, S.K. (2008). Restoration options for potential persistence of submersed aquatic vegetation: combining ecological, hydrodynamic and sediment transport modelling. *Freshwater Biology*, Vol. 53, No. 4, (April 2008), pp. 814-826, ISSN 1365-2427

Bogut, I., Vidaković, J., Palijan, G., & Čerba, D. (2007). Benthic macroinvertebrates associated with four species of macrophytes. *Biologia, Bratislava*, Vol. 62, No. 5, pp. 600-606, ISSN 1336-9563

Bousfield, E.L. (1973). *Shallow-water gammaridean Amphipoda of New England*, VII-XII 1-312, Cornell University press, ISBN 0801407265, Ithaca N. Y.

Boustany, R.G., Michot, T.C., & Moss, R.F. (2010). Effects of salinity and light on biomass and growth *of Vallisneria americana* from Lower St. Johns River, FL, USA. *Wetlands Ecology and Management*, Vol. 18, No. 2, (September 2010), pp. 203-217, ISSN 1572-9834

Brinson, M.M. (1993). *A hydrogeomorphic classification for wetland*, U.S. Army Corps of Engineers, Technical Report; WRP-DE-4, Retrieved from
http://el.erdc.usace.army.mil/wetlands/pdfs/wrpde4.pdf

Bunch, A.J., Allen, M.S., & Gwinn, D.C. (2010). Spatial and Temporal Hypoxia Dynamics in Dense Emergent Macrophytes in a Florida Lake. *Wetlands*, Vol. 30, No. 3, (May 2010), pp. 429-435, ISSN 1943-6246

Burkholder, J.M., Tomasko, D.A., & Touchette, B.W. (2007). Seagrasses and eutrophication. *Journal of Experimental Marine Biology and Ecology*, Vol. 350, No. 1-2, (November 2007), pp. 46-72, ISSN 0022-0981

Capers, R.S. (2003). Six years of submerged plant community dynamics in a freshwater tidal wetland. *Freshwater Biology*, Vol. 48, No. 9, (September 2003), pp. 1640-1651, ISSN 1365-2427

Carlson, R.E. (1977). A trophic state index for lakes. *Limnology and Oceanography*, Vol. 22, No. 2, (March 1977), pp. 361-369, ISSN 0024-3590

Carvalho, P., Thomaz, S.M., & Bini, L.M. (2003). Effects of water level, abiotic and biotic factors on bacterioplankton abundance in lagoons of a tropical floodplain (Paraná River, Brazil). *Hydrobiologia*, Vol. 510, No. 1-3 (July 2003), pp.67-74, ISSN 1573-5117

Castro-Aguirre, J.L., Espinosa-Pérez, H., & Schmitter-Soto, J.J. (1999). *Ictiofauna estuarino-lagunar y vicaria de México*, Limusa-Noriega, ISBN 968-18-5774-7, Ciudad de México.

Catling, P.M., Spicer, K.W., Biernacki, M., & Lovett Doust, J. (1994). The biology of Canadian weeds. 103. *Vallisneria americana* Michx. *Canadian Journal of Plant Science*, Vol. 74, No. 4, (May 1994), pp. 883-897, ISSN 00084220

Cetra, M., & Petrere, Jr.M. (2007). Associations between fish assemblage and riparian vegetation in the Corumbataí River Basin (SP). *Brazilian Journal of Biology*, Vol. 67, No. 2, (May 2007), pp. 191-195, ISSN 1519-6984

Collier, K.J., & Halliday, J.N. (2000). Macroinvertebrate wood associations during decay of plantation pine in New Zealand pumice-bed streams: stable habitat of trophic subsidy. *Journal of the North American Benthological Society*, Vol. 19, No. 1, (March 2000), pp. 94-111, ISSN 0887-3593

Contreras-Arquieta, A., Guajardo-Martínez, G., & Contreras-Balderas, S. (1995). *Thiara (Melanoides) tuberculata* (Müller, 1774) (Gastropoda: Thiaridae), su probable impacto ecológico en México. *Publicaciones Biológicas*, F.C.B./U.A.N.L. México. Vol. 8, No. 1 y 2, pp. 17-24, ISSN 0187-3873

Contreras-Espinoza, F., & Wagner, B.G. (2004). Ecosystem characterics and management considerations for coastal wetlands in Mexico. *Hydrobiologia*, Vol. 511, No. 1, (November 2003), pp. 233-245, ISSN 1573-5117

da Silva, H.P., Petry, A.C., & da Silva, C.J. (2010). Fish communities of the Pantanal wetland in Brazil: evaluating the effects of the upper Paraguay river flood pulse on baía Caiçara fish fauna. *Aquatic Ecology*, Vol. 44, No. 1, (March 2010), pp. 275-288, ISSN 1386-2588

Debels, P., Figueroa, R., Urrutia, R., Barra, R., & Niell, X. (2005). Evaluation of water quality en the Chillán River (Central Chile) Using physicochemical parameters and a modified water quality index. *Environmental Monitoring and Assessment*, Vol. 110, No. 1-3, (November 2005), pp. 301-322, ISSN 1573-2959

Domínguez, J.C., Sánchez, A.J., Florido, R., & Barba, E. (2003). Distribución de Macrocrustáceos en Laguna Mecoacán, al sur del Golfo de México. *Hidrobiologica*, Vol 13, No. 2, pp. 127-136, ISSN 0188-8897

Duarte, C.M., Dennison, W.C., Orth, R.J.W., & Carruthers, T.J.B. (2008). The Charisma of Coastal Ecosystems: Addressing the Imbalance. *Estuaries and Coasts*, Vol. 31, (February 2008), pp. 233-238, ISSN 1559-2731

Dudgeon, D., Arthington, A.H., Gessner, M.O., Kawabata, Z.I., Knowler, D.J., Lévêque, C., Naiman, R.J., Prieur-Richard, A.H., Soto, D., Stiassny, M.L.J., & Sullivan, C.A.

(2006). Freshwater biodiversity: importance, threats, status and conservation challenges. *Biological Reviews*, Vol. 81, No. 2, (December 2005), pp. 163-182, ISSN 1469-185X

Fernandes, R., Gomes, L.C., Pelicice, F.M., & Agostinho, A.A. (2009). Temporal organization of fish assemblages in floodplain lagoons: the role of hydrological connectivity. *Environmental Biology of Fishes*, Vol. 85, No. 2, (February 2009), pp. 99–108, ISSN 1573-5133

Fisher, B., & Christopher, T. (2007). Poverty and biodiversity: Measuring the overlap of human poverty and the biodiversity hotspots. *Ecologycal Economics*, Vol. 62, No. 1, (April 2007), pp. 93-101, ISSN 0921-8009

Florido, R., & Sánchez, A.J. (2010). Effect of the habitat complexity, mobility and density of prey by predation of the blue crab *Callinectes sapidus* (Crustacea). *Crustaceana*, Vol. 89, No. 3, (October 2010), pp. 1069-1089, ISSN 1568-5403

Froese, R., Pauly, D. (2011). FishBase. *World Wide Web electronic publication.* Available from: www.fishbase.org, version (08/2011).

García-Cubas, A. (1981). Moluscos de un sistema lagunar tropical en el sur del Golfo de México (Laguna de Términos, Campeche). *Anales del Instituto de Ciencias del Mar y Limnología, Universidad Nacional Autónoma de México*, Vol. 5, pp. 1-182, ISSN 0185-3287

García-Cubas, A., Escobar, A.F., González, L.V., & Reguero, M. (1990). Moluscos de la Laguna Mecoacán, Tabasco, México: Sistemática y Ecología. *Anales del Instituto de Ciencias del Mar y Limnología, Universidad Nacional Autónoma de México*, Vol. 17, No. 1, pp. 1-30, ISSN 0185-3287

Genkai-Kato, M. (2007). Macrophyte refuges, prey behaviour and trophic interactions: consequences for lake water clarity. *Ecology Letters*, Vol. 10, No. 2, (February 2010), pp. 105-114, ISSN 1461-0248

Gullström, M., Bodin, M., Nilsson, P.G., & Öhman, M.C. (2008). Seagrass structural complexity and landscape configuration as determinants of tropical fish assemblage composition. *Marine Ecology Progress Series*, Vol. 363, pp. 241-255, ISSN 0171-8630

Hansen, J.P., Wikström, S.A., Axemar, H., & Kautsky, L. (2011). Distribution differences and active habitat choices of invertebrates between macrophytes of different morphological complexity. *Aquatic Ecology*, Vol. 45, No. 1, (March 2010), pp. 11-22, ISSN 1386-2588

Harley, T.M., & Findlay, S. (1994). Photosynthesis-irradiance relationships for three species of sumersed macrophytes in the tidal freshwater Hudson River. *Estuaries and Coast*, Vol. 17, No. 18, (March 1994), pp. 200-206, ISSN 1559-2723

Heck, K.L., & Crowder, L.B. (1991). Habitat structure and predator-prey interactions in vegetated aquatic systems, In: *Habitat structure: The physical arrangement of objects in space*, S.S. Bell, E.D. McCoy, & H.R. Mushinsky (Eds.), 281-299, Population and community biology series, ISBN 0412322706, London

Heck, K.L., & Valentine, J.F. (2007). The primacy of top-down effects in shallow benthic ecosystems. *Estuaries and Coasts*, Vol. 30, No. 3, (June 2007), pp. 371-381, ISSN 1559-2723

Hershler, R., & Thompson, F.G. (1992). *A review of the aquatic gastropod subfamily Cochliopinae (Prosobranchia: Hydrobiidae), Malacological Review Supplement 5*, Ann Arbor Michigan

ISSG. (2011). Global Invasive Species Database, In: *Invasive Species Specialist Group*. accessed september 1st 2011, Available from: http://www.issg.org/database.

Jaschinski, S., & Sommer, U. (2008). Top-down and bottom-up control in an eelgrass-epiphyte system. *Oikos*, Vol. 117, No. 5, (May 2008), pp. 754-762, ISSN 1600-0706

Kansiime, F.M., Saunders, J., & Loiselle, S.A. (2007). Functioning and dynamics of wetland vegetation of Lake Victoria: an overview. *Wetlands Ecology and Management*, Vol. 15, No. 6, (July 2007), pp. 443-451, ISSN 1572-9834

Kiwango, Y.A., & Wolanski, E. (2008). Papyrus wetlands, nutrients balance, fisheries collapse, food security, and Lake Victoria level decline in 2000-2006. *Wetlands Ecology and Management*, Vol. 16, (January 2008), pp. 89-96, ISSN 1572-9834

Korschgen, C.E., & Green, W.L. (1988). American wildcelery (*Vallisneria americana*), In: *Ecological considerations for restoration. U.S. Fish and Wildlife Service, Fish and Wildlife Technical Report 19*. Jamestown. Date of access, Available from: http://www.npwrc.usgs.gov

Kuhlmann, M.L., & Hines, A.H. (2005). Density-dependent predation by blue crabs *Callinectes sapidus* on natural prey populations of infaunal bivalves. *Marine Ecology Progress Series*, Vol. 295, (June 2005), pp. 215–228 (June 2005), ISSN 0171-8630

Kurtz, J.C., Yates, D.F., Macauley, J.M., Quarles, R.L., Genthner, F.J., Chancy, C.A., & Devereux, R. (2003). Effects of light reduction on growth of the submerged macrophyte *Vallisneria americana* and the community of root-associated heterotrophic bacteria. *Journal of Experimental Marine Biology and Ecology*, Vol. 291, No. 2, (April 2003), pp. 199-218, ISSN 0022-0981

Lane, J.M. (1991). The effect of variation in quality and quantity of periphyton on feeding rate and absorption efficiencies of the snail *Neritina reclivata* (Say). *Journal of Experimental Marine Biology and Ecology*, No. 150, pp. 117-129, ISSN 0171-8630

Legendre, P., & Legendre, L. (1998). *Numerical Ecology*, 2nd edn. Elsevier, ISBN 978-1-4419-7975-9, Amsterdam

Leoni, V., Vela, A., Pasqualini, V., Pergent-Martini, C., & Pergent, G. (2008). Effects of experimental reduction of light and nutrient enrichments (N and P) on seagrasses: a review. Aquatic Conservation: *Marine and Freshwater Ecosystems*, Vol. 18, No. 2, (March-April 2008), pp. 202-220, ISSN 1052-7613

Lévêque, C., Balian, E.V., & Martens, K. (2005). An assessment of animal species diversity in continental waters. *Hydrobiologia*, Vol. 542, No. 1, (July 2005), pp. 39-67, ISSN 1573-5117

Li, W., Zhang, Z., & Jeppesen, E. (2008). The response of *Vallisneria spinulosa* (Hydrocharitaceae) to different loadings of ammonia and nitrate at moderate phosphorus concentration: a mesocosm approach. *Freshwater Biology*, Vol. 53, No. 11, (November 2008), pp. 2321-2330, ISSN 1365-2427

Lincoln, R.J. (1979). *British marine amphipoda: Gammaridea*, British Museum (Natural History), ISBN 1904690483, London

Liou, S.M., Lo, S.L., & Wang, S.H. (2004). A generalized water quality index for Taiwan. *Environmental Monitoring and Assessment*, Vol. 96, (June 2003), pp. 35-52, ISSN 1573-2959

Liu, W., Zhang, Q., & Liu, G. (2009). Seed banks of a river-reservoir wetland system and their implications for vegetation development. *Aquatic Botany*, Vol. 90, No. 1, (January 2009), pp. 7-12, ISSN 0304-3770

López-Ridaura, S., Masera, O., & Astier, M. (2002). Evaluating the sustainability of complex socio-environmental systems. The MESMIS framework. *Ecological Indicators*, Vol. 2, No. 1-2, (November 2002), pp. 135-148, ISSN 1470-160X

Macossay-Cortez, A., Sánchez, A.J., Florido, R., Huidobro, L., & Montalvo, H. (2011). Historical and environmental distribution of Ichtyofauna in the tropical wetland of Pantanos de Centla, southern Gulf of Mexico. *Acta Ichthyologica et Piscatoria*, Vol. XLI, No. 3, (September 2011), pp. 229-245, ISSN 1734-1515

Marceniuk, A.P., & Bentacur, R. (2008). Revision of the species of the genus *Cathorops* (Siluriformes: Ariidae) from Mesoamerica and the Central American Caribbean, with description of three new species. *Neotropical Ichthyology*, Vol. 6, No. 1, (March 2008), pp. 25-44, ISSN 1679-6225

McTigue, T.A., & Zimmerman, R.J. (1991). Carnivory versus herbivory in juvenile *Penaeus setiferus* (Linnaeus) and *Penaeus aztecus* (Ives). *Journal of Experimental Marine Biology and Ecology*, Vol. 151, No. 1, (August 1991), pp. 1-16, ISSN 0022-0981

Mendoza-Carranza, M., Hoeinghaus, D.J., Garcia, A., & Romero-Rodriguez, A. (2010). Aquatic food webs in mangrove and seagrass habitats of Centla Wetland, a Biosphere Reserve in Southeastern Mexico. *Neotropical Ichthyology*, Vol. 8, No. 1, (January-march 2010), pp. 171-178, ISSN 1679-6225

Miller, R., Minckley, W.L., & Norris, S.T. (2005). *Freshwater fishes of México*, The University of Chicago Press, ISBN 978-607-7607-20-5, Chicago, Illinois.

Minello, T.J., & Zimmerman, R.J. (1991). The role of estuarine habitats in regulating growth and survival of juvenile penaeid shrimp, In: *Developments in aquaculture and fisheries science, frontiers in shrimp research,* P. DeLoach, W.J. Dougherty, & M.A. Davidson (Eds.), 1-16, Elsevier Science Publishers, ISBN 978-0-444-51940-4, Amsterdam, The Netherlands

Minello, T.J., Zimmerman, R.J., & Barrick, P.A. (1990). *Experimental studies on selection for vegetative structure by penaeid shrimp*, NOAA Technical Memorandum NMFS-SEFC-237, Galveston, Texas.

Montalvo-Urgel, H., Sánchez, A.J., Florido, R., & Macossay-Cortez, A. (2010). Lista de crustáceos distribuidos en troncos hundidos en el humedal tropical Pantanos de Centla, al sur del Golfo de México. *Revista Mexicana de Biodiversidad*, Vol. 81, No (Octubre 2010), pp. S 121- S 131, ISSSN 1870-3453

Morris, K., Bailey, C.P., Boon, I.P., & Hughes, L. (2003a). Alternative stable states in the aquatic vegetation of shallow urban lakes. II. Catastrophic loss of aquatic plants consequent to nutrient enrichment. *Marine and Freshwater Research*, Vol. 54, No. 3, (June 2003), pp. 201-215, ISSN 1448-6059

Morris, K., Boon, I.P., Bailey, C.P., & Hughes, I. (2003b). Alternative stable states in aquatic vegetation of shallow urban lakes. I. Effects of plant harvesting and low-level nutrient enrichment. *Marine and Freshwater Research*, Vol. 54, No. 3, (June 2003), pp. 185-200, ISSN 1448-6059

Moksnes, P.O., & Heck, L.K.Jr. (2006). Relative importance of habitat selection and predation for the distribution of blue crab megalopae and young juveniles. *Marine Ecology Progress Series*, Vol. 308, (February 2006), pp. 165-181, ISSN 01718630

Nates, J.C., & Villalobos-Hiriart, J.L. (1990). Dos nuevas especies de camarones de agua dulce del género *Macrobrachium* Bate (Crustacea, Decapoda, Palaemonidae), de la vertiente occidental de México. *Anales del Instituto de Biología, Universidad Nacional Autónoma de México, Serie Zoología,* Vol. 61, (1990), pp. 1-11, ISSN 0368-8720

Ni, L. (2001). Effects of Water Column Nutrient Enrichment on the Growth of *Potamogeton maackianus* A. Been. *Journal of Aquatic Plant Management,* Vol. 39, (January 2001), pp. 83-87, ISSN 0146-6623

Orth, R.J., Scott, R.M., Kenneth, A.M., & David, J.W. (2010). Eelgrass (*Zostera marina* L.) in the Chesapeake Bay Region of Mid-Atlantic Coast of the USA: Challenges in Conservation and Restoration. *Estuaries and Coasts,* Vol. 33, No. 1, (January 2010), pp. 139-150, ISSN 1559-2723

Pelicice, F.M., & Agostinho, A.A. (2006). Feeding ecology of fishes associated with *Egeria* spp. Patches in a tropical reservoir, Brazil. *Ecology of Freshwater Fish,* Vol. 15, No. 1, (March 2006), pp. 10-19, ISSN 1600-0633

Pérez-Castañeda, R., & Defeo, O. (2001). Population Variability of Four Sympatric Penaeid Shrimps (*Farfantepenaeus* spp.) in a Tropical Coastal Lagoon of Mexico. *Estuarine, Coastal and Shelf Science,* Vol. 52, No. 5, (May 2001), pp. 631-641, ISSN 0272-7714

Pérez-Farfante, I. & Kensley, B. (1997). *Penaeoid and sergestoid shrimps and prawns of the world. Keys and diagnoses for the families and genera,* Mémoires du Muséum National d'HistoireNaturelle, ISBN 2856535100, Paris

Quiroz, C.H., Molina, I.M., Rodríguez, J.G., & Díaz, M. (2007). Los lagos Zempoala y Tonatiahua del Parque Nacional Lagunas de Zempoala, Morelos. In: *Las aguas interiores de México: Conceptos y casos,* G. de la Lanza, & S. Hérnandez (Eds.), pp. 141-168, AGT EDITOR, S.A., ISBN 978-468-463-137-3, México.

Reséndez, M.A., & Salvadores-Baledón, M.L. (2000). Peces de la Reserva de la Biosfera Pantanos de Centla. Resultados preliminares. *Universidad y Ciencia,* Vol. 15, No. 30, (January 2000), pp. 141-146, ISSN 0186-2979

Revenga, C., Campbell, I., Abell, R., de Villiers, P., & Bryer, M. (2005). Prospects for monitoring fresh water ecosystems towards the 2010 targets. *Philosophical Transactions of the Royal Society B: Biological Sciences,* Vol. 360, No. 1454 (February 2005), pp. 397-413, ISSN 1471-2970

Rocha-Ramírez, A., Ramírez-Rojas, A., Chávez- López, R., & Alcocer, J. (2007). Invertebrate assemblages associated with root masses of *Eichhornia crassipes* (mart.) Solms-Laubach 1883 in the Alvarado lagoonal System, Veracruz, Mexico. *Aquatic Ecology,* Vol. 41, No. 2, (June 2007), pp. 319-333, ISSN 1386-2588

Rozas, L.P., & Minello, T.J. (2006). Nekton use *Vallisneria americana* Michx. (Wild Celery) beds and adjacent habitats in coastal Louisiana. *Estuaries and Coasts,* Vol. 29, No. 2, (April 2006), pp. 297-310, ISSN 1559-2731

Ruiz-Carrera, V., & Sánchez, A.J. (2008). Development of in vitro culture model for *Vallisneria americana* Mitch. *Universidad y Ciencia,* Vol. 24, No. 3, (December 2008), pp. 205-218, ISSN 0186-2979

Ruiz-Carrera, V., & Sánchez, A.J. (2012). Estrategias de propagación e investigación experimental in vitro para la repoblación de angiospermas sumergidas en humedales, con énfasis en *Vallisneria americana* Michx. en el sureste de México, In: *Recursos Acuáticos Costeros del Sureste: Tendencias actuales en investigación y estado del arte,* A.J. Sánchez, X. Chiappa-Carrara, & B. Pérez (Eds.), RECORECOS, CONCYTEY, UNACAR, UJAT, ECOSUR, UNAM, Mérida, México (accepted)

Salcedo, M.A., Sánchez, A.J., de la Lanza, G., Kamplicher, C., & Florido, R. (2012). Condición ecológica del humedal tropical Pantanos de Centla, In: *Recursos Acuáticos Costeros del Sureste: Tendencias actuales en investigación y estado del arte*, A.J. Sánchez, X. Chiappa-Carrara, & B. Pérez (Eds.), RECORECOS, CONCYTEY, UNACAR, UJAT, ECOSUR, UNAM, Mérida, México (accepted)

Sánchez, A.J. (1997). Habitat preference of *Penaeus* (F.) *duorarum* (Crustacea: Decapoda) in a tropical coastal lagoon, southwest Gulf of Mexico. *Journal of Experimental Marine Biology and Ecology*, Vol. 217, No. 1, (September 1997), pp. 107-117, ISSN 0022-0981

Sánchez, A.J., & Raz-Guzman, A. (1997). Distribution Patterns of Tropical Estuarine Brachyuran Crabs in the Gulf of Mexico. *Journal of Crustacean Biology*, Vol. 17, No. 4, (November 1997), pp. 609-620, ISSN 02780372

Sánchez, A.J., Salcedo, M.A., Florido, R., Armenta, A., Rodríguez, C., Galindo, A., & Moguel, E. (2007). Pantanos de Centla, un humedal costero tropical, In: *Las aguas interiores de México: Conceptos y casos*, G. de la Lanza, & S. Hérnandez (Eds.), pp. 399-422, AGT EDITOR, S.A., ISBN 978-468-463-137-3, México

Sánchez, A.J., Florido, R., & Barba, E. (2008). Ecosistemas y recursos hídricos en Tabasco, In: *Diseño del Instituto del Agua de Tabasco. Gestión Integral del recurso hídrico*, A.J. Sánchez, & J. Barajas-Fernández (Eds.), pp. 32-45, Fondo Editorial UJAT, ISBN 978-968-9024-63-7, Villahermosa, México

Sánchez, A.J., Florido, R., Macossay-Cortez, A., Cruz-Ascencio, M., Montalvo-Urgel, H., & Garrido, A. (2012). Distribución de macroinvertebrados acuáticos y peces en cuatro hábitat en Pantanos de Centla, sur del Golfo de México, In: *Recursos Acuáticos Costeros del Sureste: Tendencias actuales en investigación y estado del arte*, A.J. Sánchez, X. Chiappa-Carrara, & B. Pérez (Eds.), RECORECOS, CONCYTEY, UNACAR, UJAT, ECOSUR, UNAM, Mérida, México (accepted)

SAS Institute Inc. (2010). *JMP Version 9.0.2, SAS Institute Inc.*, Cary, North Carolina.

Schallenberg, M., & Burns, C.W. (2004). Effects of sediment resuspension on phytoplankton production: teasing apart the influences of light, nutrients and algal entrainment. *Freshwater Biology*, Vol. 49, No. 2, (February 2004), pp. 143-159, ISSN 1365-2427

Schloesser, D.W., & Manny, B.A. (2007). Restoration of Wildcelery, *Vallisneria americana* Michx., in the Lower Detroit River of the Lake Huron-Lake Erie Corridor. *Journal of Great Lakes Research*, Vol. 33, Supplement 1, (2007), pp. 8-19, ISSN 0380-1330

Schmitter-Soto, J.J. (1998). *Catálogo de los peces continentales de Quintana Roo*, El Colegio de la Frontera Sur, San Cristóbal de las Casas, México

Schofield, P.J. (2003). Habitat selection of two gobies (*Microgobius gulosus, Gobiosoma robustum*): influence of structural complexity, competitive interactions, and presence of a predator. *Journal Experimental Marine Biology and Ecology*, Vol. 288, No. 1, (March 2003), pp. 125-137, ISSN 0022-0981

Schubart, C.D., Cuesta, J.A., & Felder, D.L. (2002). Glyptograpsidae, a new brachyuran family from Central America: larval and adult morphology, and molecular phylogeny of the Grapsoidea. *Journal of Crustacean Biology*, Vol. 22, No. 1, (February 2002), pp. 28-44, Print ISSN 0278-0372, Online ISSN 1937-240X

Schultz, S.T., & Kruschel, C. (2010). Frequency and success of ambush and chase predation in fish assemblages associated with seagrass and bare sediment in an Adriatic lagoon. *Hydrobiologia*, Vol. 649, No. 1, (July 2010), pp. 25-37, ISSN 1573-5117

SCOR-UNESCO. (1966). Determination of photosynthetic pigments in seawater. *Monographs on Oceanographic Methodology*, UNESCO, Paris, France

Seitz, R.D., Lipcius, R.N., Olmstead, N.H., Seebo, M.S., & Lambert, D.M. (2006). Influence of shallow-water habitats and shoreline development on abundance, biomass, and diversity of benthic prey and predators in Chesapeake Bay. *Marine Ecology Progress Series*, Vol. 326, (November 2006), pp. 11-27, ISSN 01718630

Sheridan, P.F. (1992). Comparative habitat utilization by estuarine macrofauna within the mangrove ecosystem of Rookery Bay, Florida. *Bulletin of Marine Science*, Vol. 50, No. 1, (January 1992), pp. 21-39, ISSN 0007-4977

Sheridan, P., Henderson, C., & McMahan, G. (2003). Fauna of natural seagrass and transplanted *Halodule wrightii* (Shoalgrass) beds in Galveston Bay, Texas. *Restauration Ecology*, Vol. 11, No. 2, (June 2003), pp. 139-154, ISSN 1526-100X

Short, F.T., Koch, E.W., Creed, J.C., Karine, M., Magalhães, K.M., Fernandez, E., & Gaeckle, J.L. (2006). SeagrassNet monitoring across the Americas: case studies of seagrass decline. *Marine Ecology*, Vol. 27, No. 4, (December 2006), pp. 277-289, ISSN 0173-9565

Smith, D.G., & Thacker, C.E. (2000). Late-stage larvae of the family Microdesmidae (Teleostei, Gobioidei) from Belize, with note on systematic and biogeography in the Western Atlantic. *Bulletin of Marine Sciences*, Vol. 67, No. 3, (November 2000), pp. 997-1012, ISSN 0007-4977

Sosa-López, A., Mouillot, D., Ramos-Miranda, J., Flores-Hernandez, D., & Chi, T.D. (2007). Fish species richness decreases with salinity in tropical coastal lagoons. *Journal of Biogeography*, Vol. 34, No. 1, (January 2007), pp. 52–61, ISSN 1365-2699

Souza-Filho, E.E. (2009). Evaluation of the Upper Paraná River discharge controlled by reservoirs. *Brazilian Journal of Biology*, Vol. 69, No. 2, (June 2009), pp. 707-716, ISSN 1678-4375

Stoner, A.W., & Lewis, F.G. (1985). The influence of quantitative and qualitative aspects of habitat complexity in tropical sea-grass meadows. *Journal of Experimental Marine Biology and Ecology*, Vol. 94, No. 1-3, (December 1985), pp. 19-40, ISSN 0022-0981

Taniguchi, H., Nakano, S., & Tokeshi, M. (2003). Influences of habitat complexity on the diversity and abundance of epiphytic invertebrates on plants. *Freshwater Biology*, Vol. 48, No. 4, (April 2003), pp. 718-728, ISSN 1365-2427

Taylor, D.M. (2003). Introduction to Physidae (Gasteropoda: Hygrophila): Biogeography classification, morphology. *Revista de Biología Tropical*, Vol. 51, No. 1, (March 2003), pp. 1-287, ISSN 0034-7744

Thomaz, S.M., Bini, L.M., & Bozelli, R.L. (2007). Floods increase similarity among aquatic habitats in rivers-floodplain systems. *Hydrobiologia*, Vol. 579, No. 1, (November 2006), pp. 1–13, ISSN 1573-5117

Touchette, B.W., & Burkholder, J.M. (2000). Review of nitrogen and phosphorus metabolism in seagrasses. *Journal of Experimental Marine Biology and Ecology*, Vol. 250, No. 1-2, (July 2000), pp. 133-167, ISSN 0022-0981

Underwood, A.J. (1997). *Experiments in ecology*, Cambridge University Press, ISBN-10: 0521553296, Cambridge, United Kingdom. 504 pp.

Varona-Cordero, F. & Gutiérrez, F.J., (2003). Estudio multivariado de la fluctuación espacio-temporal de la comunidad fitoplanctónica en dos lagunas costeras del estado de Chiapas. *Hidrobiológica*, Vol. 13, No. 3, (2003), pp. 177-194, ISSN 0188-8897

Vega-Cendejas, M.E. (2004). Ictiofauna de la Reserva de la Biosfera Celestún, Yucatán: una contribución al conocimiento de su biodiversidad. *Anales del Instituto de Biología, Universidad Nacional Autónoma de México, Serie Zoológica,* Vol. 75, No. 1, (2004), pp. 193-206, ISSN 0368-8720

Velázquez-Villegas, G. (1994). *Los recursos hidrológicos del Estado de Tabasco. Ensayo monográfico,* Universidad Juárez Autónoma de Tabasco, Centro de Investigación de la División Académica de Ingeniería y Tecnología, Villahermosa, México

Villalobos-Figueroa, A. (1983). *Crayfishes of Mexico (Crustacea: Decapoda),* Smithsonian Institution Press, ISBN 9061912954, Washington, D. C.

Waycott, M., Duarte, C.M., Carruthers, T.J.B., Orth, R.J., Dennison, W.C., Olyarnik, S., Calladine, A., Fourqurean, J.W., Heck, K.L.Jr., Hughes, A.R., Kendrick, G.A., Kenworthy, W.J., Short, F.T., & Williams, S.L. (2009). Accelerating loss of seagrass across the globe threatens coastal ecosystems. *Proceedings of the National Academy of Sciences,* Vol. 106, No. 30, (July 2009), pp. 12377-12381, ISSN 0027-8424

Wedepohl, R.E., Knauer, D.R., Wolbert, G.B., Olem, H., Garrison, P.J., & Kepford, K. (1990). *Monitoring lake and reservoir restoration,* EPA 440/4-90-007. Prepared by North American Lake Management Society for United States, Environmental Protection Agency, Washington, DC

Weilhoefer, C.L., Yangdong, P., & Eppard, S., (2008). The effects of river floodwaters on floodplain wetland water quality and diatom assemblages. *Wetlands,* Vol. 28, No. 2, (June 2008), pp. 473-486, Electronic ISSN 1943-6246

Wigand, C., Wehr, J., Limburg, K., Gorham, B., Longergan, S., & Findlay S. (2000). Effect of *Vallisneria americana* (L.) on community structure and ecosystem function in lake mesocosms. *Hydrobiology,* Vol. 418, No. 1, (January 2000), pp. 137-146, Print ISSN 0018-8158, Electronic ISSN 1573-5117

Williams, A.B. (1984). *Shrimps, lobsters and crabs of the Atlantic coast of the Eastern United States, Maine to Florida,* Smithsonian Institution Press, ISBN 0874749603, Washington DC

Zimmerman, R.J., Minello, T.J., Smith, D.L., & Kostera, J. (1990). The use of *Juncus* and *Spartina* marshes by fisheries species in Lavaca Bay, Texas, with reference to effects of floods.-NOAA Technical Memorandum. NMFS-SEFC-251: 1-40.

2

Impacts of Carbon Dioxide Gas Leaks from Geological Storage Sites on Soil Ecology and Above-Ground Vegetation

Raveendra H. Patil
Agricultural and Biological Engineering Department,
Institute of Food and Agricultural Sciences, University of Florida, Gainesville
USA

1. Introduction

The global annual carbon dioxide (CO_2) gas emissions, one of the major green house gases contributing to global warming, into atmosphere have increased to 31.5 Gt in 2008 (German renewable energy institute [IWR], 2009), of which little over two-third (~22 Gt) were emitted from manmade sources (Benson, 2005). Electricity production and transportation make up two-third of total manmade emissions while the rest is contributed from heating buildings and other industrial consumption. Our dependence on fossil fuels continues to be increasing with over 85% of the world's energy needs still coming from burning oil, coal and natural gas (Benson, 2005) and by the end of 21st century our demand for fossil fuel is projected to more than double. Some of the scenarios (e.g. with no action taken to limit emissions) suggest a doubling of CO_2 emissions by 2050. This increased energy demand is mainly driven by continued industrialization and improved quality of life in not only western countries (North America and Europe), but also in two of the largest populated countries in the world viz., China and India. This would further increase CO_2 gas emissions into atmosphere with a strong positive feedback on climate change. This demands urgent action to reduce or offset CO_2 emissions from fossil fuels. In this context, while the efforts are being made for cleaner and efficient use of fossil fuels, carbon capture and storage technology is being considered as one of the main short-term viable strategy to help mitigation climate change in coming decades (Bachu, 2000; Intergovernmental Panel on Climate Change [IPCC], 2005; Pacala et al., 2004; Van de Zwaan & Smekens, 2009).

1.1 Carbon capture and storage technology

The carbon capture and storage technology basically includes two approaches: (1) CO_2 gas is captured directly from the large and stationary source points (e.g. power plants, petroleum refineries, gas processing units & cement factories), concentrated into a nearly pure form, transported through pipelines and injected into geological storage sites (on- and off- shore) far below the ground surface (Bachu, 2000). This technology is called as "Carbon Capture and Storage" (CCS), and (2) atmospheric CO_2 gas is biologically fixed by growing vegetation (e.g. forest trees, biomass crops) and stored in above- and below-ground plant parts. This is referred to as "Carbon Sequestration" (CS) and these two approaches complement each other as CCS approach, unlike CS, cannot capture CO_2 gas directly from the atmosphere. Hence, CCS technology is increasingly seen as one of the mechanisms that can make a useful contribution to the reduction of CO_2 emissions over the next 50 years (IPCC, 2005). Some estimates predict that CCS technology has the potential to capture and store CO_2 emissions from power generation by 80−90% (Department of Trade and Industry, UK [DTI], 2002). However, in order to have CCS that significant impact, CO_2 gas will have to be injected on a large scale (in Gt year^{-1}).

1.2 The geological storage sites and their storage capacity

Geological formations suitable for CO_2 storage are located in sedimentary basins where thick sediments have accumulated over millions of years (Benson, 2005). Such sites are oil and gas reservoirs, deep saline aquifers with suitable caprocks and deep unmineable coal beds. Some estimates suggest that depleted oil and gas reservoirs can store as much as 800 Gt of CO_2 gas (Fruend et al, 2003). These sites are known to trap buoyant fluids such as CO_2 and CH_4 underground for millions of years. Hence, CCS technology intends to adopt the methods already used by oil / gas exploration and production, and enhanced oil recovery (EOR) schemes. Under EOR schemes CO_2 gas is injected into oil reservoirs to make it dissolve into the oil which reduces the viscosity of oil while increasing its volume to enhance reservoir pressure and production (Benson, 2005). For example, EOR scheme at Williston Basin oilfield, Canada started injecting ~5000 tonnes of CO_2 day^{-1} in October 2000. The CO_2 gas is being transported through 330 km long pipeline from the lignite-fired Dakota Gasification Company plant site in North Dakota, USA (DTI, 2005). The saline aquifers are the reservoirs deep underground and contain saline water (not suitable for drinking) and the global storage capacity of these sites is estimated to be between 400 and 10,000 Gt of CO_2 (Freund et al., 2003). Whereas, the unmineable coal beds, which are located too deep beneath the ground making them uneconomical to explore, can also be used to inject CO_2 gas where the injected gas is absorbed onto the coal. The global storage capacity of unmineable coal beds is estimated to be around 148 Gt of CO_2 (Freund et al., 2003).

1.3 Potential risks of failures of CCS technology

Presently existing CCS technology suggest that geological formations selected for CO_2 storage needs to be located at a minimum depth of 800 m so that CO_2 gas is stored at supercritical state to trap large amounts of gas in a small volume. However, in order to implement CCS technology on large and commercial scale, it is essential to assess all the potential risks and provide evidence to inform governments and the public that potential risks are well understood and impact assessments are studied for long-term safety and control measures (Wei et al., 2011).

As the CO_2 gas is captured from the large production sites located on land (e.g. power plants), the gas needs to be transported through pipelines over a long distance. The first source of potential leakage, therefore, would be pipeline failures or small leaks from joints. The scale of leakage and the potential impacts on surrounding environment from such leakages depends on whether the pipeline is laid over or under the ground and whether the pipeline pass through built-up area or near large drinking water sources (e.g. inland lakes or drinking water reservoirs). Pipeline corrosion over a period of time is another issue that might cause gas leakage. CO_2 gas transportation via pipelines in a supercritical state is reported to cause corrosion of steel pipes @ 0.01 mm year^{-1}. However, the corrosion rate further increases to 0.7 mm year^{-1} if free water was present in the pipeline (IPCC, 2005). If other gasses like hydrogen sulfide or hydrocarbons are mixed with CO_2 gas, then the chances of leakage and corrosion increase further (Klusman, 2003). Therefore, the pipelines require continuous surveillance for leakage or third party intrusions or encroachments. For example, the Cortez pipeline in Colorado, USA which is buried 1 m below ground, but passes through build-up areas is being air monitored once every two weeks (IPCC, 2005).

The leakage from geological storage sites may also occur due to failure of the sealing cap of injection well or migration of gas through geologic media and lead to slow but large releases either due to over-pressurization or slow releases via faults and fractures (Heinrich et al., 2003; Klusman, 2003). While the past evidence from oil and gas fields shows that the natural gases and fluids, including CO_2 gas, can be held intact underground for millions of years, incidences like the one of McElmo dome leakage (Gerlach et al., 1998; Stevens et al., 2000) demand complete evaluation of storage sites before selecting for gas injection (IPCC, 2005). Therefore, in order for CCS to be effective, the CCS technology must ensure that leakage is minimized from both sudden releases due to accidents or technical failures and slow leaks over longer period of time.

Perry (2005) reports that more than 600 natural gas storage reservoirs exist across the globe, but to-date on only 10 occasions significant leakage have occurred. These leakages were mainly attributed to failure of bore well integrity (5 times), leaks in the caprocks (4 times) and poor site selection (1 time). However, when it comes to long-term safety of CCS technology there are uncertainties as we lack the experience on the long-term fate (100 to 1000 years) and safety of large volume of CO_2 gas to be injected into geological formations (Celia et al., 2002). Some of the naturally occurring CO_2 springs and volcanic sites located across the world have been emitting CO_2 gas with no sever effect on our ecosystems or population, but the risks of CO_2 gas leakage from CCS transport pipelines or storage sites will be on larger scale as huge amounts of CO_2 gas is being handled under CCS schemes (IPCC, 2005). If leaks were to occur, the CCS technology would defy the very purpose to help mitigate climate change (Heinrich et al., 2003). While the existing CCS technology claims to reduce the risks of such leakages by applying safety systems in place and selecting safe geological storage sites (both on- & off-shore), there are chances that slow but continuous releases from CCS sites could go unnoticed as leaking CO_2 gas would quickly diffuse in the atmosphere (Heinrich et al., 2003). Therefore, it is important to understand the effects of CO_2 gas leaks on surrounding environments viz., marine life if gas were to be injected into off-shore geological sites and, vegetation and soil ecosystems if gas were to be injected into on-shore geological sites.

1.4 Effects of naturally occurring CO_2 leaks on soil geochemistry

Elevated soil CO_2 concentrations can cause changes in mineralogy composition together with changes in trace elements like As and Cr (Kruger et al., 2009; Stenhouse et al., 2009). Changes in cation exchange capacity (CEC), and the presence of oxides like CaO, MgO, Fe_2O_3, and Mn_3O_4 have also been reported elsewhere (Blake et al., 2000; Billett et al., 1987; Goulding et al., 1998). Leakage of CO_2 may also reduce groundwater pH besides affecting taste, color or smell and cause significant deterioration in the quality of potable groundwater by altering groundwater chemistry (Stenhouse et al., 2009). CO_2 leakage rising to sub-soil levels may also cause changes in subsurface microbial populations either by favoring some species or restricting others, depending on species type and site characteristics (Jossi et al., 2006; Tian et al., 2001).

1.5 Effects of naturally occurring CO_2 leaks on overlying vegetation

CO_2 is an odor less and non-toxic gas, and considered an integral part of our everyday lives. However, exposure to high concentrations of CO_2 poses danger to human beings, animals and surrounding environment, but such leaks may increase soil CO_2 concentrations in near surface and below vegetation canopies. Such a situation could have significant effects on above-ground vegetation and soil inhabiting organisms (e.g. earth worms, microorganisms), both in the short- and long-term. While plants in general are known to be more tolerant to elevated CO_2 gas, persistent leaks from geological storage sites may lead to accumulation of CO_2 gas in soil (near and sub surface). This may suppress root respiration, alter plant water / nutrient uptake capacity by altering soil pH towards acidity and ultimately affect above-ground biomass (Celia et al., 2002; Cook et al., 1998; Gahrooee, 1998; Miglietta et al., 1998; Sorey et al., 2000; Sowerby et al., 2000; Stephens & Hering, 2002). There are many studies where forest trees or perennial vegetation mortality from naturally occurring active volcanoes emitting CO_2 gas into atmosphere have also been documented (e.g. Macek et al., 2005; Stephens & Hering, 2002; Vodnik et al., 2006). In fact, to-date naturally occurring CO_2 springs and active volcanic sites are the ones quite extensively studied in Europe and America.

At Mammoth Mountain, USA, more than 30% by volume of soil CO_2 levels were measured at sites where tree mortality has occurred (Gerlach et al., 2001). At Bossoleto, Italy, soil CO_2 leaks increasing atmospheric CO_2 concentrations to as high as 75% by volume at night, but much lower during day time, were recorded (Van Gardingen et al., 1995). Low levels of CO_2 concentrations during day time, the period when plants are photosynthetically active, may have reduced the adverse effect of elevated CO_2 gas on trees and herbaceous plants or these plant species may well have adapted to elevated CO_2 gas after being exposed for long period of time (van Gardingen et al., 1995). While anaerobic conditions (anoxic or hypoxic) are harmful to plants, many species are known to adapt to such conditions, at least temporarily, by modifying their rooting system and supplying O_2 internally from leaves to roots (Vartapetian & Jackson, 1997). This may have been the case at Bossoleto, Italy where *Phragmites australis*, a wetland species adapted to temporary anoxic conditions was reported to be the most dominant plant cover (Van Gardingen et al., 1995).

Most of studies referred in above paragraphs including the ones reported elsewhere (Beaubien et al., 2008; Biondi & Fessenden, 1999; Vodnik et al., 2006;) have examined the effects of naturally occurring CO_2 leaks from active volcanoes / geothermal sites, and hot / cold CO_2 springs on ecosystems. At these environments the existing ecosystems have been exposed to elevated CO_2 for considerably long periods, thus the plant / tree species may have adapted (West et al., 2009). Therefore, the findings from these sites may not be representative of the effects of potential leakage from a CCS storage sites. Moreover, many of these sites release not only hot or cold CO_2 gas but also gas mixed with either volcanic ash or mineral particles or with other gasses like CH_4 (Bergfeld et al., 2006), which makes it difficult to isolate the effect of CO_2 gas alone. Furthermore, the effects of slow release of CO_2 from CCS underground transport pipelines or geological storage sites on plant and soil inhabiting organisms are not well studied and needs better understanding (Lewicki et al., 2005). Past studies also suggest that not all the plant species respond similarly when exposed to elevated soil CO_2 concentrations (Van Gardingen et al., 1995), hence makes it difficult to define a "lethal CO_2 concentration level" as different vegetation types respond differently to anoxic stress or anaerobic soil environments due to sever dearth of O_2 level in the soil. Added to this, variations amongst the natural ecosystems and their surrounding environment make it difficult to gauge exposure levels (Sarah & Sjogersten, 2009).

Hence, understanding the risk of CO_2 leakage from storage sites and studying the effect of leaking gas on surface ecosystem is critical for multiple reasons. First, accurate data on the quantity of CO_2 that has been injected and stored is required for trading and accounting purposes; second, any leakage will negate the original purpose of the CCS technology and; third, the leaking CO_2 might damage surface ecosystems including above-ground vegetation and soil ecology. Findings from such studies would enable CCS technology to adopt full safety measures while transporting and injecting huge amounts of CO_2 gas into on-shore geological storage sites, and alleviate the public perceptions, if any, on long-term safety of CCS technology.

In order to test the response of overlying vegetation and soil ecology to gas leaks an experimental field facility, the Artificial Soil Gassing And Response Detection (ASGARD), has been established at the University of Nottingham, UK where gas can be artificially injected into the soil to simulate build up of gas concentrations and its slow release to soil surface (Photo 1). This facility has enabled to study some impacts of a controlled injection of CO_2 gas on a non-adapted pasture grass and field crops, and on soil ecology and chemistry. This chapter describes some of the main findings from studies carried out at the ASGARD site.

Between 2002 and 2005, the ASGARD site was used to investigate the effect of elevated concentrations of soil CH_4 on pasture grass, wheat and winter bean crops (Smith et al., 2005). The natural gas, a major source of energy supply in Europe, is mainly composed of CH_4 gas (78-95% by vol.). European natural gas sub-surface transportation pipelines are under regular helicopter surveillance to detect any gas leaks from pipeline joints, third party incursion and land slippage which may physically damage pipeline and gas supply (Smith et al., 2005). According to Baggott et al. (2003) the CH_4 gas leakage from UK natural gas transportation system alone amounts to 342 kilo tonnes year^{-1}, which is equivalent to 10260 kilo tonnes of CO_2 in terms of global warming potential. Thus, early warning system to detect CH_4 gas leaks would enable the surveillance system to take immediate control measures. Therefore, Smith et al., (2005) artificially injected CH_4 gas into the soil at ASGARD site, to simulate CH_4 leakage and monitored above-ground vegetation stress symptoms using remote sensing technology.

Since 2006 the ASGARD site is being used to inject CO_2 gas, initially by the British Geological Survey, UK (West et al., 2009) and later by Patil et al. (2010) and Sarah and Sjogersten, (2009) in 2007-08. Therefore, this chapter while describing the ASGARD site in detail presents main findings of previous studies undertaken at this site. These studies were undertaken with multiple objectives: (1) to develop a field site where CO_2 gas could be artificially injected at a targeted rate into the soil to simulate build up of soil CO_2 concentrations in near-surface and its leakage into the atmosphere, (2) to monitor temporal and spatial variations in soil gas concentrations under different land cover, and (3) to study the response of vegetation(s), soil properties, and soil inhabiting organisms to elevated soil CO_2 concentrations.

Photo 1. The ASGARD site. Both pasture and fallow plots with gassing pipes and measuring tubes can be seen.

2. Methodology

2.1 The study location and the ASGARD site

The ASGARD research facility is located on a flat and open field of permanent pasture at the Sutton Bonington campus of the University of Nottingham (52° 49′ 60 N, 1° 14′ 60 E, 48 m a.s.l.), approximately 18 km south of central Nottingham, UK (Figure 1). The site was previously used as a sheep pasture and had remained grassland for over 10 years until the ASGARD facility was laid out. Long term temperature average (1971–2000) at this site shows January as the coldest month with maximum and minimum temperatures of 6.9 and 1.2 °C, respectively and July as the warmest month with maximum and minimum temperatures of 21.3 and 11.4 °C, respectively. The mean annual precipitation is 606 mm; most of it is received as precipitation, and is fairly uniformly distributed throughout the year (Sarah & Sjogersten, 2009). The ASGARD site is positioned in such a way that it is not influenced by shade from trees or fencing.

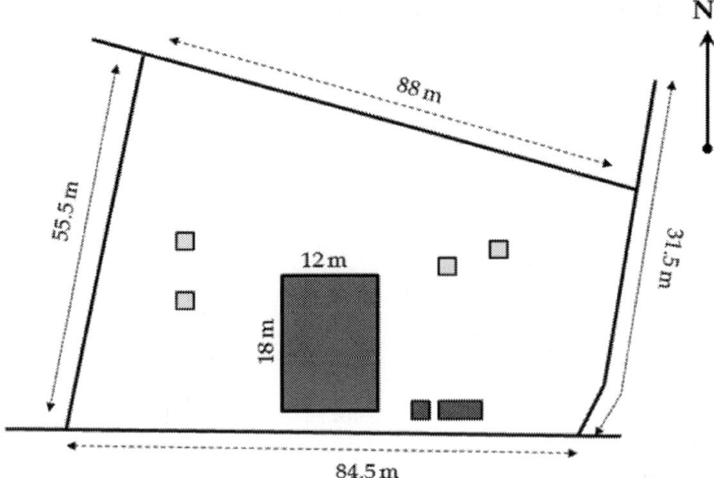

Fig. 1. Location of the ASGARD field facility and the main experimental block. Green- main experimental plots with a total of 30 plots, and Yellow – 4 remote control area located >10 m away from the main experimental plots. Blue- gas cylinders and port-cabin which accommodated automated gas supplying hardware and software system.

2.2 Site mineralogy and soil characteristics

Geologically, the study area is characterized by up to 1.5 m of head deposit overlying mudstones of the Mercia Mudstone Group. Mineralogical analyses in 2006 showed that quartz was the dominant mineral (>90% of the total dry weight) followed by K-feldspar and albite as minor along with trace amounts of mica, kaolinite, chlorite and hematite (West et al., 2009). Mineralogical composition between the A horizon (0.15–0.30 m depth) and B horizon (0.45–0.50 m depth) were found to be the same (West et al., 2009). However, readers are directed to refer to Ford (2006) for detailed geological description of this site and its surrounding area. The soil type lies within the Worcester Series and comprises 0.3 m deep sandy clay loam over 0.7 m clay and marl horizon. The top soil layer (~ 0.1 m) contains 8.91% clay, 22.89% silt and 68.2% sand (West et al., 2009).

2.3 Experimental plots and gas supply system

In the year 2006 a total of 30 plots each of 2.5 m × 2.5 m in size were laid out in a rectangular grid patterns (5 × 6 plots) with 0.5 m pathways between each plot (Figure 2). Ten plots were kept under already established pasture grass and rest were planted with agricultural crops after removing the pasture grass from those plots (West et al., 2009). CO_2 gas was delivered into the pasture plots at a constant rate of 3 liters min^{-1} for 19 weeks (May to September 2006). In the year 2007, Sarah and Sjogersten (2009) used 8 plots to plant commercial turf, previously planted with agricultural crops, but left untreated for more than 8 months before they began the study. The turf grass was composed of *Lolium perenne, Festuca rubra, Festuca rubra commutate* and *Poa pratensis* types. The CO_2 gassing was started 6 weeks after planting the turf grass. Of the total 8 turf grass plots, 4 plots were injected with CO_2 gas at a constant rate of 1 liter min^{-1} for 10 weeks (June to August 2007). The remaining 4 plots were left un-gassed (control plots). Patil et al. (2010) used a total of 16 plots: 8 pasture grass plots and 8 fallow plots which were previously cultivated with agricultural crops, but left untreated for more than 6 months. Pasture and fallow plots were chosen to represent two land use types: perennial pasture grass and fallow land (bare soil). Eight out of these 16 plots (4 pasture & 4 fallow), chosen randomly, were equipped for controlled CO_2 release at the center of the plots at a constant rate of 1 liter min^{-1} for 9 months (May 2007 through January 2008). This subsurface gas injection combined with the flat terrain of the field site prevented build up of air CO_2 concentration at the site.

During all these studies pure industrial CO_2 gas was supplied via cryogenic cylinders (supplied by British Oil Company [BOC], UK) and the gas flow to each one of those gassing plots was controlled automatically by individual mass flow controllers (Photo 2). The CO_2 gas was delivered from the cryogenic cylinder using a 32 mm polyethylene gas pipe. Automated system in turn released the gas at a pre-determined rate into 15-mm copper tubes, one tube to each gassing plot. These copper tubes carried the gas up to experimental plots, where these tubes were separately connected to 22-mm internal diameter medium density polyethylene (MDPE) gas pipes, sealed at the far end (the end which went into soil at the centre of plot). To avoid obstructing the measurement area above-ground within each gassing plot, the MDPE gas pipes were inserted into augered holes at an angle of 45° to the vertical. These pipes were drilled with twenty-six 5-mm holes at the far end 0.1 m of the tube (the end that went into soil) to deliver CO_2 gas into the soil at the center of gassing plot at 0.5–0.6 m depth (Figure 3).

Photo 2. CO_2 gas delivery hardware and software system installed at the ASGARD site. Top left: automated system to control delivery of gas at a targeted rate, bottom left: copper tubes with automatically controlled valves to supply gas, bottom right: copper tubes are connected to yellow plastic pipes which, in turn, carry gas to individual plots and top right: pasture plots at the ASGARD site with measuring tubes and gas delivery pipes in each plots, and in the back ground the cryogenic gas cylinders and port-cabin can be seen.

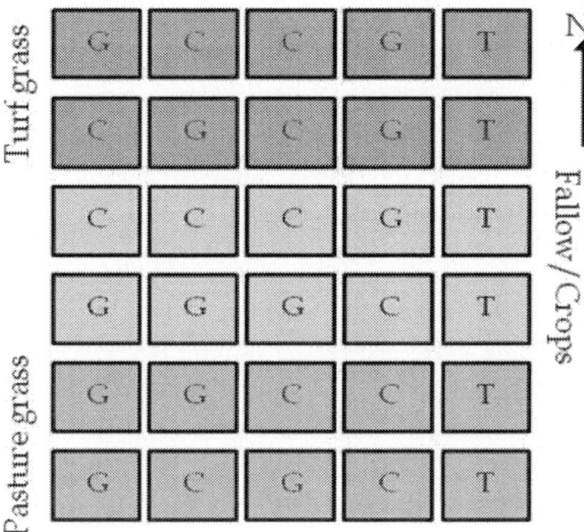

Fig. 2. Layout of 30 plots within the main experimental block. Light green plots- pasture grass, yellow plots– fallow and agricultural crops, and dark green – turf grass. G = gassed, C = un-gassed control and T = plots used only for testing the gassing system and all the T plots were not used to impose any of the treatments during study period.

The injection rate for all the gassed plots at ASGARD site since CO_2 gassing studies began varied from 1–3 liters min^{-1}, which equates to an annual injection rate of around 1–3 tonnes year^{-1}. This rate of gas injection is far less than the amount injected at some of the already existing off-shore storage sites. For example, Sleipner west field beneath the North Sea, Norway injects 1 million tonnes of CO_2 year^{-1}. Nevertheless, the injection rates selected at the ASGARD project site were constrained by the funding and the logistics put in place as well as for practical purposes.

2.4 Measurements on soil gas concentrations

Plastic tubes of 1 m long with 0.2 mm internal diameter were installed vertically into the soil to a depth of 0.3 m and at different distances from the center of plots on a diagonal transect in all the plots (gassed & un-gassed; Figure 3). The bottom end of each sampling tube (at 0.3 m depth) was sealed and the lower 0.15 m of the tube was drilled with 14 equally spaced holes (4.5 mm diameter). This portion of the tube was covered with fine messed cloth from inside the tube to prevent outside soil clogging the holes. These holes enabled free diffusion of gas from surrounding soil into the tube so as to attain equilibrium with the soil gas concentration at 0.15 to 0.30 m depth. The top end of the tube was sealed with a plastic on/off valve connectable to a portable GA2000 Landfill Gas Analyzer (Geotechnical Instruments UK Limited) to measure soil CO_2 and O_2 concentrations (in % of total 100% by vol.). In addition, Patil et al. (2010) measured soil surface CO_2 efflux from pasture grass and fallow plots (ppm hr^{-1}) using Draeger tubes (Draeger Safety AG & Co., Germany,) placed in a grid spacing of 0.5 m × 0.5 m on the surface to monitor gas diffusive pattern within each plot and its horizontal spread. The Draeger tube system is an established method for measuring and detecting contaminants in the soil, water and air.

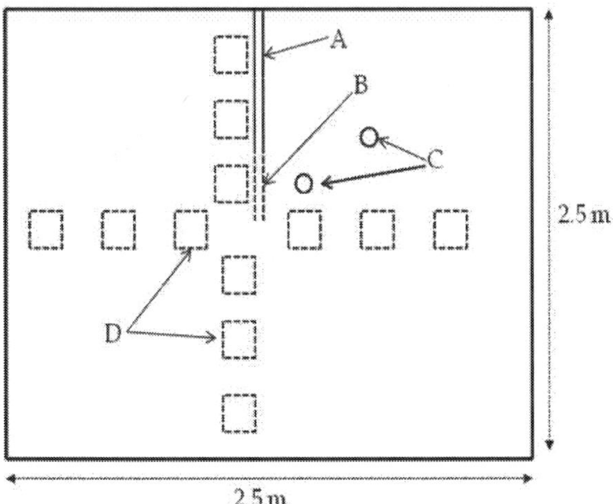

Fig. 3. Details of individual experimental plot showing the position of gas delivery pipe (A = above-ground portion, B = below-ground portion), gas measurement spots (C; one at 0.15 m & the other at 0.7 m away from center) and above-ground grass biomass sampling areas only in pasture plot (each of 0.2 × 0.2 m in size). This holds true for un-gassed plots but for the absence of soil gas measuring tube at 0.7 m away from centre of the plot.

2.5 Measurements on soil chemistry

West et al. (2009) analyzed soil samples from 0.15–0.30 m and 0.45–0.60 m depth, before and after gassing, for mineralogical (e.g. particle fraction) and geochemical (pH, organic carbon) content and changes due to gassing. Whereas, Patil et al. (2010) collected the soil samples from the top 0.3 m depth, one composite sample from each plot, three times during the gassing period at equal interval, and analyzed for pH and organic matter content. The holes left open after removing augered soil samples were re-filled immediately with local soil and marked to avoid repeated use of the same spots during subsequent sampling periods. Wei et al. (2011), on the other hand, used the top 0.12 m layer soil from the ASGARD site and incubated with CO_2 gas at varied levels of soil moisture under laboratory conditions to study the effect of CO_2 gassing on soil chemical properties and availability of mineral elements.

2.6 Measurements on soil inhabiting organisms

To study the effect of CO_2 gassing on soil-inhabiting organisms, earthworm activities were monitored by counting the number of their castings on the surface (Patil et al., 2010). Sarah and Sjogersten (2009) collected soil samples from the turf grass plots from the top 0.1 m depth and at different distance from the center of the plot. They did this on 3 occasions: before gassing began, and at 5 and 10 weeks after gassing. These soil samples were used to determine microbial biomass and microbial activity. West et al. (2009) also looked at soil bacterial population and adenosine triphosphate (ATP) concentrations, the latter as a measure of microbial activity in the soil.

2.7 Measurements on vegetation composition and growth

While West et al. (2009) did not collect destructive biomass samples, Sarah and Sjogersten (2009) collected turf grass above- and below-ground biomass samples to record dry biomass. West et al. (2009) instead monitored botanical composition (% plant species cover) of the pasture plots before and after the gassing period both in gassed and un-gassed plots. Patil et al. (2010) monitored pasture growth by collecting only its above-ground biomass from 0.2 × 0.2 m patch (0.04 m^2) above soil surface taken at distances of 0.3, 0.6 and 0.9 m from the plot center in all four directions (as shown in Figure 2), making up four samples within each plot from each distance interval. Each time after collecting biomass samples the pasture grass plots were mowed and let the grass grow again, whereas, Sarah and Sjogersten (2009) did not mow the turf grass. Between April through October 2007, the fallow plots had only bare soil with no vegetation cover and only soil gas concentrations were recorded using GA2000 Landfill Gas Analyzer as described in above paragraphs. On November 1, 2007 all the eight fallow plots were sown with winter bean (*Vicia faba* Cv. Clipper). The seeds were hand dibbled at 45 seeds per m^2. First germination count (number of seeds emerged per plot) was recorded on December 3, 2007 and the same was repeated at regular interval until the germination / emergence process was complete or no additional seeds emerged from both gassed and un-gassed plots.

2.8 Measurements on pasture grass stress responses

Pasture and turf grass stress symptoms were also monitored by recording visual appearance of grass on the surface (e.g. drying, brown / yellow coloration) both in gassed and un-gassed plots. Furthermore, Patil et al. (2010) monitored physiological stress responses to CO_2 gassing by measuring moisture content in above-ground grass biomass after each sampling (as a difference between fresh and oven dried biomass) and leaf chlorophyll content. To measure leaf chlorophyll content without any destructive sampling, the SPAD 502 Chlorophyll Meter (Spectrum Technologies Inc., USA) was used. The SPAD meter instantly measures the amount of chlorophyll content in leaves, a key indicator of plant health. The SPAD meter was clamped over pasture grass leaf part for few seconds and the meter displays an indexed chlorophyll content reading (0–99.9). Lower the value means higher stress. A more detailed description of these methodologies and measurements are given elsewhere (Patil et al., 2010; Sarah & Sjogersten, 2009; West et al., 2009).

3. Results and discussion

3.1 Soil gas leakage and migration pattern

The first objective of developing ASGARD field gassing facility was to achieve control over CO_2 gas injection into the soil at a targeted rate all through the study period. This was achieved successfully with the hardware and software logistics installed at the site. Furthermore, artificial gas injection at 0.5–0.6 m depth simulated gas diffusion and migration in the gassed plots (Patil et al., 2010; West et al., 2009). West et al. (2009), who injected CO_2 gas @ 3 liters min^{-1} for three months in 2006, reported horizontal migration of gas at a roughly similar rate in all directions in the gassed plots. They also observed that while the injected CO_2 gas had clearly migrated upwards throughout all the gassed plots, larger lateral movement was recorded at depth. This may have been probably influenced by the relative permeability of soil and the topographies of the boundaries between plots (West et al., 2009). The injected gas tended to move laterally beneath the soil beyond the boundaries of gassed plots as only the one-third amount of total injected gas was emitted at the surface from the gassed plots (West et al, 2009). However, in 2007-08 when CO_2 gas was injected at lower rates (1 liter min^{-1}) the lateral spread of diffused gas was relatively small, although the spread was much more in pasture plots than in fallow ones (Figure 4 & 5). This suggests that rate of leakage and the land use type (whether covered with pasture or left fallow) does influence the extent of surface flux and lateral migration of leaking gas. Hoeks (1972) suggests that the sub-surface injected CO_2 gas would migrate isotropically towards soil surface and creates a spherically symmetric gas plume, but that was not the case at ASGARD site (Figure 4 & 5). This may have been caused by the physical disturbance and loosening of soil along the path of inserted plastic pipe (Patil et al., 2010). Therefore, surface CO_2 flux rates recorded and its adverse effect on overlying vegetation were much higher in and around the entry point of plastic pipe (Photo 3). The diffusion and migration pattern of CO_2 gas seems to follow the pattern of CH_4 gas as reported by Adams and Ellis (1960) and Smith et al. (2005).

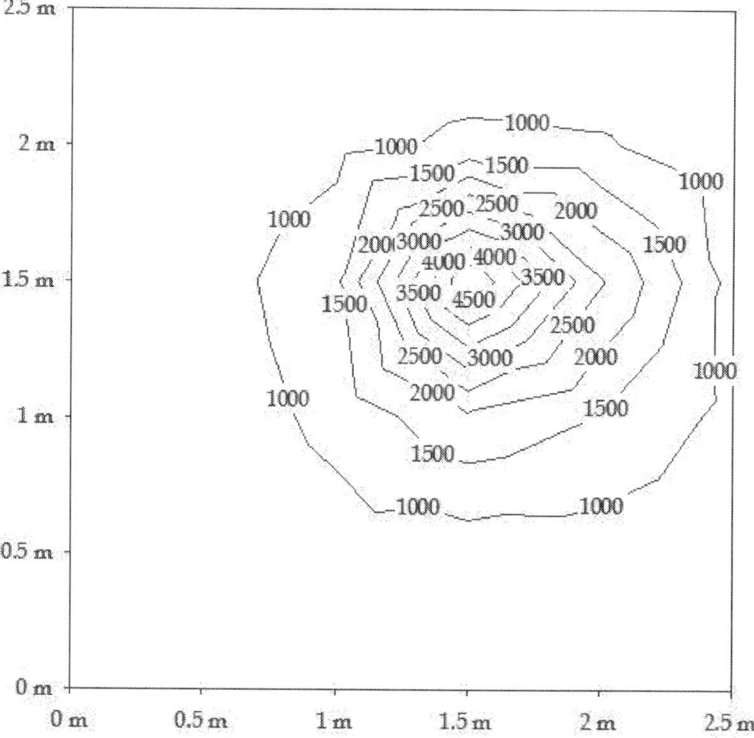

Fig. 4. Mean CO_2 efflux (ppm hr^{-1}) measured at 0.5 m × 0.5 m grid intervals at the soil-air interface in pasture plots using the Draeger tubes and shown in a 3-D contour map (source: Patil et al., 2010).

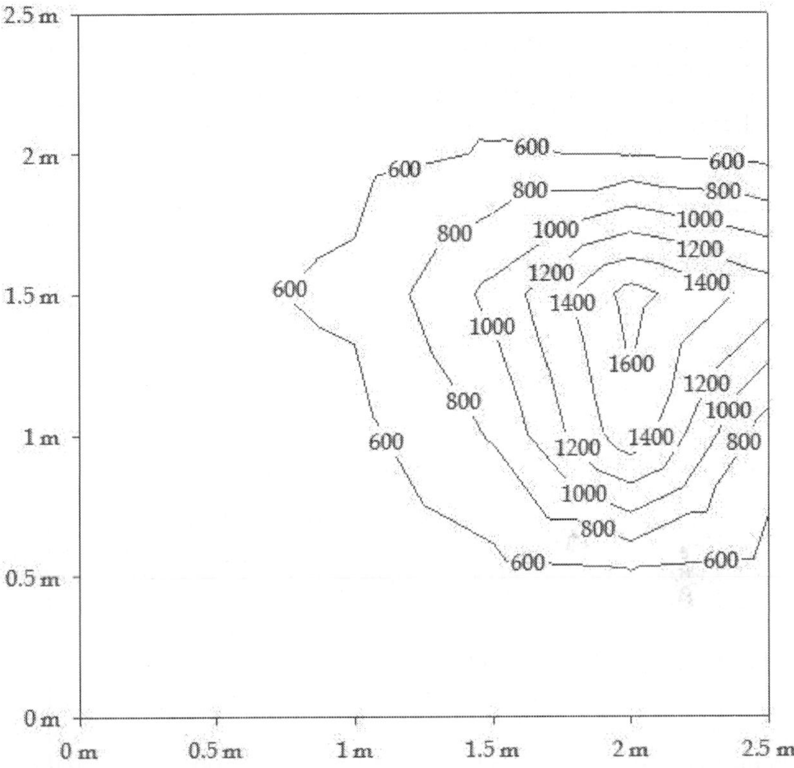

Fig. 5. Mean CO_2 efflux (ppm hr^{-1}) measured at 0.5 m × 0.5 m grid intervals at the soil-air interface in fallow plots using the Draeger tubes and shown in a 3-D contour map (source: Patil et al., 2010).

Photo 3. Vegetation in gassed pasture plots (top two) showing visual stress symptoms compared with non stressed un-gassed pasture plots (middle two), and winter bean emergence and seedlings growth pattern in gassed (bottom right) and un-gassed plots (bottom left).

3.2 Effect of land vegetation cover on soil gas concentrations

Gassing significantly increased the soil CO_2 concentrations in both pasture and fallow plots compared with their respective un-gassed pasture and fallow plots (Patil et al., 2010). In the gassed pasture plots, the maximum soil CO_2 concentrations were in the range of 19.5–76.3% with a seasonal mean of 45.7% (by vol. of the total 100%) from the measuring tubes that were 0.15 m from the center of the plots and significantly less, which ranged between 6.1–29.2% with mean of 11.5%, in measurement tubes at 0.7 m from the center as lesser amount of CO_2 migrated laterally below the soil surface. Similar trend was recorded from gassed fallow (bare soil) plots, although the seasonal high, low and mean CO_2 concentrations were on much lower side compared with gassed pasture plots. In contrast, soil CO_2 concentrations recorded from un-gassed plots were very low. From un-gassed pasture and fallow plots seasonal mean CO_2 concentrations of 1.9% and 0.7%, respectively, were recorded, but the soil O_2 concentrations reached as high as 20.8%. This clearly showed a significant negative relationship ($r^2=-0.95$) between soil CO_2 and O_2 concentrations as soil O_2 gas was displaced by injected CO_2 gas, since the latter one is heavier than former one (Patil et al., 2010). The difference in soil CO_2 concentrations between gassed pasture and gassed fallow plots also suggests that build up of injected CO_2 gas in near-surface and its diffusion / lateral migration are influenced by the land use type. Presence of vegetation tended to hold more gas due to the presence of root biomass, where as bare soil tended to release the leaking gas quickly into the atmosphere. This may have been to do with difference in soil aggregation. Pasture plots were never ploughed or tilled for over 10 years. This may have helped the pasture to develop thickly matted rooting system closing the bigger soil pores while creating larger proportion of micro pores compared with the broken soil (with more of macro pores) in fallow plots due to ploughing and cultivating with agricultural crops. Wei et al. (2011) reported that when soil was exposed to 100% CO_2, the gas is either absorbed by the soil water present in the pore space or filled the micro pores between the soil particles. This suggests that pasture soil with more of micro pores tends to hold more gas and build up CO_2 concentrations compared with loosely aggregated soil from fallow plots or the soil cultivated with seasonal crops.

When it comes to CO_2 gas migration beyond the plots of gassing, Patil et al. (2010) recorded slightly higher levels of soil CO_2 concentrations in both un-gassed pasture and fallow plots compared with the pasture plots located far away from the main treatment plots (as remote control plots). This could only be possible with sub-surface lateral migration of diffusing CO_2 gas from gassed plots to adjacent un-gassed plots (Patil et al., 2010), which again shows that even with low levels of sub-surface gas leakage (e.g. 1 liter min^{-1}) the diffusing gas can move laterally beneath the surface over quite a distance and affect the overlying vegetation and soil ecology. Therefore, if CO_2 leaks were to occur from geological storage sites located >800 m deep via cracks / faults in caprocks the gas would start diffusing upwards and spread in a funnel like shape occupying large area by the time gas reaches soil surface. This leak, if goes unnoticed, could affect large patch of land and its surrounding ecosystem.

3.3 Effect CO_2 gassing on pasture composition

The baseline botanical characterization, recorded in 2006 before gassing study was initiated, showed a range of monocotyledonous (e.g. grasses) and dicotyledonous (e.g. dandelions, thistle, plantain, chickweed, mallow, clover) plant species in all the pasture plots. However, three months after gassing the % composition changed only in gassed plots (West et al., 2009). Grass was the dominant species at the center of the plot where the soil CO_2 concentrations were much higher (> 75% of the total 100% by vol.) than at the edges of the plots. Whereas, towards the edges of the gassed plots where CO_2 concentrations were around 45% at 0.2 m depth, grass was the dominant species. In contrast, un-gassed (control) pasture plots had more species composition and higher proportions of minor plant groups than the gassed plots even at the end of the injection period. This suggests that grass species tended to be more tolerant to higher soil CO_2 concentrations than the non-grass or broad leafed species found in this perennial pasture (West et al. 2009). Similar responses were reported from a Mediterranean pasture at a natural CO_2 vent (Latera, Italy) where Beaubien et al. (2008) studied the effects of venting CO_2 gas on the shallow ecosystem.

3.4 Effect of CO_2 gassing on pasture grass stress symptoms and growth

Despite being tolerant to elevated CO_2 concentrations, the grass turned yellow or brown and patches of bare earth was exposed as no vegetation (including perennial grass) grew in and around the point of entry of gassing pipeline, the patch where the soil CO_2 concentrations reached as high as 75% at 0.2 m depth (West et al., 2009). Similar visual stress symptoms on pasture and turf grass, respectively, were observed by Patil et al. (2010) and Sarah and Sjogersten (2009) when CO_2 gas was injected @ 1 liter min^{-1} and also when CH_4 gas was injected @ 100 liters hr^{-1} in 2002-03 (Smith et al., 2005). While the grass close to the point of entry of gassing pipe turned yellow and died, a small patch of grass in a circular pattern around the dead grass showed much shorter growth, looked yellow and dry, but not dead yet. Whereas, the pasture towards the edges of the plots looked green and less affected as the soil CO_2 gas concentrations recorded towards the edges of the plot were much lower.

The observations on above-ground biomass, chlorophyll content and moisture content of pasture grass followed the same pattern; significantly lower values at the centre and higher values at the edges of gassed plots (Patil et al., 2010). This pattern of effect on overlying vegetation very much followed the pattern of soil CO_2 gas migration below the surface (Figure 4 & 5). Previous studies carried out at sites of naturally occurring CO_2 springs / vents and or volcanic sites have shown that elevated soil CO_2 concentrations reduce plant growth, disrupt plant photosynthesis, inhibit root respiration mainly due to sever dearth of soil O_2 levels, and even kill the vegetation (Cook et al., 1998; Macek et al., 2005; Miglietta et al., 1998; Pfanz et al., 2007; Vartapetian & Jackson, 1997; Vodnik et al., 2006). Previous studies on leaking CH_4 gas have also caused the same stress symptoms on above-ground vegetation as both CO_2 and CH_4 gases displace O_2 gas from the soil (Arthur et al., 1985; Hoeks, 1972; Smith et al., 2005; Smith, 2002) thus depriving the plant roots off O_2 for respiration, which in turn affects other plant functions viz., water and nutrient uptake, evapotranspiration, photosynthesis and ultimately plant growth. In fact, Adamse et al. (1972) suggested that for the proper functioning of a healthy root system, a minimum soil O_2 concentration of 12–14% is required, whereas at ASGARD site in the gassed plots this was not the case. Macek et al. (2005), Vodnik et al. (2006) and Pfanz et al. (2007) studied plant responses in relation to measured soil CO_2 concentrations at a natural CO_2 spring in Slovenia. In their studies soil CO_2 concentrations reached as high as 100% near vents, but atmospheric concentrations barely exceeded ambient concentrations (360–500 ppm) due to fast dispersion of leaking CO_2 by winds on a flat terrain. The study by Macek et al. (2005) considered root respiration in seven plant species and found that only when exposed to very high CO_2 concentrations did it inhibit root respiration, but the effects were highly variable amongst different plant species. This suggests that sensitivity of plants to elevated soil CO_2 concentrations differs with plant species as some species are more sensitive than others. The study by Vodnik et al. (2006) noted that leaf chlorophyll content was negatively correlated with soil CO_2 concentration, whereas Pfanz et al. (2007) noted that plants exposed to high soil and air CO_2 concentrations contained lower levels of nutrients in their vegetative parts and did not flower.

3.5 Effect of CO_2 gassing on winter bean crop

Smith et al. (2005) while studying the effect of leaking CH_4 gas on crops (wheat and winter bean) observed that in the plot area with the highest CH_4 concentrations most seeds germinated, but did not grow further. However, towards edges of plots, an area approximately 0.5 m in diameter, showed reduced growth with yellow leaves. Similarly, in this study significant adverse effect of CH_4 gas on chlorophyll content and leaf area of grass, wheat and bean was also reported. When it comes to CO_2 gassing, significantly lower number of winter bean seeds emerged / grew into seedlings in gassed plots compared with un-gassed plots (Patil et al., 2010). Even a very low level of leakage with 1 liter min^{-1} gas injection had lethal effect on bean seeds emergence. Plants are known to be more sensitive to anaerobic conditions especially during early growth stages (Hoeks, 1972) and that might be the reason why CO_2 gassing had lethal effect on winter bean seed emergence and seedling growth compared with the response of well established pasture grass observed by Patil et al. (2010). Schollenberger (1930) while studying the effect of CH_4 leaks on crops noted that leaking gas pipeline killed all the oat seedlings within the range of 1.0–1.3 m while stunting the seedling growth up to 4–5 m away from pipeline and beyond which no injury appeared. Godwin et al. (1990) observed wheat being more tolerant to CH_4 leaks compared with oil seed rape, which again suggests that crops differ in their sensitivity to gas leaks. In fact, this differential sensitivity could be used in early warning system of gas leaks, especially from under-ground pipeline, by growing highly sensitive crops along the path of transportation pipeline.

3.6 Effect of CO_2 gassing on soil mineralogy and chemistry

No significant changes in the mineralogical composition of soils from gassed and non-gassed plots were recorded during the 16 weeks injection period (West et al, 2009). However, soil Ca concentrations, in general, decreased, but the largest reductions were recorded in the soil samples collected close to the injection point, which also recorded the highest CO_2 concentrations (West et al, 2009). Wei et al. (2011) suggested that elevated soil CO_2 concentrations enhance weathering of minerals, thus it would be possible to assess the impact of CO_2 leaks on soil mineralogy only if the studies were carried out for longer period of time (in years). With regards to soil pH, soil CO_2 gassing @ 3 liters min^{-1} for only 16 weeks reduced the soil pH in gassed plots and the drop in soil pH was drastic near the point of entry of gas delivering pipeline (West et al., 2009). Among the soil layers, soil samples collected from gassed 'A horizon' recorded the largest reduction in pH (by 0.5 units) after only 16 weeks of gassing (West et al, 2009). Similarly, Patil et al. (2010) also recorded lower soil pH both in gassed pasture and fallow plots in comparison with their respective un-gassed control plots. This could be attributed to acidification of soil water due to dissolution of leaking CO_2 gas (Celia et al., 2002) and was corroborated by the findings of Wei et al. (2011), where the latter reported that absorption of CO_2 gas would occur either by reacting with soil pore water or by filling the pore space, thus increased soil moisture increases soil CO_2 build up in the soil. This process further leads to the formation of H_2CO_3 which, in turn, lowers the pH of soil solution. Soils in general have a buffering capacity and when pH of soil solution was lowered in the presence of absorbed CO_2, clay minerals start to weather to neutralize the lowering pH by releasing minerals (to exchange with H^+ ions) leading to an increase in the $CaCl_2$-exchangeable concentration of Al. Thus, increase in uptake of CO_2 by the soil solution leads to increased Al mobilization in moist soil (Wei et al, 2011). However, Al concentrations reported in their study were much lower than plant tolerance limits, which range between 40 µM and 60 µM depending on species (Poschenrieder et al, 2008; Taylor et al, 1998). In addition to Al, Wei et al. (2011) also reported increase in $CaCl_2$-

exchangable concentrations for Mg, K, Ti, V, Cr, Mn, Fe, Co, Cu, Rb, Sr, Mo, Cs, Ba, Pb, Th and U, while the metal concentrations for Zn and Cd decreased. Whereas, total organic carbon (TOC) concentrations in the 'A horizon' increased in both gassed and non-gassed plots by the end of the summer growing season, but at the injection depth TOC concentrations were found to be on lower side in the gassed plot compared with the un-gassed plots (West et al., 2009). This suggests that elevated soil CO_2 concentrations lower soil pH and TOC, although different soil horizons responded differently depending on their buffering capacity to changes in pH and chemical components. However, these gassing studies were carried out for a very short period (few months) and variations in soil carbon observed in these studies show only the rough indications of changes in soil chemistry which, in this case, might have been of temporary in nature influenced by factors not controlled during the study period (e.g. seasonal temperature, above-ground vegetation and other climatic conditions). Therefore, further investigation by continuing gassing over long period of time needs to be undertaken.

3.7 Effect of CO_2 gassing on soil inhabiting organisms

CO_2 gassing significantly increased the earth worm casts in gassed pasture plots compared with un-gassed pasture plots, but no earth worm activities were noticed on both gassed and un-gassed fallow plots (Patil et al., 2010). Smith et al. (2005) also reported higher worm casts in grass plots injected with natural gas. While the higher earth worm casting on gassed pasture plots indicate increased activities, but surprisingly enough the absence of earth worm casting on both gassed and un-gassed fallow plots begs question which the authors fail to clarify. Therefore, this needs to be looked at in detail to understand the causation. When it comes to soil bacterial population CO_2 gassing drastically reduced their number in gassed plots. The ATP concentrations, a measure of microbiological activity, were also reported to be below detection limits at the center of the gassed plots where CO_2 concentrations reached maximum of 87% at 0.15−0.30 m depth (West et al, 2009).

Sarah ad Sjogersten (2009) observed a tendency for reduced respiration in the soil exposed to elevated CO_2 concentrations, not significant though, compared with the soil in un-gassed plots within a period of 10 weeks. Soil moisture content seemed to have negative significant influence on microbial respiration. This negative effect of higher soil moisture on soil respiration, Sarah and Sjogersten (2009) report, was not anticipated as increased soil moisture generally results in increased microbial respiration up to the point of anoxia (Wardle and Parkinson 1990; Davidson et al. 2000). One possible explanation could be that higher soil water content reduced the O_2 supply (anoxic condition) while increasing the share of CO_2 concentrations in soil pores owing to the latter's leakage from the center of the plot. Such a soil environment would inhibit microbial respiration in the same way when soil water content reaches saturation (Davidson et al. 2000). Sarah and Sjogersten (2009) also looked at carbon source utilization rate as an estimate of change in microbial community metabolism, which however, did not differ between the soil from gassed and un-gassed plots in a 72 hr incubation study.

However, findings from the atmospheric CO_2 concentrations enrichment studies (not the soil concentrations) suggest that longer period of time is required to see any noticeable changes in microbial communities. For example, Griffiths et al. (1998) recorded no difference in soil microbial communities in a less than a month long study whereas, Grayston et al.

(1998) observed changes in bacterial community in a study carried out for four long years. Therefore, to understand the actual and long lasting effects on soil chemistry, organic matter, microbial population / their diversity, the gassing needs to be continued for longer period.

4. Conclusions

The ASGARD site developed at the University of Nottingham, UK successfully injected CO_2 gas into the soil at a targeted rate, simulated build up of soil CO_2 concentrations near soil surface and enabled measurement of soil concentrations and its effect on vegetation and soil biogeochemistry. The CO_2 gassing studies from the ASGARD site, despite being carried out for short period of time, showed the potential ecological risks of CO_2 leakage from geological storage sites on agro-ecosystems. Surprisingly enough, even the low levels of gas leakage with 1–3 liters of CO_2 gas injection min^{-1} significantly increased the soil CO_2 concentrations in a very short period of time by displacing the soil O_2 and adversely affected the growth of pasture and turf grass, severely hindered the germination and establishment of winter bean crop, lowered soil pH, TOC and microbial population while increasing the activities of earthworms. These studies also showed that different plant species have different tolerance capacity to soil CO_2 concentrations. Monocotyledonous species (e.g. grass types) tended to be more tolerant compared with dicotyledonous species (e.g. beans & broad leaf species). Therefore, some of the more sensitive plant species (e.g. non-grassy species) could be used to grow along the path of CCS transportation pipelines in early warning system to detect gas leaks.

Another limitation of studies carried out at the ASGARD site was that the CO_2 gas was injected at a very low rate (1–3 liters min^{-1}) and at a shallow depth (0.5–0.6 m) compared with CCS schemes where amount of gas injected each minutes would be in tonnes and at > 800 m deep. Therefore, while the ASGARD site based studies increased our understanding of the effects of below ground CO_2 leaks on agro-environment, long-term studies to evaluate the potential long-term consequences on ecosystem are needed to be carried out on larger scale for decision making and management. In CCS schemes the upwards movement of gas from the deep storage sites, possibly via cracks and faults in the caprocks, and its further diffusion and migration before reaching the soil surface would naturally enable the gas to spread on much larger area. Moreover, under such a situation factors like quality of selected site, amount of gas injected, the state and pressure under which it is held beneath the caprocks, geochemistry of the path of gas movement and issues related to hydrology needs to be studied in detail in addition to the impacts on overlying vegetation and environment. Therefore, further gassing studies needs to be carried out on larger scale and continued for longer period to better understand the long-term impacts.

5. Acknowledgments

The fellowship and research funding awarded to the author of this chapter by the Royal Society, London International Incoming Fellowship program 2006/R2 is duly acknowledged. The author wish to acknowledge the host institution project leaders Prof. Jeremy J. Colls (now retired while writing this chapter) and Prof. Michael D. Steven, University of Nottingham for initiating and leading the ASGARD project site. Thanks are also due to the British Geological Survey, UK for extending logistics, resources and technical support while developing the ASGARD site and for undertaking baseline studies to start with.

6. References

Adams, R.J., & Ellis, R.J. (1960). Some physical and chemical changes in the soil brought about by saturation with natural gas. *Soil Science Society of America Proceedings* 24, pp. 41–44.

Adamse, A.D., Hoeks, J., DeBont, J.A.M., & van Hessel, J.F., (1972). Microbial activities in soil near natural gas leaks. *Archive Microbiology* 83, pp. 32–51.

Arthur, J., Leone, I., & Flower, F. (1985). The response of tomato plants to simulated landfill gas mixtures. *Journal of Environmental Science and Health* 20, pp. 913–925.

Bachu, S. (2000). Sequestration of CO_2 in geologic media: approaches for site selection in response to climate change. *Energy Conversion and Management* 41, pp. 953–970.

Baggott, S.L., Davidson, I., Dore, C., Doogwin, J., Milne, R., Murrels, T.P., Rose, M., Watterson, J.D., & Underwood, B. (2003). UK Greenhouse Gas Inventory, 1990 to 2001. *Annual report for submission under the Framework Convention on Climate Change,* http://www.naei.org.uk/reports.php (accessed February 10, 2008)

Beaubien, S. E., Ciotoli, G., Coombs, P., Dictor, M.C., Kruger, M., Lombardi, S., Pearce J. M., & West J. M. (2008). The impact of a naturally-occurring CO_2 gas vent on the shallow ecosystem and soil chemistry of a Mediterranean pasture (Latera, Italy). *Journal of Greenhouse Gas Control* 2, pp. 373–387.

Benson, S.M. (2005). Carbon dioxide capture for storage in deep geological formations – results from the CO_2 capture project. *In Geological Storage of Carbon Dioxide with Monitoring and Verification* Vol. 2, Benson, S.M. (Ed.), pp. 654, Elsevier Publishing, UK.

Bergfeld, D., Evans, W.C., Howle, J.F., Farrar, C.D. (2006). Carbon dioxide emissions from vegetation-kill zones around the resurgent dome of Long Valley caldera, eastern California, USA. *Journal of Volcanology and Geothermal Research* 152, pp. 140–56.

Billett, M.F., Parker-Jervis, F., FitzPatrick, E.A., & Cresser, M.S. (1990). Forest soil chemical changes between 1949/50 and 1987. *Journal of Soil Science,* 41, pp.133–145.

Biondi, F., & Fessenden, J. E. (1999). Response of lodgepole pine growth to CO_2 degassing at Mammoth Mountain, California. *Ecology* 80(7), pp. 2420–2426.

Blake, L., Goulding, KW.T., Mott, C.J.B., & Poulton, P.R. (2000). Temporal changes in chemical properties of air-dried stored soils and their interpretation for long-term experiments. *European Journal of Soil Science,* 51 (2), pp. 345–353.

Celia, M.A., Peters, C.A., & Bachu, S. (2002). Geological storage of CO_2: leakage pathways and environmental risks. *American Geophysical Union, Spring Meeting 2002,* abstract #GC32A-03 in SAO/NASA Astrophysics Data System.

Cook, A.D., Tissue, D.T., Roberts, S.W., Oechel, W.C. (1998). Effects of long-term elevated CO_2 from natural CO_2 springs on *Nardus stricta*: photosynthesis, biochemistry, growth and phenology. *Plant, Cell and Environment* 21, pp. 417–425.

Davidson, E.A., Verchot, L.V., Cattanio, J.H., Ackerman, I.L., & Carvalho, J.E.M. (2000) Effects of soil water content on soil respiration in forests and cattle pastures in eastern Amazonia. *Biogeochemistry* 48, pp. 53–69.

DTI, Department of Trade and Industry, UK, (2002). The feasibility of carbon dioxide capture and storage in the UK. *In project definition and scoping paper.* p. 12, DTI, London, UK.

DTI, Department of Trade and Industry, UK, (2005). Geoscintific monitoring of the Weyburn CO_2 flood. A multidisciplinary approach to CO_2 monitoring. *Project summary 281* to *Cleaner fossil fuels program*, DTI, London, UK.

Ford, R. (2006). The geology of the ASGARD site, Sutton Bonington. *British Geological Survey Internal Report IR/06/049R.*

Freund, P., Gale, J., Davison J., & Riley, N. (2003). Capture and geological storage of carbon dioxide – a status report on the technology. *IEA Greenhouse Gas R &D Program.* COAL R223, British Geological Survey, London, UK.

Gahrooee, F.R., (1998). Impacts of elevated atmospheric CO_2 on litter quality, litter decomposability, and nitrogen turnover rate of two oak species in a Mediterranean forest ecosystem. *Global Change Biology* 4, pp. 667–677.

Gelrach, T., Doukas, M., McGee, K., & Lessler, R. (1998). Three-year decline of magmatic CO_2 emissions from soils of a Mammoth Mountain tree kill: Horseshoe Lake, CA, 1995–1997. *Geophysical Research Letters*, 25, pp. 1947–1950.

Gerlach, T.M., Douglas, M.P., McGee, K.A., & Kessler, R. (2001). Soil efflux and total emission rates of magmatic CO_2 at the Horseshoe Lake tree kill, Mammoth Mountain, California, 1995–1997. *Chemical Geology* 177, pp. 101–116.

Godwin, R., Abouguendia, Z., & Thorpe, J. (1990). Response of soils and plants to natural gas migration from two wells in the Lloydminster area, *Saskatchewan Research Council Report (No. E- 2510-3-E-90)*, pp. 85.

Goulding, K.W.T., Bailey, N.J., Bradbury, N.J., Hargreaves, P., Howe, M., Murphy, D.V., Poulton, P. R., & Willison, T.W. (1998). Nitrogen Deposition and its Contribution to Nitrogen Cycling and Associated Soil Processes. *New Phytologist*, 139 (1), pp. 49–58.

Grayston, S.J., Campbell, C.D., Lutze, J.L., & Gifford, R.M. (1998). Impact of elevated CO_2 on the metabolic diversity of microbial communities in N-limited grass swards. *Plant Soil* 203, pp. 289–300.

Griffiths, B.S., Ritz, K., Ebblewhite, N., Paterson, E., & Killham, K. (1998). Ryegrass rhizosphere microbial community structure under elevated carbon dioxide concentrations, with observation on wheat rhizosphere. *Soil Biology & Biochemistry* 30, pp. 315–321.

Heinrich, J.J., Herzog, H.J., & Reiner, D.M. (2003). Environmental assessment of geologic storage of CO_2. Laboratory for Energy and the Environment, MIT, presented at the *Second National Conference on Carbon Sequestration*, Washington D.C., May 5–8, 2003.

Hoeks, J. (1972). Changes in the composition of soil air near leaks in natural gas mains. *Soil Science* 113, pp. 46–54.

IPCC, Intergovernmental Panel on Climate Change, (2005). Carbon Dioxide Capture and Storage. Special report prepared by Working Group III of the Intergovernmental Panel on Climate Change, Metz, B., Davidson, O., de Coninck, H.C., Loos, M., &

Meyer, L.A. (Eds), p. 442, Cambridge, United Kingdom: Cambridge University Press; and New York, NY, USA: Cambridge University Press,.

IWR, Institute for Renewable Energy, (2009). Climate protection just a lot of hot air – Global CO_2 emissions at record levels, Press Release, August 10, 2009, Internationales Wirtschaftsforum Regenerative Energien, Muenster, Germany (http://www.iwrpressedienst.de/iwr/Global-C02-emissions-2008-renewable-energy-investment-plan.pdf; as on July 5, 2010)

Jossi, M., Fromin, N., Tarnawski, S., Kohler, F., Gillet, F., Aragno, M. & Hamelin, J. (2006). How elevated pCO_2 modifies total and metabolically active bacterial communities in the Rhizosphere of two perennial grasses grown under field conditions. *FEMS Microbiology Ecology*, 55, pp. 339–350.

Klusman, R.W. (2003). Evaluation of leakage potential from a carbon dioxide EOR/sequestration project. *Energy Conversion and Management*, 44, pp. 1921–1940.

Kruger, M., West, J., Frerichs, J., Oppermann, B., Dictor, M.C., Joulian, C., Jones, D., Coombs, P., Green, K., Pearce, J., May, F., & Moller, I. (2009). Ecosystem effects of elevated CO_2 concentrations on microbial populations at a terrestrial CO_2 vent at Laacher See, Germany. GHGT-9, *Energy Procedia* 1, pp. 1933-1939.

Lewicki, J.L., Hilley, G.E., & Oldenburg, C.M. (2005). An improved strategy to detect CO_2 leakage for verification of geologic carbon sequestration, *Geophysical Research Letters*, 32 (19), L19403.

Macek, I., Pfanz, H., Francetic, V., Batic, F., & Vodnik, D. (2005). Root respiration response to high CO_2 concentrations in plants from natural CO_2 springs. *Environmental and Experimental Botany* 54, pp. 90-99.

Miglietta, F., Berrarini, I., Raschi, A., Korner, C., & Vaccari, F.P. (1998). Isotope discrimination and photosynthesis of vegetation growing in the Bossoleto CO_2 spring. *Chemosphere* 36, pp. 771-776.

Pacala, S., & Socolow, R. (2004). Stabilization Wedges: Solving the Climate Problem for the Next 50 Years with Current Technologies. *Science*, 305(5686), pp. 968–972.

Patil, R.H., Colls, J.J., & Steven, M.D. (2010). Effects of CO_2 gas leaks from geological storage sites on agro-ecosystems. *Energy*, 35, 4587–4591.

Perry, K. (2005). Natural gas storage industry experience and technology: Potential application to CO_2 – do photosynthetic and productivity data from FACE experiments support early predictions? *New Phytologist*, 162, pp. 253–280.

Pfanz, H., Vodnik, D., Whittman, C., Aschan, G., Batic, F., Turk, B., & Macek, I. (2007). Photosynthetic performance (CO2 compensation point, carboxylation efficiency, and net photosynthesis) of timothy grass (*Phleum pratense* L.) is affected by elevated carbon dioxide in post-volcanic mofette areas. *Environmental and Experimental Botany* 61, 41-48.

Poschenrieder, C., Gunse, B., Corrales, I., & Barcelo, J. (2008). A glance into aluminum toxicity and resistance in plants. *Science of the Total Environment* 400, pp. 356-368.

Sarah, P., & Sjogersten, S. (2009). Effects of below ground CO2 emissions on plant and microbial communities. *Plant Soil*, 325, pp. 197–205.

Scholenberger, C. (1930). Effect of leaking natural gas upon the soil. *Soil Science*, 29, pp. 260-266.

Smith, K.L., Steven, M.D., & Colls, J.J. (2005). Use of hyper spectral derivative ratios in the red-edge region to identify plant stress responses to gas leaks. *Remote Sensing of Environment* 92, 207–217.

Smith, K.L., (2002). Remote sensing of leaf responses to leaking underground natural gas. PhD Thesis, p. 225, University of Nottingham, UK.

Sorey, M.L., Farrar, C.D., Gerlach, T.M., McGee, K.A., Evans, W.C., Colvard, E.M., Hill, D.P., Bailey, R.A., Rogie, J.D., Hendley, J.W., II, & Stauffer, P.H. (2000). Invisible CO_2 gas killing trees at Mammoth Mountain, California: *U. S. Geological Survey Fact Sheet* FS 0172-96, p. 2.

Sowerby, A., Ball, A.S., Gray, T.G.R., Newton, P.C.D., & Clark, H. (2000). Elevated atmospheric CO_2 concentration from a natural soda spring affects the initial mineralization rates of naturally senesced C_3 and C_4 leaf litter. *Soil Biology & Biochemistry*, 32, pp. 1323–1327.

Stenhouse, M., Arthur, R., & Zhou, W. (2009). Assessing Environmental Impacts from Geological CO_2 Storage. *Energy Procedia*, 1(1), pp. 1895–1902.

Stephens, J.C., & Hering, J.G. (2002). Comparative characterization of volcanic ash soils exposed to decade-long elevated carbon dioxide concentrations at Mammoth Mountain, California, *Chemical Geology*, 186, pp. 301–313.

Stevens, S., Fox, C., & Melzer, L. (2000). McElmo dome and St. Johns natural CO_2 deposits: Analogs for geologic sequestration. *In proceedings of the Fifth International Conference on Greenhouse Gas Control Technologies* (GHGT5), Williams, D.J., Durie, R.A., ,McMullan, P., Paulson, C.A.J., & Smith, A.Y. (Eds), pp. 328-333, CSIRO Publishing, Collingwood, VIC, Australia.

Taylor, G.J., Blamey, F.P.C., & Edwards, D.G. (1998). Antagonistic and synergistic interactions between aluminum and manganese on growth of *Vigna unguiculata* at low ionic strength. *Physiologia Plantarum*, 104(2), pp. 183–194.

Tian, S., Fan, Q., Xu, Y., Wang, Y., & Jiang, A. (2001). Evaluation of the use of high CO_2 concentrations and cold storage to control of *Monilinia fructicola* on sweet cherries. *Postharvest Biology and Technology*, 22(1), pp. 53–60.

Van der Zwaan, B., & Smekens, K. (2009). CO_2 capture and storage with leakage in an energy-climate model. *Environmental Modeling & Assessment*, 14(2), pp. 135-148.

Van Gardingen, P.R., Grace, J., Harkness, D.D., Miglietta, F., & Raschi, A. (1995). Carbon dioxide emissions at an Italian mineral spring: measurements of average CO_2 concentration and air temperature. *Agricultural & Forest Meteorology*, 73, pp.17–27.

Vartapetian, B.B., & Jackson, M.B. (1997). Plant adaptations to anaerobic stress. Annals of Botany (London) 79(Suppl. A), pp.3–20.

Vodnik, D., Kastelec, D., Pfanz H., & Macek, I. B. T. (2006). Small-scale spatial variation in soil CO2 concentration in a natural carbon dioxide spring and some related plant responses. *Geoderma*. 133, pp. 309–319.

Wardle, D.A., & Parkinson, D. (1990). Interactions between microclimatic variables and the soil microbial biomass. *Biology and Fertility of Soils*, 9, pp. 273–280.

Wei, Y., Maroto-Valer, M., & Steven, M.D. (2011). Environmental consequences of potential leaks of CO_2 in soil. *Energy Procedia*, 4, pp. 3224–3230.

West, J.M., Pearce, J.M., Coombs, P., Ford, J.R., Scheib, C., Colls, J.J., Smith, K.L., & Steven, M.D. (2009). The impact of controlled injection of CO_2 on the soil ecosystem and chemistry of an English lowland pasture. *Energy Procedia*, 1, pp. 1863–1870.

How to Keep Deep-Sea Animals

Hiroshi Miyake[1,2,3], Mitsugu Kitada[2], Dhugal J. Lindsay[3],
Toshishige Itoh[2], Suguru Nemoto[2] and Tetsuya Miwa[3]
[1]Kitasato University,
[2]Enoshima Aquarium,
[3]JAMSTEC
Japan

1. Introduction

The ocean covers 71% of the surface of the Earth. The deepest trench is the Mariana Trench, with a depth of 11000 m. The average elevation of terrestrial areas is about 840 m high, while the average depth of the oceans is about 3800 m. This means that the average elevation above the rock substrate for the Earth is -2440 m – i.e. within the seas. Moreover the volume of the ocean is 300 times the volume of all terrestrial areas combined. The coastal area of the ocean is less than 10% of the total area of the oceans. The deep sea lies under the epipelagic zone, which is the layer from the surface to 200 m depth. The volume of the deep sea is over 95% of the total volume of the oceans. The deep-sea realm is the largest biosphere on the Earth. There are three categories of habitat in the deep-sea realm. One is the deep-sea floor, the second is the bentho-pelagic layer, which is the layer from the deep-sea floor up to an altitude above the bottom of 100m, and the third is the mid-water zone, which is between the epipelagic layer and the bentho-pelagic layer. The mid-water is an extremely important realm, and holds the key to elucidate the cycles of matter in the ocean, carbon transportation from the surface layer to the deep sea, and the interaction between the behavior of oceanic circulation with global warming and the lives of deep-sea animals. However, information about mid-water biology is extremely limited because of the difficulty of sampling swimming (nekton) and floating (plankton) animals in the mid-water. The mid-water community is one of the most mysterious of all deep-sea communities.

The deep sea is a mysterious kind of Inner Space for us, even though the ocean is closer than Outer Space. High water pressure in the deep sea keeps us from easily exploring this realm. Studies of deep-sea animals have long been carried out through net sampling using tools such as dredges, trawl nets, plankton nets and line fishing. However, the development of deep-sea crewed submersibles and remotely-operated vehicles (ROVs) has drastically changed the way we study deep-sea biology. These deep-sea survey tools allow us to visit places where fishing nets cannot trawl, e.g. deep-sea valleys, outcrops of base rocks, cliffs, gaps, hydrothermal vent areas and cold seep areas. Moreover, they allow us to observe deep-sea animals and their behavior *in-situ*. This is particularly true of gelatinous zooplankton, which are vastly understudied because their fragile bodies are easily damaged and destroyed by fishing nets. Crewed submersibles and ROVs have enabled rapid progress in the study of gelatinous zooplankton because they allow us to observe their behavior and collect them in pristine condition (Miyake et al., 2001; Robison, 2004).

The most history-changing moment for deep-sea biology is undoubtedly the discovery of a deep-sea chemosynthetic ecosystem off the Galapagos Islands in 1977 (Corliss & Ballard, 1977; Corliss et al., 1979). This discovery is widely considered to be one of the greatest discoveries of the 20th century. Hydrothermal vents can spew hot water at temperatures above 300 °C at 2600 m depth. Many animals live around hydrothermal vents and most of the animals do not depend on solar energy like we do, but on heat and chemical energy from inside the Earth. Hydrothermal vent fluids can include hydrogen sulfide and methane from deep within the Earth. Bacteria use these chemicals for chemosynthesis – the making of organic molecules using chemical energy. Primary production in deep-sea chemosynthetic ecosystems relies on chemosynthetic bacteria rather than the photosynthesis of plants. Seven years after the discovery of the first hydrothermal vent, a cold seep was discovered off the coast of Florida in the Gulf of Mexico (Paull et al., 1984). Cold seeps are often located on the seafloor close to faults or the margins of oceanic plates. Chemosynthesis–based associations also occur at cold seeps. After the discovery of hydrothermal vents and cold seeps, the first whale fall community was discovered in the Santa Catalina Basin in 1987 (Smith et al., 1989). Many animals that were related to animals from hydrothermal vents and cold seeps were discovered there. Dead whales sink to the deep-sea floor and are eaten by animals such as sharks, hagfish, crabs, and so on. Remnants of the whale carcass then remain on the deep-sea floor, still containing abundant organic matter such as blubber and bone marrow. Rotten fat promotes the formation of hydrogen sulfide and methane. Therefore many animals that are related to animals from hydrothermal vents and cold seeps are able to inhabit such whale carcasses.

The study of deep-sea biology has progressed rapidly since deep-sea submersibles began to be used for science (Gage & Tyler, 1991; Van Dover, 2000; Herring, 2002; Fujikura et al., 2008). However, many aspects of deep-sea biology still need to be investigated. Surveys using deep-sea submersibles can obtain data on behavior, systematics, evolution, symbioses, and biodiversity, using video images, samples preserved using chemicals, and frozen samples. It is now possible to observe deep-sea animals *in situ* with the naked eye or with HDTV cameras. Many questions have arisen from such real time observations of deep-sea animals. However, it is difficult to observe deep-sea animals over the long term *in situ*. Our observations of deep-sea animals using these tools are limited to only a few points during the space-time of their lives. One method to connect these points over time is the rearing and observation of deep-sea animals. One of the next important and necessary steps in deep-sea biology is to study live deep-sea animals in land-based aquaria using rearing and observation methods. The keeping of deep-sea animals has been tried in many institutes and public aquaria around the world because the rearing of deep-sea animals is considered by many to be an essential development for the future. In this chapter, we would like to introduce methods for the collection and maintenance of deep-sea animals, especially midwater animals and chemosynthetic ecosystem-associated animals.

2. Environment of the deep sea

The deep sea is a realm of darkness, cold, and high water pressure. Water temperature and salinity of the surface water layer vary from one locality to another according to season, latitude, or ocean currents. However, the differences in water temperature and salinity are smaller at 600~800m depths. At these depths the water temperature and salinity is about 4 °C and 34.5 respectively in almost all ocean areas. The dissolved oxygen concentration is at its lowest at these depths – around 1.0 ml/l off Japan and 0.1 ml/l off the south-west coast of the U.S.A. As for sunlight, only blue light can penetrate to 1000m depth. There is no red light penetrating from the surface to these depths.

The environment at deep-sea chemosynthesis-based ecosystems is very different from the normal deep-sea environment. Hydrothermal vent areas are extremely different, with the temperature of the hot water being 100~300 °C. This hot water has a high chemical activity, containing high concentrations of minerals and volcanic gases from under the sea floor. This water is a reductive environment with no dissolved oxygen, high concentrations of hydrogen sulfide and carbon dioxide, low pH and containing many minerals. On the other hand, the ambient water around hydrothermal vents is normal, cold deep-sea water. Minerals dissolved in the hot water recrystallize to form chimneys when the hot water comes into contact with the cold ambient water. The low density hot water ascends quickly into the water column and mixes with ambient water. When the density of the mixed water becomes the same as the ambient water, the mixed water stops its ascent and spreads out horizontally. This water mass is called a hydrothermal plume. This environment of high hydrogen sulfide, low oxygen, high carbon dioxide and low pH is in direct contrast to the normal environment we regularly use for rearing aquatic animals. Hydrothermal vent animals live in a drastic environmental gradient with temperatures from below 4 °C to over 100 °C, both reductive and oxidative environments, and from acid to alkaline water. On the other hand, at cold seeps, hydrogen sulfide and methane seep slowly from the sea floor.

3. Sampling

Sampling of mid-water animals has been carried out using plankton nets such as the IKMT (Issacs-Kidd Mid-water Trawl), a multiple opening/closing net called the IONESS (Intelligent Operative Net Sampling System), and deep-sea submersibles and ROVs (Fig. 1). There are advantages and disadvantages to all of these sampling methods for obtaining mid-water animals. Plankton nets can collect many species and large numbers of mid-water animals, however, the animals collected in the net are usually damaged because they are buffeted around in the net during towing. Animals are also weakened by being exposed to

Fig. 1. Sampling gears of deep-sea animals. A: IONESS, B: HOV *Shinkai 6500*, C: ROV *Hyper-Dolphin*

the higher temperatures of surface waters during retrieval of the nets. To minimize these disadvantages, the duration of the tow must be as short as possible, and sampling is best done in the winter season when surface water temperatures are low.

On the other hand, deep-sea submersibles and ROVs can sample mid-water animals individually in pristine condition using slurp-gun systems, gate valve samplers, and detritus samplers (Hashimoto et al., 1992; Hunt et al., 1997). However, the number of animals able to be sampled during a single dive using present systems is at most ten or so. Plankton nets are better suited for sampling animals that swim fast and have hard bodies, such as fish and shrimp, but submersibles are better suited for sampling animals that swim slowly and have fragile bodies, such as jellyfish.

As for organisms from deep-sea chemosynthesis-based communities, deep-sea submersibles and ROVs equipped with slurp-gun systems, sample boxes, scoops or MT core samplers are used (Miyake et al., 2005).

The most important point for successfully sampling deep-sea animals to keep in an aquarium is to collect them in the best condition possible. It is essential to collect animals with minimal external wounds and to keep water temperature in the sample box or canister low. Water temperature for maintenance of deep-sea animals should be about 4 °C, but the surface water temperature off Okinawa, Japan, in summer can be over 28 °C. This high temperature is fatal for deep-sea animals. To keep temperature changes to a minimum, we use a large slurp gun canister that can hold a large volume of sea water at the ambient temperatures at which the deep-sea animals live. As for the collection of animals from the deep-sea floor, for example the giant white clam *Calyptogena*, we collect them with a scoop and place them in a sample box topped up with mud from their habitat in order to minimize the influence of warm surface waters during submersible or ROV recovery.

The next point is correct treatment of animals after retrieval to the deck of the ship. When the animals come on deck, they are quickly transferred into buckets filled with seawater at temperatures adjusted to the same temperature as their habitats (4 ~ 16°C). Animals are then transferred into rearing tanks set up in special cold-rooms on deck. Some deep-sea animals may be weakened greatly or seem to be in a moribund condition, however, these weakened animals sometimes recover in cold water aquaria overnight.

When deep-sea animals are brought to the surface, the water pressure is, as a rule, not kept at ambient deep-sea pressures, but water temperature is maintained as far as possible. The animals introduced above are usually able to endure the change of pressure between the deep-sea and the surface. We use the DEEP AQUARIUM for animals that need to be kept at deep-sea pressures for survival during retrieval (Koyama et al., 2002). The sphere of the DEEP AQUARIUM is connected to the slurp-gun set on submersibles, and animals are sucked into the sphere before the lid being closed and pressure maintained at the ambient levels. Upon retrieval the sphere is attached to the life-support system for the DEEP AQUARIUM that controls pressure, dissolved oxygen and has a filtration system for recycling the water in the sphere. The DEEP AQUARIUM is also used as a pressure chamber for animals that have been weakened by decompression sickness.

After cruises, deep-sea animals are usually transferred to the Enoshima Aquarium, to the Japan Agency for Marine-Earth Science and Technology (JAMSTEC), or to Kitasato

University in cold storage by airplane and/or by truck within 48 hours. Most animals are transported to their destinations within 36 hours.

4. Mid-water and bentho-pelagic animals

The mid-water is characterized by having no ceiling and no floor. Mid-water animals must remain suspended by floating or swimming at all times to maintain their depth positions in the mid-water. This is the most important factor to remember when trying to keep them in an aquarium.

To keep mid-water and bentho-pelagic animals, we use planktonkreisels (Hamner, 1990; Raskoff et al., 2003; Widmer, 2008) and a jellyfish tank that was developed by Kamo Aquarium, Japan (Fig. 2).

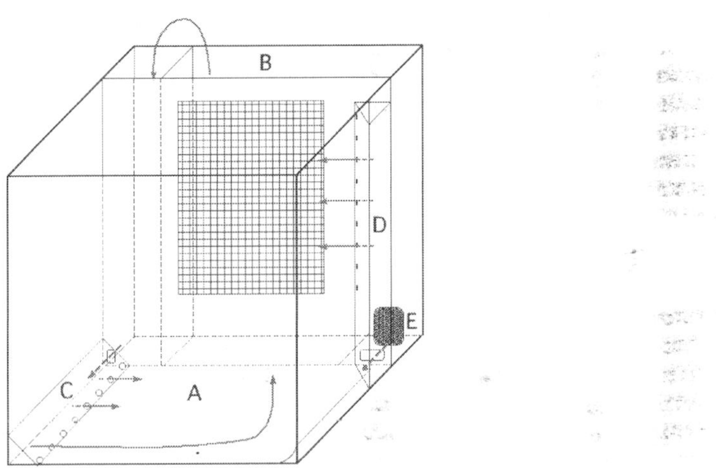

Fig. 2. Rearing tank for jellies.   A: Rearing space. B: Filtration area. C: water outlet area, designed to produce currents in the rearing area. D: water outlet area, designed to produce currents to keep jellies off the filter mesh. E: Submersible pump.

The jellyfish tank is a simple box tank that makes a slow, gentle current for jellies using differences in water level. Temperature, salinity and dissolved oxygen are controlled to conform to the habitat data recorded by the CTD-DO meter deployed on the submersible or ROV. In particular, dissolved oxygen in the deep sea is lower than in surface waters. Dissolved oxygen in land-base aquaria is very high compared to the levels that mid-water animals usually encounter and such high concentrations of oxygen may damage them. When mid-water animals show oxyecoia intoxication, dissolved oxygen is regulated to low concentrations by bubbling nitrogen gas using a Dissolved Oxygen Concentration Control System (DOCCS) (Fig. 3) (Miyake et al., 2006).

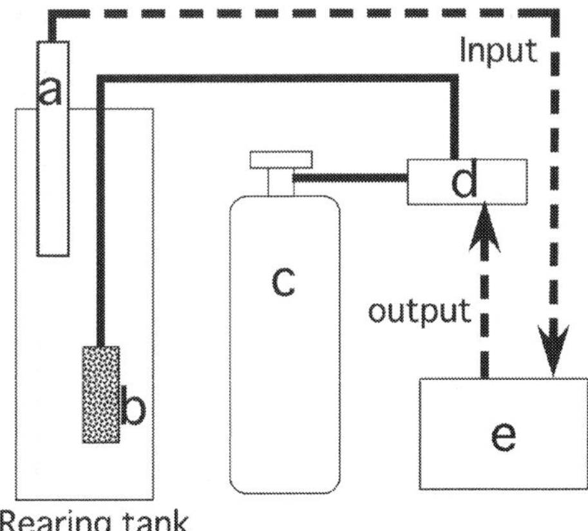

Fig. 3. Dissolved Oxygen Concentration Control System (DOCCS)
a: DO sensor, b: Air stone, c: Nitrogen tank, d: Electromagnetic valve, e: Control unit DO
electric value which is measured at DO sensor (a) is processed by control unit (e). When DO
value is higher than preset DO value, the control unit send a signal to electromagnetic valve
(d) to open the valve, and nitrogen gas added to the aquarium. When DO reaches to the
preset value, the electromagnetic valve is closed and nitrogen input stops.

Sometime an ultraviolet lamp is installed at the outlet of the filtration tank, because the mid-water generally has low abundances of bacteria. As for lighting, no statistically-proven, quantitative data exists, but experience suggests that lighting affects the survival of deep-sea animals in aquaria. Strong white light can make deep-sea animals blind (Herring et al., 1999). Some pigments of deep-sea animals such as porphyrin-derivatives can turn toxic when exposed to light (Herring, 1972). At public aquaria, artificial lighting such as by white fluorescent lamps or halogen lamps, as well as exposure to daylight over long periods, can lead to early death of deep-sea animals. We use red LED lamps to decrease mortality. An added advantage is that deep-sea animals show natural behaviors under red LED light.

4.1 Gelatinous zooplankton

Observations and experiments on live jellyfishes are necessary to understand their life history strategies in the deep sea. Therefore establishment of how to collect and keep jellyfishes that inhabit the mid-water and benthopelagic zones is necessary. However, it is difficult to collect many individuals of one species of jellyfish during a deep-sea dive. Jellyfishes have complex life cycles, including alternation of generations between a planktonic medusa stage (sexual generation), and a benthic polyp stage (asexual generation). This polyp stage has a great ability to regenerate. If we can collect a deep-sea polyp that also has a medusa generation, we can raise medusae from the polyp. The collection of polyps from the deep-sea in order to raise and keep deep-sea jellyfish in the aquarium is a very useful method.

We collect hard-bodied benthos like snails, as well as other substrates such as sunken wood, rocks and deep-sea litter, in order to find polyps. We also deploy and recover pot-plant pots with marker buoys on the deep-sea floor using manned submersibles and ROVs. Some polyps of hydrozoans and scyphozoans were successfully collected on these substrates and were subsequently kept in aquaria with temperatures regulated to the same temperature as their in situ habitats (4 ~ 12 °C) and at atmospheric pressure. These polyps were kept in small aquaria (30 cm – 20cm -25cm H) with a sponge filter and fed *Artemia* nauplii twice a week. Every species of polyp formed a colony on the substrates through asexual reproduction. Some polyps in the colony were collected using a needle under a stereoscopic microscope and transferred into a petri-dish (φ 80mm - 4cm H). After the polyps attached to the bottom of the petri-dish, *Artemia* nauplii were fed to them twice a week. About four hours after feeding, each time, all the rearing water was changed. Using this method, we have been able to observe the growth, degrowth, regeneration, colony formation, medusa bud formation, and strobilation of polyps in the laboratory.

The hydroid from a *Ptilocodium* sp. was collected on a pot-plant pot with marker buoy deployed at a depth of 1170m off Hatsushima Island, Sagami Bay (Fig. 4). This species attached to the pot-plant pot substrate itself rather than to the rope or marker. There were two types of polyp in the colony – gastrozooids and dactylozooids. Dactylozooids had four tentacles and lacked mouths. Gastrozooids ate food that the dactylozooids caught with their tentacles. One to four medusa buds were formed on the basal part of gastrozooids. Medusae just after liberation had four short tentacles. This species was kept successfully at 4 °C and was unable to be kept at temperatures above 10°C.

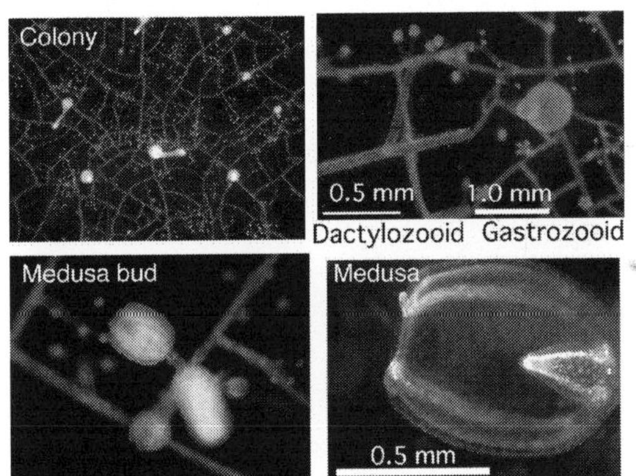

Fig. 4. *Ptilocodium* sp.

A large species of hydrozoa, *Corymorpha* sp., was collected using a suction sampler at a depth of 530m off Mera, Suruga Bay (Fig. 5a). This species inhabits sandy substrates. Just after sampling, this specimen was 15cm in height and was composed of a long hydrocaulus and a terminal hydranth. Five polyps were found at the base of the hydrocaulus. The number of aboral tentacles was about 60 and the number of oral tentacles was about 50. Blastostyles were located just above the aboral tentacles of the hydranth and had many medusa buds (Fig. 5b). Some medusae were liberated from the blastostyles (Fig. 5c). After sampling, all tentacles on the hydranth become atrophied and the hydranth dropped off from the hydrocaulus. The remaining hydrocaulus degenerated into a tissue mass. The tissue mass regenerated into a polyp after one month and started asexual reproduction(Fig. 5d). The liberated hydranth also regenerated tentacles and caught *Artemia* nauplii. One polyp degenerated into a tissue mass and was divided into two separate tissue masses. Each of these tissue masses regenerated into a new individual. Lighting had a detrimental effect on this species during rearing, causing atrophication of tentacles and degeneration into a tissue mass.

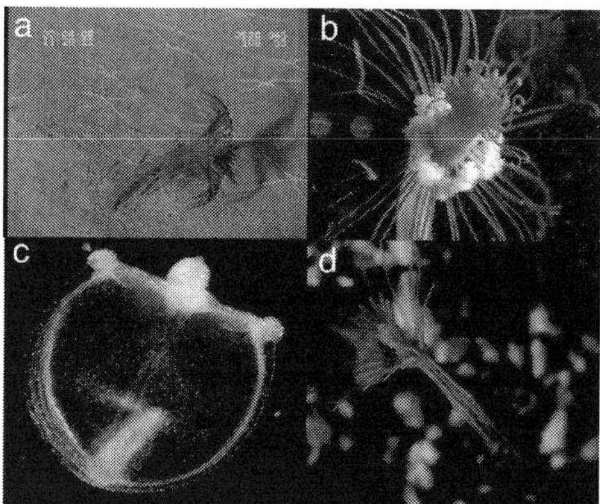

Fig. 5. *Corymorpha* sp.

It is necessary to rear the medusae that are released from such polyps until they are fully grown in order to clarify the life history of the species. Recently, some studies have been published about the life cycles and new species descriptions of deep-sea jellies from Monterey Bay (Widmer, 2007; Widmer et al., 2010; Widmer, 2011). Keeping deep-sea jellyfish in the laboratory makes it much easier to obtain detailed data about them that we would be unable to obtain in situ.

4.2 Mid-water shrimp, *Bentheogennema borealis*

Bentheogennema borealis is a mid-water shrimp referred to as Shinkai Ebi in Japanese (Shinkai means deep-sea and Ebi means shrimp in Japanese). This species is colored deep-red (Fig. 6). Sampling was done using a multiple opening/closing net – the IONESS (Intelligent Operative Net Sampling System). Many *B. borealis* were collected from 600m to 1500m depths in Sagami Bay, Japan. However, more than half of the shrimps were dead when the IONESS was retrieved onto the deck. To increase the survival rate when rearing the shrimp, it is essential to sample them in the winter season when surface water temperature is low. Lively shrimps were selected from the collected animals and transferred into buckets filled with chilled water (4°C). Selected shrimps were kept in the aquaria that were developed for jellyfish rearing (outlined above) at 4°C. Most of the shrimps had damaged antennae and could not swim normally but instead spiraled as they swam. Weakened individuals did not show any escape response. Such individuals must be eliminated from the rearing tank before the water becomes clouded and water quality worsens due to the fats and oils exuded from weakened individuals. When water became clouded, it was exchanged with fresh sea water. White light was also considered not to be good for them so red LED lights were used for observation. They could not catch food by themselves because of the damage incurred during sampling. Therefore, we fed them a piece of krill or mysid meat directly to the mouth of each individual. The survival rate was about 11% for the first month, and about 6% over three months. Maximum duration of survival was more than 575 days. Some individuals molted and regenerated antennae and/or legs.

Fig. 6. *Bentheogennema borealis*

5. Chemosynthesis-based ecosystem animals

Deep-sea biologists can observe and collect live animals from deep-sea chemosynthetic ecosystems using ROVs or manned submersibles. However, it has been difficult to keep them alive over the long-term in order to perform a large variety of biological studies. Keeping deep-sea animals from chemosynthetic ecosystems in captivity enables many useful studies of these animals, as researchers can conduct experiments at any time, without needing to participate in deep-sea diving cruises. Keeping these animals is also very useful for public aquaria or science museums as part of a deep-sea biology outreach program to the public.

The rearing of animals from deep-sea chemosynthesis-based ecosystems has been difficult due to problems in maintaining high pressure, low pH, optimizing H_2S and high CO_2 concentrations, low dissolved oxygen, and keeping low light and low temperature conditions. These conditions are far removed from the normal rearing conditions for most fishes in aquaria. To overcome these problems, artificial hydrothermal vent tank and cold seep system tanks have been developed (Miyake et al., 2006; Miyake et al., 2007) (Fig. 7).

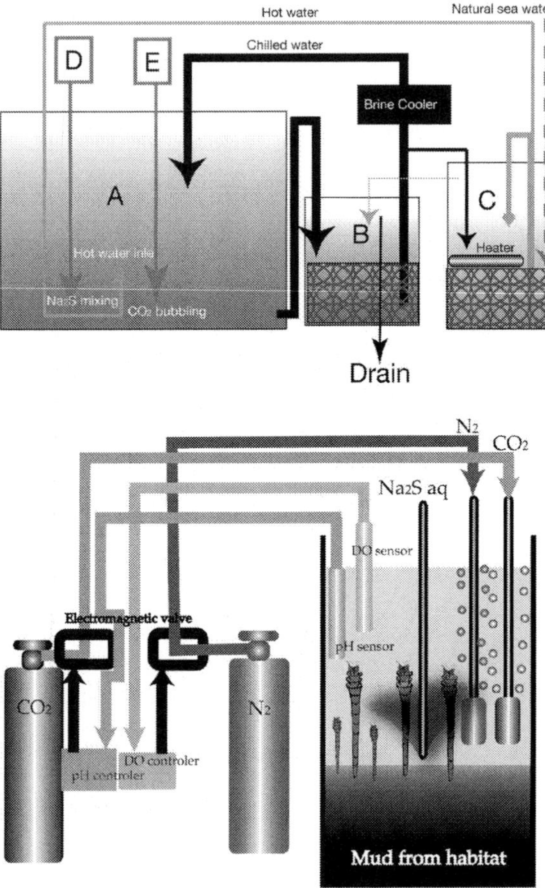

Fig. 7. Artificial hydrothermal vent and cold seep systems
Upper: Artificial hydrothermal vent system. A: Main rearing tank. Temperatures were 12 °C and 4 °C. B: Filtration tank. Filtration material was coral sand. C: Heating tank. Heating tank was also for filtration. D: Sodium sulphide tank. E: CO_2 cylinder.
Lower: Artificial cold seep systems

In 2006, the Deep-sea Chemosynthetic Ecosystem Tank (Fig. 8), with an artificial hydrothermal vent and cold seep system, was opened in the Deep-sea Corner at Enoshima Aquarium. This tank (3000WX1000-1500HX1000D (mm)) consists of three parts, a hydrothermal vent area, a cold seep area and a whale fall area. In the whale fall area, bones of a whale are displayed. In the cold seep area, the reduction zone is made using 30cm of mud mixed with decaying organic matter. The system of the tank is composed of a heating tank, a hot water outlet with added Na_2S as a source of H_2S, and added CO_2 for chemosynthetic bacteria and pH regulation. When the need arises, a DO control unit (DOCCS) is attached. There are some artificial chimneys, which act as hot water vents. The maximum temperature of hot water is 60°C. Ambient water is 2 to 6°C, with an average temperature of 4°C.

Fig. 8. The Deep-sea Chemosynthetic Ecosystem Tank, the main tank in the Deep-sea Corner in Enoshima Aquarium. Upper: Front view of the tank. Hydrothermal vent area venting hot water from imitation chimneys is located on the left side of the tank. Cold seep area is located at the right side of the tank. Whale fall area includes whale bone collected from Sagami Bay (924m). Two temperature signs (red) are displayed, the left sign is the hot water temperature, and the right sign is ambient water temperature. Below: Lighting is always red light. Hot water from imitation chimney is a maximun of 60°C.

5.1 Hydrothermal vent crabs and shrimps (Fig. 9)

Hydrothermal vent crabs, *Gandalfus yunohana*, *Austinograea alayseae*, *A. rodriguezensis* and the vent shimps, *Alvinocaris longirostris*, *Opaepele loihi* and *Opaepele* spp. have been kept in the hydrothermal vent system tank. *Alvinocaris longirostris* did not show any tendency to keep close to hot water vents, while the others have a strong tendency to keep close to hot water vents and gather around the outlet for hot water. Survival rates of these animals, except for

A. longirostris, were higher in the tank with hot water vents than in a tank without any hot water vents. Experience shows that white lighting was not good for the animals, with white light sometimes killing them, but red LEDs were a good source of illumination. They are fed krill, mysid or fish meat twice a week.

The vent crab *G. yunohana* exhibits behavior where females are guarded by the males before they molt. After molting, copulation has been observed. On the other hand, males just after molting are often cannibalized by other individuals in the aquarium. Some females spawned in the rearing tank. Gravid crabs maintained eggs on their abdomen in the hot water vent. Some eggs collected from females were kept in petri dishes with controlled temperatures at 12°C and 20°C. 12°C is the ambient water temperature in the rearing tank. Larvae hatched at 20°C, but did not hatch at 12°C. Adult crabs have no eyes, however hatched larvae have eyes. Unfortunately, rearing of the larvae has not yet been successful.

The vent shrimp graze on filaments and mats of chemosynthetic bacteria attached around the inlets for hot water, as well as eating krill, mysid or fish meat. *O. loihi* was brooding eggs when collected from the field and continued to brood them in the tank. Adult shrimps have no eyes, but larvae have eyes and were observed to swim upward in an upside-down posture. However, rearing of these larvae has not yet been successful.

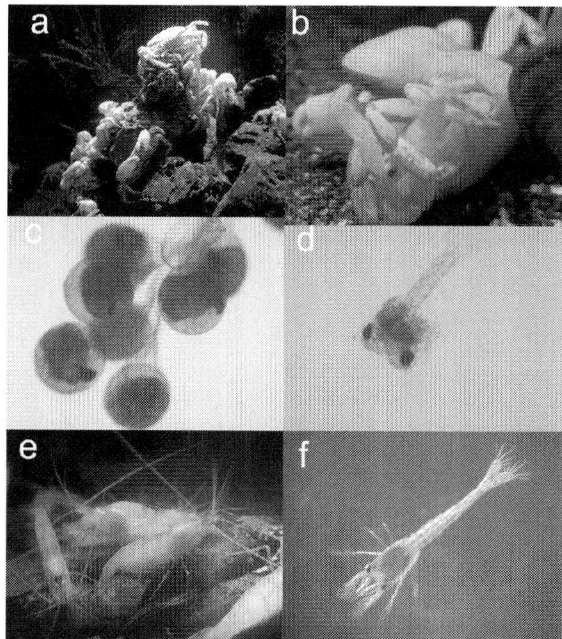

Fig. 9. Behaviors and breeding of vent crab and shrimp in the aquarium.
a: *Gandalfus yunohana* gathered around artificial hydrothermal vent, b: mating behavior of *G. yunohana*, c: eyed eggs of *G. yunohana*, d: hatched larva of *G. yunohana*, e: gravid female of *Opaepele loihi*, f: hatched larvae of *O. loihi*

5.2 Hydrothermal vent squat lobsters

The hydrothermal vent-associated squat lobster *Shinkaia crosnieri* was collected at a hydrothermal vent area at depths of 1000 ~ 1600m in the Okinawa Trough (Fig. 10a). *S. crosnieri* has a unique feeding ecology. This species has bushy white hairs, upon which dense chemosynthetic bacterial filaments are attached, on its ventral side (Fig. 10b). This species does not eat normal foods such as krill, mysid or fish meat, but instead grazes on the bacterial filaments on its bushy ventral hairs. This species lives near hydrothermal vents and farms chemosynthetic bacteria on its bushy ventral hairs. *S. crosnieri* has been kept in the vent system tank, but it showed no tendency to remain close to the hot water vent inlets like vent crabs do. Suitable rearing temperatures were 4~7 °C, and temperatures less than 3°C or more than 10°C were not suitable for long-term rearing.

S. crosnieri had bushy white hair with dense bacterial filaments attached at the time of sampling. However, a few days after sampling these dense bacterial filaments had disappeared from animals kept in a normal non-vent environment-simulating tank. In hydrothermal vent system tanks, bacterial filaments on the white hairs on the ventral side of *S. crosnieri* increased in bushiness within a few days. It has been observed that *S. crosnieri* graze on these bacteria using their mouthparts. A combination of optimum hydrogen sulfide concentrations and jetting water stream speed may be a key to emulating their *in situ* environment. A single molting was observed. The individual had no bacterial filaments on its ventral hairs after molting and died about one month later.

Fig. 10. *Shinkaia crosnieri* (a~c) and *Munidopsis myojinensis* (d~e)

Some specimens were brooding eggs when they were collected and continued brooding eggs in the vent tank. Eggs were attached to the pleopods of the females. The color of the eggs was white, and they were approximately 2 mm in diameter. The eggs of some individuals hatched (Fig. 10c). Larvae were positively buoyant and remained at the water surface without sinking. Adult individuals have no eyes, however hatched larvae had eyes.

Another species, *Munidopsis myojinensis* was collected from the Myojin Knoll (1200m depth) in the Izu-Ogasawara Arc (Fig. 10d). They did not gather together around the hot water vent inlet in the tank. This species also did not feed on krill or other meat. Individuals of this species died after a short period of rearing in a non-vent tank, but in the vent tank this species lived over a long period. This suggests that *M. myojinensis* may eat the bacterial mats around hydrothermal vents. However, in order to clarify whether their vent-dependence is due to food or some other factor, we need to do further experiments using the stable isotope tracer method, etc. Some specimens were brooding eggs when they were collected. Eggs were 2 mm in diameter and were reddish-orange in color. They were attached to the pleopods of females. Some eggs collected from females were kept in petri dishes with temperature controlled to remain at 4°C. Eggs were not positively buoyant but larvae just after hatching floated towards the surface (Fig. 10e).

5.3 Vestimentiferan tubeworms

The vestimentiferan tubeworms we have attempted to rear include *Lamellibrachia* sp. collected from Sagami Bay (850 m), and *Lamellibrachia satsuma* collected from Kagoshima Bay (80~110m) and Nikko seamount (450m) in the Izu-Ogasawara Arc. Tubeworms were kept in an improved cold seep tank. There is no DO controller and also no addition of Na_2S for making H_2S in this rearing tank. Mud collected from Ariake Bay was mixed with dog food and laid on the bottom of the tank at a thickness of 5-10 cm. Then mud from Ariake Bay was overlaid to a thickness of 10cm on top of the dogfood-mud mixture. Hydrogen sulfide, methane and ammonia for growth of the tubeworms were generated from the rotten dog food in the mud. When the dog food went rotten, large amounts of oxygen were consumed. This decreased the levels of dissolved oxygen in the tank. Carbon dioxide was added as a source of carbon for chemoautotrophic symbiotic bacteria. The pH was maintained at low levels (~ 6.8-7.0) by bubbling carbon dioxide using a pH controller. When the pH is high (> 7.0), an electro-magnetic valve connected to the pH controller automatically opens and CO_2 input starts. When the pH is low (< 6.8), the valve closes and CO_2 input stops. When the improved cold seep tank was running well, bacterial mats emerged on the surface of the mud.

If tubeworms are collected in good condition with no injuries, long-term rearing is quite easy. Tubeworms *in situ* extend their gills from the upper opening of the tube (Fig. 11a). Tubeworms in high dissolved oxygen conditions have a tendency to keep their gills within their tube. However, tubeworms in low dissolved oxygen conditions extend their gills from their tubes. Tubeworms were unable to survive in a completely anaerobic environment. The growth of the tube was faster in low oxygen and high carbon dioxide conditions than in any other combination of conditions.

A few days after sampling of tubeworms, fertilized eggs were collected by filtration of the rearing tank seawater using a 100μm mesh net or by dissection of the tubeworm egg pouches. Larvae developed to trochophores just before the settlement stage (Fig. 11b). However, we were unable to observe metamorphosis of trochophore larvae into juvenile tube-worms. Shinozaki et al. (2010) reported reproduction of *Lamellibrachia satsuma* in the aquarium. This strain of *L. satsuma* inhabited a whalebone substrate. Whalebones are an energy and nutrient source for both *L. satsuma* and its symbionts. Keeping tubeworms over the long-term can be accomplished using whalebones, but a combination of whalebones and mud would probably be better for rearing cold seep species. However, water quality must be carefully monitored because of the large amount of rotten organic matter contained within whalebones.

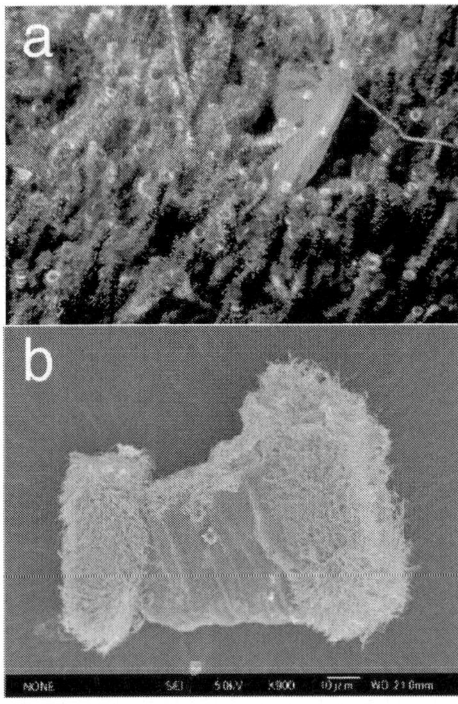

Fig. 11. *Lamellibrachia satsuma*

5.4 Deep-sea white clams

The giant white clams *Calyptogena okutanii* and *C. soyoae* were collected from cold seeps off Hatsushima, Sagami Bay (850m) and *C. kawamutai* were collected from a cold seep at Kuroshima Knoll, off Ishigaki Island (600m) using scoops, sample boxes, and MT core samplers. *Calyptogena* clams have been very difficult to keep. Before using an MT core for sampling, we were only able to keep *C. soyoae* alive for a few days. Using an MT core to collect the clam and its microenvironment simultaneously allowed us to keep *C. okutanii* and *C. soyoae* alive for a week, and *C. kawamurai* for 17 days. Spawning of *C. solidissima* has been observed on two occasions during their captivity in aquaria (Miyake et al., 2005).

Calyptogena okutanii has now been kept successfully for two months in the mud in the cold seep area of the deep-sea chemosynthetic ecosystem tank (Fig. 12).

Fig. 12 *Calyptogena okutanii* at artificial cold seep area of Chemosynthetic ecosystem tank

Dissolved oxygen was needed for the *Calyptoena* clam itself, but a low dissolved oxygen environment was needed for the symbiotic chemosynthetic bacteria to allow them to use hydrogen sulfide. Therefore, it is essential to make a gradient in the dissolved oxygen concentration so that the deeper layer of mud is an anaerobic environment and the surface layer of mud and the bottom water layer are aerobic environments. This, in principle, is the same for tubeworm from cold seeps.

5.5 Fishes from chemosynthesis-based ecosystems

Some zoarcid fish were collected from cold seeps in Sagami Bay and hydrothermal vent areas in the Okinawa Trough using a suction sampler without needing to keep pressure at ambient deep-sea levels. The species of zoarcid fishes were *Ericandersonia sagamia* (Fig. 13a), *Japonolycodes abei,* a snake-like undescribed species, and some other undescribed species (Fig. 13b, c). Zoarcid fishes collected from the deep sea were damaged during retrieval to the ship's deck because of the drastic change in pressure and temperature. After one night, almost all the specimens stayed on the bottom of the aquarium. They slowly began to feed on krill or mysids over the course of a week. The rearing method we used for zoarcid fish was the same as we use for normal deep-sea fish. Although they were collected from cold seep areas, they are unable to endure the hydrogen sulfide-rich environment that *Shinkaia crosnieri* needs to farm bacterial filaments on its ventral hairs.

The snake-like undescribed fish spawned in the aquarium. This fish was a batch-spawner that spawns large eggs. A female raised 27 eggs at a time in one egg batch. However, the eggs were not fertilized. Eggs were about 5~6 mm in diameter, were white in color and were positively buoyant (Fig. 13d). The zoarcid fishes, *Lycodes cortezianzus* and *Lycodapus mandibularis*, however, laid eggs that were negatively buoyant (Ferry-Graham et al., 2007). This species has been found within a vestimentiferan tube-worm (*Alaysia* sp.) colony, a *Bathmodiolus* mussel colony and a *Calyptogena* clam colony at cold seeps off Hatsushima, Sagami Bay. This species may lay egg batches in spaces within these colonies, e.g. in the tubeworm bush or between mussels, so as they do not rise up into the water column.

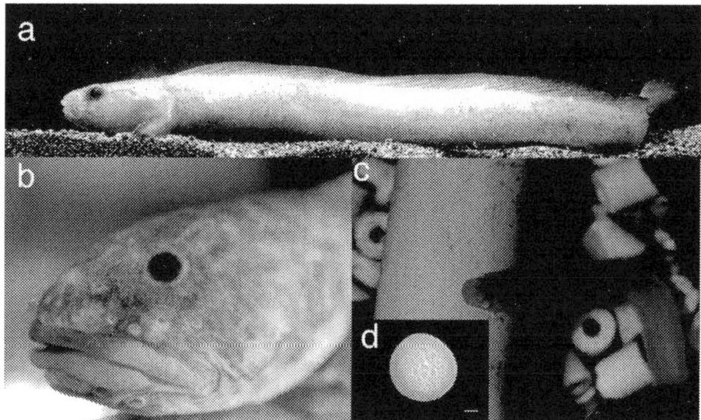

Fig. 13. Zoarcid fishes.

The vent tonguefish *Symphurus thermophiles* is also one fish that can be sampled successfully without keeping pressure at ambient levels (Fig. 14a). This species was collected from the Nikko seamount (450m) in the Izu-Ogasawara Arc. This species can also be kept in a normal aquarium as well as in the hydrothermal vent tank. *S. thermophiles* was usually either on the sand, hidden in the sand, or sometimes on the walls of the tank. *S. thermophiles* fed on any food that was dropped in front of them, especially when the food item was moving. It is supposed that, although this species inhabits deep-sea hydrothermal vent areas at depths of 400 ~ 500 m, this species finds prey items by vision.

S. thermophilis spawned in the tank. Mating occurred at approximately 30 cm above the bottom by close swimming between females and males in the aquarium. Eggs were positively buoyant, were transparent, and were approximately 1 mm in diameter (Fig. 14b). Larvae just after hatching had yolk sacs and did not have eyes or gastro- intestinal tracts (Fig. 14c). Three days after hatching, the development of the gastrointestinal tract was complete. One week after hatching, eyes and a mouth had developed and larvae could prey on small plankton.

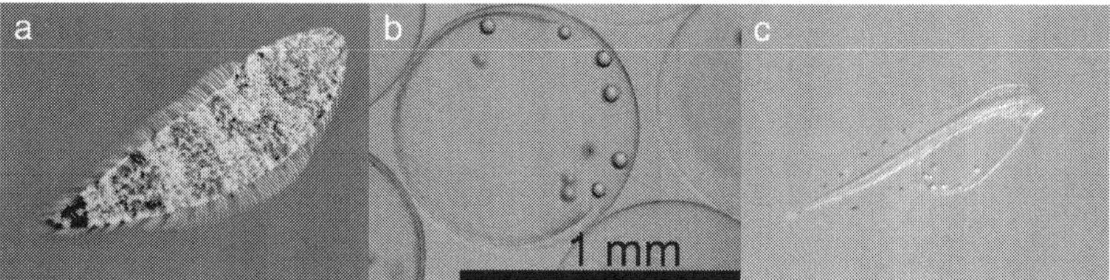

Fig. 14. *Symphurus thermophiles*

6. Summary and future studies

Recently studies on deep-sea animals using rearing methods such as those outlined above have been increasing. Mid-water hydromedusae have been raised from polyps collected from artificial substrates set on the deep-sea floor, from the bodies of mid-water shrimp, and by artificial fertilization, thereafter being described as new species (Widmer, 2007; Widmer et al., 2010; Widmer, 2011). Some other hydropolyps and scyphopolyps collected from deep-sea litter have been kept in the laboratory at Kitasato University, and a few species of them may be undescribed species according to preliminary identifications from morphology and genetics. The collection of deep-sea litter is a highly efficient way to find deep-sea sessile organisms, including jellyfish polyps. As for animals in chemosynthetic-based ecosystems, a hydrothermal vent barnacle, *Neoverruca* sp. (Watanabe et al., 2004), and hydrothermal vent crabs such as *Bythograea thermydron* (Epifanio et al., 1999; Jinks et al., 2002; Dittel et al., 2008) and *Gandalfus yunohana* (Hamasaki et al., 2010) have been kept for long periods. In particular, *Gandalfus yunohana* was observed during growth from eggs to juveniles at various temperatures. These data are very important to understand the larval ecology of vent animals and could not be obtained by *in situ* observations from submersibles. Whalebones are also useful for successfully keeping vent and seep animals in aquaria. The whale-fall mussel *Adipicola pacifica* (Fujiwara et al., 2010; Kinoshita et al., 2010) and the vestimentiferen tubeworm *Lamellibrachia satsuma* (Shinozaki et al., 2010), which were found attached to whalebones *in situ*, have been kept successfully in aquaria without adding nutrients or energy sources such as hydrogen sulfide, carbon dioxide, anmonia etc. Moreover, *L. satsuma* that was attached to a whalebone spawned in the aquarium and newly-settled individuals also appeared on this whalebone. The vent mussel *Bathymodiolus azoricus* (Kadar et al., 2005; Kadar et al., 2006; Kadar et al., 2009; Bettencourt et al., 2010; Bettencourt et al., 2011) and the thermophilic worm *Alvinella pompejana* (Pradillon et al., 2001; Pradillon et al., 2004; Pradillon et al., 2005) were kept at atmospheric pressure and *in situ* pressure using a pressurized tank and were studied immunologically, histologically, developmentally, and behaviorally. A pressurized tank may allow the rearing of deep-sea animals that cannot be collected alive at atmospheric pressure and also allow studies without the bias of different pressure regimes. Behavioral responses towards heat have been studied for the hydrothermal vent polychaete *Hesiolyra bergi* by IPOCAMP (Shillito et al., 2001). The developmental biology of the vestimentiferan tubeworm *Riftia pachyptila* has been studied in pressure vessels (Marsh et al., 2001; Brooke & Young, 2009). The vent shrimp, *Alvinocaris* sp. and the deep-sea eel, *Simenchelys parasiticus* have also been collected by the DEEP AQUARIUM system without serious damage sustained by decompression and/or exposure to the high temperatures of surface seawater (Koyama et al., 2005; Koyama, 2007).

A new pressurized recovery device, PERISCOP, has recently been developed (Shillito et al., 2008). Now we have some techniques to collect and rear deep-sea animals at atomospheric pressure, and we can also collect and keep deep-sea animals in pressure-retained aquarium systems and decrease pressure slowly to atmospheric pressure to acclimatize the animals to atmospheric pressure. It is essential to try and improve communication and technology exchanges between scientists and technicians working on rearing techniques for deep-sea animals in order for us to successfully complete life cycles and make important observations on these enigmatic organisms.

Rearing of deep-sea animals allows scientists to observe the behaviors and modes of life of these animals just as if they were *in situ* and it also allows scientists to do various experiments in the laboratory. This is an important breakthrough in biological studies of deep-sea animals. Additionally, displaying and rearing deep-sea animals allows the public to watch mysterious and curious "live" deep-sea animals without having to go on deep-sea survey missions personally and it lets us introduce deep-sea biology to the public in an easy-to-understand manner.

7. Acknowledgments

We sincerely thank the captains and crews of the R/V *Natsushima*, R/V *Kaiyo* and R/V *Yokosuka* and the commanders, pilots, and operations teams of the ROVs *Dolphin-3K* and *Hyper-Dolphin* and the crewed submersibles *Shinkai 2000* and *Shinkai 6500* for their dedicated efforts. We also thank Mrs. Yukiko Hori, the Director of the Enoshima Aquarium, and Dr. Tadashi Maruyama, the Program Director of the Marine Biodiversity Research Program, Institute of Biogeoscience, JAMSTEC, for their encouragement to us. Finally, thanks are due to many colleagues for their active support aboard ships, for laboratory assistance and for other kinds of help they have graciously provided.

8. References

Bettencourt R., V. Costa, M. Laranjo, D. Rosa, L. Pires, A. Colaco, H. Lopes & R.S. Santos. (2011). Out of the deep sea into a land-based aquarium environment: Investigating physiological adaptations in the hydrothermal vent mussel *Bathymodiolus azoricus*. *ICES Journal of Marine Science*, Vol.68. No.2: 357-364.

Bettencourt R., V. Costa, M. Laranjo, D. Rosa, L. Pires, A. Colaco, P.M. Sarradin, H. Lopes, M.J. Sarrazin & R.S. Santos. (2010). Out of the deep-sea into a land-based aquarium environment: Investigating innate immunity in the hydrothermal vent mussel i. *Cahiers de Biologie Marine*, Vol.51. No.4: 341-350.

Brooke S.D. & C.M. Young. (2009). Where do the embryos of *Riftia pachyptila* develop? Pressure tolerances, temperature tolerances, and buoyancy during prolonged embryonic dispersal. *Deep-Sea Research Part Ii-Topical Studies in Oceanography*, Vol.56. No.19-20: 1599-1606.

Corliss J.B. & R.D. Ballard. (1977). Oases of life in the cold abyss. *National Geographic Magazine*, Vol.152. 441-453.

Corliss J.B., J. Dymond, L. Gordo, J.M. Edomond, R.P.v. Herzen, R.D. Ballard, K. Green, D. Williams, A. Bainbridge, K. Crane & T.H.v. Andel. (1979). Submersible thermal springs on the galapagos rift. *Science*, Vol.203. No.1073-1083.

Dittel A.I., G. Perovich&C.E. Epifanio. (2008). Biology of the vent crab *Bythograea thermydron*: A brief review. *Journal of Shellfish Research*, Vol.27. No.1: 63-77.

Epifanio C.E., G. Perovich, A.I. Dittel & S.C. Cary. (1999). Development and behavior of megalopa larvae and juveniles of the hydrothermal vent crab *Bythograea thermydron*. *Marine Ecology Progress Series*, Vol.185. 147-154.

Ferry-Graham L.A., J.C. Drazen & V. Franklin. (2007). Laboratory observations of reproduction in the deep-water zoarcids *Lycodes cortezianzus* and *Lycodapus mandibularis* (Teleostei : Zoarcidae). *Pacific Science*, Vol.61. No.1: 129-139.

Fujikura K., T. Okutani & T. Maruyama. (2008). Deep-sea life - biological observations using research submersibles. Tokai University Press

Fujiwara Y., G. Kinoshita, M. Kawato, A. Shinozaki, T. Yamamoto, K. Okoshi, K. Kubokawa & H. Yamamoto. (2010). Protandric hermaphroditism in the whale-fall mussel *Adipicola pacifica*. *Cahiers de Biologie Marine*, Vol.51. No.4: 423-427.

Gage J.D. & P.A. Tyler. (1991). *Deep-sea biology: A natural history of organisms at the deep-sea floor*. Cambridge University Press, New York

Hamasaki K., K. Nakajima, S. Tsuchida, R. Kado & S. Kitada. (2010). Number and duration of zoeal stages of the hydrothermal vent crab *Gandalfus yunohana* from laboratory reared specimens. *Journal of Crustacean Biology*, Vol.30. No.2: 236-240.

Hamner W.M. (1990). Design developments in the planktonkreisel, a plankton aquarium for ships at sea. *Jouenal of Plankton Research*, Vol.12. No.2: 393-402.

Hashimoto J., K. Fujikura, T. Aoki & S. Tsukioka. (1992). Development of a suction sampler (slurp gun) for deep sea organisms. *Proc. JAMSTEC Symp. Deep Sea Res.*, Vol.8. 367-372.

Herring P.J. (1972). Porphyrin pigmentation in deep-sea medusae. *Nature*, Vol.238. 276.

Herring P.J (2002). *The biology of the deep ocean*. Oxford University Press, New York

Herring P.J., P. Gatent & P.M.J. Shelton. (1999). Are vent shrimps blinded by science? *Nature*, Vol.398. 1116.

Hunt J.C., J. Hashimoto, Y. Fujiwara, D.J. Lindsay, K. Fujikura, S. Tsuchida & T. Yamamoto. (1997). The development, implementation, and establishment of a meso-pelagic and bentho-pelagic biological survey program using submersibles in the seas around japan. *JAMSTEC J. Deep Sea Research*, Vol.13. 675-685.

Jinks R.N., T.L. Markley, E.E. Taylor, G. Perovich, A.I. Dittel, C.E. Epifanio & T.W. Cronin. (2002). Adaptive visual metamorphosis in a deep-sea hydrothermal vent crab. *Nature*, Vol.420. No.6911: 68-70.

Kadar E., R. Bettencourt, V. Costa, R.S. Santos, A. Lobo-Da-Cunha & P. Dando. (2005). Experimentally induced endosymbiont loss and re-acquirement in the hydrothermal vent bivalve *Bathymodiolus azoricus*. *Journal of Experimental Marine Biology and Ecology*, Vol.318. No.1: 99-110.

Kadar E., A. Lobo-da-Cunha&C. Azevedo. (2009). Mantle-to-shell CaCo3 transfer during shell repair at different hydrostatic pressures in the deep-sea vent mussel *Bathymodiolus azoricus*; (Bivalvia: Mytilidae). *Marine Biology*, Vol.156. No.5: 959-967.

Kadar E., A. Lobo-Da-Cunha, R.S. Santos & P. Dando. (2006). Spermatogenesis of *Bathymodiolus azoricus* in captivity matching reproductive behaviour at deep-sea hydrothermal vents. *Journal of Experimental Marine Biology and Ecology*, Vol.335. No.1: 19-26.

Kinoshita G., M. Kawato, A. Shinozaki, T. Yamamoto, K. Okoshi, K. Kubokawa, H. Yamamoto & Y. Fujiwara. (2010). Protandric hermaphroditism in the whale-fall mussel *Adipicola pacifica*. *Cahiers de Biologie Marine*, Vol.51. No.4: 423-427.

Koyama S. (2007). Cell biology of deep-sea multicellular organisms. *Cytotechnology*, Vol.55. No.2-3: 125-133.

Koyama S., T. Miwa, M. Horii, Y. Ishikawa, K. Horikoshi & M. Aizawa. (2002). Pressure-stat aquarium system designed for capturing and maintaining deep-sea organisms. *Deep-Sea Research Part I-Oceanographic Research Papers*, Vol.49. No.11: 2095-2102.

Koyama S., T. Nagahama, N. Ootsu, T. Takayama, M. Horii, S. Konishi, T. Miwa, Y. Ishikawa & M. Aizawa. (2005). Survival of deep-sea shrimp (*Alvinocaris* sp.) during decompression and larval hatching at atmospheric pressure. *Marine Biotechnology*, Vol.7. No.4: 272-278.

Marsh A.G., L.S. Mullineaux, C.M. Young & D.T. Manahan. (2001). Larval dispersal potential of the tubeworm *Riftia pachyptila* at deep-sea hydrothermal vents. *Nature*, Vol.411. No.6833: 77-80.

Miyake H., M. Kitada, S. Tsuchida, Y. Okuyama & K. Nakamura. (2007). Ecological aspects of hydrothermal vent animals in captivity at atmospheric pressure. *Marine Ecology-an Evolutionary Perspective*, Vol.28. No.1: 86-92.

Miyake H., D.J. Lindsay & J.C. Hunt. (2001). The study of gelatinous zooplankton using submersibles and ROVs. *Kaiyo Monthly*, Vol. Extra edition 27. 216-223.

Miyake H., J. Tsukahara, J. Hashimoto, K. Uematsu & T. Maruyama. (2006). Rearing and observation methods of vestimentiferan tubeworm and its early development at atmospheric pressure. *Cahiers de Biologie Marine*, Vol.47. No.4: 471-475.

Matsuyama & S., H. Yamamoto, M. Kitada, I. Ueda, K. Okoshi, M. Kitamura, K. Matsuyama&S. Tsuchida. (2005). Attempts to rear the deep-sea white clams calyptogena soyoae and calyptogena solidissima. *Oceanography i n Japan*, Vol.14. No.6: 645-651.

Paull C.K., B. Hecker, R. Commeau, R.P. Freeman-Lynde, C. Neumann, W.P. Corso, S. Golubic, J.E. Hook, E. Sikes & J. Curray. (1984). Biological communities at the florida escarpment resemble hydrothermal vent taxa. *Science*, Vol.226. No.965-967.

Pradillon F., N. Le Bris, B. Shillito, C.M. Young & F. Gaill. (2005). Influence of environmental conditions on early development of the hydrothermal vent polychaete *Alvinella pompejana*. *Journal of Experimental Biology*, Vol.208. No.8: 1551-1561.

Pradillon F., B. Shillito, J.C. Chervin, G. Hamel & F. Gaill. (2004). Pressure vessels for in vivo studies of deep-sea fauna. *High Pressure Research*, Vol.24. No.2: 237-246.

Pradillon F., B. Shillito, C.M. Young & F. Gaill. (2001). Deep-sea ecology. Developmental arrest in vent worm embryos. *Nature*, Vol.413. 698-699.

Raskoff K., F.A. Sommer, W.M. Hamner & K.M. Cross. (2003). Collection and culture techniques for gelatinous zooplankton. *Biological Bulletin*, Vol.204. 68-80.

Robison B.H. (2004). Deep pelagic biology. *Journal of Experimental Marine Biology and Ecology*, Vol.300. 253-272.

Shillito B., G. Hamel, C. Duchi, D. Cottin, J. Sarrazin, P.M. Sarradin, J. Ravaux & F. Gaill. (2008). Live capture of megafauna from 2300 m depth, using a newly designed pressurized recovery device. *Deep-Sea Research Part I-Oceanographic Research Papers*, Vol.55. No.7: 881-889.

Shillito B., D. Jollivet, P.M. Sarradin, P. Rodier, F. Lallier, D. Desbruyeres & F. Gaill. (2001). Temperature resistance of *Hesiolyra bergi*, a polychaetous annelid living on deep-sea vent smoker walls. *Marine Ecology-Progress Series*, Vol.216. 141-149.

Shinozaki A., M. Kawato, C. Noda, T. Yamamoto, K. Kubokawa, T. Yamanaka, J. Tahara, H. Nakajoh, T. Aoki, H. Miyake & Y. Fujiwara. (2010). Reproduction of the vestimentiferan tubeworm *Lamellibrachia satsuma* inhabiting a whale vertebra in an aquarium. *Cahiers de Biologie Marine*, Vol.51. No.4: 467-473.

Smith C.R., H. Kukert, R.A. Wheatcroft, P.A. Jumars & J.W. Deming. (1989). Vent fauna on whale remains. *Nature*, Vol. 341. 27-28.

Van Dover C.L. (2000). *The ecology of deep-sea hydrothermal vents*. Princeton University Press, Princeton

Watanabe H., R. Kado, S. Tsuchida, H. Miyake, M. Kyo & S. Kojima. (2004). Larval development and intermoult period of the hydrothermal vent barnacle *Neoverruca* sp. *Journal of the Marine Biological Association of the United Kingdom*, Vol.84. No.4: 743-745.

Widmer C.L. (2007). The hydroid and medusa of *Amphinema rollinsi* (Cnidaria : Hydrozoa), a new species of pandeid from the Monterey Bay submarine canyon, USA. *Zootaxa.* No.1595: 53-59.

Widmer C.L. (2008). *How to keep jellyfish in aquariums: An introductory guide for maintaining healthy jellies.* Wheatmark, 978-1-60494-126-5, Tucson, Arizona

Widmer C.L. (2011). The hydroid and early medusa stages of the deep sea jellyfish *Earleria purpurea* (Hydrozoa: Mitrocomidae) from the Monterey Bay submarine canyon, USA. *Zootaxa.* No.2902: 59-68.

Widmer C.L., G. Cailliet&J. Geller. (2010). The life cycle of *Earleria corachloeae* n. sp (Cnidaria: Hydrozoa) with epibiotic hydroids on mid-water shrimp. *Marine Biology*, Vol.157. No.1: 49-58.

4

An Introduced Polychaete in South America – Ecologic Affinities of *Manayunkia speciosa* (Polychaeta, Sabellidae) and the Oligochaetes of Uruguay River, Argentina

Laura Armendáriz, Fernando Spaccesi and Alberto Rodrigues Capítulo
Instituto de Limnología "Dr. Raúl A. Ringuelet"
(ILPLA- CONICET La Plata- UNLP),
Facultad de Ciencias Naturales y Museo,
UNLP, Consejo Nacional de Investigaciones
Científicas y Técnicas (CONICET),
Argentina

1. Introduction

1.1 The Uruguay River and its alluvial valley

Rivers are very dynamic and complex ecosystems. The elements forming their structure are the riverbed, the bank area and the alluvial valley. They can change in size and complexity according to their geological features and the climate of the region they cross. This variation in the horizontal axis, the flow regime, the physicochemical characteristics of the water and the type and density of the vegetation determine, in fluvial systems, a spatial heterogeneity, in agreement with the temporal dynamism (Sabater et al., 2009).

South America is characterized by the vast extension of its river basins, among which the Amazon, Orinoco and Del Plata can be cited. Due to the geologic structure of the continent, the rivers of these basins, generally long and with an abundant flow drain into the Atlantic Ocean. Among the main rivers forming the Del Plata Basin are the Paraguay, Paraná and Uruguay (Bonetto & Hurtado, 1998).

Uruguay River flows along 1800 km, starting in Brazil, and passing through Argentina and Uruguay before draining in the estuarial system of the Río de la Plata. The Uruguay River basin comprises some 365,000 km² with a complex geology, forming a mosaic of volcanic (basaltic flows) and sedimentary rocks and quaternary alluvial sediments (Bonetto & Hurtado, 1998; López Laborde, 1998). These Holocene formations occurred during a long period of some 5000 years under a tropical/subtropical humid climate relatively homogeneous in the whole basin, giving rise to partially erosioned soils covered by young sediments, mainly eolian (Iriondo & Kröhling, 2004). According to its hydrographic features it can be divided into three sections: upper, middle and lower. The Upper Uruguay, fast and hardly navigable, starts in Brazil at the confluence of Pelotas and Canoas rivers extending to the drainage of the Piratiní River, along some 800 km with a slope of approximately 43 km; the middle section is ca. 600 km long reaching the locality of Concordia (Entre Ríos Province, Argentina) where the Salto Grande hydroelectric dam is found; the lower section is 350 km long and presents a slope of only 3 km.

The average discharge of Uruguay River is 4500m³/s with extremes varying from 800 to 14,300 m³/s and although its discharge level is lower than the Paraná River's, its variability is higher (from 11 to 17 %). Its waters are transparent with much less solids in suspension than in the Paraná River (López Laborde, 1998; Nagy et al., 1998). The swelling of Uruguay River takes place in winter, with a secondary maximum in spring and a minimum flow in summer-autumn. An increase in the discharge has been observed since 1950 and the modifications in its flow have been due to the Salto Grande hydroelectric dam (Nagy et al., op. cit.). Both Uruguay and Paraná rivers show a strong oscillation in phase with the ENSO events (El Niño/La Niña Southern Oscillation), with remarkable swellings during the mild events (or El Niño) and normal to low discharges during the cold events (or La Niña) (Pasquini & Depetris, 2006). Its alluvial valley is formed by ecosystems with wetlands of relative extension and importance where the gallery forest connected to the drainage system stands out. The basin is found in the phytogeographic territory corresponding to the Amazonic ("Paranaense" province) and Chaqueño ("del Espinal" and Pampean provinces) domains. It presents numerous edaphic communities and marginal forests formed by species from the Paranaense province, although impoverished in a northern-southern direction (Cabrera & Willink, 1973). These marginal or gallery forests are narrow strips, densely populated by fairly tall species like "ibirapitá" (*Peltophorum dubium* (Spreng.), "lapacho rosado" (*Tabebuia impetiginosa* ((Mart. ex DC.) Standl.), "azota caballo" (*Luehea divaricada* Mart.), and "pitanga" (*Eugenia uniflora* L.). Towards the Lower Uruguay in the riparian zone of Argentina and in the islands there is an almost absolute predominance of "sarandí blanco" (*Phyllanthus sellowianus* Müll.) (Bonetto & Hurtado, 1998; Cabrera, 1971). This abundant riverside vegetation plays an important role in the intake of coarse particulate allochthonous material which is then used by the benthic organisms associated to the riverside, determining a fauna characteristic of these environments.

With respect to the uses of the river, navigation is concentrated mainly in the lower section towards the mouth of the Río de la Plata River, in the middle section it is navigated by ships of reduced draft while upstream, in Brazilian territory, navigation is in small boats. The main use of its waters is connected with the generation of hydroelectric energy (e. g. the Salto Grande hydroelectric dam, already mentioned, and two other dams in its upper section).

The recreational use of the river includes fishing, sports practice and bathing resorts in riverside Argentinian cities such as Gualeguaychú and Colón. The impact on the river is due to an inadequate management of the resources, such as agricultural, industrial and transport activities, the production of energy, and the habitat fragmentation (Bonetto & Hurtado, 1998).

1.2 Benthic invertebrates

Invertebrates are among the most adapted organisms in fluvial ecosystems, playing a fundamental role in the transference of energy from the inferior levels (macrophytes, algae and detritus) to the superior consumers (aquatic vertebrates and birds). As they have evolved in an unpredictable environment, many of them have developed opportunistic ways of life and generalistic diets. Their communities often show a great plasticity with respect to their vital features and reproductive strategies which they optimize according to the environmental characteristics (Barbour et al., 1999; Rodrigues Capítulo et al., 2009).

The community of organisms living at the bottom of a water body constitutes the benthos. The characteristics of the physical habitat, the water quality and the availability and quality of the food determine, to a great extent, the presence and abundance of benthic organisms. In fluvial environments the current and type of substratum also determine the benthic structure. The current produces a homogenizing effect of the environment, diminishing diversity, removing particles and favoring the gaseous exchange. The type of substratum can be formed by sediments of different granulometry, such as gravel, sand, silt and clay, and by riparian vegetation. Spatial and temporal variations of the fluvial systems determine that the communities of organisms inhabiting the upstreams usually differ from those inhabiting the middle and downstreams. Besides, it is possible to distinguish a littoral benthos, with habitat heterogeneity, greater diversity, and coarse particulate organic matter, and a profoundal benthos with less available habitats, moderate diversity and thin particulate organic matter (Margalef, 1983).

The study of the benthic community of Uruguay River has been carried out by Ezcurra de Drago & Bonetto (1969) and Di Persia & Olazarri (1986) concerning fouling and arborescent sponges; bivalves and gastropod mollusks; and caddisflies, mayflies and true flies insects (Bonetto & Hurtado, 1998). Nevertheless, and differently from other great South American rivers as the Paraná, Paraguay and Río de la Plata, where the oligochaetes have been extensively studied (Armendáriz et al., 2011; Ezcurra de Drago et al., 2004, 2007; Marchese & Ezcurra de Drago, 1992; Marchese et al., 2005; Montanholi-Martins & Takeda, 1999; Takeda, 1999), the Uruguay annelid fauna has received, so far, scarce attention.

1.3 The introduced species

The introduction of species is one of the processes which produce major alterations in the biodiversity of ecosystems. Even though it is not a problem caused only by human activity, the number of species involved and the frequency of their relocation has grown enormously as a result of the expansion of transportation and trade (Glasby & Timm, 2008; Penchaszadeh, 2005). The continental aquatic environments are highly susceptible to the accidental or deliberate introduction of exotic species. Besides, the natural connection of the basins and the dispersion capacity of the aquatic organisms together with the human activity determine the possibility of invasion of new species (Ciutti & Cappelletti, 2009). In these aquatic environments, the transfer of exotic species occurs through a wide variety of ways and means, for example, in maritime transportation, in the ballast water of ships (El Hadad et al., 2007).

In South America, most of the coastal ecosystems between the Río de la Plata Estuary and central Patagonia have been modified due to the introduction of exotic organisms. From 31 identified cases, 6 have caused a considerable ecological impact (Orensanz et al., 2002). In this region, the introduction of crustaceans such as *Balanus glandula* Darwin, 1854 and

B. amphitrite Darwin, 1854; mollusks, such as *Crassostrea gigas* (Thumberg, 1793), *Corbicula fluminea* (Müller, 1774), *C. largillierti* (Philippi, 1844), *Limnoperna fortunei* (Dunker, 1857), and *Rapana venosa* (Valenciennes, 1846), and polychaetes such as *Ficopomatus enigmaticus* (Fauvel, 1923) have been recorded in the last 40 years (Penchaszadeh, 2005). Recently, the presence of *Manayunkia speciosa* Leidy, 1858 (Armendáriz et al., 2011) has been registered in Uruguay River.

M. speciosa is a small polychaete strictly from freshwater, first recorded in Schuylkill River (USA) in 1858. Some decades later its distribution extended through the whole Nearctic Region, being recorded in the eastern and western lotic systems of North America (Brehm, 1978; Hazel, 1966; Holmquist, 1967; Krecker, 1939; Leidy, 1883; Mackie & Qadri, 1971; Meehean, 1929; Spencer, 1976). A hundred and fifty years after its first record, it is found for the first time in the Neotropical Region, enlarging its limit distribution from the United States of America to Argentina. Although it could not be determined whether its introduction had been recently, it can be assumed that *M. speciosa* arrived at the Uruguay River through the ballast water of commercial ships (Armendáriz et al., 2011).

In recent years, the studies on *M. speciosa* have been focused in some aspects of its population dynamics and in its role as intermediate host of Myxozoa parasites of Salmonidae (Bartholomew et al., 1997; Bartholomew et al., 2006; Stocking & Bartholomew, 2007). Due to the lack of information on the ecological requirements of *M. speciosa* together with the scarce knowledge on the annelid fauna of the Lower Uruguay River, we propose, in this chapter, to analyze the main ecological variables which could have determined the establishment of *M. speciosa* with the oligochaete annelids, forming the benthic assemblage of the Lower Uruguay River.

2. Methodology

2.1 Study area

The study was carried out in the Lower Uruguay River (Figure 1). Eight sampling sites were selected between 33° 05′ S 58° 12′W, 33° 5.94′ S 58° 25′ W. Sites S1 to S5 were established on the main river basin, site S6 at the centre of Bellaco Bay, site S7 inside Ñandubaysal Bay, and site S8 in Inés Lagoon. Sites S1 to S3 were located upstream of a pulp mill industry on the Uruguayan coast. Site S4 is an area of sediment accumulation, located downstream of the mentioned industry. Site S5 is located at the centre of the main riverbed. It is characterized by a high current velocity and a pronounced erosion effect. Hence, it constitutes a clean water zone with a limited offer of refuge and food for the benthic fauna. Site S6, located in a zone with calm waters, with bulrush (*Schoenoplectus californicus* (C. A. Mey) Palla) and riverside vegetation is influenced by Gualeguaychú River and Gualeguaychú city. Site S7 is also located in a zone with a low current velocity, Ñandubaysal Bay, with low waters and abundant littoral vegetation. Depending on the seasonal hydrological regime of the river, this site is connected with site S8 that shows lentic characteristics.

The sediments of the study area were formed by plastic silty clay with small proportions of scarcely sorted quartz sand (Iriondo & Kröhling, 2004).

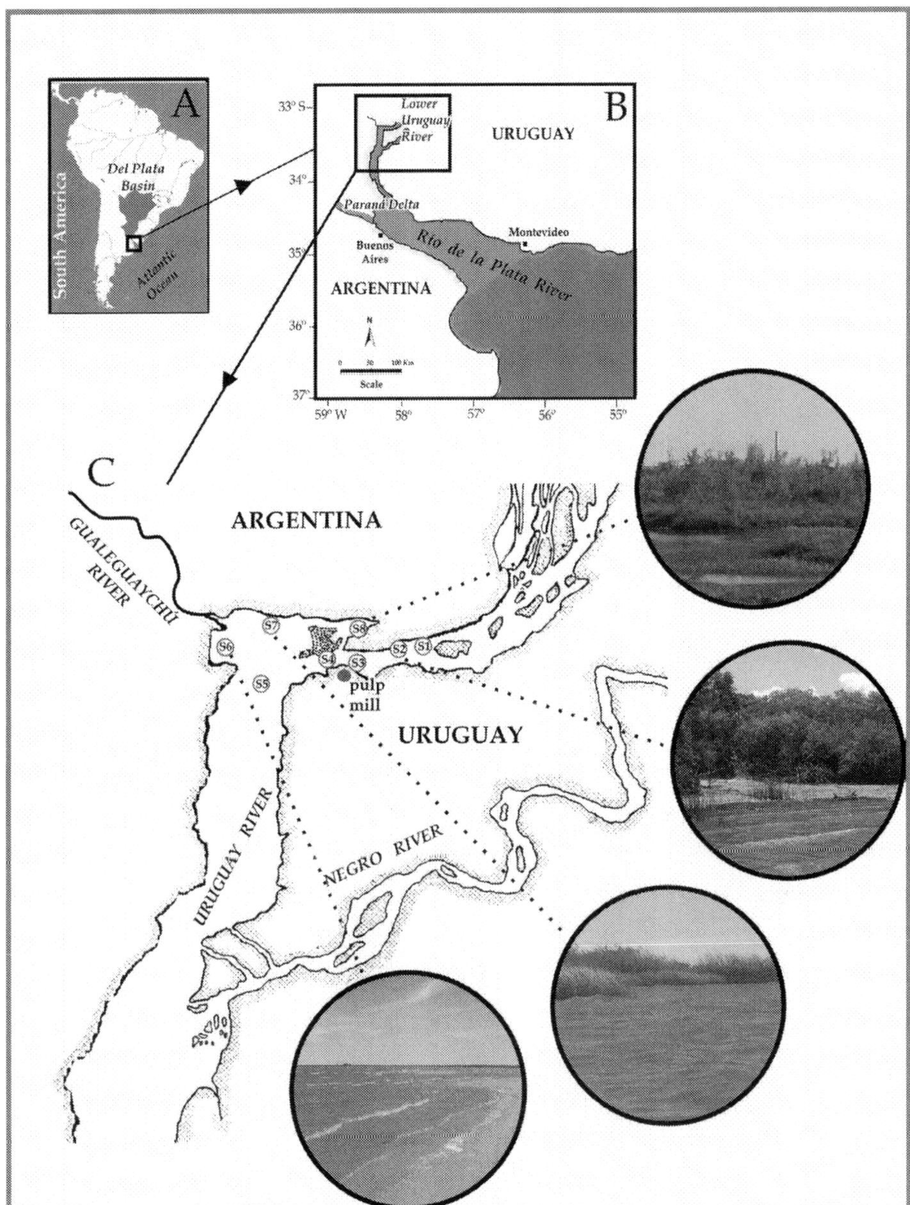

Fig. 1. A) Map of Del Plata Basin in South America. B) Study area in the Lower Uruguay River. C) Sampling sites: sites S1 through S5 at the main course of the Uruguay River, S6 at Bellaco Bay, S7 at Ñandubaysal Bay, and S8 at Inés Lagoon. Photos: A. Rodrigues Capítulo.

2.2 Sampling of the annelid assemblage

The work methodology included the taking of samples with a van Veen dredge (470 cm²) from a boat (Figure 2). Seasonal samples, from November 2007 through March 2009, were extracted in the 8 mentioned sites, distinguishing, when possible, the main riverbed and the bulrushes from the coast (Fig. 1). Site S8 was not sampled during autumn and winter due to the scarce depth of the river and the abundant aquatic vegetation which made navigation impossible. Six replicates in each site and season were collected. The samples were fixed *in situ* with formaldehyde 5%. The following physicochemical variables were measured: Temperature in °C and pH (Hanna HI 8633), dissolved oxygen in mg/l (Ysi 52), conductivity in μS/cm (Lutron CD-4303) and turbidity in UTN (Turbidity meter 800-ESD). Due to difficulties with the equipment, depth could only be measured in spring, summer and autumn. In March 2009, water samples for the analysis of nutrients (nitrates, NO_3^-; nitrites, NO_2^-; ammonium, NH_4^+; and soluble reactive phosphorus PO_4^{3-}) were also taken.

Fig. 2. Van Veen dredge used for collecting the benthic samples in Lower Uruguay River, Argentina.

In the laboratory, the samples were washed with a 500 μm mesh sieve, dyed with erythrosine B, and the organisms were separated manually from the sediment under stereoscopic microscope. Annelids were identified under optical microscope according to keys and specific bibliography (Brinkhurst & Marchese, 1992; Glasby & Timm, 2008;

Opinion 2167, ICZN, 2007; Pettibone, 1953; Struck & Purschke, 2005). Final preservation of the organisms was made in alcohol 70%.

Indirect measurements of the content of organic matter (OM) in sediment of the collected samples were measured using the loss organic ignition method (LOI: loss organic ignition) at 500°C during 4 h, with a previous drying of 48 h at 60°C (APHA, 1998).

2.3 Statistical analysis

The annelid assemblage was evaluated through the application of ecological indexes: diversity (H') (Shannon-Wiener, in $\log_e$), evenness (E') and specific richness (R: number of species or taxa). The frequency of occurrence of each species was calculated as FO (%) = 100 m_i/M, where m_i: is the number of samples containing species i, and M: the total number of samples. According to this, the species were classified in: Constant (C): with a FO ≥ 75 %; Frequent (F): with 75% < FO ≤ 50%; Accesorial (A): 50% < FO ≤ 25%; and Incidental (I): FO < 25%.

A Pearson correlation was performed in order to analyze the relationship between: 1) percentage of the organic matter sediment vs. individual abundance (N) of *M. speciosa* and 2) percentage of the organic matter sediment vs. total abundance of oligochaetes.

The relationship between the sampled sites and the environmental variables recorded was explored by a Principal Components Analysis (PCA). A Detrended Correspondence Analysis (DCA) was applied to the biotic data in order to determine if the species responded linearly to gradients or passed through some environmental optimum. Due to the fact that the maximum length of gradients in standard deviation units in this analysis was 3.328, a model of unimodal response for the species has been assumed. It was decided to apply a Canonical Correspondence Analysis (CCA) (Ter Braak & Smilauer, 2002) to explore the relationships between the annelid abundance and the environmental variables recorded in the sampling sites. The abundances of the species were $\log_e$ (x+1) transformed, and all the species presenting a frequency of occurrence higher than 3 % were included (Table 1). The physicochemical variables were standardized. Their inflation factors were smaller than 3. Depth was excluded from the analysis due to the lack of records corresponding to winter 2008. Significance of all canonical axes was evaluated through the Monte Carlo test (499 permutations under the reduced model, P < 0.05). The two first axes of the ordination were selected for the graphical representation.

3. Results

3.1 Physicochemical characteristics of the water

The physicochemical characteristics of the water in the section of the Uruguay River studied varied according to the following values: temperature fluctuated from 9.8 (S6, June 08) to 30.6 °C (S6, December 08), conductivity from 42 (S8, March 09) to 177 µS/cm (S6, February 09), pH from 6.57 (S1, March 09) to 9.22 (S7, February 08), turbidity from 10 (S5, February 08) to 193 UTN (S1, December 08), depth from 0.5 (S1, June 08) to 18 m (S3, March 09), dissolved oxygen from 4 (S6, March 09) to 12.8 mg/l (S7, Sep 08), and organic matter from 0.083 (S3, in the main stream of the river, February 08) to 23.67 % (S7, in closed systems with vegetation, March 09). Nutrients were measured in only one sampling and varied according to the following values: nitrates, 0.082 (S8) to 0.564 mg/l (S5); nitrites, 0.001 (S3) to 0.027 mg/l (S7); ammonium, 0.004 (S8) to 0.486 mg/l (S4); and phosphates, 0.035 (S8) to 0.766 mg/l (S6).

3.2 The annelid assemblage and its ecological affinities

The annelid assemblage studied in the Uruguay River during the period comprised a total richness R=40 taxa, most of which belonged to the oligochaete Naididae, being significant the quantity of species of Naidinae and Pristininae (Table 1). According to their frequency of occurrence and taking into account all the sampling sites, no species was constant. Only two taxa, *Aulodrilus pigueti* and Megadrili, were frequent; 11 taxa, among them *Limnodrilus hoffmeisteri*, *Bothrioneurum americanum*, *Nais variabilis*, *Pristina longidentata*, Enchytraeidae, *Narapa bonetoi*, and *Manayunkia speciosa* were accessories, while the rest of the taxa were incidental. Nevertheless, considering sites S3 and S7, *M. speciosa* was constant during the studied period.

TAXA				FO	
Oligochaeta					
Naididae	Tubificinae	*Aulodrilus pigueti* Kowalewski, 1914		Ap	F
		Limnodrilus hoffmeisteri Claparède, 1862		Lh	A
		Limnodrilus udekemianus Claparède, 1862		Lu	I
	Rhyacodrilinae	*Bothrioneurum americanum* Beddard, 1894		Ba	A
		Branchiura sowerbyi Beddard, 1892		Bs	I
	Naidinae	*Stylaria fossularis* Leidy, 1852		Sf	I
		Dero (Dero) pectinata Aiyer, 1930		Dp	I
		Dero (Dero) digitata (Müller, 1773)		Dd	I
		Dero (Dero) righii Varela, 1990		Dr	I
		Dero (Aulophorus) furcata (Müller, 1773)		Df	I
		Dero (Aulophorus) costatus (Marcus, 1944) emm. Harman, 1974		Dc	I
		Allonais lairdi Naidu, 1965		Al	I
		Nais variabilis Piguet, 1906		Nv	A
		Nais communis Piguet, 1906		Nc	I
		Nais pardalis Piguet, 1906		Np	A
		Slavina appendiculata (d'Udekem, 1855)		Sa	I
		Slavina isochaeta Cernosvitov, 1939		Si	I
		Slavina evelinae (Marcus, 1942)		*	I
		Stephensoniana trivandrana (Aiyer, 1926)		St	I
		Bratislavia unidentata (Harman, 1973)		Bu	I
		Chaetogaster diastrophus (Gruithuisen, 1828)		Cdt	I
		Chaetogaster diaphanus (Gruithuisen, 1828)		Cdp	I
	Pristininae	*Pristina osborni* (Walton, 1906)		Po	A
		Pristina proboscidea Beddard, 1896		Pp	I
		Pristina jenkinae (Stephenson, 1931)		Pj	A
		Pristina acuminata Liang, 1958		Pa	I
		Pristina americana Cernosvitov, 1937		Pam	A
		Pristina leidyi Smith, 1896		*	I
		Pristina longidentata Harman, 1965		Plo	A
		Pristina aequiseta Bourne, 1891		Pae	I
		Pristina sima (Marcus, 1944)		Ps	I
		Pristina longisoma Harman, 1977		*	I
		Pristina macrochaeta Stephenson, 1931		Pm	I

TAXA			FO
Opistocystidae	*Trieminentia corderoi* (Harman, 1970)	Tc	I
	Crustipellis tribranchiata (Harman, 1970)	Ct	I
Enchytraeidae		EN	A
Narapidae	*Narapa bonettoi* Righi y Varela, 1983	Nb	A
Megadrili		ME	F
Polychaeta			
Aeolosomatidae	*Aeolosoma* sp.	*	I
Sabellidae	*Manayunkia speciosa* Leidy, 1858	Ms	A

Table 1. List of annelid species collected at the sampling sites in the Uruguay River
(Argentina), with the abbreviation of each species showed in CCA biplot and the frequency
of occurrence category: F, frequent (75 % < FO ≤ 50 %), A, accessory (50 % < FO ≤ 25 %); I,
incidental (FO < 25 %).* Taxa not included in the CCA (frequency of occurrence < 3%).

The ecological indexes applied showed certain fluctuations during the year in the sampling
sites (Figure 3). Diversity (H') varied between 0.43 and 2.12. Site S5 presented the lowest H'
values, while site S6 showed the highest ones. Evenness (E') varied between 0.24 and 0.87,
although it was rather regular in all the sites throughout the period of study. Taxonomic
richness (R) varied between 2 and 16, and it was very fluctuating in almost all the sampling
sites, except in site S7 that showed similar values during the year (Fig 2).

Annelid density (N) varied between 66 (S5, spring) and 3985 ind./m² (S8, summer) (Figure
4). *M. speciosa* was present in sites S3 and S7 in all the seasons, while it was never recorded
in site S8; in the rest of the sampling sites its presence was variable during the year. Its
density varied between 4 (S6, spring) and 517 ind./m² (S3, autumn). Two egg-bearing
females were found in March 2009.

Figure 5 shows an individual of *M. speciosa* —with part of its tube built by fine sediments—,
and another annelids representative of the assemblage found in Lower Uruguay River, such
as the naidid oligochaetes Naidinae and Tubificinae.

The Pearson correlation coefficient demonstrated a positive relationship between the
abundance of *M. speciosa* and the percentage of organic matter in the sediments ($r= 0.508$; $P< 0.001$), as well as between the total abundance of oligochaetes in the samples and the
percentage of organic matter in the sediments ($r= 0.278$; $P< 0.05$).

According to the PCA results, the first two axes explained 57.4 % of the cumulative variance
of the data set (Table 2).

	Axis 1	Axis 2	Axis 3	Axis 4	p-value
PCA-eigenvalue	0.337	0.237	0.180	0.134	
PCA-cumulative % variance	33.7	57.4	75.4	88.9	
CCA-eigenvalue	0.592	0.340	0.242	0.171	
CCA-cumulative % variance explained (species-environment relation)	38.6	60.7	76.4	87.6	0.004

Table 2. Summary results from the PCA and CCA ordination analysis.

Figure 6 represents the ordination of the sampling sites studied determined by the two first
axes (axis 1, eigenvalue: 0.337; axis 2, eigenvalue: 0.237). The variables that best correlated

with axis 1 were temperature and dissolved oxygen, while pH and conductivity correlated with axis 2. A marked seasonality in the ordination of the sites was observed.

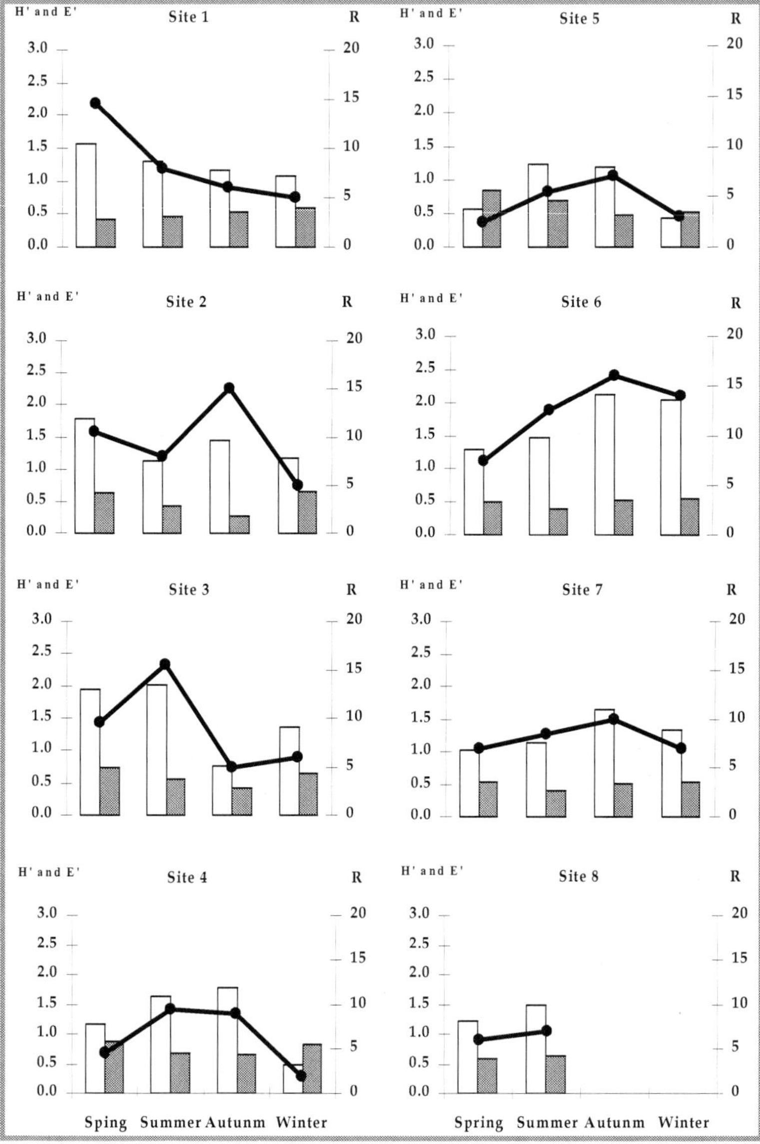

Fig. 3. Ecological Indexes at the sampling sites in Lower Uruguay River. The Taxonomic Richness (R) is shown in black line, Evenness (E') in gray bars and Shannon-Wiener Diversity (H') in white bars.

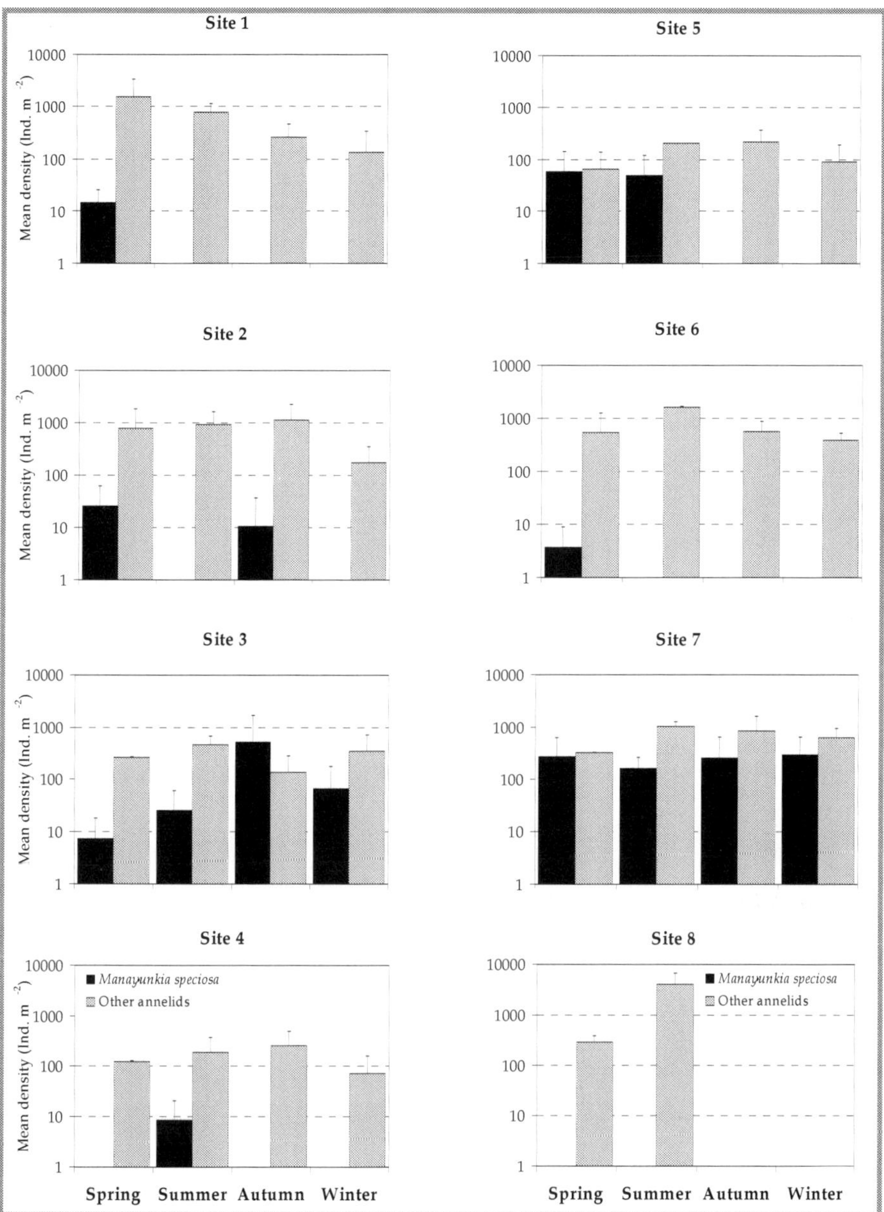

Fig. 4. Mean abundances of the introduced polychaete *Manayunkia speciosa* and the other annelids (Oligochaeta and Polychaeta Aeolosomatidae) collected during the study at the Lower Uruguay River, Argentina.

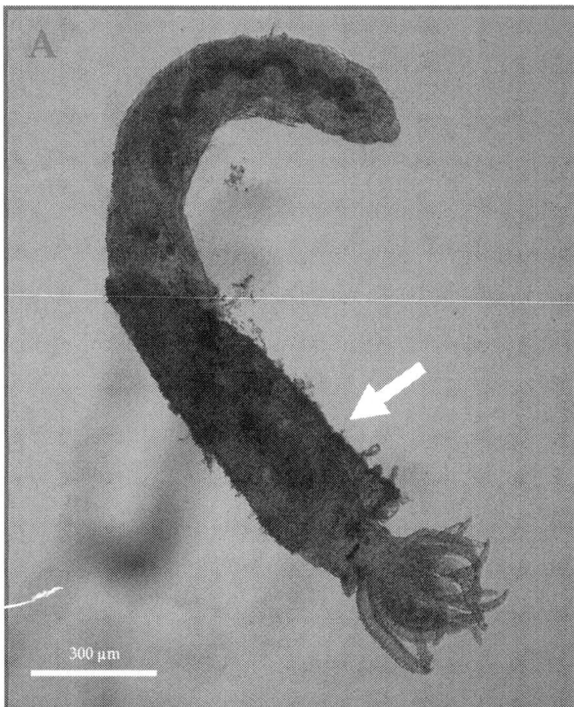

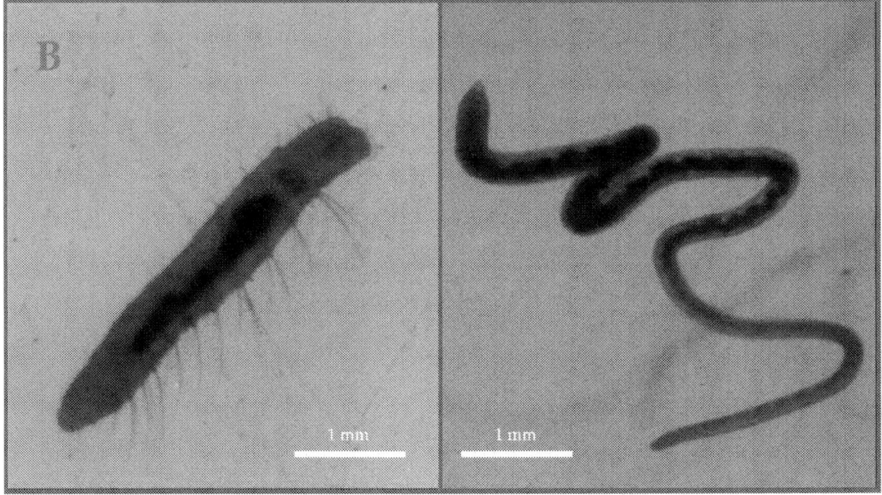

Fig. 5. A. *Manayunkia speciosa* specimen collected in the Lower Uruguay River, Argentina. The arrow indicates part of the tube formed by fine sediments agglutinated. B. Oligochaeta Naididae: Naidinae (left) and Tubificinae (right).

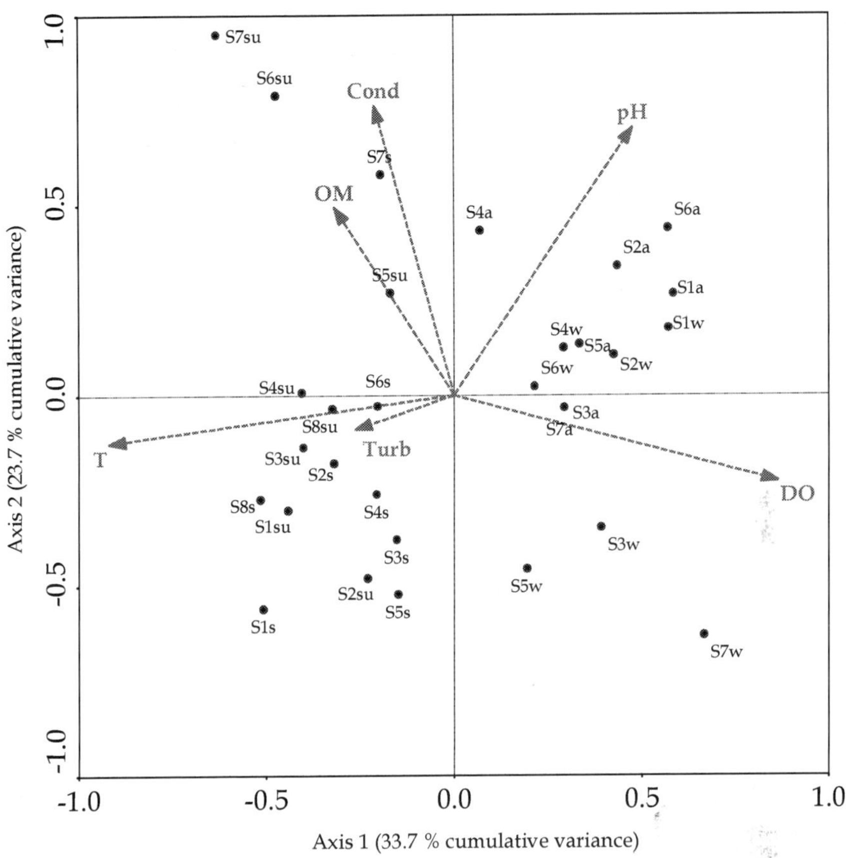

Fig. 6. Ordination (PCA) diagram of the first two axes showing the variation in the sampling sites (dots) among habitat related variables (arrows). S1- S8: sampling sites; su: summer; a: autumn; w: winter; s: spring.

All the sampling sites corresponding to autumn and winter were grouped in the right (top and bottom) quadrants of the biplot, while the sampling sites corresponding to spring and summer were grouped in the left quadrants.

The first two axes of CCA explained 60.7 % of the cumulative variance from the data set (axis 1, eigenvalue: 0.592; axis 2, eigenvalue: 0.340) (Figure 7). The variables that best correlated with axis 1 were pH (r= 0.613) and conductivity (r= 0.533), while turbidity (r= 0.525) and temperature (r= 0.409) were the variables that best correlated with axis 2. The annelid species were distributed along a gradient associated to turbidity, for example *Dero digitata*, *Stephensoniana trivandrana*, *Trieminentia corderoi*, and *Slavina appendiculata*, among others. *Bratislavia unidentata*, *Limnodrilus hoffmeisteri*, *Branchiura sowerbyi*, and *Pristina osborni* were associated to organic matter. *Stylaria fossularis* markedly separated from the rest of the species and presented a strong association with conductivity and pH. Otherwise, *M. speciosa*, did not show any particular association with the physicochemical variables and grouped together with other species as *Narapa bonettoi*, *Limnodrilus udekemianus*, *Aulodrilus pigueti*, *Pristina jenkinae* and the Enchytraeidae, close to the intersection of the axes.

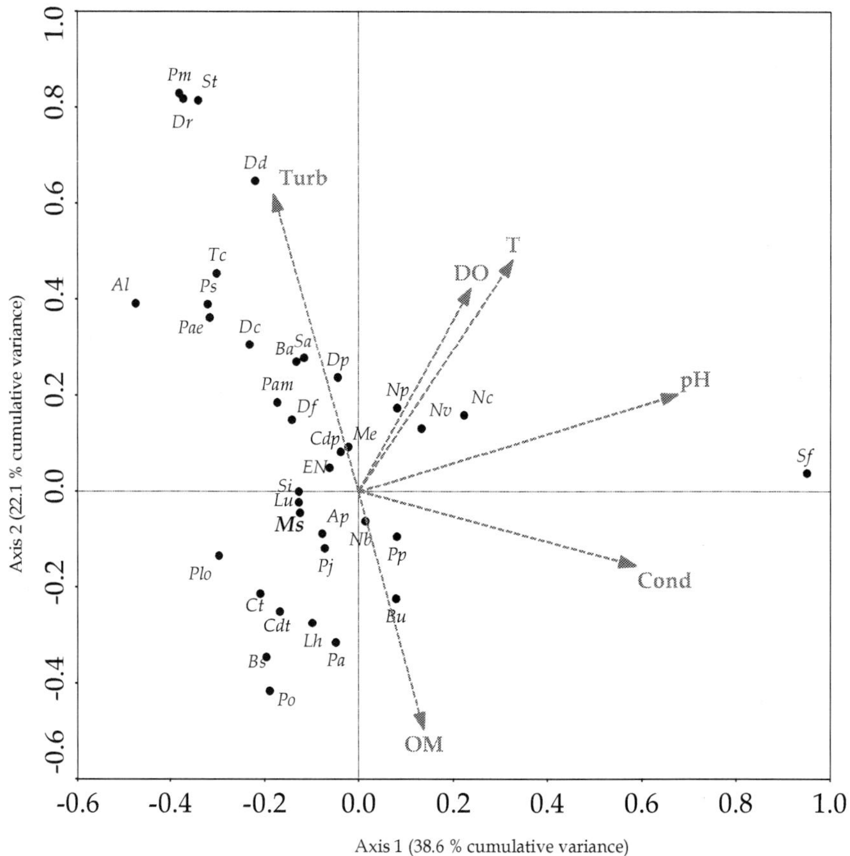

Fig. 7. Ordination diagram (CCA) of the first two axes showing the taxa (dots) and environmental variables (arrows). Taxa abbreviations are as in Table 1.

4. Discussion and conclusions

- Macroinvertebrates are good indicators of the ecological quality of rivers; they integrate the changes occurring over time in the environmental conditions. The species respond to the environmental variables through the gradients and the communities of specific assemblages can be used as entities in order to describe the biological state of rivers under certain environmental conditions. Ordination techniques are used to describe the species-environment associations and to identify the factors which could influence the habitat preference of each taxon. It is a way of understanding the range of tolerance of a particular species to certain environmental variables (Adriaensses et al., 2007). In this study, ordination techniques were used in order to understand the environmental preferences of each species. Nevertheless, as the ordination only analyzes the correlations between environmental variables and species distribution, and does not reveal causal relationships, it cannot be used as a tool to determine the real factors within the many variables which could be affecting the species distribution (Lin & Yo, 2008).

The Principal Components Analysis (PCA) showed that the tendency in the habitat distribution between the sampling sites was affected, to a great extent, by seasonality. The principal environmental variables acting in the ordination of the sites were temperature, conductivity and dissolved oxygen. The Canonical Correspondence Analysis (CCA) evaluated the relationship between the species populational abundance and the environmental variables. The length of the arrows representing certain variables, such as turbidity, pH and conductivity shows the impact they have had on the annelid assemblage. Some species showed a strong association with turbidity (e.g. *Dero digitata* and *Stephensoniana trivandrana*), others to organic matter (e.g. *Bratislavia unidentata* and *Limnodrilus hoffmeisteri*), while others, like *Manayunkia speciosa, Aulodrilus pigueti, Narapa bonettoi* did not show a particular association with the physicochemical variables and grouped together close to the intersection of the axes. *L. hoffmeisteri* is a species frequently associated to eutrophic environments (Dumnicka, 2002; Krodkiewska & Michalik-Kucharz, 2009; Särkkä, 1987) and together with *A. pigueti* dominates in the alluvial plain environments of Paraná River (Ezcurra de Drago et al., 2007). *N. bonettoi* is a species commonly associated to coarse sediment and sand, generally found in the main riverbed of Paraná and Paraguay rivers, where it can reach high densities (Ezcurra de Drago et al., 2007; Marchese et al., 2005; Takeda et al., 2001). Nevertheless, we found it mainly in shallow zones (bays) with abundant aquatic vegetation.

- The oligochaetes are important organisms in the aquatic systems due to their impact on the sediment structure and water-sediment exchanges. Their role in the trophic nets and as intermediate hosts for several myxozoan parasites of fishes of economic importance is significant, as well as their application in pollution monitoring programs and their potential use to reduce sludge volumes in sewage treatment systems (Martin et al., 2008). Current studies on the zoobenthos of Uruguay River only deal with the fauna of its middle and upper sections (Ezcurra de Drago & Bonetto, 1969; Amestoy et al., 1986; Di Persia & Olazarri, 1986; Pintos et al., 1992). The annelid fauna of the lower section has not received much attention so far. In our work, the freshwater polychaete *Manayunkia speciosa* was associated with the oligochaete assemblage formed by cosmopolitan organisms, like Naididae (mainly Tubificinae and Naidinae) and Enchytraeidae (Martin et al., 2008), and an endemic family of the Neotropical region, Narapidae (Ezcurra de Drago et al., 2007).

Such assemblage showed certain similarities with the oligochaete fauna recorded in other large South American rivers as Paraguay and Paraná (Ezcurra de Drago et al., 2007; Marchese et al., 2005; Monthanoli & Takeda, 1999; 2001), although a higher number of species have been recorded in our study in the Lower Uruguay River. Among the especies registered in all these rivers, *A. pigueti, L. hoffmeisteri, L. udekemianus, Bothrioneurum americanum, Pristina americana, P. longidentata, Stephensoniana trivandrana, Trieminentia corderoi* and *Narapa bonettoi* can be mentioned. Ezcurra de Drago et al., (2007) noticed that the benthos of large rivers (with sandy-mobile beds) has a high similarity world-wide, with high abundances, low biomass and low species richness. In the lateral dimension of the alluvial plain, these authors (Ezcurra de Drago, op. cit.) observed a gradient from the main channel to the temporary marginal wetlands, with an increase in the species richness, diversity and biomass, and a decrease in abundance. In the alluvial river-plain transect they found a higher occurrence of oligochaetes in channels and lakes. The spatial heterogeneity, habitat structure, temporal instability and high productivity determine the highest importance of the transversal dimension over the longitudinal in the large rivers with alluvial plain. Vertical dimension also plays an important role in these systems, as it offers refuge for the hatching of eggs and the eclosion of juveniles, providing a habitat for these first stages which represent a faunistic reserve of the system under adverse conditions. Many invertebrate species, the oligochaetes among them, can move to the sediments during the drought phases until favorable conditions are restored. In the Parana River system, the benthic assemblage exhibits more variation in an extended spatial scale, where geophysical processes occur, rather than in smaller scales.

We observed a similar complex as pointed out by Marchese et al., (2005), and Ezcurra de Drago et al., (2007) in Paraná River, in which the highest richness and diversity were recorded in the marginal habitats (or floodplain lakes), that represent the highest spatial and temporal heterogeneity of habitats. Likewise, these authors recorded the highest densities in the main riverbed with values much higher than the ones registered by us. In our study, the highest abundances of individuals have been found in sites with high values of richness and diversity (as bays and lagoon), but not in the main riverbed. In our case, the low values of abundance, diversity and richness observed in the main riverbed (S4), may be due to the influence of the pulp mill industry placed upstream. It is known that the benthic invertebrates are exposed to xenobiotic chemicals in several ways, like the direct contact with sediment through the tegument or the ingestion of contaminated sediment, as the pulp mill effluents and their sediments contain possibly bioavailable and toxic amounts of resin acids, β-sitosterol and chlorophenolics (Merilaïnen & Oikari, 2008).

In the large South American Rivers (e.g. Paraná and Paraguay), the composition of the benthic assemblage and the total density show changes determined by the annual pulse of floods and the degree of connectivity of the alluvial environments, and by the long term phenomena involving ecological successions. For the benthic communities, the low water periods represent a more significant stress factor than the floods (Ezcurra de Drago et al., 2004; 2007; Montanholi & Takeda, 2001). Nowadays, the ENOS (El Niño/ Southern Oscillation) is one of the most widely studied weather phenomena. During the sampling period in the Lower Uruguay River, the ENOS (El Niño/La Niña) events underwent a cold phase or "La Niña" with transition to a neutral phase (information of the Departamento Climatológico, Servicio Meteorológico Nacional, Argentina, www.smn.gov.ar), related with periods of normal to low discharges (Pasquini & Depetris, 2007). Montaholi & Takeda (2001) recorded *Aulodrilus pigueti* in the Upper Paraná River during the low water phase, where it was dominant. In our study, this species appeared in every site and sampling occasion. Other studies associated *A. pigueti* with fine sediment and a high content of organic matter (Ezcurra de Drago et al., 2007; Lin & Yo, 2008). Montaholi & Takeda (op. cit.) considered that its highest abundances in the alluvial lagoons during the low water phase and its low density in lotic environments is an adaptation to lentic environments and muddy substrate with a high quantity of organic matter and low velocity of current. Besides, its asexual reproductive mode would be intensified during these low water periods.

- Many studies indicate that the anthropogenic pressure in natural ecosystems favors the invasion and persistence of non-native species when the habitat conditions are modified (Gabel et al., 2011). Polluted areas could accumulate a higher quantity of invasive species than the less impacted areas (Crooks et al., 2011). Freshwater ecosystems are particularly affected by biological invasions. The artificial navigation routes, when connecting the freshwater systems previously separated by natural biogeographic boundaries, constituted important ways of invasion for non-native species, especially with maritime ports as entrance gates from overseas. Freshwater species can be transported attached to the hulls of ships (in continental aquatic systems) or in their ballast water (Gabel et al., 2011). The anthropogenic alteration of the abiotic factors (pollution, physic disturbances, artificial navigation routes, etc.) facilitates biological invasions (Crooks et al., 2011; Gabel et al., 2011). Estuaries, coastal lagoons and inner waters generally present more species of introduced invertebrates than the marine environments (Zaiko et al., 2011). This can be due to the fact that they are more vulnerable environments (variety of habitats, salinity gradient, nutrient enrichment, etc.) and with a higher anthropogenic pressure —extensive shipping activities, proximity to invasive corridors, e.g. channels, artificial substrata, degradation of habitat, over-fishing—. Some authors even consider that these ecosystems should be treated as "hot spots" for the introduction of species (Zaiko et al., 2011) and therefore the management strategies and monitoring programs tending to prevent the invasion impacts should be focused on them. The efforts to improve the environmental conditions and to prevent the mechanisms which favor potential invaders should be encouraged.

- The biogeographic aspects of the genus *Manayunkia* arises certain interest because it includes species which have colonized freshwater environments (Rouse, 1996). Benthic and littoral organisms can present a high degree of endemism, being an example *M. speciosa* (Margalef, 1983). Its distribution has been restricted to North America, with a rather disperse and also intriguing distribution pattern. Nevertheless, and as with other polychaete species, it is supposed that its distribution has been widened due to human activities, both intentionally (aquiculture) and inadvertently (basins interconnection and maritime transportation activities) (Glasby & Timm, 2008). As early as in the first decades of the 20th century, Meehan (1929) suggests its entrance in Lake Superior through the ballast water of maritime ships.

- *Manayunkia speciosa* is a tube-dwelling organism, with poorly developed parapods and therefore a limited swimming capacity (Croskery, 1978). It builds its tubes with fine particles clustered by mucus secretions (Leidy, 1883). Under certain conditions —when disturbed— it can leave the tube and roam freely in the sediment (Rouse, 1996; Stocking & Bartholomew, 2007). Like other freshwater polychaetes, it has modified its reproductive strategies for protecting their larvae from osmotic stress. Although the knowledge on its reproductive ecology is scarce, it reproduces sexually or asexually inside the tube, forming typically large and yolky eggs. The juveniles develop within the tubes and when they become young adults, they leave the refuge to build their own ones (Croskery, 1978; Glasby & Timm, 2008). Like other representatives of the Fabricinae subfamily, it is believed that they are suspension feeders, and in a facultative way they are surface deposit feeders, consuming fine organic detritus and microalgae. As they require great quantities of organic matter and in order to optimize their ingesta, these polychaetes are able to select the size of the particles. They either reject or use those bigger particles in the building of their tubes (Stocking & Bartholomew, 2007). Successful non-native invertebrates often present ecological characteristics which result beneficial in disturbed environments, e.g. short generation time, early sexual maturity, high fecundity, a wide food range, euryhalinity and tolerance to pollution and habitat degradation (Gabel et al., 2011). Some of these characteristics would have allowed *M. speciosa* the settlement in the Lower Uruguay River. As regards its ecological requirements, Brehm (1978) found this polychaete associated to silt-clay sediments, scarce sands, abundant organic matter, salinities up to 11 ‰, and temperatures between 7 to 31 °C. Besides, Margalef (1983) mentioned *M. speciosa* as a resistant species to polluted water. In the Lower Uruguay River, we have found this species mainly in calm water sites with aquatic vegetation and abundant organic matter. Despite the very low abundance of ovigerous females collected in March 2009, we suggest that the population of *M speciosa* could be well established in the Lower Uruguay River.

During biological invasion, "lag times" are frequently observed between the first introduction and the consequent spread of the species, and ussually extend from decades to centuries. Several species have been present for long periods without an apparent major impact before rapidly increasing their populations and becoming highly consequential (Simberloff, 2011). We hypothesize that a similar event might be occurring with *M. speciosa*. Furthermore, the only record of this polychaete on the South American Uruguay River highlights that it goes unnoticed due to its small size, infaunal lifestyle and low abundances. According to Brehm (1978), who reported that most records are based on a few individuals, we found ca. 500 ind./m². Only Hiltunen (1965) recorded more than 45,000 ind./m² at the mouth of Detroit River (USA). Nevertheless, this does not imply that introduced species that now have scarce or no repercussion could eventually have a greater impact on the involved ecosystems in the future.

5. Acknowledgments

We would like to thank Jorge Donadelli from Laboratorio de Química, ILPLA for the analyses of nutrients of the water samples and Mrs. Mónica Caviglia for the translation of the manuscript. This is scientific contribution N° 914 of Instituto de Limnología Dr Raúl A. Ringuelet (ILPLA).

6. References

Adriaensses, V.; Verdonschot, P. F. M.; Goethals, P. L. M. & De Pauw, N. (2007). Application of clustering techniques for the characterization of macroinvertebrate communities to support river restoration management. *Aquat Ecol*, vol 41, pp. 387–398, ISSN: 1573-5125.

APHA (American Public Health Association). (1998). *Standard Methods for Examination of Water and Wastewater*, 20th ed. American Public Health Association, American

Water Works Association and Water Pollution Control Federation, ISBN: 087-5532-35-7, Washington DC.

Amestoy, F.; Fabiano, G. & Spinetti, M. (1986). Comunicación preliminar sobre la presencia de *Corbicula spp.* (Mollusca, Pelecypoda) en contenidos estomacales de peces de importancia comercial. IS en seminario "El río Uruguay y sus Recursos Pesqueros". Comisión Administradora del Río Uruguay. Pub. N° 4: 49-53.

Armendáriz, L. C.; Paola, A. & Rodrigues Capítulo, A. (2011). *Manayunkia speciosa* Leidy (Polychaeta: Sabellidae): introduction of this nonindigenous species in the Neotropical Region (Uruguay river, South America). *Biol Invasions*, vol 13, pp. 281–284, ISSN: 1573-1464.

Armendáriz, L. C.; Rodrigues Capítulo, A. & Ambrosio, E. S. (2011). Relationships between the spatial distribution of oligochaetes (Annelida, Clitellata) and environmental variables in a temperate estuary system of South America (Río de la Plata, Argentina). *NZ J. Mar. Freshwater Res*, vol 45, pp. 263-279, ISSN: 0028-8330.

Barbour, M. T.; Gerritsen, J.; Snyder, B. D.; Stribling, J. B. (1999). *Rapid Bioassessment Protocols for use in streams and wadeable rivers. Periphyton, benthic macroinvertebrates and fish* 2nd edn. EPA 841- B-99-002. U. S. Environmental Protection Agency; Office of Water, Washington D.C.

Bartholomew, J. L.; Whipple, M. J.; Stevens, D. G. & Fryer, J. L. (1997). The life cycle of *Ceratomyxa shasta*, a myxosporean parasite of salmonids, requires a freshwater polychaete as an alternate host. *J. Parasitol*, vol 83, pp. 859–868, ISSN: 0022-3395.

Bartholomew, J. L.; Atkinson, S. D. & Hallett, S. L. (2006). Involvement of *Manayunkia speciosa* (Annelida: Polychaeta: Sabellidae) in the life cycle of *Parvicapsula minibicornis*, a myxozoan parasite of Pacific salmon. *J Parasitol*, vol 92, pp. 742–748, ISSN: 0022-3395.

Brehm, W. T. (1978). First Gulf of Mexico coast record of *Manayunkia speciosa* (Polychaeta: Sabellidae). *Northeast Gulf Sci*, vol 2, pp. 73-75. ISSN: 0148-9836.

Bonetto, A. & Hurtado, S. (1998). Región 1: Cuenca del Plata. In: *Los humedales de la Argentina: clasificación, situación actual, conservación y legislación*, Canevari, P.; Blanco, D.; Bucher, E.; Castro, G.; Davidson, I. (Eds), 31- 72, Wetland International Publ, ISBN: 987-97187-1-2, Buenos Aires, Argentina.

Brinkhurst, R. & Marchese, M. (1992). *Guía para la identificación de oligoquetos acuáticos continentales de sud y centroamérica*. Asociación de Ciencias Naturales del Litoral, Colección Climax, vol 6, ISBN: 950-9267-07-4, Santo Tomé, Argentina.

Cabrera, A. L. (1971). Fitogeografía de la República Argentina. *Bol. Soc. Argent. Bot.*, vol 14, pp. 1-42. ISSN: 1851-2372.

Cabrera, A. L. & Willink, A. 1973. *Biogeografía de América Latina*. Programa regional de desarrollo científico y tecnológico. Secretaría General de la Organización de los Estados Americanos, Washington, D.C.

Ciutti, F. & Cappelletti, C. (2009). First record of *Corbicula fluminalis* (Müller, 1774) in Lake Garda (Italy), living in sympatry with *Corbicula fluminea* (Müller, 1774). *J Limnol*, vol 68, pp. 162-165, ISSN: 1723-8633.

Crooks, J. A.; Chang, A. L. & Ruiz, G. M. (2011). Aquatic pollution increases the relative success of invasive species. *Biol Invasions*, vol 13, pp. 165–176, ISSN: 1573-1464.

Croskery, P. (1978). The freshwater co-occurrence of *Eurytemora affinis* (Copepoda: Calanoida) and *Manayunkia speciosa* (Annelida: Polychaeta): possible relicts of a marine incursion. *Hydrobiologia*, vol 59, pp. 237– 241, ISSN: 0018-8158.

Di Persia, D. & Olazarri, J. (1986). Zoobenthos of the Uruguay system. In: *The ecology of river systems*, Danes, B & Walker, K (Eds), pp. 623-631, Junk Publishers, Dordrecht, The Netherlands. ISBN 906-1935-40-7.

Dumnicka, E. (2002). Upper Vistula River: response of aquatic communities to pollution and impoudment. X. Oligochaete taxocens. *Pol. J. Ecol*, vol 50, pp. 237-247, ISSN: 0420-9036.

El Haddad, M.; Capaccioni Azzati, R. & García-Carrascosa, A. M. (2007). *Branchiomma luctuosum* (Polychaeta: Sabellidae): a non-indigenous species at Valencia Port (western Mediterranean Sea, Spain). *J Mar Biol Assoc*, vol 5660, pp. 1-8, ISSN: 0025-3154.

Ezcurra de Drago, I. & Bonetto, A. (1969). Algunas características del bentos en los saltos del río Uruguay, con especial referencia a la ecología de los poríferos. *Physis*, vol 28, pp. 359-369.

Ezcurra de Drago, I.; Marchese, M. & Wantzen, K. M. (2004). Benthos of a large neotropical river: spatial patterns and species assemblages in the Lower Paraguay and its floodplains. *Arch. Hydrobiol.* 160, 347-374, ISSN: 0003-9136.

Ezcurra de Drago, I.; Marchese, M. & Montalto, L. (2007). Benthic Invertebrates. In: *The Middle Paraná River: Limnology of a Subtropical Wetland,* Iriondo, M. H.; Paggi, J. C. & Parma, M. J. (Eds.), pp. 251-275, ISBN 9783540706236, Springer-Verlag, Berlin Heidelberg New York.

Gabel, F.; Pusch, M. T.; Breyer, P.; Burmester, V.; Walz, N. & Garcia, X. F. (2011). Differential effect of wave stress on the physiology and behaviour of native versus non-native benthic invertebrates. *Biol Invasions*, vol 13, pp. 1843–1853, ISSN: 1573-1464.

Glasby, C. & Timm, T. (2008). Global diversity of polychaetes (Polychaeta; Annelida) in freshwater. *Hydrobiologia*, vol 595, pp. 107–115, ISSN: 0018-8158.

ICZN. (2007). Bulletin of Zoological Nomenclature, Opinion 2167.

Hazel, C. R. (1966). A note on the freshwater polychaete, *Manayunkia speciosa* Leidy, from California and Oregon. *Ohio J Sci*, vol 66, pp. 533–535, ISSN: 0030-0950

Hiltunen, J. K. (1965). Distribution and abundance of the polychaete *Manayunkia speciosa* Leidy, in western Lake Erie. *Ohio J Sci*, vol 65, pp. 183- 185, ISSN: 0030-0950.

Holmquist, C. (1967). *Manayunkia speciosa* Leidy—a freshwater polychaete found in Northern Alaska. *Hydrobiologia*, vol. 29, pp. 297–304. ISSN: 1573-5117

Iriondo, M. & Kröhling, D. (2004). The parent material as the dominant factor in Holocene pedogenesis in the Uruguay River Basin. *Rev Mexicana Cienc Geol*, vol 21, pp. 175-184, ISSN: 1926-8774.

Krecker, F. H. (1939). Polychaete annelid worms in the Great Lakes. *Science*, vol 89, pp. 153, ISSN: 0036-8075.

Krodkiewska, M. & Michalik- Kucharz, A. (2009). The bottom Oligochaeta communities in sand pits of different trophic status in Upper Silesia (Southern Poland). *Aquat Ecol*, vol 43, pp. 437- 444, ISSN: 1386-2588.

Leidy, J. (1883). *Manayunkia speciosa. Proc Acad Nat Sci Phil*, vol 35, pp. 204–212, ISSN: 0097-3157.

Lin, K. J. & Yo, S. P. (2008). The effect of organic pollution on the abundance and distribution of aquatic oligochaetes in an urban water basin, Taiwan. *Hydrobiologia*, vol 596, pp. 213–223, ISSN: 1573-5117.

López Laborde, J. (1998). Marco geomorfológico y geológico del Río de la Plata. En: *El Río de la Plata. Una revisión ambiental. Un informe de antecedentes del ProyectoEcoPlata*, Wells, P. G. & Daborn, G. R. (Eds), 1-16, ISBN: 077-0328-52-0, Dalhousie University, Halifax, Novo Scotia, Canada.

Mackie, G. L. & Qadri, S. U. (1971). A polychaete, *Manayunkia speciosa*, from the Ottawa River, and its North American distribution. *Can J Zool*, vol 49, pp. 780–782, ISSN: 0008-4301.

Marchese, M. & Ezcurra de Drago, I. (1992). Benthos of the lotic environments in the middle Paraná River system: transverse zonation. *Hydrobiologia* 237, 1-13, ISSN 0018-8158.

Marchese, M. R.; Wantzen, K. M. & Ezcurra de Drago, I. (2005). Benthic invertebrates assemblages and species diversity patterns of the Upper Paraguay River. *River Res Appl*, vol 21, 485- 499, ISSN: 1535-1467.

Margalef, R. (1983). *Limnología*. ISBN: 84-282-0714-3, Omega, Barcelona, España.

Martin, P.; Martinez- Ansemil, E.; Pinder, A. & Timm, T. (2008). Global diversity of oligochaetous clitellates (Oligochaeta; Clitellata) in freshwater. *Hyrobiologia*, vol 595, pp. 117-127, ISBN: 878-1-4020-8258-0.

Meehean, O. (1929). *Manayunkia speciosa* (Leidy) in the Duluth Harbor. *Science*, vol 70, pp. 479–480, ISSN: 0036-8075.

Merilaïnen, P. & Oikari, A. (2008). Uptake of organic xenobiotics by benthic invertebrates from sediment contaminated by the pulp and paper industry. *Wat Res*, vol 42, pp. 1715–1725, ISSN: 0043-1354.

Montanholi-Martins, M. C. & Takeda, A. M. (1999). Communities of benthic oligochaetes in relation to sediment structure in the Upper Paraná River, Brazil. *Stud Neotrop Fauna Environ*, vol 34, 52- 58, ISSN: 0165-0521.

Montanholi-Martins, M. C. & Takeda, A. M. (2001). Spatial and temporal variations of oligochaetes of the Ivinhema River and Patos Lake in the upper Paraná River Basin, Brazil. *Hydrobiologia*, vol 463, pp. 197-205, ISBN: 1-4020-0090-1.

Nagy, G. J.; Martínez, C. M.; Caffera, R. M.; Pedrosa, G.; Forbes, E. A.; Perdomo, A. C. & López Laborde, J. (1998). Marco hidrológico y climático del Río de la Plata. En: *El Río de la Plata. Una revisión ambiental. Un informe de antecedentes del ProyectoEcoPlata*, Wells, P. G. & Daborn, G. R. (Eds), 17-70, ISBN: 077-0328-52-0, Dalhousie University, Halifax, Novo Scotia, Canada.

Orensanz, J. M.; Schwindt, E.; Pastorino, G.; Bortolus, A.; Casas, G.; Darrigran, G.; Elías, R.; López Gappa, J. J.; Obenat, S.; Pascual, M.; Penchaszadeh, P.; Piriz, M. L.; Scarabino, F.; Spivak, E. D. & Vallarino, E. A. (2002). No longer the pristine confines of the world ocean: a survey of exotic marine species in the southwestern Atlantic. *Biol Invasions*, vol 4, pp. 115-143, ISSN: 1387-3547.

Pasquini, A.I.; Depetris, P. J. (2007). Discharge trends and flow dynamics of South American rivers draining the southern Atlantic seaboard: An overview. *J Hydrol.*, vol 333, pp. 385-399, ISSN: 0022-1694.

Penchaszadeh, P. E. (Coord). (2005). *Invasores: Invertebrados exóticos en el Río de la Plata y región marina aledaña*. 1° edición, ISBN: 950-2313-88-7, Eudeba, Buenos Aires, Argentina.

Pettibone, M. (1953). Fresh-water polychaetous annelid, *Manayunkia speciosa* Leidy, from Lake Erie. *Biol Bul*, vol 105, pp. 149–153, ISSN: 0006-3185.

Pintos, W.; Conde, D. & Norbis, W. (1992). Contaminación orgánica en el río Uruguay (Paysandú, Uruguay). *Rev Asoc Cienc Nat Litoral*, vol 23, n° (1 y 2), pp. 21-29, ISSN: 0325-2809.

Rodrigues Capítulo, A.; Muñoz, I.; Bonada, N.; Gaudes, A. & Tomanova, S. (2009). La biota de los ríos: los invertebrados. En: *Conceptos y técnicas en ecología fluvial*, Elosegi, A. & Sabater, S (Eds), 23-37, ISBN: 978-84-96515-87-1, Rubes Editorial, España.

Rouse, G. W. (1996). New *Fabriciola* and *Manayunkia* species (Fabriciinae: Sabellidae: Polychaeta) from Papua New Guinea. *J Nat History*, vol 30, pp 1761-1778, ISSN: 0022-2933.

Sabater, S.; Donato, J.; Giorgi, A. & Elosegi, A. (2009). El río como ecosistema. En: *Conceptos y técnicas en ecología fluvial*, Elosegi, A. & Sabater, S (Eds), 23-37, ISBN: 978-84-96515-87-1, Rubes Editorial, España.

Särkkä, J. (1987). The occurrence of oligochaetes in lake chains receiving pulpl mill waste and their relation to eutrophication on the trophic scale. *Hydrobiologia*, vol 155,pp. 259-266, ISSN: 0018-8158.

Simberloff, D. (2011). How common are invasion-induced ecosystem impacts? *Biol Invasions*, vol 13, pp. 1255–1268, ISSN: 1573-1464.

Spencer, D. R. (1976). Occurrence of *Manayunkia speciosa* (Polychaeta: Sabellidae) in Cayuga Lake, New York, with additional notes on its North American distribution. *Trans Am Micr Soc*, vol 95, pp. 127–128.

Stocking, R. W. & Bartholomew, J. L. (2007). Distribution and habitat characteristics of *Manayunkia speciosa* and infection prevalence with the parasite *Ceratomyxa shasta* in the Klamath River, Oregon–California. *J Parasitol*, vol 93, pp. 78–88, ISSN: 0022-3395.

Struck, T. & Purschke, G. (2005). The sister group relationship of Aeolosomatidae and Potamodrilidae (Annelida: "Polychaeta") - a molecular phylogenetic approach based on 18S rDNA and cytochrome oxidase I. *Zool Anzeiger*, vol 243, pp. 281–293, ISSN: 0044-5231.

Takeda, A. M. (1999). Oligochaete community of alluvial Upper Paraná River, Brazil: spatial and temporal distribution (1987- 1988). *Hydrobiologia*, vol 412,pp. 35- 42, ISSN: 0018-8158.

Takeda, A. M.; Stevaux, J. C. & Fujita, D. S. (2001). Effect of hydraulics, bed load grain size and water factors on habitat and abundance of *Narapa bonettoi* Righi & Varela, 1983 of the Upper Paraná River, Brazil. *Hydrobiologia*, vol 463, pp. 241-228, ISBN: 1-4020-0090-1

Ter Braak, C.J. F.; Smilauer, P. (2002). *CANOCO reference manual and CanoDraw for Windows user's guide: software for canonical community ordination (version 4.5).* Microcomputer Power, Ithaca.

Zaiko, A.; Lehtiniemi, M; Narščius, A. & Olenin, S. (2011). Assessment of bioinvasion impacts on a regional scale: a comparative approach. *Biol Invasions*, vol 13, pp. 1739–1765, ISSN: 1573-1464.

5

Evaluation of Aquatic Ecosystem Health Using the Potential Non Point Pollution Index (PNPI) Tool

Camilla Puccinelli[1,*], Stefania Marcheggiani[1,*],
Michele Munafò[2], Paolo Andreani[3] and Laura Mancini[1]
[1]Italian-National Institute of Health Dept. Environment and Primary Prevention
[2]Italian National Institute for Environmental Protection and Research
[3]Province of Viterbo
Italy

1. Introduction

The aquatic ecosystem health is a topic that has been developed by scientific world and local authorities. This study shows numerous aspects of environmental management: one of the most interesting fields is the that investigating the relationships between ecosystem and human health.

An healthy aquatic ecosystem is able to preserve and recovering quickly its structure and functionality against adverse effects due to natural, like floods and landslides, or human causes, like pollution and urbanization.

The degradation of aquatic ecosystems can have important impacts on human health: The ways that water can damage people are different: consumption of contaminated waters and fishes, infections by vectors related to these environments and algal blooms in inland and coastal waters.

The aquatic ecosystem protection come to prominence after the emanation of Water Frame Directive 2000/60/EC (European Union, 2000). The innovative point of WFD is the assessment of water quality entrusted to biological communities and their relationship with human pressures and impacts.

The analysis of four biological elements required, phytobenthos, macrophytes, benthic invertebrates and fishes, describe the ecological status, that represents the functionality of the ecosystems. The concept of "ecological status" is another new point of view in the preservation activities of natural systems, that before took into account only the biodiversity and the preservation of rare species.

Water Frame Directive main objective is to reach a good ecological status for all water bodies until 2015. To achieve this goal, two key steps are needed: the first is the assessment

---

* The authors equally contributed to this work

of ecological status, then the river management plan in order to preserve healthy ecosystems and restore the damaged ones.

The evaluation of ecological status, as wrote before, required the study of the structures (composition and abundance of species) of the biological elements, giving a global view of different pressures that affect aquatic ecosystem. Main adverse effect are sourced from: nutrients load, organic pollution, hydromorphological alterations.

Biotic communities detect the effect of anthropogenic pressure; evaluate the negative externalities that human activities may have on the water bodies, but cannot be enough for a right management planning of these environments.

The river management plans should include: objectives for each water body; reasons for not achieving objectives where relevant; and the program of actions required to meet the objectives. The restoring actions for the achievement of good ecological status should begin, first of all, by identifying the pressures and the impacts that insist on the water bodies.

Between pressures that affect aquatic communities, eutrophication and organic pollution, generally are not easy to assess. The sources of these pressures could be classified in point pollution sources, and non point pollution sources.

Point pollution sources are linked to those sources "were originally defined as pollutants that enter the transport routes at discrete identifiable locations and that can usually be measured", while nonpoint source pollution was "everything else"(Loague & Corwin, 2005). Point sources of pollution are represented by industrial or municipal wastewater discharges.

Nonpoint pollutants are defined as "contaminants of air, and surface and subsurface soil and water resources that are diffuse in nature and cannot be traced to a point location" (Corwin & Wagenet, 1996). They are linked to agricultural activities (e.g. irrigation and drainage, applications of pesticides and fertilizers, runoff and erosion); urban and industrial runoff; pesticide and fertilizer applications; nitrogen and phosphorus atmospheric deposition; livestock waste; and hydrologic modification e.g. dams, diversions, channelization, over pumping of groundwater, siltation, (Loague & Corwin, 2005, Haycock *et al.*, 1993; European Environment Agency, 1999a; Crouzet, 2000; Schilling & Libra, 2000).

Point pollution source are easily identified, and also restoring activities can directly control work on them. Concerning non point pollution sources, they cannot be directly identified, assessed and controlled. One way to evaluate diffuse pollution is by a description of the hydrologic rainfall-runoff transformation processes with attached quality components (Notovny & Chesters, 1981). The useful methods, collecting hydrologic information rainfall-runoff, other environmental parameters are those based on Geographical Information System (Solaimani *et al.*, 2005; Hellweger & Maidment 1999; Olivera & Maidment 1999; Kupcho, 1997).

In this context a Geographical Information System (GIS) based index was developed, the Potential Non-Point Pollution Index (PNPI), in order to describe the global pressure exerted on water bodies by different land uses across the catchment areas (Munafò *et al.*, 2005). The chapter showed three case studies of the application of PNPI: the Trasimeno lake (Baiocco *et al.*, 2001) the Tiber River basin (Munafò *et al.*, 2005) and the province of Viterbo.

2. Potential Non Point Pollution Index

The Potential Non-Point Pollution Index (PNPI) is a tool designed to assess the global pressure exerted on rivers and other surface water bodies by different land use areas and infrastructures across the catchments. PNPI is a GIS-based watershed scale tool designed to give fast and reliable support to decision makes and public opinion about potential environmental impact of different land management scenarios. PNPI doesn't need a great amount of input data nor highly skilled operators and its high communication potential makes it particularly interesting for a participatory approach to land management. PNPI values is based on land use, geological features and distance from the water body of each land unit (Munafò *et al.*, 2005, Cecchi *et al.*, 2007). The information required are land use (like Corine Land Cover) maps, geological maps and digital elevation models (DEM).

The potential pollution is the expression of three indicators:

Land Cover Indicator, LCI: refers to the potential generation of non-point pollution due to the land uses of the parcel; it depends on the pollution potential load of the single cell mainly due to management practices (i.e. fertilizers and manure application for agricultural areas).

Run-Off Indicator, ROI,: takes into account pollutant mobility and possible filtering with respect to terrain slope, land cover and geology, it depends on the physical features of the entire path from cell to drainage network, features that affect flow velocity and subsequently pollution filtering.

Distance Indicator, DI, is the distance from the water body into a pollution dumping coefficient.

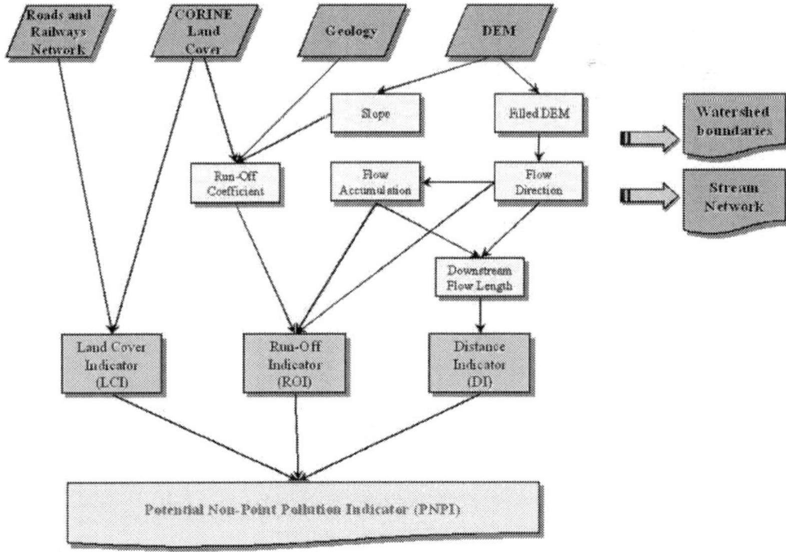

Fig. 1. Calculation pathway of PNPI (Munafò *et al.*,2005).

PNPI gives the potential contribution of each cell of the DEM (as land unit) to the non-point pollution of the studied aquatic ecosystem. PNPI values can then be used to build up five quality classes: natural and unaltered zones gave the lowest values and represented the first class (Munafò *et al.*, 2005), whereas densely populated areas and intensively cultivated crop lands highest values and corresponded to the fifth class. The output of the calculation can be presented in the form of maps showing areas that are more likely to produce pollution and the aquatic ecosystem section that is more affected by these areas.

2.1 Calculation of PNPI

The PNPI for every cell of the river basin is then calculated as a combination of the three indicators described above (LCI, ROI, DI) following the formula:

$$PNPI = 5 * LCI * 3 * DI * 2 * ROI$$

Most important of the three indicator is the LCI. For its values, experts, provided to each land use a coefficient depending on the polluting potential Land use types and their geographic distribution were taken from CORINE land cover (CLC) digital maps. Densely built areas and intensively cultivated fields were given the highest coefficients whereas natural and unaltered zones were placed at the opposite end of the scale (Munafò *et al.*, 2005; Cecchi *et al.*, 2007). Experts such as biologists, engineers, naturalists assessed for each land use type a score from 0, minimum pollution, to 10, maximum pollution (Fig. 2).

For the calculation of both the ROI (run off indicator) and DI (distance indicator), data on terrain elevation are needed. DEM is used to draw the basin shape and drainage network.

A 75-meter grid DEM but higher or lower resolution can be used. 75-meter resolution was chosen to keep computing time under control and because it is congruent with the reference scale of other layers (ex. 1:100.000 for Corine Land Cover maps). As a consequence of this choice, the grid size of PNPI computation is 75 m.

The method used to model the outflow area (Jenson & Domingue, 1988) consisted of:

- Filling of the depressions
- Calculation of the flow directions
- Calculation of flow accumulation
- Definition of the boundary of the river basins and the hydrographical network

The filling of the depressions is essential since DEM always contains some cells working as accumulation areas, which are at a lower level than surrounding ones. The goal of this phase is to produce a modified DEM in which any cell is part of a decreasing monotone route, leading to the outlet of the watershed. A route is made up of adjacent hydraulically linked cells, the assumption being that the route is always downhill or flat. The second step is to recognize the directions of flow coming out from any cell, assuming that the water leaving a cell enters only one of the eight adjacent cells; the receiving cell is the one on the steepest slope. Such information is used to draw a new grid where every cell is assigned a value (flow accumulation, FA) corresponding to the number of cells that flow into it. Since all the cells are part of a route always leading to the outlet of the study area, by selecting the cells exceeding a given FA, a river network is obtained. Similarly the boundaries of the drainage basins are easily identified. Additional information such as a vector river network can be used to refine the automatic delineation (Munafò *et al.*, 2005).

The DI (distance indicator) can be calculated from the hydraulic distance between each point of the basin and the receiving water body. Distances are calculated along the theoretical route taken by water on ground. DI is such that longer distances from the river give lower DI values to take into account the pollution dumping effect of the distance.

Corine land cover class	Score - Average value (0 - 10)	Score - standard deviation
Continuous urban fabric	8.22	2.22
Discontinuous urban fabric	6.89	1.36
Industrial or commercial units	7.78	2.49
Road and rail networks and associated land	5.67	2.55
Port areas	7.00	3.10
Airports	5.56	1.67
Mineral extraction sites	7.78	1.72
Dump sites	8.11	2.32
Construction sites	7.22	2.54
Green urban areas	2.33	1.66
Sport and leisure facilities	3.00	1.66
Non-irrigated arable land	6.33	2.50
Permanently irrigated land	8.89	2.03
Rice fields	7.67	1.80
Vineyards	7.00	2.24
Fruit trees and berry plantations	7.89	2.26
Olive groves	5.22	1.99
Pastures	4.00	2.35
Annual crops associated with permanent crops	7.44	2.24
Complex cultivation patterns	6.89	1.96
Land principally occupied by agriculture, with significant areas of natural vegetation	5.67	1.94
Agro-forestry areas	2.89	2.03
Broad-leaved forest	0.56	1.13
Coniferous forest	0.56	0.88
Mixed forest	0.44	0.88
Natural grasslands	1.94	2.27
Moors and heathland	0.56	1.01
Sclerophyllous vegetation	0.22	0.44
Transitional woodland-shrub	0.78	1.09
Beaches, dunes, sands	0.78	1.64
Bare rocks	0.00	0.00
Sparsely vegetated areas	0.89	1.96
Burnt areas	2.67	2.24
Glaciers and perpetual snow	0.11	0.33
Inland marshes	0.89	1.17
Peat bogs	1.00	1.50
Salt marshes	0.44	0.88
Salines	0.43	1.13
Intertidal flats	0.43	1.13
Water courses	0.14	0.38
Water bodies	0.88	1.81
Coastal lagoons	0.14	0.38
Estuaries	0.43	1.13
Sea and ocean	0.14	0.38

Fig. 2. Estimated diffuse pollution generation of Corine land cover classes as resulted from an experts' consultation. The table reports the average values of the consultation and the relevant standard deviations (Cecchi *et al.*, 2007).

The ROI (run off indicator) is calculated as the average of the run-off coefficient along the entire path from cell to river. The run-off coefficient for every cell is a function of soil permeability, land use and slope (Fig. 3). In this way, the effects of velocity and flow rate on pollution can be taken into account (Fig 4). Before using the LCI, ROI and DI for the calculation of the PNPI, they are normalized between their maximum and minimum values in order to have indicators ranging between 0 and 1.

Land use class (Corine land cover)	Permeability Classes			
	A	B	C	D
Continuous urban fabric	0.77	0.85	0.90	0.92
Discontinuous urban fabric	0.57	0.72	0.81	0.86
Industrial or commercial units	0.89	0.90	0.94	0.94
Road and rail networks and associated land	0.98	0.98	0.98	0.98
Port areas	0.89	0.92	0.94	0.94
Airports	0.81	0.88	0.91	0.93
Mineral extraction sites	0.46	0.69	0.79	0.84
Dump sites	0.46	0.69	0.79	0.84
Construction sites	0.46	0.69	0.79	0.84
Green urban areas	0.39	0.61	0.74	0.80
Sport and leisure facilities	0.39	0.61	0.74	0.80
Non-irrigated arable land	0.70	0.80	0.86	0.90
Permanently irrigated land	0.70	0.80	0.86	0.90
Rice fields	0.90	0.90	0.90	0.90
Vineyards	0.45	0.66	0.77	0.83
Fruit trees and berry plantations	0.45	0.66	0.77	0.83
Olive groves	0.45	0.66	0.77	0.83
Pastures	0.30	0.58	0.71	0.78
Annual crops associated with permanent crops	0.58	0.73	0.82	0.87
Complex cultivation patterns	0.58	0.73	0.82	0.87
Land principally occupied by agriculture with significant areas of natural vegetation	0.52	0.70	0.80	0.85
Agro-forestry areas	0.45	0.66	0.77	0.83
Broad-leaved forest	0.36	0.60	0.73	0.79
Coniferous forest	0.36	0.60	0.73	0.79
Mixed forest	0.36	0.60	0.73	0.79
Natural grasslands	0.49	0.69	0.79	0.84
Moors and heathland	0.49	0.69	0.79	0.84
Sclerophyllous vegetation	0.49	0.69	0.79	0.84
Transitional woodland-shrub	0.36	0.60	0.73	0.79
Beaches. dunes. sands	0.76	0.85	0.89	0.91
Bare rocks	0.77	0.86	0.91	0.94
Sparsely vegetated areas	0.49	0.69	0.79	0.84
Burnt areas	0.77	0.86	0.91	0.94
Glaciers and perpetual snow	1.00	1.00	1.00	1.00
Inland marshes	1.00	1.00	1.00	1.00
Peat bogs	1.00	1.00	1.00	1.00
Salt marshes	1.00	1.00	1.00	1.00
Salines	1.00	1.00	1.00	1.00
Intertidal flats	1.00	1.00	1.00	1.00
Water courses	1.00	1.00	1.00	1.00
Water bodies	1.00	1.00	1.00	1.00
Coastal lagoons	1.00	1.00	1.00	1.00
Estuaries	1.00	1.00	1.00	1.00
Sea and ocean	1.00	1.00	1.00	1.00

Fig. 3. Run-off coefficients by land use class (Corine Land Cover) and by permeability class: A: high permeability, D: low permeability (Cecchi *et al.*, 2007).

Slope classes (degrees)	Correction coefficient
< 2°50'	0
2°50'-3°41'	0.1
3°41'-4°32'	0.2
4°32'-5°23'	0.3
5°23'-6°14'	0.4
6°14'-7°05'	0.5
7°05'-7°56'	0.6
7°56'-8°47'	0.7
8°47'-9°38'	0.8
9°38'-10°29'	0.9
> 10°29'	1.0

Fig. 4. Slope correction coefficients for the calculation of the Run-off indicator(Cecchi *et al* *2007*). The coefficient is added to the Run-off coefficient as derived from Figure 3.

LCI, ROI and DI might not seem completely independent. For example both LCI and ROI depend on land cover. Nevertheless there is no double counting because different features of the land cover are taken into account. Input to the DB comes from the processing of basic maps performed by ESRI ArcView GIS 3.2 (Environmental System Research Institute, 1999b), together with its extensions 3D Analyst (Environmental System Research Institute, 1999a), Spatial Analyst (Environmental System Research Institute, 1999c) and Hydrologic Modelling (Environmental System Research Institute , 1999d).

3. Potential Non Point Pollution Index case studies

Potential Non Point Pollution Index (PNPI) has been evaluated on the aquatic ecosystems of Central Italy, at three different scales. It was applied on the Trasimeno lake, Tiber river basin, on all water bodies of Viterbo Province (Fig. 5).

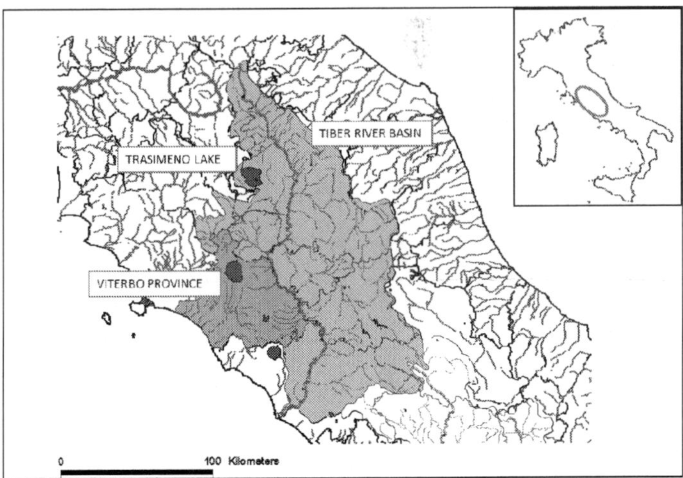

Fig. 5. Localization of PNPI three case studies.

3.1 Trasimeno Lake

3.1.1 Study area

The first study case of Non Point Pollution Index reported is on a volcanic lake ecosystem in Central of Italy, Trasimeno Lake (Baiocco *et al.*, 2001). It has a catchment area is 309 km², the surface area is 128 km² The main tributary is Fosso Anguillara, and the effluent is the Emissario of Trasimeno, an artificial waterbody.

3.1.2 Potential Non Point Pollution Index of Trasimeno Lake

Potential non Point Pollution Index was applied on the Trasimeno lake catchment area

The PNPI application described the potential contribution of each land unit to the non-point pollution on Trasimeno lake (Fig. 6). This basin, mainly, resulted affected by slightly or moderate pollution (second and third class of PNPI) due to agricultural and farm activities. The most critical areas resulted, those near the banks of the lake, characterized by urban areas.

Fig. 6. Potential non Point Pollution Index on Trasimeno Lake (Ciambella *et al.*, 2005).

3.2 Tiber River

3.2.1 Study area

The Tiber river is the third-longest river in Italy, rising in the Apennine Mountains, in Monte Fumaiolo, and flowing 406 kilometres through six region Emilia Romagna Toscana

Marche, Umbria Abruzzi, Lazio. It flows into the Tyrrhenian Sea near the City of Rome. The Tiber River basin covers a territory of 17.156 km² and the length is 409 km. The Tiber River basin's population accounts for 4 344 000 inhabitants (population census from 2001), of which 70% lives in the metropolitan area of Rome, about 10% in five of the main cities (Rieti, Perugia, Terni, Tivoli, Spoleto), and the remaining in the other small municipalities.

The Tiber River Basin is identified as Pilot Basin for testing of the implementation of the Water Framework Directive 2000/60/EC by the Italian Government.

3.2.2 Potential Non Point Pollution Index on the Tiber River basin

Potential Non Point Pollution index was applied on the whole catchement area of Tiber River and on the rivers (Figs.7-8).

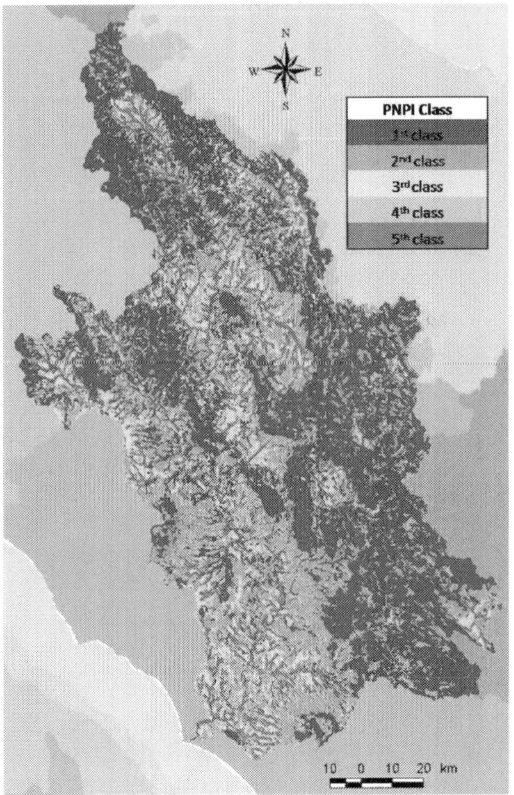

Fig. 7. Potential non Point Pollution Index results on Tiber River Basin.

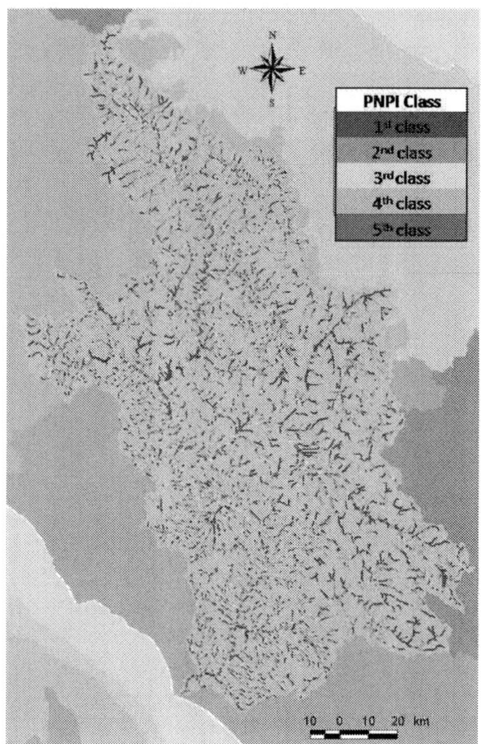

Fig. 8. Potential non Point Pollution Index results on Tiber River Basin, applied on watercourses.

The application of PNPI of Tiber River Basin described the non-point pollution of the Tiber river main course and its tributaries.

As described in figure (Fig. 7), it is possible to recognize slightly non point pollution only in the northern part of the basin, near the spring and in the left part of river basin. The natural habitats of the Apennines obtained the highest class of PNPI. Upstream to downstream there is a progressively increasing of impact and pressures, such as agricultural activities and urban areas (Fig. 8).

All the cities showed high level of pollution. A critical situation was described for the flood plain around Rome showing that only can be classified as low pollution driver (Cecchi *et al.*, 2005).

3.3 Viterbo Province

3.3.1 Study area

Viterbo Province is characterized by two groups of rivers: one that flows directly into the Tyrrhenian Sea and the second right tributaries of Tiber river. The main sub-basins of the first group are represented by Arrone , Fiora River, Mignone River; and for the second group Treja River, right tributary of the Tiber.

The geology of the area is volcanic due to the activity of the Vulsini, the Vicano and the Cimino.

3.3.2 Potential Non Point Pollution Index of Viterbo Province

The data obtained on PNPI of river basin of, Arrone, Fiora, Mignone e Treja, were merged in order to describe the non point pollution of the study area within Viterbo Province (Fig 9).

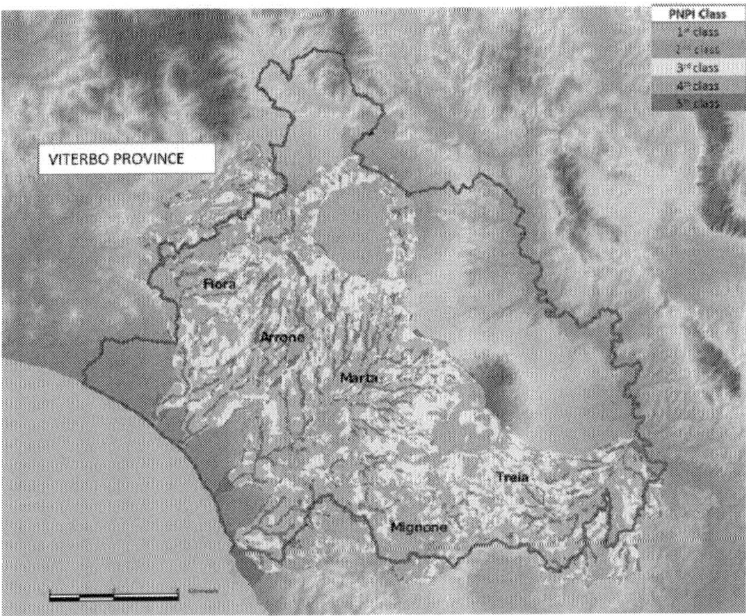

Fig. 9. Potential non Point Pollution Index of Viterbo Province.

The natural areas, where are located the river springs, are represented by non polluted or minimum polluted land units. Significant pollution was detected for the large part of the Province (Fig. 8). Strong impacts are located near the urban areas, and near the mouths of these rivers.

Fiora River basin has natural and less impact areas around its watercourse at the boundary between Tuscany and Latium; this basin is affected by significant diffuse pollution due to agricultural activities. The most critical area resulted the city of Montalto di Castro, where the river flows into the sea. The fourth class of PNPI is due to the presence of hydroelectric power plant and to the urban areas.

Arrone basin is characterized generally by significant non point pollution. The catchment of this River is set in an area with predominantly agricultural activities , which are uncommon civilian settlements, and fewer still industrial (Andreani *et al.*, 2003; Ciambella *et al.*, 2005; Mancini *et al.*, 2008). Near the mouths, except for a small area, represented by a pinewood, there are presence of significant pollution effects.

Marta River presented two different pollution degrees: the northern part of its basin has moderate diffuse pollution and the southern part, no or slightly pollution. The fourth class of PNPI is due to the presence of industrial activities, farms and from urban areas near its course and the fifth class, near the mouth of the river is due to the city of Tarquinia (Mancini *et al.*, 2008). The first and the second class of PNPI in the southern part are due to Tuscania Regional protected area and the surrounding areas.

Mignone River basin, by application of PNPI, resulted without pollution or shligtly pollution in the upper part of the basin (first and second class of PNPI), indicating natural areas around the water courses. It worst downstream, when the water course is surrounded by agricultural activities and urban areas (third and fourth class). In critical situation of pollution resulted the area near the mouth, due to the presence of the harbour, industries of the city of Civitavecchia.

River Treja is the major tributary of the right side of Tiber river; its course is characterized by slightly or moderate impact come from non pollution sources. Its quality could be influenced by urban areas and agricultural activities near its flow in the Tiber River (Andreani *et al.*, 2005).

4. Conclusion

Potential Non Point Pollution Index was set up with the view to give a reading of the diffuse pollution, using the Geographic Information Systems and the knowledge on the biological communities and their relationship with environmental parameters.

This index, as others systems based on GIS describes complex environmental phenomena, using a large amount of parameters and their results can be easily interpreted.

The results of three case studies showed the application of Non potential Point Index at three different scales: the aquatic ecosystem of Trasimeno Lake, the Tiber River basin and the river basins of Viterbo Province.

The information obtained in the first study case are obviously more detailed than those achieved in the others, and useful to identify where are the vulnerable areas. On the other hand, the Tiber River case study showed a global view of diffuse pollution at interregional level, indicating the main pressures that damage the whole catchment area. The analyses of PNPI values of Viterbo Province described the first experimental approach of the assessment of non point pollution, done by a local authority. PNPI was resulted able to give useful information at all three scales.

Compared with other models (Di Luzio *et al.*, 2002), the PNPI approach showed some advantages: a low need for input data, simplicity of use and a direct reference to the pollution driving forces. The strength point of this index, is the combination of environmental data and expert judgment of biologists, naturalists, and engineers, giving a qualitative ecological component.

The evaluation of each land use type could be improved with the more information achieved from biological communities researches. The implementation activities of Water Frame Directive at Italian level have focused the attention of scientific world to improve the knowledge of the biological indicators and their response to environmental parameters.

In particular the acquired knowledge about the four biological elements, phytobenthos, macrophytes, benthic invertebrates and fishes will improve the evaluation of land use type making the PNPI more suitable for the achieving a good ecological status, after restoring activities of the river management plans, as required at national and European level (Italy, 2006; European Union, 2000).

5. References

Andreani, P., Bernabei, S., Cecchi, G., Charavgis, F., D'Angelo, A.M., Formichetti, P., Mancini, L., Martinelli, A., Notargiacomo T., Pace G., Pierdominici E., Tancioni L., & Venanzi D. (2005). "Perennial Water Courses Types". Tevere pilot river basin article 5 report pp. 40-54. Gangemi Editore, Roma p 175.

Andreani, P., Cecchi, G., Sforza M.R., Pecorelli, A., Dello Vicario, E., Ciambella, M., Venanzi, D. & Mancini L. Ed. (2006). *Indicatori e modelli per la gestione ed il monitoraggio delle acque superficiali*. Rapporto Istisan Roma: Istituto Superiore di Sanità.

Baiocco, F., Munafò, M., Bernabei, S. & Mancini L. (2001). Un'applicazione GIS per la valutazione dell'inquinamento diffuso: il caso del Lago Trasimeno. *Conference Proceedings. Conferenza italiana utenti ESRI*. Atti.

Cecchi G., Fabiani C., Mancini L., Munafò M., 2005. "Pressures of human activity on water status". Tevere pilot river basin article 5 report. pp. 79-87. Gangemi Editore, Roma. pp. 175.

Cecchi, G., Munafò, M., Baiocco, F., Andreani, P. & Mancini L. (2007). "Estimative river pollution from diffuse sources in the Viterbo province using the potenzial non-point pollution index". *Annali dell'Istituto Superiore di Sanità 2007*, Vol. 43, No. 3pp. 295-301 ISSN 0021-2571

Ciambella, M., Venanzi, D., Dello Vicario, E., Andreani, P., Cecchi, G., Formichetti, P., Munafò M. & Mancini L. (2005). *Convenzione Istituto Superiore Di Sanità -Provincia Di Viterbo: messa a punto di uno strumento conoscitivo per la gestione delle acque superficiali della Provincia di Viterbo. Available on line*
http://www.provincia.vt.it/ambiente/Stato_amb03/PDF/227-238_Convenzione.pdf.

Crouzet, P. (2000). *Calculation of Nutrient Surpluses from Agricultural Sources*. EEA, Copenhagen.

Di Luzio, M., Srinivasan, R. & Arnold, J.G. (2002). Integration of watershed tools and SWAT model into BASINS. *Journal of American Water Resources Association*. Vol. 38, No. 4, pp. 1127–1141.

Environmental System Research Institute (1999a). *3D Analyst Version 3.0*. ESRI, Redlands, CA.

Environmental System Research Institute, (1999b). *ArcView GIS Version 3.2*. ESRI, Redlands, CA.

Environmental System Research Institute (1999c). *Spatial Analyst Version 3.2*. ESRI, Redlands, CA.

Environmental System Research Institute (1999d). *Hydrologic Modelling Version 1.1.* ESRI, Redlands, CA.

European Environment Agency (1999b). *Nutrients in European Ecosystem*, Environmental Assessment Report, Vol. 4. EEA.

European Union 2000. -Directive 2000/60/EEC of 23 October 2000 establishing a framework for community action in the field of water policy. Official Journal of European Communities, L327/1.

Haycock, N.E., Gilles, P. & Wlker, C. (1993). Nitrogen retention in river corridors: European perspective. *Ambio* Vol. 22, pp. 340–346.

Hellweger F. & Maidment D. R. (1999). Definition and Connection of Hydrologic Elements Using Geographic Data, *Journal of Hydrologic Engineering.* Vol. 4, pp. 10-18.

Italy, (2006). Decreto Legislativo 3 aprile 2006, n. 152. Norme in materia ambientale. *Gazzetta Ufficiale - Supplemento Ordinario* n. 96 del 14 aprile 2006.

Jenson, S.K. & Domingue, J.O. (1988). Extracting topographic structure from digital elevation data for geographic information system analysis. *Photogrammetric and Remote Sensing* Vol. 54, N11, pp. 1593–1600.

Kupcho K., (1997). Obtaining SCS Synthetic Unit Hydrograph By GIS Techniques, 12 http://gis.esri.com.

Loague K. & Corwin, D.L. (1996) Uncertainty in regional-scale assessments of non-point source pollutants. In *Applications of GIS to the Modeling of Non-Point Source Pollutants in the Vadose Zone,* Corwin D.L. and Loague K. (Eds.), Special Publication 48, Soil Science Society of America: Madison, pp. 131–152.

Loague, K. & Corwin, D.L. (2005). Point and nonpoint source pollution. In: M.G. Anderson (Ed.) Encyclopedia of Hydrological Sciences. John Wiley & Sons, Ltd. Chichester, UK. Chapter 94: 1427-1439.

Mancini, L., Andreani, P. (Eds.) (2008) *Guida agli indicatori biologici dei corsi d'acqua della provincia di Viterbo..* Roma: Istituto Superiore di Sanità. (Rapporti ISTISAN 08/34).

Mohammadi H. (2001). *Hydrological analysis of flood forecasting: Kasilaian watershed.* MSc thesis, University of Mazandaran, 152 p.

Munafò, M., Cecchi, G., Baiocco, F., Mancini L., (2005). River Pollution from non-point source: a new simplified method of assessment. *Journal of Environmental Management Vol* 77pp.93-98.

Novotny, V. & Chesters, G. (1981). Handbook of Nonpoint Source Pollution: Sources and Management. Van Nostrand Reinhold Co, New York, N.Y. ISBN 0-442-00559-8. 1

Olivera F. & Maidment D. (1999). Developing a Hydrologic Model of the Guadalupe Basin/ Center for Research in Water Resources, Austin, Texas.

Schilling, E. & Libra, R.D. (2000). The relationship of nitrate concentrations in streams to row crop land use in Iowa. *Journal of Environmental Quality,* Vol. 29 pp. 1846–1851.

Solaimani, K., Mohammadi, H., Ahmadi, M. Z., & Habibnejad, M. (2005). "Flood Occurrence Hazard Forecasting Based on Geographical Information System". *International Journal of Environmental Science and Technology,* Vol. 2, No. 3, pp 253-258.

6

Long Term Changes in Abundance and Spatial Distribution of Pelagic *Agonidae, Ammodytidae, Liparidae, Cottidae, Myctophidae* and *Stichaeidae* in the Barents Sea

Elena Eriksen[1,*], Tatyana Prokhorova[2] and Edda Johannesen[1]
[1]Institute of Marine Research, Bergen,
[2]Polar Research Institute of Marine Fisheries
and Oceanography (PINRO), Murmansk,
[1]Norway
[2]Russia

1. Introduction

The Barents Sea is a high-latitude, arctoboreal shelf sea, situated between 70° and 80°N. Its topographical features include Spitsbergen to the North, Novaya Zemlya to the East, the mainland coast of Russia and Norway to the South, and the continental shelf break to the West (Figure 1). In the Barents Sea, warm and saline Atlantic water flows in from the Southwest and meets the cold Arctic water with low salinity, or the mixed waters in the North and East, along the polar front. The interannual variation in sea temperature in the Barents Sea is, to a large extent, determined by the inflow of Atlantic water: in years with strong inflow of Atlantic water the sea temperature is higher (Loeng 1991). This variability in ocean climate strongly influences the abundance of commercially exploited fishes in the Barents Sea ecosystem (e.g. Hamre 1994).

The Barents Sea fish fauna is dominated by 8-10 very abundant, commercially exploited species, and among them some of the world's largest fish stocks: the Northeast Arctic cod *Gadus morhua*, Barents Sea capelin *Mallotus villosus*, Northeast Arctic haddock *Melanogrammus aeglefinus*, and the juvenile Norwegian spring-spawning herring *Clupea harengus*.

The Institute of Marine Research (IMR), Norway, and the Polar Research Institute of Marine Fisheries and Oceanography (PINRO), Russia, have carried out routine surveys to assess stock size of exploited species as basis for advice in the Barents Sea for several decades. Since 1965 an international 0-group fish survey in the Barents Sea has provided pelagic trawl data to give an early indication of year class strength of targeted fish species: capelin, herring, cod, haddock, saithe (*Pollachius virens*), redfish (*Sebastes spp.*), Greenland halibut

---

* Corresponding Author

(*Reinhardtius hippoglossoides*), long rough dab (*Hippoglossoides platessoides*), and polar cod (*Boreogadus saida*) (Haug and Nakken, 1973, 1977; Dingsør 2005; Eriksen *et al.* 2009). Since 1980 standard trawling procedures have been used (Anon. 1980). Since 2005 standard procedures (stratified sample mean method) have been used to estimate abundance and biomass indices of 0-group fish (Dingsør 2005; Eriksen *et al.* 2009, 2011). On the same survey by-catch of non-targeted species (invertebrates and fishes) has been recorded but these data have never been analysed and published. Non-targeted fishes are much less studied than the exploited species in the Barents Sea. However, these fishes are potentially important components of the food web and thus the functioning of the Barents Sea ecosystem. Also they are of interest because they contribute to the biodiversity of the Barents Sea.

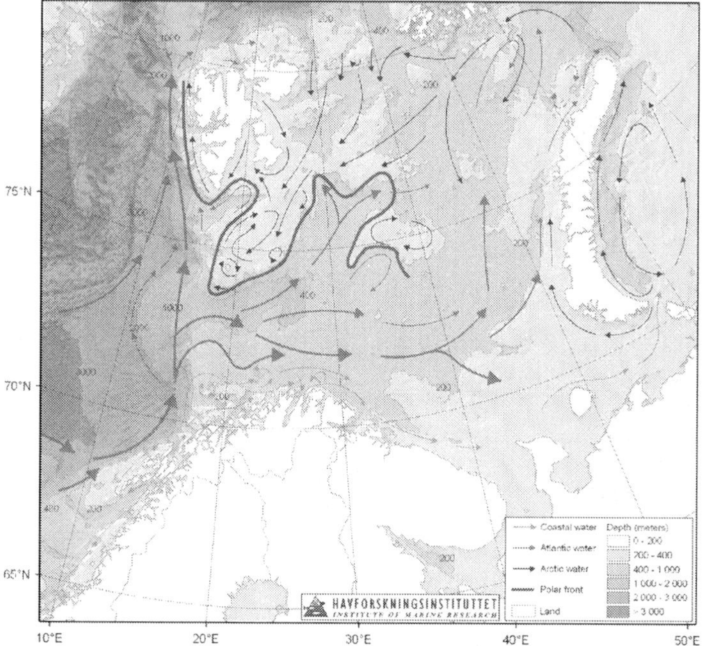

Fig. 1. Overview of the Barents Sea geography and current system.

Several authors, e.g. Andriashev and Chernova (1995) have grouped the Barents Sea fishes into zoogeographical groups that are associated with different water temperatures and water masses (Arctic or Atlantic, above, see material and methods for details on the biology of the studied fishes). Species from different zoogeographical groups are likely to respond differently to climatic change and fluctuations, and thus can serve as indicators of the effect of climate change.

In this paper, we study six families of non-targeted fishes (called "small fishes" here) caught in the 0-group survey from 1980-2009: *Agonidae, Ammodytidae, Cottidae, Liparidae, Myctophidae* and *Stichaeidae*.

We calculate indices of biomasses of these fish families and compare the estimates to the biomass estimates of targeted pelagic fish species (0-group and adult, Eriksen *et al.* 2011). We then discuss the potential ecological role of small fishes based on their biomass and distribution.

Furthermore, we study how variation in the distributions and biomasses of small fishes are related to variation in climate. We use time series of temperature of 0-50 m from Fugløya-Bear Island (FB) oceanographic section (70°30' and 20°00' to 74°15' and 19°10') as an indicator of climate variation. The FB section is located in the main entrance of the inflow of the Atlantic water masses into the Barents Sea. We predict that the biomass of fishes from families mostly associated with warmer, Atlantic water (*Ammodytidae, Myctophidae* and *Stichaeidae*) should increase in years with high temperatures associated with high inflow of Atlantic water, due to more favourable environmental conditions according to their requirements. We furthermore predict that the distribution of all fish families should be shifted northwards in warm years and southwards in cold years.

2. Materials and methods

2.1 Survey data

Our data was the by-catch from the standardised international 0-group fish survey in the Barents Sea. The survey has been carried out annually during August-September since 1965. Since 2004 the 0-group survey has been a part of the Joint Norwegian-Russian ecosystem survey in the Barents Sea in August-September (Anon. 2004). In 1980 a standard trawling procedure was recommended and applied (Anon. 1980). The standard trawling procedure consists of pelagic tows at predetermined positions 25-35 nautical miles (nm) apart. A "Harstad trawl" trawl having 7 panels and a cod end was used. The panels have mesh sizes (stretched) varying from 100 mm in the first panel to 30 mm in the last panel, and 7 mm in the cod end. The tows are done at three depths: head-line at 0 m, 20 m and 40 m, each tow is 0.5 nautical miles (nm) with a trawling speed of 3 knots. Additional depths are towed (60 and 80 m), at dense concentration of fish recorded deeper than 40 m depth on the echo-sounder (Anon. 1980, 2004). To study pelagically distributed small fishes we used pelagic catches from the 0-group survey (1980-2003) and the ecosystem survey (2004-2009) from a revised and updated 0-group database (Eriksen *et al.* 2009).

We also mapped the demersal distribution of the fishes studied here, using demersal trawl catches from the ecosystem survey (2004-2009). Note that the area covered by the bottom trawl extended further north than that of the pelagic trawl (see Figure 2 and 3). Bottom catches were taken with a "Campelen 1800" shrimp trawl with 80mm (stretched) mesh size in the front and cod end with 22 mm mesh size. Standard towing time was 15 minutes after the trawl had made contact with the seabed and the towing speed was 3 knots. A subsample of each fish species/group, was length measured from all demersal and pelagic trawl stations all years. The average lengths from demersal and pelagic trawls were calculated separately. Individual weights for small, non-targeted fishes were recorded only occasionally. During the ecosystem survey in 2008, fishes from the families that we studied here were sampled from some pelagic and bottom trawl stations, and their otoliths were mounted in epoxy and cut to precision into units of 0.1 mm for later age determination with binocular, the results of age determination are presented in the results section.

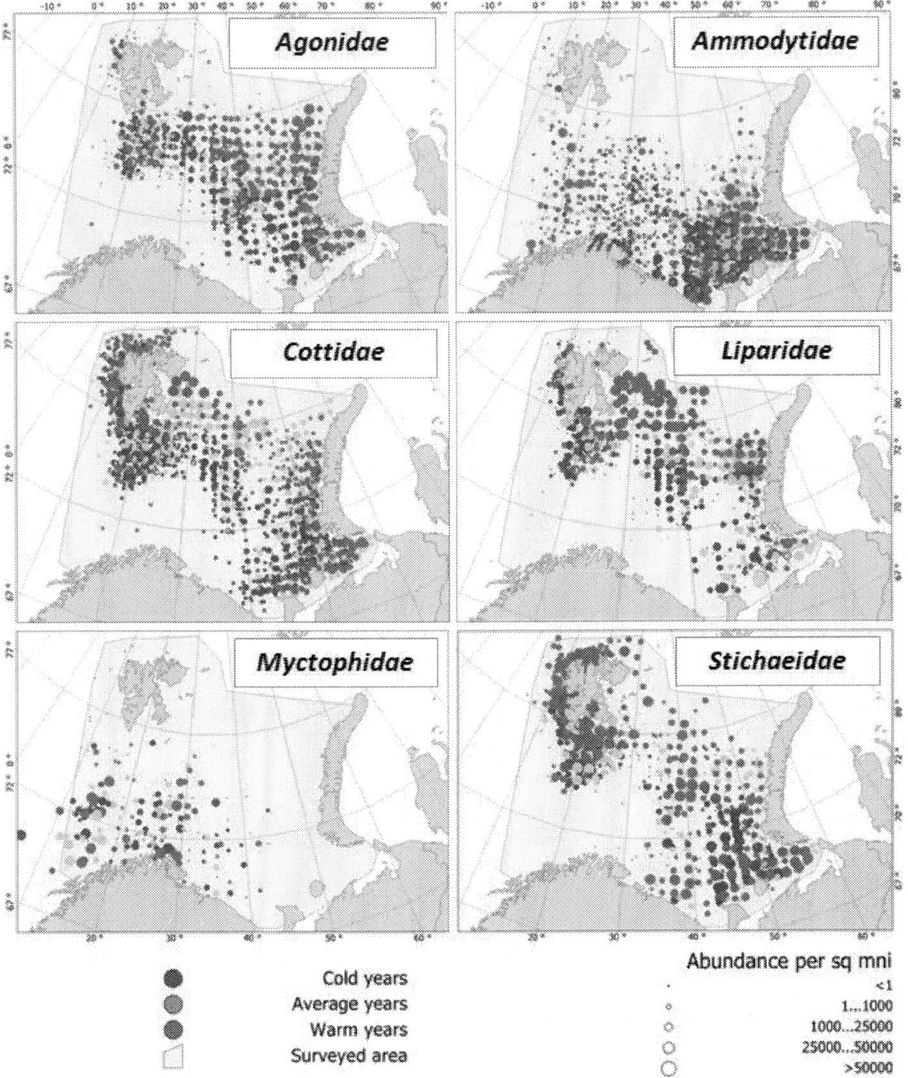

Fig. 2. Spatial distribution of pelagically distributed *Agonidae*, *Ammodytidae*, *Cottidae*, *Liparidae*, *Myctophidae* and *Stichaeidae* in the Barents Sea during 1980-2009. Abundance of small fishes per mn^2 is shown by circle, where size indicates density of fishes, while colour of circle indicates temperature condition (years which were categorized as cold shown in blue, average in yellow and warm years in red).

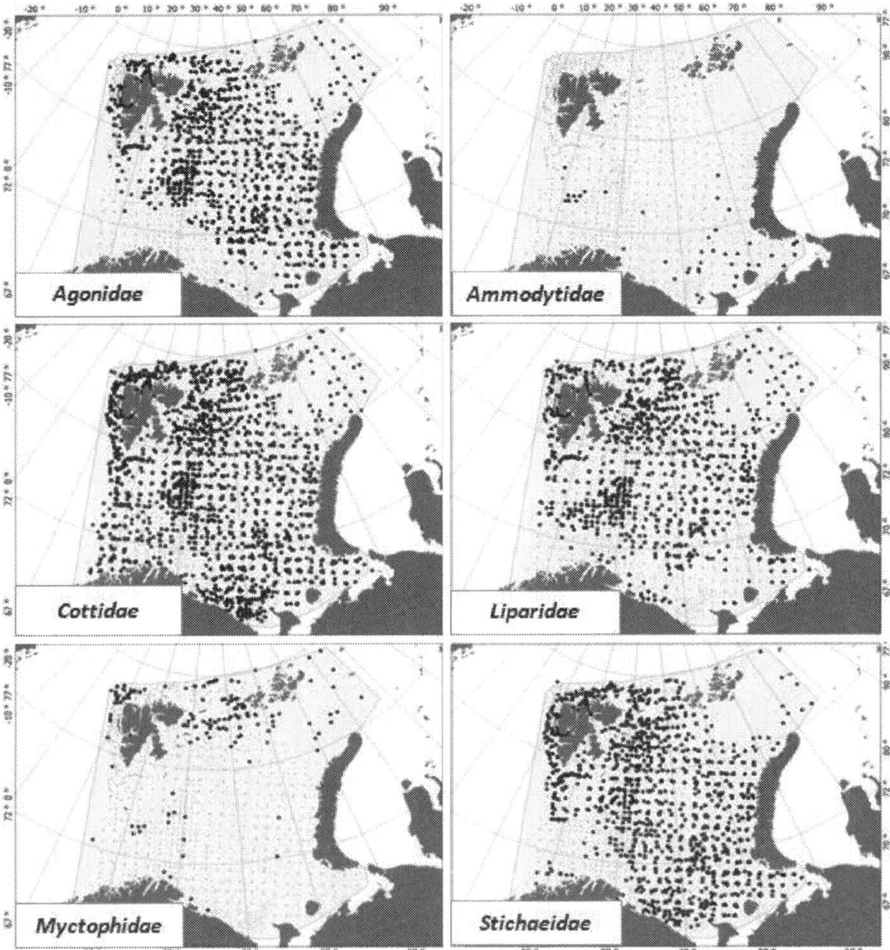

Fig. 3. Spatial distribution of demersal *Agonidae, Ammodytidae, Cottidae, Liparidae,*
Myctophidae and *Stichaeidae* in the Barents Sea during 2004-2009. Occurrence of small fishes
in bottom catches is shown by black circles, while grey and small circles showed bottom
trawl stations.

With the exception of redfish (genus level), the targeted fish species (see introduction) were
identified to species level during the 0-group survey. The non-targeted small fishes, on the
other hand, were at times only identified to family level; in some years up to 50% of the
small fishes were identified to the family level only. This was due time constraints on board
and difficulties in species identification. We therefore decided to combine the small fishes

into larger groups (families). We excluded the family *Paralipidae*, tusk (*Bromse brosme*), Norway pout (*Trisopterus esmarkii*) and *Phycis sp.*, because they were only present in very low numbers and only in some years. We then analysed the data on the remaining six non-targeted fish families recorded in the survey. Their biology is described below.

2.2 Fish families examined and their biology

Agonidae are small bottom-dwelling cold-water marine fish that are widely distributed in the northern part of the Atlantic and Pacific Oceans. There are three species in the Barents Sea (Karamushko 2008). *Leptagonus decagonus* (Bloch and Schneider 1801) is the most abundant and is widely distributed in both the northern and central areas, however it is rare in the south-eastern part of the Barents Sea and along the Finnmark coast (Byrkjedal and Høines 2007). It is generally found in depths of 120-350 m and at a temperature of between -1.7°C and 4.4°C, while the juvenile fish are found in the upper 100-meter layer (Andriashev 1986; Ponomarenko 1995). The larvae of *L. decagonus* drift from the North and North-east with the colder arctic waters that enter the Barents Sea, and are not observed in the Atlantic waters (Rass 1949). *Agonus cataphractus* (Linnaeus 1758) is mostly distributed along the Norwegian coast and in the White Sea (Andriashev 1986). *Ulcina olriki* (Lütken 1876) is a small arctic fish, distributed in the northern parts of the Atlantic and Pacific Oceans, and found on sand and mud bottoms, mostly at temperatures that are below 0°C, and rarely up to 2-3°C (Andriashev 1954, 1986). *Agonidae* generally feed on small crustaceans and worms that are found on the sea-bed (Andriashev 1954, 1986; Pethon 2005).

Ammodytidae are small schooling benthopelagic fish, which are usually found at depths of 10-150 meters, and in the brackish and marine waters of the Atlantic, Pacific and Indian Oceans, mainly in the northern hemisphere. In the Barents Sea, *Ammodytidae* are represented mostly by *Ammodytes marinus* (Raitt 1934) which is distributed along the Norwegian coast, in the Southeast and between Novaya Zemlya and Bear Island (Andriashev 1954). *Ammodytes tobianus* (Linnaeus 1758) and *Hyperoplus lanceolatus* (Le Sauvage 1824) were reported by Andriashev and Chernova (1995) along Murman and the Norwegian coasts, but have not been observed more recently. *A. marinus* spawns during November-February (Rass 1949) at depths of 25-100 m, and feeds on plankton. During low light intensity (night and winter) they bury themselves in the bottom sediment (Muus and Nielsen 1999). In periods with strong tidal currents they leave their bottom hides and form large shoals. Several fish species (cod, flatfish and others), sea birds and seals have been observed to be feeding on *Ammodytidae* in the Barents Sea (Adriashev 1954; Viller 1983; Krasnov & Barret 1995; Borisov *et al.* 1996; Yezhov 2008, 2009; Sakshaug *et al.* 2009).

Cottidae is a large family consisting of about 800 species (Pethon 2005). They are small bottom-dwelling fish that are found in marine, brackish and fresh waters, and most species occur in Arctic or temperate waters. In the Barents Sea 11-13 species were found (Pethon 2005; Byrkjedal and Høines 2007; Karamushko 2008). *Artediellus atlanticus* (Jordan & Evermann 1898), *Myoxocephalus scorpius* (Linnaeus 1758), *Taurulus bubalis* (Euphrasen, 1786), and *Triglops murrayi* (Günther 1888) are mostly boreal species and linked to Atlantic water masses. Others such as *Artediellus scaber* (Knipowitsch 1907), *Gymnocanthus tricuspis* (Reinhardt, 1830), *Icelus bicornis* (Reinhardt 1840), *Triglops nybelini* (Jensen, 1944) and *Myoxocephalus quadricornis* (Linnaeus 1758) are Arctic, while *Icelus spatula* (Gilbert & Burke

1912) and *Triglops pingelii* (Reinhardt 1937) are Arctoboreal species. Earlier studies have reported occurrences of pre-spawned *Cottidae* females, egg and larvae (Rass 1949; Kazanova and Perzeva-Ostroumova 1960; Ponomarenko 1995; Mukhina 2005), indicating spawning in the Barents Sea. *Cottidae* feed on small fish, bottom crustaceans and worms. Cod have been observed to be feeding on *Cottidae* in the Barents Sea (Borisov *et al.* (1996)).

Liparidae are small bathypelagic or bottom fish, and is one of the most diverse and abundant fish families that dwell in polar and deep-sea habitats (Chernova 1991, 2005a). The biology of many of these species is poorly studied. In the Barents Sea, *Liparidae* are represented by approximately 12 species, however, their taxonomy is under revision (e.g Chernova 2005b). In the Barents Sea, *Liparidae* from the *Liparis* and *Careproctus* (Pethon 2005; Byrkjedal and Høines 2007) are abundant and widely distributed throughout the North and Northeast. Species from the genera *Paraliparis* and *Rodichthys* are very rare. Species identification is complicated due to their wide morphological variation (Fevolden *et al.* 1989; Knudsen *et al.* 2007). Several species of *Liparis* spawn during winter, while some species of *Careproctus* spawn during late summer (Ponomarenko 1995). However, the spawning biology is unknown for several species (Pethon 2005). *Liparidae* feed on small pelagic and bottom crustaceans and worms (Stein and Able 1986). Gadoids, skates, seals and auks prey on *Liparidae* (Berestovskij 1990; Borisov *et al.* (1996); Pethon 2005).

Myctophidae include more than 200 circumglobal, high-oceanic, mesopelagic fish species (Hulley 1990). Almost all *Myctophidae* undergo vertical migration associated with foraging on planktonic crustaceans (euphausiids and copepods). They have been observed at depths between 10-200 m at night, and at 375-800 m during the day (Hulley 1990). Not all individuals undertake diel vertical migration (Pearcy *et al.* 1979). *Benthosema glaciale* (Reinhardt 1837), which is very abundant in the North Atlantic, is the only species found in any quantity in the Barents Sea (Kristoffersen and Salvanes 2009). Both males and females of *B. glaciale* mature at age 2-3, when they are between 45-50 mm in length. *B. glaciale* spawns in fjords and along the Norwegian coast, and is an important food resource for several fish, especially for salmon along the coastal area (Giske *et al.* 1990; Salvanes and Kristoffersen 2001; Salvanes 2004; Kristoffersen and Salvanes 2009).

Stichaeidae are circumpolar elongated small bottom fish that are found in northern oceans: Arctic, North Pacific and Northwest Atlantic (Coad and Reist 2004). The Barents Sea *Stichaeidae* include *Lumpenus lampraetaeformis* (Linnaeus 1758), *Leptoclinus maculatus* (Fries 1838) and *Anisarhus medius* (Reinhardt 1837) (Byrkjedal and Høines 2007). *Lumpenus fabricii* (Reinhardt 1836) are very rare. *L. lampraetaeformis* spawns during the winter, but the spawning biology of the other species is not known. However, their larvae and younger fish have been found pelagically. *Stichaeidae* feed on small crustaceans, worms, clams and fish. Cod have been observed to be feeding on *L. lampraetaeformis* in the Barents Sea (Borisov *et al.* (1996)).

2.3 Data analyses

The capture efficiency of the sampling trawl differs between species and decreases with the fish length (Godø *et al.* 1993; Hylen *et al.* 1995). Species specific catch correction functions dependent on length are established and used in the annual computations of 0-group abundances for cod, haddock, herring and capelin (Anon. 2004; Dingsør 2005;

Eriksen *et al.* 2009). The species specific and length dependent catch efficiencies are unknown for all our study fishes. We assume that the small fish were only captured by the panels with less than 50 mm mesh size which was the last two panels and the cod end. The mouth opening of the 50 mm mesh panel is 4 meters, so this is the effective wingspread of trawl used here.

Abundance indices (AI) were calculated for the studied fish families by the standard procedures (equations 1, 2 and 3, below) used for commercial fish species. The stratified sample mean method described by Dingsør (2005) and Eriksen *et al.* (2009) was employed, using the Barents Sea 0-group strata system, consisting of 23 strata.

First the fish density (p_s individuals per square nautical mile) at each station, *s*, was calculated from the following equation

$$p_s = \frac{n_s \cdot 1852}{wsp * (td_s / dl_s)} \tag{1}$$

where n_s is the observed number of fish at station *s*, *wsp* is the effective wingspread of the trawl (4 m), td_s (nm) is the total distance trawled at station *s*, and dl_s is the number of depth layers at station *s*.

The average density y_i in stratum *i* was then calculated from

$$\bar{y}_i = \frac{1}{n_i} \sum_{s=1}^{n_i} p_s \tag{2}$$

where n_i is the number of stations in stratum *i*.

Finally, the abundance estimate (AI) was then calculated from

$$AI = \sum_{i=1}^{N} A_i \bar{y}_i \tag{3}$$

where *N* is the number of strata and A_i is the covered area in the *i*-th stratum.

Yearly estimates of the relative biomasses indices of *Agonidae, Ammodytidae, Liparidae, Cottidae, Myctophidae* and *Stichaeidae* were calculated as a product of the yearly relative abundance indices and average fish weight. Data on individual weight were however, scarce (see above). Therefore, we pooled all available weights from pelagic hauls and calculated average individual weights across all years: *Agonidae* (0.30g), *Ammodytidae* (0.50g), *Liparidae* (0.35g), *Cottidae* (0.30g), *Myctophidae* (0.45g) and *Stichaeidae* (0.50g).

To visualise a huge amount (30 years) of spatial distribution of the small fishes in relation to temperature we separated the time series into three categories. We categorised the temperature conditions in the Barents Sea based on the mean annual temperature at the Kola section (average surface to bottom). The temperature data from the Kola section is the longest time series in existence from the Barents Sea (1900-), and is regarded as a good indicator of the temperature variation in the Barents Sea. The mean annual temperature from 1980 to 2009 was 4.2°C, with minimum of 3.2°C and maximum of 5.1°C. Years with the mean annual temperatures at the Kola section being less than 4°C were classified as cold (1980-82, 1985-88, 1994, 1996-98, a total of 11 years); temperatures between 4 and 4.5°C were categorised as average (1984, 1989, 1991, 1993, 1995, 1999 and 2001-03, a total of 9 years) and temperatures higher than 4.5°C were described as warm (1983, 1990, 1992, 2000, 2004-09, a total of 10 years). Maps of pelagic fish distribution that separated warm, cold and intermediate years were produced using the manifold software (Manifold System 8.0.15.0 Universal Edition).

In addition, catches from the ecosystem survey bottom trawls were mapped in order to compare distributions of the fish families caught in demersal trawls. All years of the ecosystem survey were classified as warm (2004-09).

3. Results and discussion

3.1 Spatial variation

Agonidae and *Liparidae,* mostly arctic fishes, were observed over large area in the north and in the east, with some scattered occurrences in the central part of the Barents Sea. The densest concentrations were found around the Svalbard archipelago and along the western side of Novaya Zemlya (Figures 2 and 3).

Ammodytidae, Myctophidae and *Stichaeidae,* mostly boreal fishes, were recorded in the south-eastern and central parts of the Barents Sea, some of *Stichaeidae* were also found along southern and western side of Svalbard archipelago. The main findings for each family group is summarized in Figures 2 and 3. We also had some data at the species level recorded during the 0-group survey (pelagic trawl) and at the ecosystem survey (pelagic and demersal trawl), described below under their respective families.

3.1.1 *Agonidae*

Pelagically distributed *Agonidae* were mostly found on the banks south and east of the Spitsbergen archipelago to the Novaya Zemlya (Figure 2). Demersally distributed *Agonidae* occupied almost the entire Barents Sea, except the south-western area (Figure 3). At the species level, we found that *L. decagonus* was widely distributed in the central and northern areas of the Barents Sea, while *U. olriki* was distributed along the western part of Novaya Zemlya. A few individuals of *Agonus cataphractus* were found in the colder coastal waters of the eastern part of the Barents Sea on the survey in the pelagic trawl.

The mean fish length of pelagic *Agonidae* in the period of 1980-09 varied from 2.8 to 4.6 cm. The long term average length of pelagically distributed fish was 3.7 cm, while the average length of bottom distributed fish was much higher (12.6 cm, Figure 4). The higher mean length of demersal *Agonidae* supported the assumption that the *Agonidae* at age 0 live pelagically, while the distribution of adults is near the bottom. There were no otoliths sampled from the *Agonidae.*

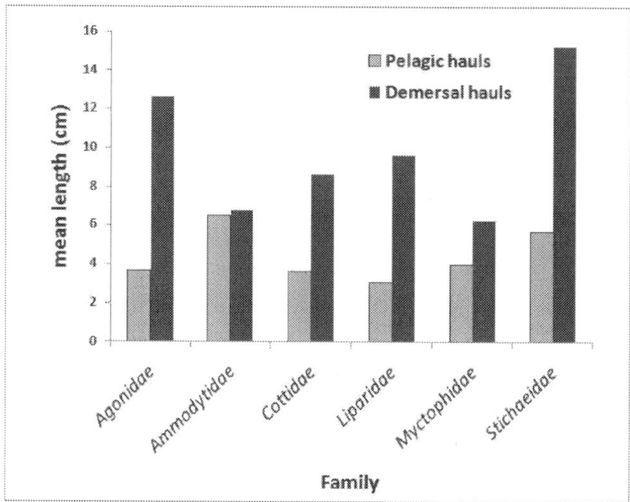

Fig. 4. Mean length of small species from pelagic and bottom trawl catches.

3.1.2 *Ammodytidae*

Pelagically distributed *A. marinus* was observed mainly in the south-eastern area of the Barents Sea (Figure 2). Small catches were also registered in the central and western areas. Demersally distributed fish were observed scattered throughout the south-east, while some were found in the western part of the Barents Sea (Figure 3). The average lengths of pelagically and bottom distributed fish were similar, 6.5 and 6.8 cm, respectively. Therefore, we believe that individuals from the pelagic and demersal hauls were part of the same component of the stock, and that the variation in their vertical distribution reflects feeding migrations for planktonic crustaceans. A total of 20 otoliths were studied, and showed that the 0-group fish were between 5.2-6.9 cm in length, while fish at age 1 were between 6.0-7.6 cm and fish at age 2 and older were longer than 8.9 cm. Therefore we believe that our abundance estimates represented both young and adult fish.

3.1.3 *Cottidae*

Cottidae were observed in shallow areas in the whole area (Figure 2). They showed a clear shift in their area of occupation during different temperature conditions. In colder years they were observed in North Kanin, Central Bank and the Svalbard bank, while in warmer years they were also observed further north in the Great Bank and Novaya Zemlya Bank. Demersally distributed *Cottidae* were observed widely during the warm years when the Barents Sea ecosystem survey was conducted (Figure 3).

The mean length of the fish observed pelagically varied between 3.0 and 4.2 cm, with much lower average lengths than of bottom distributed fish (3.6 cm and 8.7 cm, respectively, see Figure 4). Otoliths were taken from both pelagic and demersal trawl catches from the following species: *A. atlanticus* (24 fish), *I. bicornis* (3 fish) and *T. nybelini* (5 fish). The length of the one-year-old *T. nybelini* varied between 6.1 and 7.3 cm. Two year old *A. atlanticus* (5.4-6.5 cm) were smaller than *T. nybelini* (10.4-11.4cm), 3-5 years old *A. atlanticus* were between 5.9 and 9.7cm, while 3-5 year old *I. bicornis* were 4.8-6.8. As expected, this showed that different species exhibit different growth patterns. We believe that our results on abundance, biomass and distribution from pelagic trawl represented mostly juvenile fish.

3.1.4 *Liparidae*

The following species was recorded during the 30 years of observations: *Liparis liparis, L. cf. fabricii, L. cf. gibbius, Paraliparis bathybius* and *Careproctus reinhardti* (species complex under taxonomic revision). However, the taxonomy in this family is relatively unknown because of the rarity of many species and difficulties in identifying distinct morphological characteristics. Therefore, we believe that our results apply mostly at the genus level, with *Liparis* and *Careproctus* being much more abundant than the others. Pelagically distributed *Liparidae*, like *Cottidae*, were generally found over shallow areas and banks (Figure 2), while demersally distributed *Liparidae* were more widely observed, but not found in coastal areas (Figure 3).

Three otoliths of *Careproctus sp.* were taken from fish larger than 8 cm. Age determinations showed that the fish of 8.3 cm were 3 years old, while two fish of 9.8 and 10 cm were 4 years old. The calculated mean length varied between 2.1 and 4.0 cm for pelagically distributed fish (Figure 4). The average length of pelagically distributed fish was much lower than of bottom distributed fish, (3.1 cm and 9.6 cm, respectively, Figure 4), therefore our results represent juvenile fish.

3.1.5 *Myctophidae*

The Barents Sea *Myctophidae* are largely represented by *B. glaciale*. Pelagic *B. glaciale* were found in the south-western part of the Barents Sea (Figure 2). The demersal distribution is shown in Figure 3, with most fish found in the northern parts of the Barents Sea, where there were only a few pelagic catches. Records of such a northerly distribution of *B. glaciale* are absent in the literature. *B. glaciale* is a good swimmer, and catchability is low due to high avoidance of the pelagic trawl (own experience), resulting in strong underestimation of abundance. In this study the mean length of pelagically distributed fish varied between 3.1 and 5.2 cm, and their average length was somewhat lower than the demersally distributed fish (4.0 cm and 6.0 cm, respectively, Figure 4). *Myctophidae* mature at age 2-3, when the fish are 4.5-5.0 cm in length, therefore we believe that our result mainly represents adult fish that are migrating to the surface to feed on small crustaceans.

3.1.6 *Stichaeidae*

The pelagically distributed *Stichaeidae* was found in shallow waters (Figure 2). *L. maculatus* was most abundant and was widely distributed, while *L. lampraetaeformis* and *A. medius* were mostly found near Spitsbergen archipelago and Novaya Zemlya, although *L. lampraetaeformis* was also frequently found in the northern Barents Sea. Demersally distributed *Stichaeidae* were found widely in the Barents Sea (Figure 3). During the studied period (1980-09) the mean length of fish varied between 3.4 and 8.1 cm. The average length of pelagic fish was much lower than that of the bottom distributed fish (5.7 cm and 15.3 cm, respectively, Figure 4). In addition otoliths studies (a total of 27 fish) showed that the growth rate of *L. lampraetaeformis* was higher than that of *L. maculatus,* and the fish length of those of age 1 and older was 12.8 cm and 9.1 cm, respectively. Our results from pelagic trawls represented younger fish.

3.2 Biomass and abundance

Relative abundance and biomass indices were calculated for *Agonidae, Ammodytidae, Liparidae, Cottidae, Myctophidae* and *Stichaeidae* for the period of 1980-2009. The biomasses varied greatly between fish families, and were dominated by boreal fishes, *Ammodytidae* and *Stichaeidae* (Table 1).

The total biomass of small fishes varied between from 2 to 90 thousand tonnes, with long term average of 23 thousand tonnes for the 1993-2009 (Table 1), The average biomass of the most abundant 0-group fishes (cod, haddock, herring and capelin) 1993-2009 was ca 1 million tonnes, so the small fishes was only 2.3% the biomass of the most abundant 0-group fish.

Year	Agonidae		Ammodytidae		Cottidae		Liparidae		Myctophidae		Stichaeidae		Total biomass	FB 0-50, August
	AI	B	AI	B	AI	B	AI	B	AI	B	AI	B		
1980	758	227	133169	66584	1640	492	1216	426	521	234	4180	2090	70053	8.34
1981	855	256	302	101	535	160	521	182	30	14	3482	1741	2455	7.78
1982	1048	314	56872	28436	154	46	290	101	8	3	0	0	28902	7.69
1983	276	83	24049	12024	762	229	151	53	430	194	39	20	12602	7.41
1984	199	60	4030	2015	1337	401	61	21	595	268	4	2	2768	7.30
1985	456	137	1733	866	1515	454	697	244	70	32	4576	2288	4022	8.10
1986	652	196	51172	25586	1824	547	380	133	77	35	1	0	26497	8.29
1987	339	102	103686	51843	1142	343	139	49	153	69	0	0	52405	7.18
1988	341	102	39482	19741	3248	974	345	121	1236	556	3877	1939	23433	8.78
1989	145	43	48330	24165	14980	4494	471	165	683	307	1878	939	30114	8.17
1990	226	68	10819	5409	938	281	1	0	14	6	6193	3096	8862	9.62
1991	888	266	8766	4383	17992	5398	2115	740	31	14	10262	5131	15933	7.95
1992	425	128	6833	3417	1155	346	178	62	1367	615	3276	1638	6206	8.93
1993	58	17	17607	8803	415	125	77	27	7679	3456	1	0	12428	8.98
1994	3224	967	165192	82596	18171	5451	55	19	66	30	0	0	89064	7.62
1995	188	57	22560	11280	432	130	8	3	1	1	15	8	11477	7.84
1996	585	176	40791	20395	1606	482	179	63	1372	617	0	0	21733	8.25
1997	178	53	18652	9326	1611	483	1008	353	52	24	11460	5730	15969	8.02
1998	564	169	2283	1141	2336	701	778	272	69	31	8707	4354	6668	9.27
1999	1794	538	8877	4439	12859	3858	4582	1604	2	1	5312	2656	13095	8.31
2000	1671	501	43244	21622	5077	1523	2426	849	1094	492	10036	5018	30007	7.99
2001	410	123	6316	3158	1161	348	598	209	767	345	4740	2370	6554	8.21
2002	98	29	16180	8090	265	79	330	115	85	38	0	0	8352	9.56
2003	150	45	3048	1524	8389	2517	45	16	10	5	278	139	4245	8.90
2004	804	241	21472	10736	1127	338	1023	358	0	0	378	189	11862	9.49
2005	1823	547	75874	37937	2458	738	19304	6756	209	94	2984	1492	47564	8.33
2006	2183	655	110823	55411	465	139	10096	3533	159	72	5995	2997	62808	9.10
2007	1237	371	4158	2079	2073	622	423	148	123	55	1430	715	3991	9.23
2008	548	164	886	443	48	14	122	43	338	152	1823	912	1728	8.33
2009	1787	536	53719	26859	16580	4974	5079	1778	2165	974	23012	11506	46627	8.50
Mean	797	239	36694	18347	4077	1223	1757	615	647	291	3798	1899	22614	8.38
CV	0.95		1.16		1.42		2.23		2.22		1.32		0.99	
% of total biomass	1		58		11		4		2		25			
Spearman r	0.25		-0.01		-0.00		0.17		0.15		0.20			
and *p* value	0.18		0.96		0.98		0.36		0.42		0.29			

Table 1. Abundance indices (AI) (in millions), biomass (B) (in tonnes) of pelagically distributed *Agonidae, Ammotydae, Liparidae, Cottidae, Myctophidae* and *Stichaeidae* and water temperature (0-50 m) at FB-section in August over the period 1980-2009. In addition, the long term mean of fish amount (AI and B), coefficient of variation (CV) and Spearman rank correlation (r and p value) between fish abundance and temperature are given.

However, small fishes can be locally important in some areas. In the North, especially near the banks around Spitsbergen, the abiotic conditions are very variable. The waters are thoroughly mixed throughout the year yielding high primary production (Sakshaug and Slagstad 1991). Only ca 5 % of the most abundant 0-group fish were distributed in the northern areas of the Barents Sea, where smaller fishes are abundant (Figure 5). Therefore, in some years (e.g. 1994, 2001 and 2009) with low 0-group fish abundance, the relative importance of small fishes may increase in the northern areas. Polar cod occupies northern Barents Sea, adult capelin occupies the central and northern areas, and thus both these species are important components of ecosystem in the north. The polar cod stock is a comparatively large (average 1986-2009, 0.8 million tonnes), while the capelin stock is largest pelagic stock in the Barents Sea (average 1973-2009, ca 3 million tonnes (Anon. 2009)). In this study, *Ammodytidae* constituted 58 % of the total biomass of small fishes in the pelagic layer. This corresponds to approximately 18 thousand tonnes annually, and the majority of this biomass is found in the south-eastern area. In this area, juvenile herring is also found, with its biomass varying greatly according to the year class strength. In the Barents Sea *Ammodytidae* biomass is low in comparison to e.g. the North Sea, where the sandeel spawning stock varied between 475 thousand and 1 million tonnes from 1983 to 1998 (Lewy *et al.* 2004). The ecology of Barents Sea *Ammodytidae* is insufficiently known. One problem is that younger *Ammodytidae* have very small otoliths (0.5-1.2 mm length) with large shape variation, therefore otoliths are often impossible to identify due to digestion and this substantially increases uncertainties in diet analyses (Eriksen Svetocheva, 2011). However, several studies from other areas, ranging from the North Sea to the Bering Sea have reported that *Ammodytidae* are important prey for a long list of predators, including fish, seals,

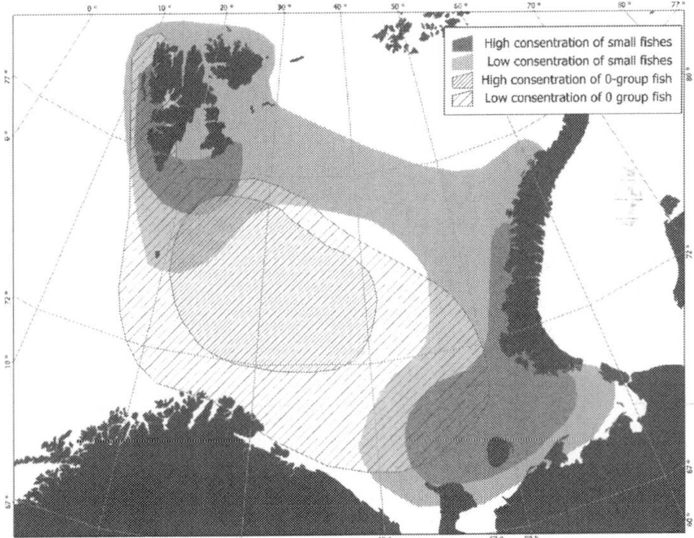

Fig. 5. Distribution of 0-group (cod, haddock, herring and capelin) and small fishes. The distribution of 0-group fish is modified from Eriksen and Prozorkevich 2011 and Eriksen *et al.* (2011).

whales, dolphins, porpoises, and many species of seabirds in different areas (Andriashev 1954; Viller 1983; Wright & Begg 1997; Wanless *et al.* 1998, Yezhov 2008, 2009). The south eastern part of the Barents Sea is known as the migration routes of seals and whales, and being close to large seabird colonies, *Ammodytidae* is likely an important food source there, especially in years with low abundance of juvenile herring.

In some years the rich fauna of small fishes may be a locally important addition to the large pelagic stocks and 0-group of commercially important fish, and make a significant contribution to the energy transport between the different trophic levels in the Barents Sea. This is likely to occur in years with low abundance of the large pelagic stocks, and in areas bordering the main distribution of the pelagic fish and 0-group fish, where their abundance varies greatly.

3.3 Variation in abundance and distribution in relation to temperature

Agonidae and *Liparidae*, mostly arctic fishes, showed a shift in their occupation area towards the North in warm years (Figure 2). *Cottidae*, including boreal, Arcto-boreal and Arctic fishes, showed, as *Agonidae* and *Liparidae*, a shift in the occupation area towards the North in warm years. *Stichaeidae* in the Barents Sea is dominated by mostly boreal species and was more widely distributed during warmer years. *Ammodytidae* and *Myctophidae* are mostly boreal fishes. Their occupation areas were mostly unchanged in cold vs. warm years. When comparing the small fish abundances indices with temperature from the FB oceanographic section (0-50m), no significant Spearman's rank correlations were found (Table 1).

Warmer temperature conditions associated with the increased inflow of Atlantic water are commonly considered an indicator of good feeding conditions in the Barents Sea (Sætersdal and Loeng 1987; Loeng and Gjøsæter 1990; Ottersen and Loeng 2000). Most of the studied fish feed on small crustaceans during their pelagic juvenile stages. The increased inflow may transport Atlantic plankton further north-eastwards, and this redistribution of food resources may have an impact on the survival of fish in specific areas, changing the overall distribution area.

Distribution of the boreal fishes *Ammodytidae* and *Myctophidae* were mostly unchanged as temperature varied. This is probably due to specific habitat requirements. *Ammodytidae* depends on bottom sediment (sand), where they bury themselves in the bottom to avoid predators; whereas *Myctophidae* is mostly restricted to areas deeper than 300 m. These restrictions might limit the potential for distributional changes in response to changes in temperature and prey availability.

We have used many assumptions in our analysis and interpretation, therefore our results are only crude approximations. The length distribution differences found between fish caught by pelagic and demersal trawls might reflect a difference in catchability related to size rather than age specific differences in vertical distribution. Most of small fishes are bottom fish with have pelagic life stages, and they are probably also distributed deeper than the surveyed layer. Hence, we acknowledge that the abundance and biomass indices are most probably underestimated. Despite these shortcomings, this unique long term data set provides new information on poorly known fish species, and their long term fluctuations in pelagic distribution, abundance and biomass.

4. Acknowledgments

We thank G.E. Dingsør for help with SAS programming, Eike Rachor and Anne Gro Vea Salvanes for valuable comments on the manuscript. We thank Sven Lemvig (IMR) for otolith preparation and age determination and Daniel Howell for proofreading. We also thank The Barents Sea Program at the Institute of Marine Research and the Norwegian Research Council project ("BarEcoRe", ref nr 200796) for funding.

5. References

Anonymous (1980). Preliminary report of the international 0-group fish survey in the Barents Sea and adjacent waters in August–September 1980. *Annales biologiques, Conseil international pour l'Exploration de la Mer, 37*, pp. 259–266

Anonymous (2004). Proceedings of the international 0-group fish survey in the Barents Sea and adjacent waters in August-September 1965-1997. IMR/PINRO Joint Report Series, No. 2/2004, PINRO Press, Murmansk. 551 pp

Anonymous (2009). Survey report from the joint Norwegian/Russian ecosystem survey in the Barents Sea August-October 2009 (adopted vol.). IMR/PINRO Joint Report Series 12/2009, 120 pp, ISSN 1503-6294

Andriashev, A.P. (1954). Fishes of the Arctic seas of the USSR. *Acad Sc Press*, 566 pp (in Russian)

Andriashev, A.P. (1986) .*Agonidae*. In: Whitehead PJP *et al.* (ed) Fishes of the North-eastern Atlantic and the Mediterranean. UNESCO, Paris, 3, pp. 1265-1268

Andriashev, A.P. & Chernova, N.V. (1995). Annotated list of fishlike vertebrates and fish of the Arctic seas and adjacent waters. *Journal of Ichthyology*, 34, 4, pp. 435-456 (in Russian)

Berestovskij, E.G. (1990). Feeding in the skates, Raja radiate and Raja fyllae, in the Barents and Norwegian Seas. *Journal of Ichthyology*, 29, 8, pp. 88-96

Boitsov, V.D., Lebed, N.I., Ponomarenko, V.P., Ponomarenko, I.Ya., Tereschenko, V.V., Tretyak,V.L., Shevelev, M.S. & Yaragina, N.A. (1996). The Barents Sea cod (biological and fisheries outline). PINRO, Murmansk

Byrkjedal, I. & Høines, A. (2007). Distribution of demersal fish in the south-western Barents Sea. *Polar Research*, 26, pp. 135-151

Chernova, N.V. (1991). Snailfishes of the Eurasian Arctic. Apatity, *Kola Scientific Center Press*, 111 pp (in Russian)

Chernova, N.V. (1998). A new species *Gymnelus andersoni sp.nova*, from the Arctic Seas with refinement of the species status of *G. retrodorsalis* LeDanois and *G. pauciporus* Anderson (Fam. *Zoarcidae*). *Journal of Ichthyology*, 38, 6, pp. 737-744

Chernova, N.V. (2005a). Review of *Careproctus* (*Liparidae*) of the North Atlantic and adjacent Arctic, including the generic type *C. reinhardti*, with rehabilitation of *C. gelatinosus* (Pallas) from Kamchatka. *Journal of Ichthyology*, 45, pp. 1-22

Chernova, N.V. (2005b). New species of *Careproctus* and *Liparidae* from the Barents Sea and adjacent waters. *Journal of Ichthyology*, 45, pp. 689-699

Coad, B.W. & Reist. .JD. (2004). Annotated list of the arctic marine fishes of Canada. *Canadian Manuscript Report of Fisheries and Aquatic Sciences*, 2674, 112 p, ISSN0706-6473

Dingsør, G.E. (2005). Estimating abundance indices from the international 0-group fish survey in the Barents Sea. *Fisheries Research*, 72, pp. 205-218

Eriksen, E., Prozorkevich, D. & Dingsør, G.E. (2009). An evaluation of 0-group abundance indices of Barents Sea fish stocks. *The Open Fish Science Journal*, 2, pp. 6-14

Eriksen, E. and Prozorkevich, D. 2011. 0-group survey. In: Jakobsen T and Ozhigin V (ed) The Barents Sea ecosystem, resources, management. Half a century of Russian-Norwegian cooperation. Tapir Academic press, 557-569

Eriksen, E., Bogstad, B. & Nakken, O. (2011). Fluctuations of 0-group fish biomass and role of 0-group fish in the Barents Sea ecosystem. *Polar Biology*, 34, pp. 647–657

Fevolden, S.E., Haug, T. & Vader, W. (1989). Intra- and interspecific allozymic variation in Liparis fabricii and Liparis gibbus (Teleostei, *Liparididae*) from Spitsbergen waters. *Polar Biology*, 10, pp. 107-111

Giske, J.D., Aksnes, B.M., Balino, M. *et al.* (1990) Vertical distribution and trophic interaction of zooplankton and fish in Masfjorden, Norway. *Sarsia*, 75, pp. 65-81

Godø, O.R., Valdemarsen, J.W. & Engås, A. (1993). Comparison of efficiency of standard and experimental juvenile gadoid sampling trawls. *ICES Journal of Marine Science*, 196, pp. 196–201

Hamre, J. (1994). Biodiversity and exploitation of the main fish stocks in the Norwegian-Barents Sea ecosystem. *Biodiversity and Conservation*, 3, pp. 392-473

Haug, T. & Nakken, O. (1973). Echo abundance indicies of 0-group fish in the Barents Sea 1965-1972. *Proceedings of the ICES/FAO/ICNAF Symposium on Acoustic methods in Fisheries Research*, Bergen, pp. 1-13

Haug, A. & Nakken, O. (1977). Echo abundance indices of 0-group fish in the Barents Sea 1965-1972. *Rapports et Procès-Verbaux des Réunions du Conseil International pour l' Exploration de la Mer.*, 170, pp. 259-64

Hulley, P.A. (1990). *Myctophidae*. In: Quero JC *et al.* (ed) Check-list of the fishes of the eastern tropical Atlantic (CLOFETA). *UNESCO*, Paris, 1, pp.398-467

Hylen, A., Korsbrekke, K., Nakken, O. & Ona, E. (1995). Comparison of the capture efficiency of 0-group fish in the pelagic trawls. *In: Precision and Relevance of Pre-recruit Studies for Fishery Management Related to Fish Stocks in the Barents Sea and Adjacent Waters*. Hylen A (ed), *Proceedings of the Sixth IMR-PINRO Symposium*, Institute of Marine Research, Bergen, Norway, 1994, pp. 145–156

Karamushko, O.V. (2008). Species composition and structure of the ichthyofauna of the Barents Sea. *Journal of Ichthyology*, 48, 4, pp. 277-291

Kazanova, I.I. & Pertseva-Ostroumova, T.A. (1960). Investigations of the reproduction and development of fishes in the Nordic Seas. In: Soviet fisheries research in the seas of the European north. VNIRO, *Fisheries*, pp 233-251

Knudsen, S.W., Møller, P.R. & Gravlund, P. (2007). Phylogeny of the snailfishes (Teleostei: *Liparidae*) based on molecular and morphological data. *Molecular Phylogenetics Evolution*, 44, 2, pp. 649-66

Krasnov, J.V. & Barrett, R.T. (1995). Large-scale interactions among seabirds, their prey and humans in the southern Barents Sea. In: *Ecology of Fjords and Coastal Waters*. Skjoldal HR *et al.* (ed), Elsevier Science B.V., Amsterdam, pp 443-456

Kristoffersen, J.B, and Salvanes, A.G.V. (2009). Distribution, growth, and population genetics of the glacier lanternfish (*Benthosema glaciale*) in Norwegian waters:

Contrasting patterns in fjords and the ocean. *Marine Biology Resesarch*, 5, 6, pp. 596-604

Lewy, P., Nielsen, A. & Gislason, H. (2004). Stock dynamics of sandeel in the North Sea and sub-regions including uncertainties. *Fisheries Research*, 68, 1-3, pp. 237-248

Loeng, H. & Gjøsæter, H. (1990). Growth of 0-group fish in relation to temperature conditions in the Barents Sea during the period 1965-1989. ICES CM 1990/G:49, 9 pp

Loeng, H. (1991) Features of the physical oceanographic conditions of the Barents Sea. *Polar Research*, 10, 1, pp. 5-18

Mukhina, N.V. (2005). Distribution of egg and fish larvae in the Norwegian and Barents Seas. *PINRO Press*, Murmansk

Muus, B.J. & Nielsen, J.G. (1999). Sea fish. In: *Scandinavian Fishing Year Book*. Hedehusene, Denmark

Ottersen, G. & Loeng, H. (2000). Covariability in early growth and year-class strength of Barents Sea cod, haddock, and herring: the environmental link. *ICES Journal of Marine Science*, 57, pp. 339-348

Pearcy, W.G., Lorz, H.V. & Peterson, W. (1979). Comparison of the feeding habits of migratory and non-migratory *Stenobrachius leucopsarus* (*Myctophidae*). *Journal of Marine Biology*, 51, 1, pp. 1-8

Pethon, P. (2005). Aschehoug's big fish book – Norway's fishes in colour. (Aschehougs store fiskebok - Norges fisker i farger) Aschehoug, Oslo (in Norwegian)

Ponomarenko, V.P. (1995). On the frequency of occurrence and fecundity of some species of the families *Agonidae, Cottunculidae, Cottidae, Lumpenidae, Cyclopteridae*, and *Zoarcidae* in the Barents Sea. *Journal of Ichthyology*, 35, 8, pp. 245-247

Rass, T.S. (1949). Composition of the Barents Sea ichthyofauna and taxonomic characteristics of fish egg and larvae. *Trydu VNIRO*, 17, pp. 7-65 (in Russian)

Sakshaug, E. & Slagstad, D. (1991). Light and productivity of phytoplankton in polar marine ecosystems: a physiological view. *Polar Research*, 10, pp. 69-86

Sakshaug, E., Johnsen, G. & Kovacs, K. (2009). Ecosystem Barents Sea. Tapir Academic Press, Trondheim

Salvanes. A.G.V. & Kristoffersen, J.B. (2001). Mesopelagic Fish (life histories, behaviour, adaptation). In: *Encyclopedia of Ocean Sciences*, Steele J *et al.* (ed), Academic Press Ltd, pp. 1711-1717

Salvanes, A.G.V. (2004). Mesopelagic fish. In: *The Norwegian Sea*, Skjoldal HR (ed) Ecosystem, Tapir, Trondheim, Norway, pp. 247-260

Stein, D.L. & Able, K.W. (1986). Liparididae. In: *Fishes of the North-eastern Atlantic and the Mediterranean*, Whitehead PJP *et al.* (ed). UNESCO, Paris, 3, pp. 1275-1283

Svetocheva, O., Eriksen, E. & Haug, T. (submitted) Barents Sea *Ammodytidae* and their ecological significance for the top predators during summer feeding. *Proceeding of the 15th Russian-Norwegian Symposium in Longyearbyen*, Svalbard, Norway, 7-8 September 2011

Sætersdal, G. & Loeng, H. (1987). Ecological adaptation of reproduction in Northeast Arctic cod. *Fisheries Research*, 5, pp. 253–70

Viller, A. (1983). Key of fishes of the seas and former seas of the north-eastern basins. *Moscow Press* (in Russian)

Wanless, S., Harris, M.P. & Greenstreet, S.P.R. (1998). Summer sandeel consumption by seabirds breeding in the Firth of Forth, south-east Scotland. *ICES Journal of Marine Science,* 55, pp. 1141-1151

Wright, P.J. & Begg, G.S. (1997). A spatial comparison of Common Guillemots and sandeels in Scottish Waters. *ICES Journal of Marine Science,* 54, pp. 578-592

Yezhov, A.V. (2008). Feeding condition for sea birds along Murman coast. All-Russian conference of young scientist. Murmansk, *PINRO press,* pp. 82-86 (in Russian)

Yezhov, A.V. (2008). Trophic conditions in the bird colonies of the Murman Coast in the Modern Period. *Proceedings of the Russian Conference of Young Scientist and Specialists on the 90th anniversary of the K.G. Konstantiniv,* Murmansk, PINRO, pp. 82-88

Yezhov, A.V. (2009). Assesment of the status of the lagest marine bird colonies of the Murman Coast in 2008; outlook for their existence. *Proceedings of the XXVII Conference of Young Scientists,* Murmansk Marine Biological Institute, Murmansk, MMBI KSC RAS, pp. 54-58

7

Rangelands in Arid Ecosystem

Selim Zedan Heneidy
Department of Botany and Microbiology,
Faculty of Science, Alexandria University
Egypt

1. Introduction

The world's rangelands constitute an important global resource. Range has been defined by the society for range management as land which supports vegetation useful for grazing on which routine management of that vegetation is through manipulation of grazing rather than cultural practices. Grazing systems are biological systems controlled by: a- the biotic factors of climate and site, b- the management inputs and decisions of man, and 3- internal regulatory mechanisms involving feedback.

General introduction for dry lands in which, most of rangelands are distributed, refer to areas with primary productivity limited by water. They cover about 40% of the land surface and contain about one fifth of the human population. This population is typically convergent on areas with relatively lower aridity, further intensifying the stresses on these marginal lands. The concept of marginality applies to these dry lands in a socio-economic sense, where the inhabitants commonly suffer poverty and lack of resources. Anthropogenic activities are almost entirely responsible for these factors. Included with, over-utilization of vegetative cover through improper rangeland management, poorly planned conversion of rangelands to croplands through irrigation schemes, and degradation of soil quality through salinization and input of chemical pollutants. The impacts of these factors on the human society are quite profound and often lead to trans-migration to other eco-regions as well as social and political strife. Thus approaches for management of dry lands resources must be viewed in the broader socio-economic context by providing the opportunities for local communities to explore viable, alternative livelihoods while maintaining their own cultural and societal fabric. Many approaches for managing the scarce water resources are available and form the underpinnings of overall resource management in drylands. These include water harvesting techniques, safe re-use of treated wastewater for irrigation, improving ground water recharge and deficit irrigation. Newer approaches of drylands aquaculture using brackish water and ecotourism also hold considerable promise for the future (UNU/MAP/UNESCO, ACRDA, 2003).

1.1 The Western Mediterranean coastal region

The western coastal desert of Egypt and its hinterland is renowned by its wealth of natural resources. This region has been a point of attraction for development projects due to this richness in natural resources, fine location, good weather and pleasant conditions. Most of the north western desert falls in the arid region except for the coastal strip. Water resources are scarce and variable. As a result the local community has developed a wide range of strategies for managing water resources in this region. Traditionally, they move around for water, pasture and croplands, based on the rainfall pattern. But recently and after being sedentary, together with population growth, overuse of water resources, over grazing and uprooting of indigenous vegetation. Climate changes, and other political and social forces, there has been an increased pressure on land resources that affected its performance and provision of goods and services.

One of the most common forms of land use in the Western Mediterranean coastal region of Egypt is animal husbandry. Its contribution to the livelihood of Bedouin increases from Alexandria to the westward direction. Grazing material and fire-wood as energy sources are two major basic needs of the inhabitants of the western desert of Egypt. The western Mediterranean desert land is one of the richest phytogeographical region in Egypt for natural vegetation because of its high rainfall. The natural vegetation includes many species of annuals, mostly herbs and a few grasses, perennial herbs, shrubs, sub-shrubs, and a few trees. These species represent 50% of the total flora of Egypt. The botanical composition is spatially heterogeneous depending on soil fertility, topography, and climatological conditions but with sub-shrubs dominating the vegetation (Ayyad and Ghabbour, 1977).

The Mediterranean desert west of Alexandria is referred to as Mareotis (or Mariut) and It is one of the arid regions which has a long history of intensive grazing and rain-fed farming. It is seldom found that any one family is engaged either in cultivation or in herding usually there is a mixture of these activities. In such region the yield and nutritive value of consumable range plants are highly dependent upon rainfall (Le Houérou and Hostte 1977; Heneidy 1992). However, the conditions of the pasture and vegetation performance also depend upon rainfall.

The total area of the rangeland is about 1.5 million hectare. The number of animals per folks varies among the Bedouin, and between regions. A herd size of about 200 head of sheep and 60 head of goats, may be reaches to 300 heads. The total number of animals which graze the rangelands is about 1.5 million head, therefore the average stocking rate is one head /ha. (Heneidy, 1992).

There is no consistent range management strategy in control of the season long grazing in the area, which varies mainly with climatic condition, availability of watering points and availability of supplementary feed. Absolutely the stocking rate leads to the problem of overgrazing. Overgrazing has been one of the main factors causing the deterioration of ecosystem productivity in the Mediterranean coastal region. It has resulted in severe reduction of perennial cover, soil erosion and formation of mobile dunes.

Grazing and rain fed farming provide a clear example of the impact of man's disruptive action on arid and semiarid regions. Continued uncontrolled grazing, wood cutting and farming have, however, induced a process of degradation. This is coupled with severe environment, uncertainty of rainfall and a change in socio-economic circumstances, has resulted in an advanced stage of desertification. As a result, the region now is producing at a rate less than its potential, and is continuing to lose the productivity it had. The degradation of renewable natural resources in arid and semi-arid areas has become a matter of great concern. This region includes several types of habitats (e.g. sand dunes, saline depressions, coastal ridges, non-saline depressions, wadies, inland plateau, inland ridges, etc....).The causes of degradation of rangeland in the Mediterranean coastal region of Egypt, are mixes of environmental, socio-political and socioeconomic conditions (Ayyad, 1993; Heneidy and Bidak, 1998). In recent years reclamation and cultivation of desert areas in Egypt became a necessity. The coastal Mediterranean semi-arid stripe and the inland new valley areas are possible promising fields for such a purpose.

2. Methodology

2.1 Selection of stands for vegetation analysis

Several stands were distributed within the study area extending from the non saline depression to the inland plateau. These stands were selected to represent major apparent variations in the physiogonomy of vegetation and in topographic and edaphic features. For sampling of the vegetation, about fifty 1 X 2 m^2 quadrates were located randomly in each stand. The presences of each species in each quadrate were recorded, and the number of its individuals counted. The number of individuals and occurrences of each species in the quadrates was then used to calculate its density and frequency respectively. The line-intercept method was used for estimation of cover. The lengths of intercept of each species in a stand was measured to the nearest centimeter along five, 50 m long transects. This length was then be summed and expressed as a relative value of the total length of the five transects. Relative density, frequency and cover for each perennial species were summed up to give an estimate of its importance.

2.2 Consumption

Consumption is the amounts taken by grazing animals- Consumption depends on many factors and it is differ according to types of livestock. Estimation of the consumption rate was carried out using the Bite method as follows. The flock, which normally consists of about 100 heads, was observed for 24 hours twice a week in each habitat. A particular animal in the flock was observed for five inputs, using field glasses and a tape recorder. During each hour of the daily grazing period, ten animals were observed in a cycle of 50 minutes to represent the hourly consumption behavior to record the observations. Estimation of the consumption of different plant species by the grazing animals is based on three main measures: (a) The number of times each plant species was used in the diet of a single animal (number of bites per unit time). (b) Average size of material removed from each species in one bite. (c) The location on the canopy from which this material was removed. At the end of each month, 20 samples of each species simulating the average bite size were clipped and used to estimate the average of fresh and dry weights of the bite of that species. The weight and the number of bites were used to calculate the fresh and dry weight of consumed material and the amount of water. This technique was applied also by Jamieson & Hodgson, 1979; Chamber et al. 1981; Hodgson & Jamieson 1981; Illius & Gordan 1987; and Heneidy (1992, 1996). The total number of bites per animal per day of each plant species multiplied by the average weight of material removed in each bite will express the amount of material removed per animal per day. This estimate will then be multiplied by the value of the stocking rate of animals in the area to provide an assessment of the amount consumed from different plant species per hectare per day.

2.3 Palatability

Palatability of the range plants was determined and classified into four categories according to the palatability index reported by Heneidy (2000) and Heneidy & Bidak, (1999).

2.4 *Phytomass* (standing crop phytomass, accessible parts, and necromass or litter)

Homogeneity of a stand was judged according to edaphic and physiographic features. The direct harvest method was used for phytomass determination according to Moore and Chapman (1986). All above-ground parts of different life-forms of the most common palatable species were excavated in each stand, and directly weighed in the field. Representative individuals of each species were collected in each stand during two seasons, spring and summer (representing the wet and dry season), for standing crop phytomass determination and also to determine the vegetative and accessible parts (available parts) depending upon the morphology and configuration of the plant species (Heneidy, 1992; 2003a). Nomenclature and identification were carried out according to Täckholm (1974) and the Latin names of species were updated following Boulos (1995).

2.5 Economic value

Economic value of plant species were estimated based on: 1- direct observation in the field; 2- preparing a questionnaire form to the inhabitants. The questionnaire was prepared to obtain information about fuel wood, traditional uses of some plant species; 3- the previous experience in the field of study.

3. Omayed biosphere reserve as a case study

Omayed Biosphere Reserve (OBR) is located in the western Mediterranean coastal region of Egypt (29° 00' - 29° 18' E and 30° 52' - 20° 38' N). It extends about 30 km along the Mediterranean coast from west El-Hammam to El-Alamin with a width of 23.5 km to the south (Fig.1). Its N-S landscape is differentiated into a northern coastal plain and a southern inland plateau. The coastal plain is characterized by alternating ridges and depressions running parallel to the coast in E-W direction. This physiographic variation leads to the distinction of 6 main types of ecosystems:

1- Coastal ridge, composed mainly of snow-white oolitic calcareous rocks, and overlain by dunes. 2- Saline depressions, with brackish water and saline calcareous deposits (i.e. salt marshes). 3- Non-saline depressions, with a mixture of calcareous and siliceous deposits of deep loess. Rainfed farms are a pronounced man made habitat within the non-saline depressions in this region. 4- Inland ridges, formed of limestone with a hard crystallized crust, and less calcareous than the coastal ridge. 5- Inland plateau, characterized by an extensive flat rocky surface and shallow soil. And, 6- Inland siliceous deposits, sporadically distributed on the inland plateau. The deposits of this ecosystem vary in depth and overly heavier soil with calcareous concretions at some distance from the soil surface. This soil is poor in organic matter and nutrients, particularly in the siliceous sand horizon (Heneidy, 2003b).

Fig. 1. Location of OBR in the western coastal desert of Egypt (After UNU/MAP/UNESCO, ACRDA, 2003)

Major economic activities: Raising herds by grazing, intensive quarrying and rainfed cultivation of grain crops, vegetables and orchards depending on water availability (mainly rain). Irrigated land agriculture is another potential acitivity that is introduced due to the extension of an irrigation canal from the Nile delta to the region.

Major environmental/economic constraints: Land degradation, habitat fragmentation, overgrazing, loss of biodiversity, salinization of soil, over exploitation of mineral and water (ground water) resources are the major environmental constraints The economic constraints stems from the general economic status of the local inhabitants that ranges from low to moderate, due to the following:

- Absence of permanent source of income (revenue),
- Lack of skills
- Major activities are seasonal (agriculture and grazing)
- spread of unemployment and thus poverty

Integration of environmental conservation and sustainable development can perfectly be implemented in OBR as the area is an ideal site for the sustainable development of 18 natural resources, by rationalizing ecotourism, rangeland management, propagation of multipurpose woody species, and promoting local industries based on native knowledge and experience of inhabitants. Tourism oriented towards natural history and traditional lifestyle is expected, especially with the spreading of nearby tourist villages. Nature watching (vegetation in spring, birds and wild animals, etc.), ancient devices of rain water harvesting and historical buildings (Zaher Bebars citadel from the thirteenth century) are types of tourist attractions in this reserve.

3.1 Vegetation - Environment relationships

Five major types of habitat are recognized in the study area: coastal sand dunes, inland ridges, non-saline depressions, wadis and inland plateau. Each of these habitats is characterized by local physiographic variations which effectuate variations in vegetation composition and species abundance. The application of the polar ordination, tabular comparisons and principal component analysis, resulted in the classification of vegetation into groups related to environmental and anthropogenic factors.

Dunes are fashioned by the influence of onshore winds which are predominantly north-western. Sand moves inland in a series of transverse dunes. Close to the shore, dunes are relatively small and active. Further inland they become larger and, being more heavily covered by vegetation, tend to become more or less stabilized. They exhibit a typical dune from with gentle windward slopes and steep leeward slopes. In the shelter of stabilized dunes, active deposition of sand in front of steep leeward slopes results in the formation of sand shadows Ayyad (1973). Kamal (1988) distinguished seven main physiographic categories:

i. Sand dunes comprise 5 categories:
    1. Very active baby dunes, lying close to the shore, where erosion and deposition take place are dominated by *Ammophila arenaria,*
    2. Active, partly stabilized dunes, codominated by *Ammophila arenaria. Euphorbia paralias,* and *Lotus polyphyllos,*
    3. Stabilized dunes with typical dune form, and where almost no erosion or deposition is taking place are codominated by *Pancratium maritimum* and *Hyoseris lucida.*
    4. Deep protected sand shadows are codominated by *Crucianella maritima, Ononis vaginalis, Echinops spinosissimus* and *Thymelaea hirsuta.*
    5. Exposed barren rock and escarpment of the coastal ridge, co-dominated by *Deverra triradiata, Echiochilon fruticasum , Gymnocarpos decandrum.*and *Salvia lanigera.*
        All vegetation groups of both eastern and western provinces of the coastal dunes indicates that the vegetation is co-dominated by *Ammophila arenaria, Euphorbia paralias, Zygophyllum album and Deverra triradiata, sincethese groups* (Kamal, 1988).
ii. The salt marsh is characterized by high density of salt tolerant shrubs (e.g. *Atriplex halimus, Suaeda spp., and Salsola spp.*) .
iii. The rocky ridge habitat is dominated by the subshrub (e.g. *Gymnocarpos decandrum, Anabasis oropediorum* and *A. articulata,* and *Artemisia herba alba).*
iv. The non-saline depression habitat is dominated by *Anabasis articulata Asphodelus ramosus, Plantago albicans, and Thymelaea hirsute.*
v. The inland plateau habitat is dominated by *Thymus capitatus,Globularia arabica, Asphodelus ramosus, Herniaria hemistemon, Scorzonera undulata* and *Deverra tortuosa Plantago albicans,Thymelaea hirsuta. Noaea mucronata,* and *Echinops spinosissimus.*

3.1.1 Effect of ploughing on species abundance and diversity

This study focuses on the effect of ploughing on plant abundance, vegetation cover, species richness, and taxonomic diversity during the growing seasons (winter and spring) of 1992 and 2000 in the habitat of inland plateau. Ninety-five species belonging to 27 families were recorded High percentages of life-forms and a large number of species were recorded in ploughed and unploughed stripes in the winter and spring of 2000 (Fig.2). Higher averages of importance values (IVs) and absolute frequencies were recorded for most perennial and annual species in the unploughed stripes compared to the ploughed ones. This may be attributed to crop failure and consequently unfavourable soil conditions. On the other hand, some shrubby species (e.g. *Noaea mucronata* and *Haloxylon coparium*) and perennial herbs

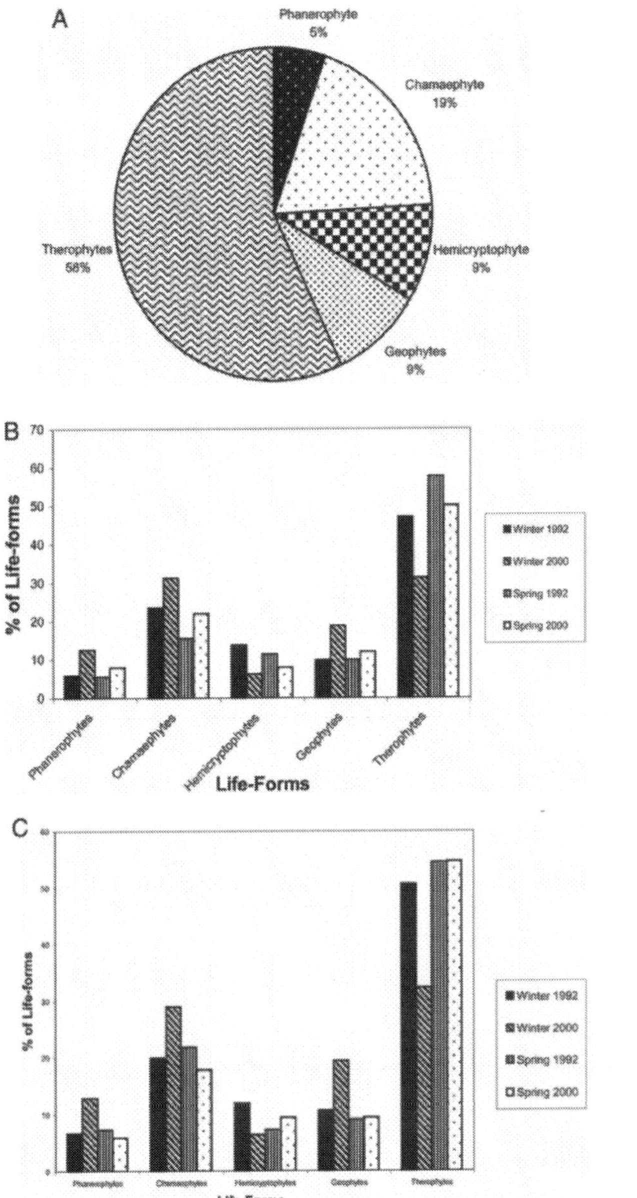

Fig. 2. (A) Different percentages of life forms in the study area. (B) Life form spectrum of the two seasons winter and spring (1992 and 2000) for the ploughed stripes. (C) Life form spectrum of the two seasons winter and spring (1992 and 2000) for the unploughed stripes.

(e.g. *Gynandriris sisyrinchium*) attained higher IVs in the ploughed stripes compared to unploughed ones. This may be attributed to the cultivation of *Prosopis juliflora* trees in the elevated part of the ploughed stripes, which have an ecological role in protecting and enriching the soil with organic matter, thus favouring the growth of these shrubs and perennial herbs. Higher species richness and diversity were associated with low concentration of dominance and low taxonomic diversity in the spring of 2000 in ploughed and unploughed stripes compared to the winter of 1992, for both perennials and annuals. The lowest taxonomic diversities were exhibited in the spring of 2000 for ploughed and unploughed stripes where the vegetation had the largest number of congeneric species and confamilial genera. Higher species richness and diversity characterized the vegetation of the unploughed stripes, especially in winter and spring 2000, as compared to those of ploughed ones. This study also reveals low species richness and diversity of therophytes in winter for both ploughed and unploughed stripes (Kamal et al., 2003).

Species richness and other diversity measures of ploughed and unploughed stripes showed that species richness (Figure 3) was usually high for perennials in the year 2000 in both ploughed and unploughed stripes. It also indicated that the unploughed stripes are substantially the most diverse, as estimated by Shannon's index (H' = 3). The same trend was also notable for annual species (Figure 4). Accordingly the lowest dominance was attained in spring 2000 for ploughed and unploughed stripes.

4. Grazing as a major activity

4.1 Existing state

Rangelands constitute about 47% of the total of 332 million hectares. The vegetation resource on these lands is often sparse and droughty. Therefore these lands often require extra measures of management in order to provide forage and wildlife habitat as well as uses on a sustained yield basis (Tueller, 1988).

Generally, one of the most common forms of land use in the western Mediterranean coastal region of Egypt is animal husbandry. Its contribution to the livelihood of Bedouin increases from west of Alexandria to Sallum (border of Egypt). The natural vegetation includes many species of annuals, mostly herbs and a few grasses, perennial herbs, shrubs, sub-shrubs, and a few trees. These species represent 50% of the total flora of Egypt. The botanical composition is spatially heterogeneous depending on soil fertility, topography, and climatologically conditions but with sub-shrubs dominating the vegetation (Ayyad and Ghabbour, 1977). The composition of growth forms in the region expresses a typical desert flora. The majority of species are either annuals (ephemerals) or geophytes (perennial ephermeroid herbs and grasses). Both are "drought evaders" in the sense that the whole plant or the greater part of the photosynthetically active and transpiring (above ground) organs are shed during the rainy season. The majority of other perennials in the study area are evergreen shrubs or sub-shrubs (Chamaephytes).

There are three major land uses in the dry lands: rangelands, rain- fed cropland and irrigated land. The following table shows generally about 88% of the dry lands are used as rangelands, 9% are rain-fed croplands, and 3% are irrigated lands in the arid, semi-arid and dry sub-humid climatic zones.

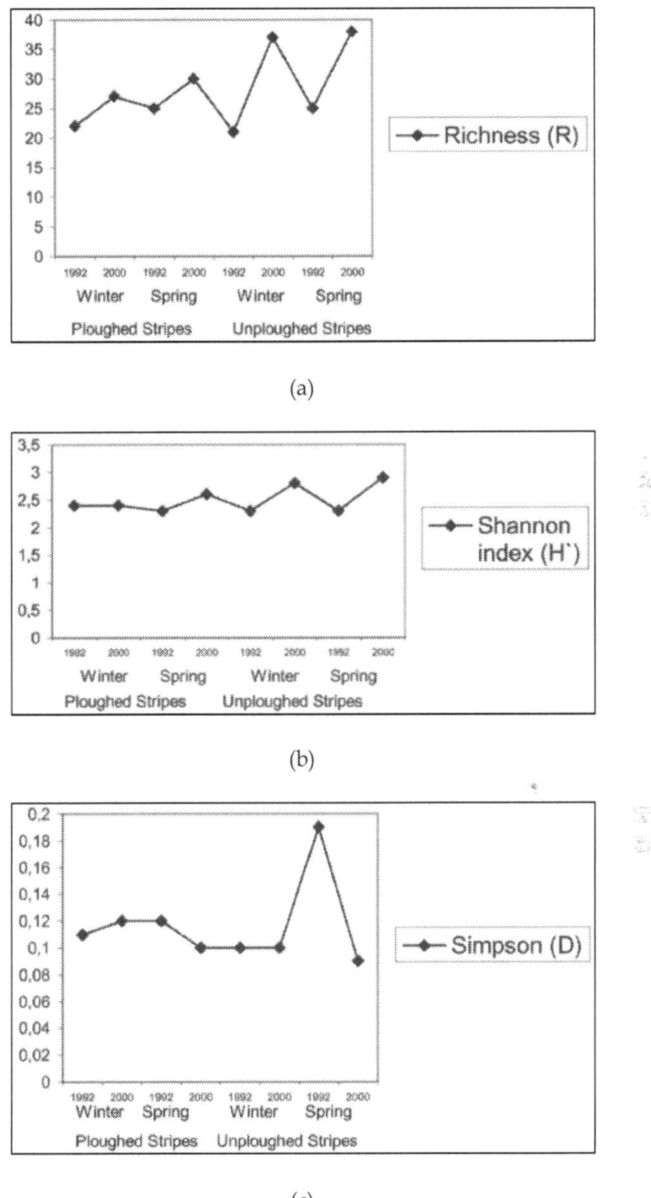

(a)

(b)

(c)

Fig. 3. Diversity indices for ploughed and unploughed stripes (winter and spring of 1992 and 2000) for perennial species. (a) Species richness; (b) Shannon Index; (c) Simpson Index.

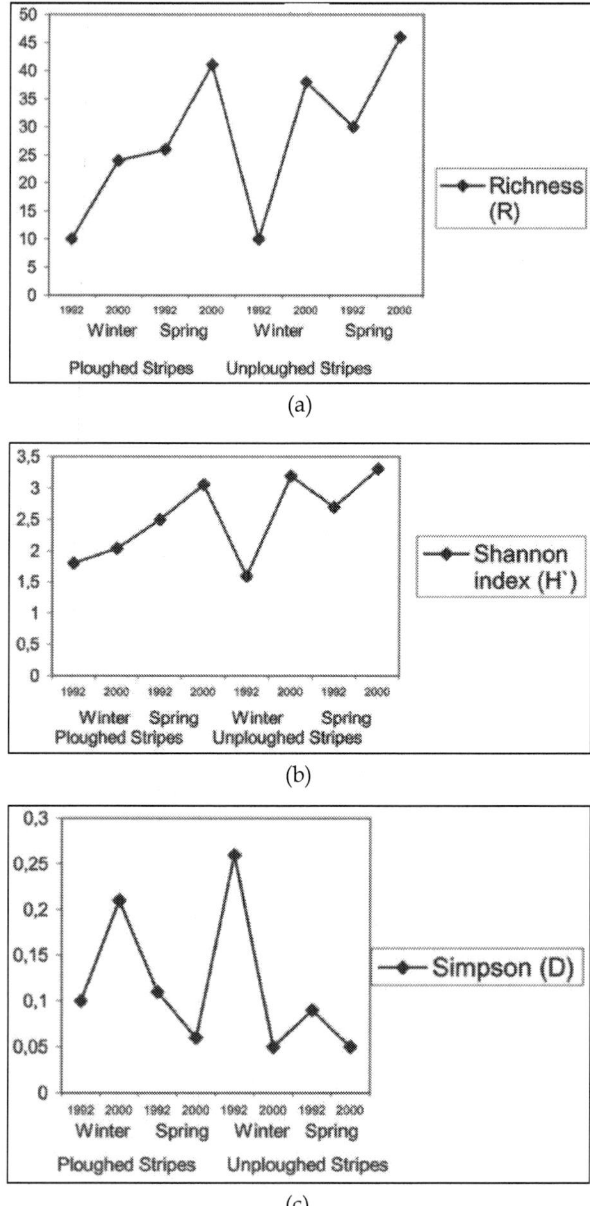

Fig. 4. Diversity indices for ploughed and unploughed stripes (winter and spring of 1992 and 2000) for annual species. (a) Species richness; (b) Shannon Index; (c) Simpson Index

Land use area in global dry lands

Type of land use	Area (Hectare)
Rangeland	4,550,000,000
Rain-fed cropland	457,000,000
Irrigated land	145,000,000
Total Area	5,152,000,000

Some of most common rangeland species in the Mediterranean coastal region are *Anabasis articulata, A. oropediorum, Artemisia monosperma, A. herba-alba, Asphodelus ramosis. Convolvulus lanatus, Carduncellus eriocephalus, Eciochilon fruticosum, Echinops spinosissimus, Gymnocarpos decandrum, Helianthemum lippii, H. kahiricum, Lycium europaeum, Noaea mucronata. Deverra triradiata. Periploca aphylla, Scorzonera alexandrina,* and *Thymelaea hirsuta.*

Table 1 shows a list of plant species with different life-forms, in El-Omayed Biosphere Reserve (Heneidy, 1992).

Species	Life-form	Pala.	Species	Life-form	Pala.
1- *Achilla santolina*	Ch.	P	87- *Hippocrepis cyclocarpa*	Th.	P
2- *Adonis dentatus*	Th.	LP	88- *Hordium marinum*	Th.	HP
3- *Aegialophia pumila*	Ch.	HP	89- *Hyoscyamus muticus*	Ch.	NP
4- *Aegilops kotschyi*	Th.	HP	90- *Hyoseris lucida*	Th.	P
5 — *Ajuga iva*	Geo.	P	91- *Hyoseris scarba*	Th.	P
6- *Aeluropus lagopoides*	Ch.	HP	92- *Ifloga spicata*	Th.	HP
7- *Alhagi graecorum*	Ch.	HP	93- *Imperata cylindrica*	Geo.	P
8- *Alkanna tinctoria*	Ch.	P	94- *Iris sisyrinchium*	Geo.	LP
9- *Allium sp.*	Geo.	LP	95- *Juncus acutus*	Geo.	HP
10- *Ammi visnaga*	Th.	P	96- *Juncus rigidus*	Geo.	HP
11- *Ammophila arenaria*	Ch.	NP	97- *Kickxia aegyptiaca*	Ch.	HP
12- *Anabasis articulata*	Ch.	P	98- *Kochia indica*	Th.	P
13- *Anabasis oropediorum*	Ch.	HP	99- *Lactuca serriola*	Bi.	P
14- *Anacyclus alexandrinus*	Th.	P	100- *Launaea nudicaulis*	Ch.	HP
15- *Anagalis arvensis*	Th.	P	101- *Launaea resedifolia*	Th.	HP
16- *Anchusa azurea*	Th.	P	102- *Limoniastrum monopetalum*	Ch.	LP
17- *Anthemis retusa*	Th.	P	103- *Limonium tubiflorum*	Ch.	LP
18- *Argyrolobium abyssinicum*	Ch.	HP	104- *Lobularia arabica*	Th.	P
19- *Arisarum vulgare*	Geo.	LP	105- *Lobularia maritima*	Ch.	P
20- *Artemisia herba-alba*	Ch.	HP	106- *Lolium perenne*	Geo.	HP
21- *Artemisia monosperma*	Ch.	P	107- *Lotus creticus*	Ch.	P
22- *Arthrocnemum glaucum*	Ch.	P	108- *Lotus polyphyllos*	Ch.	P
23- *Asparagus stipularis*	Ch.	LP	109- *Lycium shawii*	Ch.	HP
24- *Asphodelus ramosus*	Geo.	HP	110- *Lygeum spartum*	Geo.	HP
25-*Astenatherum forsskalii*	Ch.	P	111- *Malva parviflora*	Th.	P
26- *Astragalus spinosus*	Ch.	HP	112- *Marrubium vulgare*	Ch.	LP

Species	Life-form	Pala.	Species	Life-form	Pala.
27- *Atractylis carduus*	Ch.	P	113- *Melilotus indicus*	Th.	P
28- *Atriplex halimus*	Ph.	P	114- *Moltkiopsis ciliata*	Ch.	HP
29- *Avena barbata*	Th.	P	115- *Narcissus tazetta*	Ch.	LP
30- *Avena sativa*	Th.	P	116- *Noaea mucronata*	Ch.	HP
31- *Bassia muricata*	Th.	LP	117- *Ononis vaginalis*	Ch.	LP
32- *Brassica tournefortii*	Th.	P	118- *Onopordum alexandrinum*	Bi.	LP
33- *Bromus rubens*	Th.	HP	119-*Ornithogalum trichophyllum.*	Ch.	LP
34-*Bupleurum semicompositum*	Th.	P	120- *Otanthus maritimus*	Ch.	LP
35- *Cakile maritima*	Th.	P	121- *Pancratium maritimum*	Geo.	LP
36- *Carduncellus eriocephalus*	Ch.	HP	122- *Pancratium sickenbergeri*	Geo.	LP
37- *Carrichtera annua*	Th.	P	123- *Papaver rhoeas*	Th.	LP
38- *Carthamus lanatus*	Ch.	P	124- *Paronchia argentea*	Ch.	P
39- *Centaurea alexandrina*	Th.	P	125- *Peganum harmala*	Ch.	NP
40- *Centaurea calcttrapa*	Th.	P	126- *Phagnalon schweinfurthii*	Ch.	P
41- *Centaurea glomerata*	Th.	P	127- *Phlomis floccosa*	Ch.	LP
42- *Chenopodium murale*	Th.	NP	128- *Phragmites australis*	Geo.	P
43-*Chrysanthemum coronarium*	Th.	LP	129- *Deverra triradiata*	Ch.	HP
44- *Citrullus colocynthis*	Ch.	NP	130- *Plantago albicans*	Geo.	HP
45- *Cleome africana*	Ch.	NP	131- *Plantago crypsoides*	Th.	P
46- *Colchicum ritchii*	Ch.	NP	132- *Plantago major*	Ch.	P
47- *Convolvulus althaeoides*	Ch.	P	133- *Polygonum equisetiforme*	Ch.	P
48- *Convolvulus arvensis*	Ch.	LP	134- *Polygonum maritimum*	Ch.	P
49- *Convolvulus lanatus*	Ch.	HP	135- *Prasium majus*	Ch.	HP
50- *Conyza linifolia*	Th.	P	136- *Reaumuria hirtella*	Ch.	LP
51- *Cressa cretica*	Ch.	LP	137- *Reseda decursiva*	Bi.	P
52- *Crucianella maritima*	Ch.	NP	138- *Retama raetam*	Ch.	LP
53- *Cuscuta planifolora*	Th.	NP	139- *Rumex pictus*	Th.	HP
54- *Cutandia dichotoma*	Th.	HP	140- *Rumex vesicarius*	Th.	P
55- *Cynodon dactylon*	Geo.	HP	141- *Salicomia fruticosa*	Ch.	LP
56- *Cyperus rotundus*	Geo.	LP	142- *Salsola longifolia*	Ch.	LP
57- *Dactylis glomerata*	Geo.	HP	143- *Salsola tetragona*	Ch.	HP
58- *Ebenus armitgei*	Ch.	HP	144- *Salsola tetrandra*	Ch.	P
59- *Echinops spinosissimus*	Ch.	HP	145- *Salsola volkensii*	Ch.	LP
60- *Echiochilon fruticosum*	Ch.	HP	146- *Salvia aegyptiaca*	Ch.	HP
61- *Echium sericeum*	Ch.	LP	147- *Salvia lanigera*	Ch.	HP
62- *Elymus farctus*	Geo.	P	148- *Scorzonera alexandrina*	Geo.	HP
63- *Emex spinosus*	Th.	P	149- *Senecio vulgaris*	Th.	P
64- *Eminium spiculatum*	Geo.	NP	150- *Silybum marianum*	Ch.	P
65- *Ephedra alata*	Ch.	NP	151- *Solanum nigrum*	Th.	LP
66- *Erodium cicutarium*	Th.	P	152- *Stipa capensis*	Th.	P
67- *Eryngium campestre*	Ch.	LP	153- *Stipa-grosts ciliata*	Geo.	HP

Species	Life-form	Pala.	Species	Life-form	Pala.
68- *Euphorbia bivonae*	Ch.	P	154- *Suaeda pruinosa*	Ch.	LP
69- *Euphorbia granulata*	Th.	NP	155- *Suaeda vera*	Ch.	LP
70- *Euphorbia helioscopia*	Th.	LP	156- *Tamarix nilotica*	Ph.	LP
71- *Euphorbia paralias*	Ch.	NP	157- *Teucrium polium*	Ch.	P
72- *Euphorbia peplis*	Th.	LP	158- *Thymelaea hirsuta*	Ch.	P
73- *Fagonia arabica*	Ch.	NP	159- *Thymus capitatus*	Ch.	P
74- *Filago desertorum*	Th.	HP	160- *Tribulus terrestris*	Th.	LP
75- *Foeniculum vulgare*	Ch.	P	161- *Trigonella stellata*	Th.	HP
76- *Frankenia revoluta*	Ch	LP	162- *Typha domingensis*	Geo.	NP
77- *Fumana thymifolia*	Ch.	LP	163- *Urginea undulata*	Geo.	NP
78- *Fumaria parviflora*	Th.	LP	164- *Urtica urens*	Th.	LP
79- *Globularia arabica*	Ch.	P	165- *Vaccaria pyramidata*	Th.	LP
80- *Gymnocarpos decandrum*	Ch.	HP	166- *Varthemia candicans*	Ch.	P
81- *Halocnemum strobilaceum*	Ch.	LP	167- *Verbascum tetoumeurii*	Bi.	LP
82- *Haplophyllum tuberculatum*	Ch.	LP	168- *Vicia sativa*	Th.	P
83- *Helianthemum kahiricum*	Ch.	HP	169- *Zilla spinosa*	Ch.	P
84- *Helianthemum lippii*	Ch.	HP	171- *Zygophyllum album*	Ch.	NP
85- *Herinaria hemistemon*	Ch.	LP			
86- *Hippocrepis bicontorta*	Th.	P			

Table 1. Wild plant species recorded in Omayed Biosphere Reserve.
Ch. = Chymophytes, Th. =Therophytes, Ph. Phanerophytes, and Geo. = Geophytes.
Pala.= Palatability, P. = Palatable, HP. = High palatable, LP.= Low palatability,
NP.= unpalatable, Bi.= Biennial, and *Asphodelus microcarpus* = *A. ramosus* = *A. aestivus*
(After Heneidy, 2002a).

4.1.1 Palatability and preference

Palatability of range land species is a very complex notion, very difficult to generalize as it is
linked to many factors that vary in time and place. Some of these variables are linked to the
plant, other to the animal, while a third category depends on various environmental factors
(Heneidy, 1992, 1996: Le Houéou, 1994). In the north western Mediterranean region,
Heneidy (1992) recorded that the temperature is the most effective factors on the animal
behaviour. One approach to recording food selectively is based on observation of feeding
animals (Bjustad et al., 1970). Grazing animals change their behaviour, and hence their food
preferences in relation to differences in temperature and rainfall (Castle and Halley, 1953).
Given a free choice, herbivores exhibit marked preferences amongst the available foods but
will rarely eat one food to the exclusion of all others. Selectivity of herbage expresses the
degree to which animals harvest plants or plant parts differently from random removal. Van
Dyne and Heady (1965) use ratios between the proportion of any species, part of plant, or
group of plants in the diet, and the proportion of that item in the herbage available to the
animal as an expression of relative preference (or selectivity ratio). Using such ratios for the
vegetation of the present study has resulted in ranking of perennial species growing in the
different habitats during spring as follows (Table 2).

It is interesting to note that species with the highest ranks of relative preference are those with relatively lower availability or abundance. For example, *P. albicans* is the species with the highest selectivity ratio in the habitats where it grows. It's share in the diet selected by animals in season of maximum consumption is more than fourfold its share in the available forage for the animals. This species represents less than 5% of the grazeable phytomass in the range, but as high as about 22% of the diet consumed. This is also true in the case of *H. lippii* which provides available forage not more than 7%, while it contributes up to 19% of the animal diet especially in the ridge habitat. On the other hand, the species ranked as of lower selectivity ratio are those which are more available or abundant in the ridge. Fore example, *E. fruticosum* has a proportion in the animal diet (about 13%) of one third its proportion in the available forage (about 40%). Le Houéou (1980a) discusses such relationship between palatability and abundance, and states that all other attributes being equal, the palatability of a given taxon is inversely related to its abundance in the range, except for a few species which are specially relished in all circumstances. *A. microcarpus* in the present study represents such species whose selectivity by animals is independent of their abundance on the range. It's selectivity is related more to the phenological development of the plant. Consequently, its highest selectivity ratio is attained during winter at the phase of flower bud development. At this phase, the animals consume much of these flower buds (up to 40% of the diet), while they represent not more than 4% of the available forage. On the other hand, when the above-ground shoot of *A. microcarpus* dries up during and summer and autumn, it provides about 30% of the material available for grazing and constitutes more than 50% of the animal diet.

Rank	Non-saline depression		Ridge		Inland plateau	
	Species	Selectivity ratio	Species	Selectivity ratio	Species	Selectivity ratio
1	*Plantago albicans*	4.69	*H. lippii*	2.69	*P. albicans*	5.02
2	*Helianthemum lippii*	3.81	*G. decandrum*	2.36	*A. microcarpus*	3.09
3	*Asphodelus ramosus*	2.23	*A. ramosus*	1.78	*H. lippii*	1.50
4	*Gymnocarpos decandrum*	0.82	*A. articulata*	0.87	*G. decandrum*	1.27
5	*Anabasis articulata*	0.47	*T. hirsuta*	0.19	*E. fruticosum*	0.34
6	*Eiochilon fruticosum*	0.30	--		*A. articulata*	0.29
7	*Thymelaea hirsuta*	0.16	--		*T. hirsuta*	0.11

Table 2. Relative preference (or selectivity ratio) in OBR (After Abdel-Razik et al., 1988a).

Many studies (e.g. Cook, 1972; Dicko-Toure, 1980; Le Houéou, 1980b) refer to the relationship between palatability of a given species or taxon, its stage of development (phenology), its relative abundance on the rangeland, and its chemical composition. Cowlishow and Alder (1960) assert that palatability of a given species changes because of changing characteristics that an animal can recognize by its senses of sight, taste and smell. Table 2 shows a list of plant species and their palatability in Omayed Biosphere Reserve. 63% of these species are palatable while, 42% of them are highly palatable.

5. Biomass and productivity

El-Kady (1980) estimated the effect of protection and controlled grazing on the vegetation composition and productivity as well as the rate of consumption of phytomass by domestic animals and their grazing behavior, in the ecosystem of non-saline depression. Plots were fenced and different grazing pressures were applied in the vegetation in these plots. Changes in the density, cover, frequency, phytomass and the phenological sequence of species were recorded and compared to those of the same species outside the fenced plots. Remarkable increases were recorded in total density and cover of perennials, in frequency and presence of animals, and in phytomass as a result of protection and controlled grazing. It was concluded that partial protection and controlled grazing could be of better consequence than full protection: light nibbling and removal of standing dead by domestic animals may promote vigor and growth of defoliated plants, and the availability of nutrients may be enhanced by the passage of herbage through their guts and out as feces.

Heneidy (1992) reported that the annual above-ground dry matter production of the rangeland at Omayed area in the different habitats were about 2833 kg/ha in the non-saline depression, 1448 kg/ha in the ridge, and 4416 kg/ha on the inland plateau habitats. In general preliminary field observations on the behaviour of grazing animals indicated that almost all the consumed forage is contributed by sixteen perennial species (most common) and annuals (Table 3) throughout the year. In general the total phytomass of new growth was highest in the habitat of the inland plateau, and lowest in the ridge habitat.

Habitat	Non-saline depression		Ridge			Inland plateau		
	Above-ground	Below-ground	Above-ground	Below-ground	ground	Above-ground		Below-ground
Asphodelus ramosus	71.43 **28.6**	35.29	7.94 **16.2**	14.13		87.11 **39**		43.15
Plantago albicans	7.82 **4.8**	3.08	--	--		12.47 **9**		3.11
Carduncellus eriocephalus	3.53 **0.8**	0.89	--	--		--		--
Echinops spinosissimus	--	--	0.89 **0.6**	0.61		0.66 **0.4**		0.28
Scorzonera alexandrina	--	--	10.95 **2.9**	7.98		3.37 **0.8**		2.53
Echiochilon fruticosum	107.08 **21.1**	15.91	--	--		69.6 **21**		10.83
Helianthemum lippii	7.5 **4.5**	1.47	6.6 **3.9**	1.3		10.22 **7.3**		1.58
Gymnocarpos decandrum	51.87 **5.0**	5.75	53.8 **6.9**	5.76		60.74 **8.0**		7.49
Deverra triradiana	2.89 **0.59**	0.45	10.61 **3.6**	2.93		8.52 **2.1**		0.63
Convolvulus lanatus	12.69 **2.3**	0.18	--	--		16.2 **2.2**		1.66
Noaea mucronata	4.29 **1.3**	0.85	20.0 **5.6**	3.74		21.6 **1.9**		2.89
Artemisia monosperma	28.59 **9.5**	5.39	--	--		--		--
Artemisia herba alba	--	--	5.98 **0.24**	1.84		4.38 **0.1**		0.88
Anabasis articulata	411.9 **92.4**	74.32	111.6 **23**	26.39		1011 **251**		96.79
Anabasis oropediorm	--	--	13.53 **3.1**	2.73		15.9 **3.8**		2.94
Thymelaea hirsuta	826.7 **104**	414.8	382.73 **39**	231.67		1160 **99**		77.5
Annuals	2.3 **2.3**	--	0.9 **0.9**	--		1.13 **1.13**		--

Table 3. The total standing crop phytomass (old organs + new-growths) and below-ground biomass (gm/dr.wt/m²) of different organs of species in the different habitats at Omayed. Above-ground (old organs + new growths), Bold value = New growths (grazeable parts).

In general the total phytomass accessible to grazing animals (kg/ha} varied remarkably with season and habitats, from a minimum of 308 kg/ha in the ridge habitat to maximum of 1543 kg/ha in the habitat of inland plateau. Inland plateau habitat is characterized by (1) Remarkably higher annual production (Accessible); (2) Remarkably higher nutritive value; and

(3) Presence of palatable species (browse species) e.g. *Astragallus spinosis, Anabasis oropediorum, Salsola tetragona, Lygeum spartum* and *Lycium europaeum* (Plate 1). Therefore, it is recommended that the shepherds should extend the grazing area further to the inland habitat.

Plate 1. *Overgrazing of highly palatable species* (After Heneidy, 2003b)

The highest contribution by perennial herbs in all habitats was attained during winter and spring (13%, and 10.9% in the none-saline depression, 12.7% and 13.5% on the ridge, and 9.5% and 8.3% on the inland plateau). On the other hand, the maximum contribution by most subshrubs was attained during spring. The contributions by subshrubs in the habitats of non-saline depression, the ridge and the inland plateau in spring were 15%, 14.5% and 10.1% respectively.

5.1 Accessible (consumable) biomass

The accessible dry matter production of the subshrubs and forbs together (maximum values of the different seasons) is estimated by Heneidy (1992) as about 741 kg/ha/yr in the non-saline depression, 371 kg/ha/yr on the ridge and 745 kg/ha/ yr on the inland plateau (average of 677 kg/ha/yr, equivalent to about 4.5 kg/ha/yr/mm rainfall). For sub-shrubs te accessible dry matter production represents about 4,5, and 2% of total above-ground phytomass in the non-saline depression, on the ridge and on the inland plateau respectively, while they contribute about 30,20, and 19% to the amount of consumable forage in the tree habitats respectively. On the other hand, The accessible dry matter production of forbs represents about 4,6 and 3% of total above-ground phytomass in the non-saline depression, on the ridge and on the inland plateau, while they contribute about 31, 34 and 27% to the amount of consumable forage in these habitats respectively. The accessible dry matter production of shrubs is about 491, 253, and 879 kg/ha/yr in the three habitats respectively (average of 554 kg/ha/yr, equivalent to about 3.7 kg/ha/yr/mm rainfall). The accessible dry matter production of the large shrubs represents about 5,7, and 6% of total above-ground phytomass in the non-saline depression, on the ridge and on the inland plateau respectively, although they contribute about 10, 39 and 11% to the amount of consumed forage in the three habitats respectively (the higher percent of contribution of shrubs to the diet on the ridge is due to the presence of the highly palatable species (*Anabasis oropediorum*). This indicates that the shrubs form the skeletal structure of the community, provide the needs for the long term conservation, and share in the annual cycling of material through the food chain. The above mentioned results indicate that a relatively high annual primary production is provided by the rangeland under study (about 3016 kg/ha, equivalent to 20.0 kg/ha/yr/mm rainfall), most of which is used to build up the ligneous structural component of the plant community. Only about 18.4% of the total above–ground annual production is accessible to the grazing animals. Heneidy (1995) reported that the accessible biomass in the different habitats in the east of Matrouh (rocky plateau as about 436, flat plateau is 292, rocky ridge is 620, non-saline depression is 776, and saline depression is 393 kg/ha.). Table 4 summarizes the biomass and accessible parts in the different habitats during the two seasons. The highest total above-ground phytomass of the vegetation is produced in the rocky ridge during the wet season (1897 ± 843 kg ha^{-1}) where, the contribution of perennials is 86% while the lowest is in the saline depression (675 ± 22.9 kg ha^{-1}) where the contribution of perennials is 91%. However, the highest accessible parts are produced in the non-saline depression 864 ± 35 and 400 ± 64.3 kg ha^{-1} during wet and dry seasons respectively. Table 5 showed the RUE for primary production level as an average and accessible production of different habitats.

Habitat		Rocky plateau	Flat plateau	Rocky ridge	Non-saline depression	Saline depression
T.A.G	Spring	1053 ± 110	1347 ±176	1897 ± 843	1427 ± 68	675 ± 22.9
	Summer	637 ± 92.8	733 ± 66	796 ± 186	881 ±126	557 ± 53
Accessible biomass	Spring	510 ± 41	708 ± 69	744 ± 188	864 ± 35	356 ± 23.6
	Summer	199 ± 22.6	218 ± 22	231 ± 34	400 ± 64.3	208 ± 26.2

Table 4. Total above-ground (TAG) biomass and accessible biomass (mean ± standard error (SE) kg ha^{-1}) in different habitats during the two seasons.

Habitat	Rocky plateau	Flat plateau	Rocky ridge	Non-saline depression	Saline depression
NPP (kg ha^{-1}yr^{-1})	412.7 ± 78.5	668.9 ± 185.2	1083 ± 674	547.4 ± 73.4	169 ± 5.30
Rainfall (mm)	77.8 ± 13.7	59.9 ± 13	75.2 ± 11	74.7 ± 14.5	107.3 ± 9.2
Accessible (kg ha^{-1}yr^{-1})	309.8 ± 32.2	488 ± 68.2	499 ± 165	471 ± 69.7	149.5 ± 1.8
RUE to primary production	5.3	11.2	14.4	7.3	1.6
RUE to accessible	3.9	7.5	6.6	6.3	1.4

Table 5. The net primary production (NPP), accessible production (mean ± SE as kg ha^{-1} yr^{-1} mm^{-1}) and rain use efficiency (RUE) of different habitats.

The primary productivity of the pasture in different habitats ranges from 169 ± 5.3 to 1083 ± 674 kg ha^{-1} yr^{-1} in habitats of the saline depression and rocky ridges respectively. The accessible production level reached its maximum in the habitats of rocky ridge (449 ± 165 kg ha^{-1} yr^{-1}) while the minimum is attained in the saline depression (149.5 ± 1.8 kg ha^{-1} yr^{-1}).

Measurement of plant biomass or productivity has been of interest to range workers and ecologists for some times because herbivores depend directory upon plant biomass for their food (Milner and Hughes, 1970). On the other hand, any ecological argument in land use planning in arid rangelands should be based on a through knowledge of the harvestable primary productivity. Study of the vegetation by Heneidy, (2002a) indicated that it consists of 39 perennial species and 43 annuals. The grazeable (mostly new-growth) phytomass is either accessible or non-accessible to the grazing animals due to morphological configuration of the plant. A portion of the accessible phytomass is actually grazed and another portion is left over (Heneidy, 1992). The study also reveals that the study area is vegetitionaly rich and the woody species are the most abundant life-forms. The woody species are considered the skeletal part of the grazing system in arid ecosystem (Abdel – Razik et al., 1988a). The grazing system represents 60 to 80% of the utilization land of North Africa area in terms of economic output (Le Houèrou, 1993). In the study area most of the land is used as grazing land.

Most perennial species exhibit their greatest vegetative activity during winter and spring, and they are less active or dormant during summer. This observation agrees with that noticed by Abdel-Razik et al. (1988a) at Omayed in the northwestern region. However, some shrubs and sub-shrubs are active throughout the whole year. These species are more conservative in the use of their own resources, especially soil moisture and have developed a root system that is capable of exploiting soil moisture and minerals from a large volume of soil and at depth that is permanently wet, which in-turn enables them to extend their activities under conditions of moisture stress Ayyad et al. (1983). This behavior of plant species occurs in some species in the study area. This type of species plays an important role in the sustainable production of the natural forage of the pasture.

(Noy-Meir, 1973) suggested that the bulk of primary standing biomass of the community in semi-arid regions is made up of woody life-forms. This means that accessible parts do not depend upon the primary above-ground phytomass, but depend upon the configuration and morphological shape of the species and their life-forms (Heneidy, 1992). Fritzi and Bradley, (1992) recorded that the production level of herbivores may depend more upon plant architecture than on the particular species of natural enemies present.

The annual average of the primary production in the western desert was 590 ± 117 kg ha^{-1} yr^{-1}, while the accessible production was 410 ± 39 kg ha^{-1} yr^{-1}, compared with that of the woody steeps in arid zones which ranges from 300 to 600 kg ha^{-1} yr^{-1} (Le Houérou, 1972). This value is less than that obtained by Abdel-Razik et al. (1988a) and Heneidy (1992) at Omayed in the coastal region (668 and 720 kg ha^{-1} respectively). However, the average of annual forage yield in the saline depression habitat in coastal region was 1560 kg/ha (Heneidy and Bidak 1996) which is three times higher .

The RUE factor is the quotient of annual primary production by annual rainfall. RUE tends to decrease when aridity increases together with the rate of useful rains, and as potential evapotranspiration increases. But it also strongly depends on soil condition and, more than anything, on vegetation condition particularly on its dynamic status. It thus greatly relies on human and animal impact on the ecosystems. The RUE is a good indicator of ecosystem productivity (Le Houérou, 1984).

The average of accessible dry matter production per mm rainfall ranges from 7.5 to 1.4 kg ha^{-1} yr^{-1} in the habitats of flat plateau and the saline depression respectively. The average Rain Use Efficiency (RUE) was 5.1 kg ha^{-1} yr^{-1} mm^{-1} for accessible production while for primary production was 10 kg ha^{-1} yr^{-1} mm^{-1}. In Comparison, the average of the grazeable dry matter production per mm rainfall at Omayed area is 4.8 kg ha^{-1} yr^{-1} (Abdel-Razik et al., 1988a) while that average of RUE was 10.4 kg ha^{-1} yr^{-1} mm^{-1} at salt marshes in the coastal region (Heneidy and Bidak 1996). Actual RUE figures throughout the arid zones of the world may vary from less than 0.5 in depleted subdeseric ecosystems to over 10 in highly productive and well managed stepped (Le Houérou, 1984).

Generally, the great variations in the productivity levels in different sites are due to variations in soil, climate, vegetation types, and grazing pressure. Consequently Coefficient of Variation (CV) was taken as a measure of this relative variation in production responsiveness at different sites. Calculated CV of the primary production (1.2) indicates that there is a great variation between sites. This variation does not depend upon rainfall and only may hide other factors as mention above (e.g. life-forms, soil condition, topography and human impact ..etc). This result agrees with results obtained by Le Houérou (1988) where he assessed that variability in primary production does not depend only on rainfall, but also on ecosystem dynamic, soil surface condition and texture. The following Table summarizes the average of the primary, production and accessible production ± SE on one hand, and RUE, carrying capacity (CC), Coefficient of Variation (CV), Production to Rain Variability Ratio (P/RVR) on the other hand.

Item	Average value	Rain Use Efficiency (RUE)	P/RVR	CV	Carrying capacity ha/ head	
					Range	Mean ± S.E
Primary production (kgd.wt.ha-1 yr-1) ± S.E	590 ± 117	8.7	2.4	1.2	0.5 – 4.4	1.7 ± 0.17
Accessible production (kgd.wt.ha-1 yr-1) ± S.E	388 ±35	5.1	1.1	0.54		
The rainfall (mm) ± S.E	75.4 ± 6.7	--	--	0.5		

Carrying capacity ranged from 0.5 to 4.4 ha head-1. The variability of annual production was 1.2 times that of the variability of annual precipitation. Production to Rain Variability Ratio (P/RVR) averages 2.4 world-wide in primary production than in accessible. Conversely, P/RVR increases when rainfall decreases and with ecosystem degradation. Finally natural vegetation is affected by rainfall variability in its composition, structures, morphology, ecophysiological adaptation and physiological processes. RUE decreases with rainfall and with the depletion status of the ecosystem (biomass, permanent ground cover, organic matter, microbial activity), (Le Houérou, 1988).

The relative highly annual production of the range land under study as compared to the average production of North Africa which varies between 200 ton ha-1 yr-1 (Le Houérou, 1975 and Sarson & Salmon, 1977) may be attributed to the fact that its vegetation is composed mainly of perennials that can exploit moisture substantial at deep layers of the soil. It is also represents a good production level if compared with the production level at Omayed area 80 km west of Alexandria (Heneidy, 1992). However, this area needs more studies and a good plan for improvement as rangeland within the carrying capacity of the ecosystem, for a sustainable development.

5.2 Biomass and production in the three main habitats

Heneidy (1992) reported that the annual above-ground dry matter production of the rangeland at Omayed area in different habitats was about 3833 kg/ha in the non-saline depression, 1448 kg/ha in the ridge, and 4416 kg/ha on the inland plateau habitats. In general preliminary field observations on the behaviour of grazing animals indicated that almost all the consumed forage is contributed by sixteen perennials (Table 3) throughout the year. In general the total phytomass of new growth is highest in the habitat of the inland plateau, and lowest in the ridge habitat are summarized in the following Table:

Production kg/ha/yr.		
Habitat		
None-saline depression	Ridge	Inland Plateau
3833	1448	4416
Accessible		
----	308	1543

5.3 Necromass

Generally, pasture production depends, among various factors, on the type and intensity of management, e.g. grazing pattern and stocking rate. Such practices can significantly influence species, life-form and growth-form of plants. Herbivory can also affect the structure and functioning of the community. Therefore, a relatively large proportion of annual productivity is either incorporated in ligneous parts of standing dead material near the end of growth period, since most of species, except the evergreen shrubs, are dormant during the dry summer season. Besides, intense defoliation during the growing season may conserve sufficient soil moisture to allow for adequate growth during the late season even during years with below average precipitation. Therefore, it may be concluded that the accessible forage in the area can support at least three fold the present stocking rate, provided that the supplementary feed be also tripled. In general, the contribution of shrubs to the consumed diet on the ridge habitat is higher than that in the non-saline depression and the inland plateau habitats. Conversely, the contribution of annuals in the non-saline depression is much higher compared to those in the other habitats. In detail the average necromass (Kg/ha) exhibited notable variations with habitat and season (Table 6). For example in the non-saline depression it varied between 46.2 kg/ha in summer and 262.2 kg/ha in autumn, on the ridge it varied between 14.4 kg/ha in spring and 159 kg/ha in autumn, and on the inland plateau it varied between 20.7 kg/ha in soring and 156.6 kg/ha in summer. It may be noted that most of the litter was contributed by shrubs and subshrubs. The annual rate of litter production is about 465, 174 and 177 kg/ha in the non-saline depression on the the ridge and the inland plateau respectively. Thus the average litter loss in the three habitats was about 333 kg/ha/yr, which represents less than 11% of annual production of the area. This may confirm the notion that substantial proportion of production of dry matter is invested in the ligneous structural component of the community.

Season	Autumn			Winter			Spring			Summer		
Habitat	I	II	III	I	II	III	I	II	III	I	II	III
Litter	262	159	131.5	87.8	39.5	92.4	50.3	14.4	20.7	46.2	28.1	156.6

Table 6. Seasonal variations in the necromass (Kg dry wt./ha) in different habitats, I = Non-saline depression, II= Ridge, and III = Inland plateau.

5.4 Ecological stress on the grazing ecosystem

The desert ecosystem is exposed to different types of stress as the following:

1- Overgrazing, 2- Woodcutting, 3- Aridity, 4 – Salinity, and different types of human activities, and 5- Erosion of soil surface (e.g. Plates 2A, B, C and D). Rangelands constitute approximately 90% of the country's land surface area. The products are renewable; thus the ranges are capable of providing continuous goods and services such as forage, fiber, meat, water and areas for recreation. These resources are considered by many people to be an integral part of their traditional Arab heritage, which adds special importance to their value. Concomitantly, rangelands are now in a poor condition due to pressures that have either altered or destroyed them as a result of overgrazing, uprooting of plants and off-route use by vehicles. These factors have resulted in an almost complete removal of vegetation cover, a speeding up of the desertification process and the destruction of wildlife habitats.

Vegetation and land degradation is still widespread in the Mediterranean region. Degradation results from various kinds of mismanagement of the land. This include the introduction or expansion of agro forestry systems with multiple-use of the land to develop tourism, wildlife, hunting and sports, combined with extensive grazing of livestock and game and timber production from elite clones of selected high yielding or highly valued species.

5.5 Causes of land degradation

These causes are essentially linked, either directly or indirectly, to soil denudation in areas, which, because of their topography, slope, geological substratum and soils are erosion-prone. They also may be subjected to sedimentation, flooding, water logging, salinisation and alkalization for reasons of the same nature.

Soil surface denudation, in turn, results from a number of causes that may be natural or man-induced.

a.   Natural causes play a minor role in Mediterranean region.
b.   Man-induced causes play by far a major role. These are the following and may act in isolation or in combination of two or more causes:

1- Deforestation due to over-exploitation either for timber, firewood, charcoal, often combined with over-browsing; 2- Long standing overstocking and overgrazing; 3- Forest and shrub land wildfires, combined or not with over-browsing; 4- Inappropriate grazing systems and patterns: heavy stocking combined with continuous grazing; 5- Clearing for cultivation of land that is inappropriate to cropping without adequate precautions of water and soil conservation; 6- Inadequate tillage practices (down slope, steep ground, utilization of disc plough in sandy soils etc.); 7- Unsound cultivation practices and crop distribution patterns; 8- Chemical exhaustion of soil nutrients involving lack of fertility, inappropriate crop rotation practices favouring the leaching of soil nutrients, or export of nutrients consecutive to wild fires. The coastal belt of the western desert of Egypt has some potential of natural productivity and is the site of a variety of landuse and development programme. The main activities include grazing, agriculture, and woodcutting and over collection especially Matruh to Salloum. Woody plants are, in many instances, the only source of fuel for the desert inhabitants. Batanouny (1999) reported that along the coastal line from Alexandria to Alamein, the sand dunes represent a landscape with special characteristics and features. For more than two decades, due to the conspicuous socio-economic changes, privation, open-door policy in economy and other political changes in Egypt, a great part of the coastlines, has been destroyed. This is due to the continuous construction of summer resort villages. The consequences of the human activities in the area are numerous. These include impacts on the soil, water resources, the flora and fauna, migration birds, trends of the indigenous people, and the cultural environment.

Plate 2A. Grazing of diffent livestock (WMCD) (After Bidak et al., 2005).

Plate 2B. Habitat lost along the western Mediterranean coast (remnants of Abu-sir ridge) (After Bidak et al., 2005).

Plate 2C. Wood collecting as a fuel (After Bidak et al. , 2005).

Plate 2D. Wind erosion (After Bidak et al. , 2005).

Human activity over thousands of years on the Mediterranean landscape has resulted in major management problems. This long period of intensive grazing, fire cycles, and cutting caused degraded or drastically changes in the natural vegetation of the area (Cody, 1986) and have created a mosaic ecosystems which represent degradation stage (Di Castri, 1981).

Bedouins in Maruit region depend on fuel wood collection from the natural vegetation as an essential source of energy. The daily fuel wood collected by a household was about 29 kg. The quantity used was about 24 kg/ha, which equal 8.8 ton/year. The excess represents a waste that should be controlled (Heneidy & El-Darrir 1995). All these activities have their effects on the ecological balance of the ecosystem especially if they are carried out at a rate higher than the rate of regeneration of vegetation cover. Heneidy et al. (2002c) recorded, in Omayed Biosphere Reserve that the human activities have more impact on the vegetation than the over grazing by livestock, where various activities (e.g. wood cutting and collection) have different impacts on the plant diversity.

The causes of degradation of the vegetation in the western desert are mixes of environmental, socio-political and socio-economic conditions. Three main lines may be suggested for the initiation of a long-term strategy for the restoration and conservation of degraded vegetation:

a. Establishment of pilot areas for protection and controlled grazing in each of the main habitats and communities.
b. Initiation of a cooperative system for grazing management between the main social sectors (tribes).
c. Formation of an extensive programme me for propagation of endangered species, it is necessary that the decisionmakers and land-users participate in the planning and execution of the activities along these three lines, and that extension services and incentives be ensured in order to encourage their participation. Ayyad (1993) initiated a test programme me for propagation of multipurpose species in the Mediterranean desert and put the main items for executive (1- seed bank, 2- establishment of seed banks, 3- establishment of nurseries, and 4- demonstration of field experiments).

5.6 Plants and their defenses

The role of physical defensive factors by range specie against livestock has been interest to many range management researchers. In natural ecosystem, plants are associated with a great number of potential predators and pathogens. Nearly all ecosystems contain a wide variety of bacteria, fungi, nematodes, mites, insects, mammals and other herbivores. By their nature, plants cannot avoid these enemies simply by moving away, but they protect themselves in other ways. The cuticle, periderm, thorns, stinging hairs, and tough, leathery leaves help deter herbivores feeding (Taiz & Zeiger, 1991). Heneidy and Bidak (1998) studied the type of defenses in the range species in the coastal Mediterranean region of Egypt. The vegetation of the region is composed of rangeland species of different life-forms . However, some of these plants may avoid or resist grazing animals by different mode of defenses (spiny leaves or branches, compact woody structure, taste, and fragrant smell.

This study depends upon recorded notes through 12 years including daily field observation of different flocks (e.g. grazing behaviours, walking, grazing time, resting time, drinking, number of bites from each species and preference) (Abdel- Razik, et al. 1988a Heneidy, 1992). Heneidy and Bidak (1999) reported that some of the rangelands species have ability for adaptation that avoid grazing activity and how these defenses affect on the palatability, aversion, and consumption.

Fifty-two plant species(44 species perennials and 8 annuals) belonging to 19 families have different defensive mechanisms were recorded in the study area. The type ,degree and the percentage of defense, life-forms, and abundance of each species are presented in Table (7).

Species	Type of defense	Defense%	Family	Growth Form	Abundance
1. Spiny organs	Spiny branches	100	Leguminosae	P. herb	Commn
Alhagi graecorum					
Anacyclus alexandrinus	Weak spiny plant	80	Compositae	Annual	Common
Asparagus stipularis	Spiny plant	80	Liliaceae	Sub-shrub	Few
Astragllus sieberi	Spiny branches	100	Leguminosae	Shrub	Few
Astragllus spinosus	Spiny plant	80	Leguminosae	Shrub	Few
Atractylis carduus	Spiny plant	100	Compositae	P. herb	Commn
Carduncellus eriocephalus	Spiny plant	80	Compositae	P. herb	Commn
Carthamus lanatus	Spiny plant	80	Compositae	P. herb	Commn
Centaurea alexandrina	Spiny plant	100	Compositae	P. herb	Commn
Centaurea glomerata	Woolly, spiny	60	Compositae	Annual	Commn
Centaurea pumilio	Woolly, spiny	50	Compositae	P. herb	Rare
Echinops spinosissimus	Spiny fruits	80	Compositae	P. herb	Commn
Emex spinosa	Spiny stipules	25	Polygenaceae	Annual	Few
Fagonia arabica	Spiny plant	100	Zygophyllaceae	Sub-shrub	Commn
Juncus rigidus	Spiny branches	100	Jancaceae	Geophyte	Commn
Noaea mucronata	Spiny plant	100	Chenopodiaceae	Sub-shrub	Commn
Onopordum alexanrinum	Spiny branches	80	Compositae	P. herb	Rare
Salsola vermiculata	Spiny plant	70	Chenopodiaceae	Sub-shrub	Commn
Traganum nudatum	Spiny plant	100	Chenopodiaceae	P. herb	Commn
Xanthimum spinosum	Spiny fruits	50	Compositae	Annual	Few
Zilla spinosa	Spiny plant	100	Cruciferae	Shrub	Commn
2. Woody and spine -like					
Convoloulus lanatus	W. spine branches	60	Convolvulaceae	Sub-shrub	Commn
Kickxia aegyptiaca	W. pine branches	70	Scrophulariaceae	Sub-shrub	Few
Lycium shawii	W. spine branches	80	Solanaceae	Tall shrub	Commn
Moricandia nitens	W. spine branches	70	Cruciferae	Sub-shrb	Few
Periploca aphylla	W. spine branches	100	Asclepiadaceae	Tall shrub	Few
3. Lathery leaves					
Adonus dentatus	Odour	100	Compositae	Annual	Common
Artemisia herba-alba	Woolly, odour	100	Compositae	Sub-shrub	Common
Chenopodium murale	Odour	100	Chenopodiaceae	Annual	Common
Cleome arabica	Sticky, odour	100	Cleomaceae	Sub-shrub	Rare
Haploophyllum tuberculatum	Sticky, odour	100	Rutaceae	P. herb	Few
Peganum harmala	Sticky, odour	100	Zygophyllaceae	P. herb	Common
Pulicaria incisa	Hairy, odour	100	Compositae	P.herb	Common
Thymus capitatus	Odour	100	Compositae	Sub-shrub	Common
Varthemia candicans	Odour	100	Compositae	Sub-shrub	Common

Species	Type of defense	Defense%	Family	Growth Form	Abundance
4. Leathery and hairy					
Alkana lehmanii	Stiff hairs	100	Boraginaceae	Sub-shrub	Few
Artemisia monosperma	Woolly hairy	100	Compositae		Common
Echichilon fruticosum	Woody, hairy	100	Boraginaceae	Sub-shrub	Common
Echium sericeum	Stiff hairy	100	Boraginaceae	Sub-shrub	Common
Marrubium vulgare	Woolly	100	Labiatae	P. herb	Few
Moltkiopsis ciliata	Stiff hairs	100	Boraginaceae	P. herb	Few
Phlomis floccosa	Stiff woolly	100	Labiatae	P. herb	Few
Plantago albicans	Woolly leaves	80	Plantagonaceae	P. herb	Common
Verbascum letourneuxii	Hairy acute branches	80	Scrophylariaceae	Bi-herb	Few
5. Leathery and latex					
Euphorbia paralias	Latex	100	Euphorbiaceae	P. herb	Common
Citrullus colocynthis	Leathery leaves	100	Cucurbitaceae	P. herb	Common
Hyoscyamus muticus	Sticky odour	100	Solanaceae	P. herb	Common
6. Weaky plants defense by protection					
Didesmus aegyptius	Hinding	80	Cruciferae	Annual	Comon
Euphorbia hierosolymitana	Hinding	80	Euphorbiaceae	Sub-shrub	Rare
Launaea nudicaules	Hinding	80	Compositae	P. herb	Common
Prasium majus	Hinding	100	Labiatae	Sub-shrub	Rare
Rumex dentatus	Hinding	80	Polygonaceae	Annual	Rare

Table 7. Types of defense, defense percentage, family, growth-form and abundance of the some modified range species in the western coastal region (After Heneidy and Bidak, 1998), W = Woody and P = Perennual.

Aversion factor (AF) was estimated ratio of the relationships between palatability and the degree of protection. The degree of protection acquired by the studied species distinguished into three classes: 100%, 80% and less than 80%. Each of these classes includes three of palatability rates (HP, P, and LP or NP).Table (8) shows that 19 species (31%) are of AF =zero, 53% of which is due to protection and 11% is due to palatability while, 36% is due to both. 46% of these species are AF = 1 where, 58% is due to protection, 25% is due to palatability, and 17% is due to both.

Species	Palatability	Consumed part	Stock	Aversion factor
Alhagi graecorum	Hp	Leaves, branches	CG	Zero*
Anacyclus alexandrinus	P	Leaves, inflorescence	GS	0.5***
Asparagus stipularis	P	Young branches	CG	0.5*
Astragllus sieberi	HP	Leaves, young branches	CG	Zero***
Astragllus spinosus	HP	Leaves, young branches	CG	Zero*
Atractylis carduus	P	Leaves, inflorescence	SCG	0.5***
Carduncellus eriocephalus	HP	Leaves, inflorescence	SG	Zero***
Carthamus lanatus	HP	Leaves, inflorescence	CGS	Zero*

Species	Palatability	Consumed part	Stock	Aversion factor
Centaurea alexandrina	P	Leaves, inflorescence	CGS	Zero**
Centaurea glomerata	P	Leaves, inflorescence	GS	1**
Centaurea pumilio	LP	Young branches	GS	1**
Echinops spinosissimus	HP	Young branches, inflorescence	CG	Zero*
Emex spinosa	LP	Above-ground	GS	1**
Fagonia arabica	NP	--	--	1*
Juncus rigidus	HP	Spiny branches	C	Zero*
Noaea mucronata	HP	Leaves, branches	SGC	Zero***
Onopordum alexanrinum	P	Leaves, inflorescence	CG	0.5***
Salsola vermiculata	P	Leaves, branches	SGC	0.5*
Traganum nudatum	LP	Young branches	GC	1*
Xanthimum spinosum	NP	--	--	1**
Zilla spinosa	P	Young branches	CG	0.5*
Convoloulus lanatus	HP	Leaves, young branches	SGC	Zero**
Kickxia aegyptiaca	P	Leaves, branches	SGC	0.5**
Lycium shawii	HP	Leaves, young branches	SGC	Zero***
Moricandia nitens	LP	Leaves, young branches	SG	1**
Periploca aphylla	HP	Leaves, branches	SCG	Zero*
Adonus dentatus	LP	inflorescence	GS	1**
Artemisia herba-alba	VHP	Leaves, young branches	SG	Zero*
Chenopodium murale	NP	--	--	1**
Cleome arabica	NP	--	--	1**
Haploophyllum tuberculatum	LP	Dead parts	GS	1*
Peganum harmala	NP	Dead parts	SG	1*
Pulicaria incisa	LP	Leaves, inflorescence	GSC	1*
Thymus capitatus	LP	Leaves, branches	SGC	1*
Varthemia candicans	NP	--	--	1*
Alkana lehmanii	P	Leaves, branches	SG	0.5*
Artemisia monosperma	P	Leaves, young branches	SGC	Zero*
Echichilon fruticosum	HP	Leaves, young branches	SG	Zero*
Echium sericeum	LP	Leaves, young branches	GS	1*
Marrubium vulgare	LP	Leaves, branches	SG	1*
Moltkiopsis ciliata	P	Leaves, branches	SG	0.5*
Phlomis floccosa	LP	leaves	GS	1*
Plantago albicans	HP	All above-ground	SG	Zero*
Verbascum letourneuxii	LP	Leaves, young branches	SG	1***
Euphorbia paralias	NP	--	--	1*
Citrullus colocynthis	NP	--	--	1*
Hyoscyamus muticus	NP	--	--	1*
Didesmus aegyptius	LP	All plant	SG	1***
Euphorbia hierosolymitana	NP	--	--	1***
Launaea nudicaules	HP	All above-ground	SG	Zero***
Prasium majus	P	Leaves, branches	SG	Zero***
Rumex dentatus	HP	All above-ground	SG	Zero***

Table 8. Palatability rate, consumed parts, stock and aversion factor of defensed plant species. S = Sheep, G.= Geots, C. = Camels, HP= High palatable LP.= Low palatable and NP.= Unpalatable.

5.7 Browse in grazing lands and types of livestock in Egypt

The role of browse in natural grazing lands varies in importance according to ecological zone. Under temperate climates and in the humid tropics, browse is of limited use in animal production. However, in the Mediterranean isoclimatic zone, the arid and semi-arid tropics and in montane areas, browse plays an essential role in the animal production, and thus greatly contributes to the protein supply of mankind in Africa. For instance, over 250 million heads of domestic animals live in arid, semi- arid and montane zones where browse is an important quantitative component in livestock diets. In the arid zone the main browse species are dwarf shrubs which constitute the bulk of animal feed (Le Houérou, 1980a).

Rangelands in Egypt are dominated by shrubs; most of those are browsed. These "browselands" cover the following surface in 10^3 km^2 (Table 9). In Egypt the main components of rangelands are shrubs, and dwarf shrubs (subshrubs) Heneidy (1992). The area of rangelands in Egypt is about 125.000 km^2. Table 10 shows the livestock populations (10^3).

Country	Total	Semi-arid to humid rangelands zone shrublands (R > 400 mm)	Arid rangelands chamaephytic steppe (400 >R> 100)	Desert rangelands chamaephytic steppe (100 >R> 50)
Egypt	125	0.0	25.0	100.0

Table 9. Egypt browse lands in 10^3 km^2, (After Le Houérou, 1980b).

Country	Cattle	Equines	Sheep	Goats	Camels	Total
Egypt	2150	1600	1940	1400	100	7190

Table 10. Livestock populations (Egypt) in 10^3 (After Le Houérou, 1980b).

The total area of the rangeland in the western Mediterranean region is about 1.5 million hectare. The number of animals per folks varies among the Bedouin, and between regions in the coastal zone. A herdsize of about 200 head of sheep and 60 head of goats may be reaches to 300 heads (Heneidy 1992). The total number of animals which graze the rangelands is about 1.5 million head where, in Omayed biosphere reserve is about 15,000 head (ratio 3 sheep to 1 goat) Heneidy (1992). Therefore the average stocking rate is one head /ha. Duivebooden (1985) reported that the total number of animals is about 1.2 million head of sheep, 0.9 million head of goats, and 0.2 million head of camels. The stocking rate of Omayed was estimated as about 0.5-0.6 head/ha (Ayyad & Ghabbour, 1977; Ayyad & El-Kady 1982), which is much lower compared to the average for the whole area. The average rate averaged throughout the year at Burg El-Arab varies from 0.75 head/ha to 1.0 or 1.5 head/ha (El-Kady, 1983 and Abdel-Razik et al., 1988a). Further more, the calculated stocking rate in the Omayed by Heneidy (1992) is 0.286 head/ha, is far below the regional average practices. It should be mentioned, however, that the calculated stocking rate holds true only for those animals grazing yearlong on the pasture, while twice as much as the resident animals is removed to the areas of irrigated cultivation for most of the year. Consequently, spatial variation in stocking rate occurs. In addation, variation in time occurs, due to the migration of livestock in summer. Stocking rate may be reaches to 1.2 head/ha at

Omayed (Van Duivenbooden, 1985). There is no consistent range management strategy in control of the season long grazing in the area, which varies mainly with climatic condition, availability of watering points and availability of supplementary feed. Absolutely the stocking rate leads to the problem of overgrazing. Overgrazing has been one of the main factors causing the deterioration of ecosystem productivity in the Mediterranean coastal region. It has resulted in severe reduction of perennial cover, soil erosion and formation of mobile dunes.

5.7.1 Grazing pressure

The pressure of grazing at any given moment is defined as a relationship between demand for by animals and a comination of daily herbage increment and standing crop of vegetation. This function is related only indirectly to number of animals and area of pasture (Heady ,1975). El-Kady (1980) recorded that partial protection and controlled grazing, might be of better consequences than full protection.

5.7.2 Stocking rate

The total area of the rangeland in the western Mediterranean region is about 1.5 million hectare. Therefore the average stocking rate is one head /ha. The stocking rate of Omayed was estimated as about 0.5-0.6 head/ha (Ayyad & Ghabbour, 1979; Ayyad & El-Kady 1982), which is much lower compared to the average for the whole area. The average rate averaged throughout the year at Burg El-Arab varies from 0.75 head/ha to 1.0 or 1.5 head/ha (El-Kady, 1983 and Abdel-Razik et al., 1988a). Further more, the calculated stocking rate in the Omayed by Heneidy (1992) is 0.286 head/ha, is far below the regional average practices. It should be mentioned, however, that the calculated stocking rate holds true only for those animals grazing yearlong on the pasture, while twice as much as the resident animals is removed to the areas of irrigated cultivation for most of the year. There is no consistent range management strategy in control of the season long grazing in the area, which varies mainly with climatic condition, availability of watering points and supplementary feed. Absolutely the stocking rate leads to the problem of overgrazing. Overgrazing has been one of the main factors causing the deterioration of ecosystem productivity in the Mediterranean coastal region. It has resulted in severe reduction of perennial cover, soil erosion and formation of mobile dunes.

5.7.3 Forage consumption

Heneidy (1992) reported that the annual consumption of forage by the grazing animals is about 619 kg/ head (average of 1.696 kg/ head/ day). If this value of annual consumption is multiplied by the calculated stocking rate of grazing animals, the consumption rate would be about 0.485 kg/ha/day which is well below the total accessible forage production of different habitats in Omayed. Normally, browse (grazing on shrubs) is supplementary to a herbage based system. It may serve to raise the protein intake for all or part of the year, or it may be essential that animals subsist on browse during cold or dry periods, so that they survive to utilize and breed on the herbage at other times of the year. This indicates the importance of browse species in the area. Taking into consideration that subshrubs constitute the bulk of feed of grazing animals. Such shrubs are therefore valuable in tiding over the periods of limited forage supply. Combining this with the high production levels of woody species, may reflect the importance of such species in arid rangelands. It should be added that the relatively high production levels of some of the perennial herbs ensure a good supply of standing dead material during the dry period that is selected by grazing animals with even higher priority than the available green. The consumption rate of dry matter in the three habitats of the study area (non-saline depression, the ridge, and the inland plateau) is about 0.53, 0.36, and 0.47 kg/ha/day respectively.

The total phytomass accessible to grazing animals (gm dy weight/m^2) varied remarkably with season and habitat, from 30.8 gm/m^2 in the ridge habitat in autumn to 154.3 gm/m^2 in the habitat of inland plateau in spring. The total livestock offtake is estimated to be on the order of 6% of primary production in the study area. Therefore, it seems unlikely that livestock exert a major control on plant biomass.

The monthly variations in the species composition of the daily diet of animals in different habitats are presented in Table 11 (After Heneidy 1992).

Species	Annual mean of daily consumption		
	Non-saline depression	Ridge	Inland plateau
Asphodelus ramosus	694.32	167.27	574.1
Plantago albicans	84.19	--	171.65
Carduncellus eriocephalus	17.58	--	--
Echinops spinosissimus	--	5.73	3.64
Scorzonera alexandrina	--	273.69	62.68
Echiochilon fruticosum	161.21	--	206.15
Helianthemum lippii	111.07	80.08	113.39
Gymnocarpos decandrum	120.66	87.24	163.33
Deverra triradiana	12.34	30.31	20.36
Convolvulus lanatus	26.26		23.86
Noaea mucronata	78.56	71.87	89.7
Artemisia monosperma	200.11	--	--
Artemisia herba alba	--	39.25	18.52
Anabasis articulata	86.03	169.26	38.61
Anabasis oropediorm	--	203.66	66.73
Thymelaea hirsuta	98.35	90.9	63.12
Lycium europaeum	8.1	16.81	15.45
Annuals	168.86	11.44	11.75

Table 11. Variations in the daily consumption rate (gm dry wt/head/day) in the different habitats at Omayed.

Seasonality in consumption behavior of grazing animals in the Omayed site is mainly a function of: (1) variation in the availability of consumable parts of species and their relative palatability, (2) the phenological stages of plants having different nutritional status and characterized their specific chemical composition in energy and mineral elements, (3) the relative need for drinking water, (4) climatic factors, and (5) structure and size of the flock.

The interplay between the previously mentioned three factors let the grazing animal spend periods of voluntary walk over distances that vary between 20 km/day during winter and 9 km/day during summer. These walking experiences represent a transect of about 15 km in length in winter and 7 km in length during summer (Heneidy, 1992). The corresponding grazing time averages about 9 hours/day during winter and 7 hours/day during summer(at the rate of 2.22 and 1.29 km/hour during winter and summer respectively)

Ayyad, (1978) reported that in the Mediterranean desert ecosystem variable input of rain would be matched, under relatively light usage, by the supply of forage, but heavy stocking might so damage the grazing that there would be an enfprced decrease in cattle numbers when the rains failed. If the damage were severe, neither forage no cattle numbers could recover and there would be a sustained fall in yield. Grazing tends to damage the most palatable first and then progressively less palatable species, so reducing their frequency in the community. Not all palatable species, however, need be decreases: some may actually thrive on grazing. In the present study some species exhibited a negative response to protection. These species may have been either of low play abilities and were deprived of light nibbling and removal of standing dead shoot, a practice which usually promotes vigour and growth or they may have been unpalatable and due to competition by the protected palatable species became of lower potential for productivity.

5.7.4 Grazing behavior of animals

Grazing animals do not graze continuously . They have specific periods during the 24 hours daily cycle, when ingestive intake is very high and others when grazing is punctuated by ruminating resting and idling. In the coastal Mediterranean region the grazing times varied between 7 and 9 hours with the increase occurring from summer to winter, and with small differences between animals in their annual grazing time spent on different life forms.)

In general, there are two periods of grazing activity of domestic animals in Omayed area one in the morning and one in the afternoon, with a resting period in between. The peak of grazing activity differs with season and habitat, but it usually occurs either in the early morning or the late afternoon, particularly during summer. In spring and winter, the peak activity may occur towards mid-day.

In general, the grazing activity during the day stats at six in the morning, and ends at six in the afternoon during autumn and spring, while, in summer it starts at eight, and in winter it starts one hour later and ends one hour earlier. The resting period for grazing animals starts at 12 noon, and lasts for two hours during winter and spring, four hours during autumn, and seven hours during summer.

The annual pasture can be divided into two periods: the wet grazing period (winter, spring) and the dry grazing period (summer, autumn). The effective wet grazing period is that period when green pasture availability is sufficient for the animals to satisfy their appetite and meet all nutritional requirements (about 4 or 5 months) depending on the amount and distribution of rainfall. Seligman and Spharim (1987) indicate that, this green period in arid rangelands of the Middle East lasts only 3 months from December till March. In the rainy season most species are in vegetative stage and furnish an adequate supply of forage to the grazing animals. Some species start activity for a few months when the soil moisture is available, after which they go to dormancy or inactivity.

5.7.5 Selectivity of plant species by grazing animals (animal visiting)

The frequency distribution of the number of visits per hour or per day to a plant species may be used as a measure of the non-randomness of its selection by domestic animals Heneidy, 1992). The number of visits per hour to each plant species, irrespective of the number of bites, during one day representing each season was recorded. This number was then classified into classes representing low, medium and high frequency of visits. The χ^2 – test of goodness-of-fit was applied (Table 12). It indicated the probabilities that the difference between the observed and the expected (random) distribution may be attributed to the sampling error.

The goodness –of-fit of the observed ferquency distribution to the expected random distribution of the number of visits per hour during one day representing each season varied with species, season and habitat. The largest number of significant deviations from randomness was that exhibited by *Anabasis articulata*, followed by that of *Deverra triradiana*, *Thymelaea hirsuta* and *Noaea mucronata*. It is notable that the number of these significant deviations was lowest during summer. The largest number of such deviations was recorded for plants of the non-saline depression, taking into consideration the species common to the three habitats.

Habitat	Non-saline		Ridge		Inland plateau	
Species \ Season	Spring	Summer	Spring	Summer	Spring	Summer
Asphodelus ramosus	<0.005	0.25-0.5	0.05-0.1	0.1-0.25	<0.005	0.1-0.25
Plantago albicans	0.25-0.5	0.1-0.25	<0.005	--	0.01-0.025	0.1-0.25
Carduncellus eriocephalus	<0.005	--	--	--	<0.005	--
Echinops spinosissimus	--	--	<0.005	--	--	--
Scorzonera alexandrina	--	--	<0.005	--	--	-
Echiochilon fruticosum	<0.005	0.1-0.025	--	--	0.025-0.5	0.25-0.5
Helianthemum lippii	0.1-0.25	<0.005	<0.005	<0.005	0.1-0.25	0.25-0.5
Gymnocarpos decandrum	<0.005	0.0-0.1	0.25-0.5	0.1-0.25	<0.005	0.05-0.1
Deverra triradiana	<0.005	0.5-0.75	<0.005	0.1-0.25	<0.005	0.025-0.05
Convolvulus lanatus	<0.005	0.5-0.75	--		0.025-0.05	0.01-0.025
Noaea mucronata	<0.005	0.5-0.75	<0.005	0.1-0.25	<0.005	0.01-0.025
Artemisia monosperma	0.1-0.25	0.25-0.5	--		--	--
Artemisia herba alba	--	--	0.005-0.01	0.1-0.25	<0.005	<0.005
Anabasis articulata	<0.005	<0.005	0.01-0.025	0.025-0.05	<0.005	0.01-0.025
Anabasis oropediorm	--	--	<0.005	<0.005	0.05-0.1	0.01-0.025
Thymelaea hirsuta	0.05-0.1	0.01-0.025	<0.005	0.5-0.75	<0.005	0.1-0.25
Lycium europaeum	--	--	<0.005	--	<0.005	--
Annuals	0.005-0.001	--	0.1-0.25	--	<0.005	--

-- These species are not recorded.

Table 12. Probabilities of random distribution of the number of visits/hour by domestic animals to each plant species during one day representing each of the two seasons in the three habitats (After Heneidy, 1992).

Generally the average number of plants visited by each animal was about 1318 per day (Heneidy, 1992) (representing about 12% of the total number of plants if the animal is confined to one hectare). The average numbers of bites was 8235 bite/head/day, with the maximum number during spring.

5.7.6 Water consumption

The temporal variations in the consumption of water by the grazing animals from the rangeland forage in different habitats, as well as the supplementary amounts during the dry period (liter/head/day) are presented in Table (13) assuming that the flock spends the whole grazing day in one habitat.

	October			November			December			January			February			March		
Habitat	I	II	III	I	II	III	I	II	III	I	II	III	I	II	III	I	II	III
Amount of water from forage	0.4	0.2	0.7	0.7	0.3	1.4	4.6	1.9	4.5	4.5	5.1	4.3	4.7	4.5	5.1	8.7	10.6	6.2
Amount of drinking water	3.0-3.5			2.0-3.0			0.5-0.7			0.0			0.0			0.0		

	April			May			June			July			August			September		
Habitat	I	II	III	I	II	III	I	II	III	I	II	III	I	II	III	I	II	III
Amount of water from forage	8.6	1.7	3.8	5.9	0.9	1.5	1.1	0.4	1.2	0.4	0.4	0.8	0.3	0.3	0.5	0.2	0.2	0.6
Amount of drinking water	0.0			0.8-1.5			3.5-4.0			4.5-5.5			5-5.5			4.4-4.8		

Table 13. Monthly variations of the amount of water (liter/head/day) consumed by the grazing animals from the rangeland forage in different habitats, and that of drinking water from October till September at Omayed (After Heneidy, 1992).

It is notable that the species composition and time period, affect the amount of water available for animal from the natural forage. For example, the amount of water of the rangeland forage on the ridge habitat is higher than in the other two habitats, probably due to the abundance of the more or less succulent species *Scorzonera alexandrina* on the ridge where it is highly consumed by grazing animals. In the non-saline depression the amount of water is greater than on the inland plateau, which may also be due to the high abundance of the succulent species *Rumex pictus* in the former habitat.

In the dry grazing period, the subsidies of concentrates, wheat, barley and remains of onion are supplied to the animals, and supplementary drinking water becomes more necessary. Water consumption by grazing animals depends on five main factors: (1) air temperature, (2) amount of dry matter ingested, (3) water content of the feed, (4) salinity of drinking water, and of feed (Le Houéou et al., 1983; Heneidy, 1992). Heneidy (1992) recorded that sheep and goats in the coastal region do not supply of drinking water during winter and spring, since the water content of the vegetation is sufficient in the study area. However, in the other seasons watering of the flock is necessary. The total annual water intake by the grazing animals was calculated as about 1876 kg/head. Drinking water represents 43% of the total water intake, while the remainder was the water content in the consumed fresh forage.

5.7.7 Faeces, necromass and water loss

The monthly variations of fresh and dry weight of faeces (gm/head/day), its moisture content and the animal water loss are presented in Table (14). The maximum amount of faeces is deposited during February (473.4 gm dr. wt./head/day), while the maximum fresh weight is attained in March (about 1103.7 gm fresh wt./head/day).

The highest moisture contents of faeces are attained during the wet grazing period. The lowest amounts of feces are deposited during the dry grazing period, particularly in October. The maximum water loss in the faeces of animals per day occurs during the rainy period, when the natural forage is enough to meet their feed requirements.

Month	Fresh wt./ (gm/day)	Dry wt. (gm/day)	Moisture content%	Amount of water loss (gm/day)
October	350.0	159.3	54.5	190.7
November	548.0	255.7	53.3	292.4
December	774.6	353.9	54.3	420.8
January	1027.0	302.0	70.6	725.0
February	1080.0	473.4	56.2	607.4
March	1102.7	324.7	70.6	779.0
April	981.6	288.7	70.6	692.9
May	487.5	183.5	62.4	303.9
June	539.0	184.7	65.7	354.3
July	556.0	191.5	65.6	364.5
August	446.6	204.8	54.1	241.7
September	365.3	166.4	54.5	198.9

Table 14. Monthly variations in fresh and dry weight, moisture content, and water loss (gm/head/day) of faeces, (October 1987-88), at Omayed (After Heneidy, 1992).

5.8 Palatability and nutritive value

Palatability of range plant species is a very complex notion, very difficult to generalize as it is linked to many factors that vary in time and place. Some of these variables are linked to the plant, others to the animals, while a third category depends on various environmental factors (Le Houérou, 1980; Heneidy 1996). Based on several years of field observation, chemical composition and energy content help us to know the palatability information of the plant species. In the coastal Mediterranean region numerous factors were considered when classifying a plant as palatable or not: phenological stage, morphological form, odour, taste, chemical composition and its abundance relative to associated species. The topography of the habitat, climate and state of animal itself are all contributing factors (Heneidy and Bidak, 1999).

Abdel-Razik et al. (1988a) reported that the grazing domesticated mammals show highest selectivity ratios (preferences) towards plants with relatively low availability. However, the palatability of a few species of plants is related to their phenological development. Woody plants represent the principle source of most nutrients to grazing animals, including, with the exception of phosphorus, more than adequate minerals and are available throughout the year.

Many physical and chemical factors influence palatability of plant species, of which animal behavior and chemical composition are the most important. The contents of chemical constituents in plants and their digestibility vary independendently between different organs and according to plant age (Le Houéou, 1980a and Heneidy, 1992). Of these, crude protein (CP) is viewed classically as an indicator of the nutritional value of food for ruminants.Abdel-Razik et al. (1988 a,b) reported that the analysis of samples which represent the diet selected by the herbivores indicated that the grazeable proportion of the phytomass of most species had in general, higher nutrient concentration than the non-grazeable proportion. Crude protein was a highest concentration early in the growing season, and was lowest at the end of the dry season. The concentration of total nonstructural carbohydrates (TNC) and in the grazeable parts varied also with life-forms and habitats and crude fibers was lower in the wet season in the young parts. The concentration of lipids in the grazeable parts of different species slightly higher in winter and does not exhibit notable differences between habitats (Heneidy, 1992).

Heneidy (1992) reported that the percentage of digestible crude protein(DCP) in Omayed area is 4.9%, while Heneidy and Bidak (1996) reported that it reached 6.9%. DCP content of the forage in the coastal region at Omayed area is ranked good according to the scale suggested by Boudet & Riviere (1980) who consider a fair DCP percentage between 2.5 and 3.4.

Total digestible nutrients (TDN) is reported by Abdel-Salam (1985) in the coastal Mediterranean region as about 66% DM, while Abdel-Razik et al (1988b) reported annual average value of TDN for consumable forage as 75% DM and Heneidy (1992) reported the annual average of TDN in coastal region as about 67% DM. Heneidy and Bidak (1996) reported that TDN in the saline depression habitat in the coastal region as about 66.2%. Accordingly the energy content in the coastal Mediterranean region equals 0.76 SFU of the halophytes species, while it was estimated by El-Kady (1983) at Omayed as 0.3 SFU. In general, the nutritive value of forage decreases with advancing maturity of plants. This decrease is mainly due to the decrease in protein content together a concomitant increase in the fiberous components and general lignification.

6. Acronyms

AF: Aversion factor
CV: Coefficient of Variation
OBR: Omayed Biosphere Reserve
RUE: Rain Use Efficiency
P/RVR: Production to Rain Variability Ratio
WMCD: Western Mediterranean Coastal Desert

7. Recommendations and some suggestions

1. Degradation of vegetation and even the whole ecosystem due to overgrazing is manifest in almost all the localities in the protectorate. The communal system of grazing

is followed in the protectorate, as well as the other arid and semi-arid ecosystem in the region. While the communal grazing system has certain advantages in an arid range area, it can be an obstacle to rational utilization of the range and to range improvement practices. Individuals try to increase the size of their flocks to obtain the highest possible income, and there is no interest to protect the range by limiting the number of animals or to improve the range by any means.

2. There is a strong view that only a holistic approach should be used to address sustainable rangelands management and pastoral development. The challenge is to address policy issues and to enhance national and regional collaboration to promote exchange of information and experiences as well as to facilitate coordinated national and regional programmes.

3. Various factors affecting plant and animal production in a high potential desert system (arid and semi- arid). Use of cultural treatment for increased forage production, effect of protection, use of native species for rehabilitation work, utilization of supplemented irrigated shrubs in the feed calendar for semi intensive systems, flock management for reduced grazing pressure are among the different aspects studied for integrated management of rangeland resources which have considerable implication in combating desertification.

4. Range Improvement several trials all aimed at the improvement of vegetation for range-livestock production using principles of restoration ecology. Techniques used include protection, seed propagation of indigenous range for reseeding degraded rangelands, use of cultural treatments and establishment of supplementary irrigated vegetation.

8. Acknowledgments

I warmly thank to my wife Dr. Laila M. Bidak, Professor of Plant Ecology, Faculty of Science, Alexandria University and my daughter Nada, for their encouragement, valuable suggestions and patience throughout the preparation of this chapter.

9. References

Abdel-Razik, M.; Ayyad, M. A. & Heneidy, S.Z. 1988a. Preference of grazingmammals for forage species and their nutritive value in a Mediterranean desert ecosystem (Egypt). *J. of Arid Environments* 15: 297-305.

Abdel-Razik, M.; Ayyad, M. A. & Heneidy, S.Z. 1988b. Phytomass and mineral composition in range biomass of a Mediterranean arid ecosystem (Egypt). *Oecol. Plant.* 9 (4): 359-370.

Abdel-Salam, H. 1985. Grazing Capacity per Faddan at Omayed Grazing Area the Northern Coastal Zone Western to Alexandria. Ph. D. Thesis. University of Alexandria. 166 pp.

Ayyad, M. A. 1973. Vegetation and environment of the western Mediterranean coast of Egypt. I. The habitat of sand dunes. J. Ecol., 61:509-523.

Ayyad, M. A. 1978. A preliminary assessment of the effect of protection on the vegetation of the Mediterranean desert ecosystems. *Taeckholmia* 9: 85-101 p.

Ayyad, M. A. 1993. The terrestrial Ecosystems of Fuka-Matruh Area (Egypt). Mediterranean Action Plan- UNEP. RAC/SPA. pp. 33.

Ayyad, M. A. & Ghabbour, S. I. 1977. System analysis of Mediterranean desert ecosystems of Northern Egypt "SAMDENE". Environ. Conse. Vol.4., no. 2, summer. 91-102 pp.

Ayyad, M. A. & Ghabbour, S. I. 1979. Phytosociological and environmental variation in a sector of the western desert of Egypt. *Vegetatio*, 31: 93-102.

Ayyad, M. A. & El-Kady, H. F. 1982. Effect of protection and controlled grazing on the vegetation of Mediterranean desert ecosystem s in Northern Egypt. *Vegetatio*, 49: 129-139.

Ayyad, M. A.; Abdel-Razik, M., and Ghali, N. 1983. On the phenology of desert species of western Mediterranean coastal region of Egypt. *Int. J. Ecol. Environ. Sci.* 9: 169-183.

Batanouny, K.H. 1999. The Mediterranean coastal dunes in Egypt: an endangered landscape, Estuarine, *Coastal and Shelf Science*, 49 (Supplement A), 3-9. Bjustad, A. J., Crawford, H. S. & Neal, D. L. 1970. Determining forage consumption by direct observation of domestic grazing animals. *U.S.Dep. Agri. Misc. Pub.* 1147: 101-104

Bidak, L.M; Heneidy, S.Z; Shaltout, K.H; Al Sodany, Y. & Seif El-Naser, M. 2005. Conservation and sustainable use of Medicinal Plants. Vol.(4). pp 120 pp.

Boulos, L. 1995. *Flora of Egypt: Checklist*. Al-Hadra Publishing, Cairo. 286 pp.

Boudet, G. & Riviere, R. 1980. Emploi practique des analyses fouregeres pour l' apprecietion paturages tropicauz. *Rev. Elev.Med. vet. Pays Trop.* 2 (21):227-266.

Castle, M. E. & Halley, R. J. 1953. The grazing behaviour of dairy cattle of the National Institute for Research in Dairying. *Br. J. Ani. Behav.* 1: 139-143.

Chambers, A. R.; Hodgson, J. and Milne, J. 1981. The development and use of equipment for the automatic recording of ingestive behaviour in sheep and cattle. *Grass and Forage Science,* 36: 97-105.

Chapman, V.J. 1960. Salt Marches and Salt Deserts of the World. Interscience publishers Inc.., New York.

Cody, M. L. 1986. Diversity rarity and conservation in Mediterranean – climate regions. In: Soule, M. E. (ed.) Conservation Biology, the Science of Scarcity and Diversity. Sinaure, Sunderland, M. A. 122-152.

Cook, C. W. 1972. Comparative nutritive values of forbs, grasses and shrubs. In: *Wildland shrubs, their Biology and Utilization*. Ogden, Utah, USDA Forest Service. Gen. Tech. Rep. INT-1: 303-310.

Cowlishow, S. J. & Alder, F. E. 1960. The grazing preferences of cattle and sheep. *J. Agr. Sci.* 54:257-265.

Di Castri, F. 1981. Mediterranean type shrublands of the world. In; dicastri, F., Goodall, D. W. and Specht, R. L. (eds.), Ecosystem of the world. Mediterranean type shrub lands. Elsevier, Amsterdam, 2:1-52.

Dicko-Toure, M.S. 1980. Measuring the secondary production of pasture: an applied example in the study of an extensive production system in Mali. In: *Browse in Africa*. (Le Houéou, ed.) ILCA, Addis Ababa: 247-258.

Duivenbooden, N. V. 1985. The effect of grazing on the growth of subshrubs and the soil-watr balance in the North-western Coastal Zone of Egypt. Student Report, Theoretical Production Ecology, Agriculture University Wageningen, The Netherlands.

El-Kady, H. F. 1980. Effect of Grazing Pressure and Certain Ecological Parameters on Some Fodder Plants of the Mediterranean Coast of Egypt. M.Sc. Univ. Tanta. 97 pp.

El-Kady, H. F. 1983. Animal resources. In: Ayyad, M. A.; Le Floc'h, E. (eds.). An Ecological Assessment of Renewable Resources For Rural Agricultural Development in the Western Mediterranean Coastal Region of Egypt. Pp. 77-79.

Mapping Workshop of C.N.R.S./C.E.P.E. Montpellier: CNRS: 104 pp. Fritzi, S. G. and Bradley, W. K. 1992. The influence of plant architecture on the foraging efficiencies of a suite of Ladybird beetles feeding on aphids. *Oecologia*, 92: 399-404.

Heady, H.F. 1975. *Rangeland Management*. Mc. Graw-Hill, Inc. 480 pp.Heneidy, S. Z. 1992. An Ecological Study of the Grazing Systems of Mariut. Egypt. Report submitted to UNESCO. 51 pp.

Heneidy, S.Z. 1995. A study of rangeland species, biomass and chemical composition in different rainguage sites of eastern Matrouh, Egypt. Submitted to faculty of Agriculture through IDRC Project (Canada and University of Guelph).

Heneidy, S. Z., 1996: Palatability and nutritive value of some common plant species from the Aqaba Gulf area of Sinai, Egypt. Journal of Arid Environments 34,115– 123.

Heneidy, S. Z., 2000. Palatability, chemical composition and nutritive value of some common range plants from Bisha, Region, Southwestern of Saudi Arabia *Desert Inst. Bull. Egypt*, 2: (50):345-360.

Heneidy, S. Z. 2002a: Browsing and nutritive value of the most common range species in Matruh area, a coastal Mediterranean region, Egypt. Ecologia Mediterranea 2: 39–49.

Heneidy, S. Z. 2002b. Role of indicator range species as browsing forage and effective nutritive source, in Matruh area, a Mediterranean coastal region, NW- Egypt. Journal of Biological Sciences, 2: 136-142.

Heneidy, S. Z. 2002c. On the ecology and phytosociology of El-Omayed area, Egypt. *Desert Inst. Bull. Egypt*, 52 (1): 85-104.

Heneidy, S. Z. 2003a. Accessible forage biomass of browse species in Matruh area, a Mediterranean coastal region, Egypt. Pakistan-Journal Of Bilogical Sciences, 6 (6): 589-596

Heneidy, S. Z. 2003b. "Rangelands" In: Sustainable Management of Marginal Dry Lands National UNESCO Commission, Cairo, Egypt. SUMAMAD, pp. 97.

Heneidy, S. Z. & El-Darier, S. M. 1995. Some ecological and socio-economic aspects of bedouins in Mariut rangelands, Egypt. Journal of Union of Arab Biologists Cairo, vol. 2(B): 121-136.

Heneidy, S. Z. &, L. M. Bidak. 1996. Halophytes as forage source in the western Mediterranean coastal region of Egypt. Desert Inst. Bull., Egypt. 49, 2: 283-304.

Heneidy, S. Z. &, L. M. Bidak. 1998. Diversity of the wadi vegetation in Matrouh region, Egypt. *J. of Union of Arab Biologists* Cairo, vol. 6 (B). 13-28.

Heneidy, S. Z. &, L. M. Bidak. 1999. Physical defenses and aversion factor of some forage plant species in the western Mediterranean coastal region of Egypt. *J. of Union of Arab Biologists* Cairo, vol. 9 (B):15- 30. 1999.

Heneidy, S. Z.; Halmy, M., 2007: Rehabilitation of degraded coastal Mediterranean rangelands using *Panicum turgidum* Forssk. *Acta Botanica,* Croatia 66, 161–176.

Heneidy, S. Z.; Halmy, M., 2009: The nutritive value and role of *Panicum turgidum* Forssk. In the arid ecosystems of the Egyptian desert", *Acta Botnica,* Croatia, 68 (1): 127– 146.

Hodgson, J. & Jamieson, W. S. 1981. Variations in herbage mass and digestibility and the grazing behaviour and herbage intake of adult cattle and weaned calves. *Grass and Forage Science*, 36, 39-48.

Illius, A. W. & Gordan, I. J. 1987. The allometry of food intake in grazing ruminants. *J. of Animal Ecol.* 56: 989-999.

Jamieson, W. S & Hodgson, J. 1979. The effects of variation in sward characteristics upon the ingestive behaviour and herbage intake of calves and lambs under continouous stoching management. *Grass and Forage Science*, 34: 273-282.

Kamal, S.A. 1988. A of Vegetation and Land Use in the Western Mediterranean Desert of Egypt. Ph. D. Thesis Alexandria University, 193 pp.

Kamal, S.A.; Heneidy, S. Z. & El-Kady, H. F. 2003. Effect of ploughing on plant species abundance and diversity in the northern western coastal desert of Egypt, Biodiversity and Conservation, *12: 749-765*.

Le Houéou, H. N. 1972. Continental aspects of shrubs distribution, utilization and potentils: Africa- The Mediterranean region. In: Wildland Shrubs, their Biology and Utilization (McKell, C. M. Blaisde Blaisdell, J.P. and Goodin, J. P. eds.). USDA, US for sero. Gen. Tech. Repo. INT.01: Ogden, Utah.

Le Houèrou, H. N. 1975. The natural pastures of North Africa, types, productivity and development In: *Proceeding of International Symposium on range survey and Mapping in tropical Africa*, Bamako. Addis Ababa, ILCA.

Le Houéou, H. N. 1980a. The role of Browse in the management of natural grazing lands. In: Le Houéou, H. N. (ed.). *Browse in Africa.* pp. 320-338. Addis Ababa, ILCA. 491 pp.

Le Houérou, H. N., 1980b: Chemical composition and nutritive value of browse in tropical West Africa. In: Le Houérou (ed.), Browse in Africa, 261–289. Addis Ababa, ILCA.

Le Houérou,H. N. 1984. Rain use efficiency: a unifying concept in arid-land ecology. *Academic Press Inc.* (London). pp. 213-247.

Le Houérou, H. N. 1988. Interannual variability of rainfall and its ecological and managerial consequences on natural vegetation, crops and livestock. In: di Castri F., Floret Ch., Rambal S., Roy J. (Eds.) *Time Scale and Water Stress- Proc. 5th Int. On Mediterranean Ecosystems.* I.U.B.S., Paris.

Le Houèrou, H. N. 1993. Grazing lands of the Mediterranean Basin. In: Corpland. R. T. (ed.). *Natural grassland, Eastern Hemisphere and Resume. Ecosystem of the World,* Vol. 8b: pp.171-196. Amsterdam: Elsevier Science Publisher.

Le Houéou, H.N. 1994. Forage halophytes and salt-tolerant fodder crops in the Mediterranean basin. In: Squires, V. R. & Ayoub, A. T. (eds). Halophytes as a resource for livestock husbandry: problems and prospects, pp. 123-137. Bordrecht. The Netherlands: Academic Publisher.

Le Houéou, H.N & Hostte, C. H. 1977. Rangeland production and annual rainfall relations in the Mediterranean basin and in the African Sahelo-Sudanian Zone. *J. of R. Manag.* 30 (3): 181-189.

Milner, C. and Hughes, R. 1970. Methods for the management of primary production of grassland. IBP Handbook no. 6, 70 pp.

Moore, P. D. and Chapman, S. R. 1986. *Methods in Plant Ecology.* 588 ppNoy-meir, I. 1973. Desert ecosystems: environment and producers. *Annual Review of Ecology and Systematics,* 4: 25-51.

Sarson, M., and Salmon, P. 1977. Eleuage, Paturage et donnees de bass pour un amenayment sylvo-pastural dans la Zone no. 2. MOR 73/016, Rabat FOA/MIN. Agric. Ref. Agric.

Täckholm, V. 1974. *Student's Flora of Egypt.* Cairo university press. 888 PP.Taiz, L. & Zeiger, E. (1991). *Plant Physiology.* The Benjamin Cummings Publishing Compony. INC. California USA. Pp 565.

Tueller, P.T. 1988. *Vegetation Science applications for rangeland analysis and management.* Kluwer Academic Publishers, Dordrecht. 642pp.

UNU/MAP/UNESCO, ACRDA. 2003. Site Assessment Methodology for "Omayed Biosphere Reserve" In: Sustainable Management of Marginal Dryland, SUMAMAD. 97 pp.

Van Dyne, G. M. ; and Heady,H. F. 1965. Botanical composition of sheep and catle iets on a mature annual range. *Hilgardia* 36: 465-492.

8

Desertification-Climate Change Interactions – Mexico's Battle Against Desertification

Carlos Arturo Aguirre-Salado[1], Eduardo Javier Treviño-Garza[2],
Oscar Alberto Aguirre-Calderón[2], Javier Jiménez-Pérez[2],
Marco Aurelio González-Tagle[2] and José René Valdez-Lazalde[3]
[1]*Universidad Autónoma de San Luis Potosí*
[2]*Universidad Autónoma de Nuevo León*
[3]*Colegio de Postgraduados*
Mexico

1. Introduction

Until a few centuries, Earth enjoyed a fragile balance in the ecosystem. With the development of mankind this balance has started to lose. More than one is starting to believe that the imbalance is modifying Earth's climate. This is evident in the increasingly frequent news on natural disasters. Both forest and soil degradation have contributed to anger the weather, which has responded directly or indirectly through destruction processes, fostering biological organisms to congregate on continuing the chaos (Allen et al., 2010).

It is a common idea that droughts cause desertification. While the lack of rain is a contributing factor, the root causes are related to the overexploitation of the environment by humans. Droughts are common in arid and semiarid lands, but lands with good management can recover when rains return (Runnström, 2000, Rasmussen et al., 2001). However, the persistent abuse of the land plus drought increases land degradation. Larger population and agricultural pressure on marginal lands accelerates desertification (Viglizzo & Frank, 2006). In some areas of the world such as Africa and Asia, nomads migrate to less arid regions and disrupt the local ecosystem by increasing the rate of soil erosion. While trying to escape from desert, the wilderness pursues them with their practices in land use (UNCCD, 2004; Henson, 2008, Barua et al., 2010). Thus, poor people become both the cause and the victims of land degradation (Günter et al., 2009).

For Mexico, the regulation of the General Law for Sustainable Forest Development defines desertification as the loss of productive capacity of land caused by nature or man in any ecosystem (DOF, 2005). Although this definition assumes the concept completely, it is worth contrasting it with the one used by the United Nations Convention to Combat Desertification (UNCCD) which reads: "Land degradation in arid, semiarid and subhumid lands, is the loss of biological or economic productivity and complexity of agriculture (irrigation and temporal), pastures, woods and forests that results from a process or combination of processes including those promoted by the man such as: a) water or wind erosion, b) deterioration of physical, chemical and biological soil properties and) the permanent loss of vegetation." This definition is used worldwide to describe desertification and its impacts (Grainger et al., 2000).

Causes of desertification vary worldwide. These can be divided into two groups (Geist & Lambin, 2004), 1) those that predispose desertification called underlying causes and 2) those that physically start a desertification process, called proximate causes. Among underlying causes are climate, economics, institutions, national policies and remote influences of population growth. While proximate causes include the expansion of agriculture, overgrazing and infrastructure development. Climate variability represented by meteorological phenomena acts concomitantly and synergistic with other drivers. magnitude of storms or duration of droughts can also be drivers to start desertification process. Economic factors such as market conditions can induce abandonment of arid lands that were used someday for agriculture leading land degradation and thus desertification. Social factors like in- or out-migration determine the potential amount of people available to work in field. Both situations can precede degradation, while immigration could exert higher pressure on land, out-migration would start abandonment of agriculture land. Policy and institutional related factors can also foster degradation. These drivers are associated to formal development policies such as market liberalization, money injection to farming sector as subsidies, incentives or credits. Others have stated that property rights of land tenure have not worked well, weakening conservation of ecosystems in arid lands and predisposing land to excessive grazing, best illustrated by the Tragedy of Commons.

Earth system works following a series of rules or consequences called feedbacks or interactions. When one climatic variable changes, it alters another in a way that influences the initial variable that fostered the change. There are positive and negative feedbacks. A positive feedback stimulates an increase on the initial variable response. For example, global warming reduces snow cover in winter, increases both surface albedo and absorption of solar energy by ground, thus average temperature becomes higher. Or when, land use change reduces retention and availability of soil moisture, hindering the establishment of flora and leading to more soil degradation until a desert area appears. By other hand, negative feedbacks include an amelioration of effects in this circular process. For example, global warming could induce more water vapour in the air, thus more cloud formation which can reflect more sunlight to space, thereby reducing the amount of heating of the surface and with that reversing the initial warming process. Depending on the systems involved, these types of chain reaction can lead to a variety of responses to any ecosystem perturbation (Borroughs, 2007).

Further describing these relationships, at local and regional scales, involves understanding how human activities alter the Earth's surface and influence its atmospheric behaviour. In turn, it will be interesting to evaluate the opposite, i.e. how desertification affect weather conditions in soils, ecosystems, water balance and land use in arid regions (Sivakumar et al., 2007; Bied-Charreton, 2008). These interactions generate a vicious circle which if we do not stop it with specific actions of conservation and restoring, the productive capacity of a particular field could be completely lost. In this sense, the objective of this Chapter is to address these interactions and pinpoint the actions that both the Mexican government and civil society are tackling to fight every day against desertification, mitigating these two-directional effects between Earth ecosystem and climate.

2. Influences of desertification on climate

The awareness that climate influences both the development and distribution of plants and animals, comes from prehistoric times. While recognizing that ecosystem changes have an influential effect on climate change is a recent concern. Vegetation degradation includes a temporal or permanent reduction in the density, structure and species composition. These changes may accelerate climate impacts and exacerbate climate variability (Grainger et al., 2000; Ustin et al., 2009).

2.1 Land use and land cover change

Land degradation caused by land use changes is one of the main themes of the environmental research agenda. This process normally starts with the replacement of forest vegetation for subsistence farming, while the productive potential is retained; otherwise, the reduction of nutrients shore to drop. As a consequence, grazing is the next activity (Viglizzo & Frank, 2006). The introduction of cattle reduces the likelihood for ecosystem restoration. Forest seedlings are part of the diet of animals; besides soil compaction, created by the grazing, eliminates any possibility of water retention and hence growth and development of new forestation (Arnalds et al., 2004). Another possible cause of land degradation is deforestation propelled by forest fires. Although fire may be used as a tool for forest management and even some pine species require high temperatures to open and download off the seeds (Juárez & Rodríguez, 2003), forest fires are uncontrolled events that eliminate vegetation by burning and emit into the atmosphere large amounts of greenhouse gases.

Terrestrial ecosystems contain three or four times more carbon than atmospheric CO_2, and more than 1/8 of atmospheric CO_2 is exchanged within a year by ecosystems through the process of photosynthesis and respiration (Burroughs, 2007). These natural flows of CO_2 are approximately 10 times larger than the flux of anthropogenic additions which are blamed for global warming. In Mexico, it is common the occurrence of fires of various magnitudes during the dry season from December to August. The average annual occurrence of forest fires for the period 2000-2005 was 7880, affecting an average annual area of 208,000 ha. The most common causes of forest fires were agricultural activities (42%), campfires of tourists (10%), smoking (9%), forestry (4%), burning of garbage (3%), poaching (2%), near roads (2%) and unspecified causes (27%). This shows the impact of emissions of greenhouse gases by land use change on global climate (SEMARNAT & INE, 2006).

Another point-of-view related to deforestation and climate change is that tree cutting in urban areas affects so much more global warming than the same activity occurring outside-city. This is explained by the increasing need of air-conditioning in urban areas, which leads to a higher microclimate temperature, thereby raising the use of more and more energy resources such as electricity, especially in warm and dry climates (Rappaport & Hammond, 2007).

2.2 Albedo and energy exchange

The amount of solar radiation that can be reflected or scattered into space is defined as the surface albedo. Surface albedo is closely related to Earth's energy flux. Desert sand reflects much more solar energy than any wooded ecosystem, so the consequences of desertification are an alteration of the radiation balance of the areas affected by this degradation process. The amount of solar energy available at the surface for heating the atmosphere depends on the surface albedo, which varies by location, season and land use. The worldwide average albedo is around 0.3, but ranges from 0.9 to less than 0.05. Bare-soil albedo ranges from 0.1 to 0.6, while albedo of forested areas varies from 0.08 to 0.15. Large changes in albedo occur in regions suffering desertification and deforestation. Depending on soil albedo, the reduction in vegetation cover can lead to an increase in the albedo. An increase in surface albedo could decrease surface temperature (Burroughs, 2007; Rafferty, 2011).

2.3 Evapotranspiration and relative moisture

The removal of vegetation decreases the reduction of water vapor flowing into the atmosphere by evapotranspiration, this process will additionally impact on reducing the relative humidity and on the possibility of cloud formation (Andréassian, 2004; Geist & Lambin, 2004; Burroughs et al., 2007). Other climate impact of evapotranspiration is the regulation of temperature. Thus lower evapotranspiration rates could induce a warmer effect on environmental temperature, and vice versa (Davoudi et al., 2009).

2.4 Surface roughness, wind speed and dust storms

The sun generates energy at about 345 watts/m². Approximately 30% of this energy is reflected back into space and is never used in the atmosphere-land system. Of the remainder, slightly less than 1% (3.1 watts/m²) accelerates the air and creates the wind. Without tree cover, the continents would offer less friction to the wind, and wind speeds in un-vegetated landscapes would be about twice as fast compared to the places covered with forest vegetation, increasing the potential for wind erosion. Therefore, as desertification creates vegetation of lower size will therefore lower values of surface roughness and as a result, increased wind speed over previous patterns observed in landscapes with greater vegetation height (Rafferty, 2011). The opposite also occurs, a greater wind speed can contribute to an increase on soil erosion (Lu & Shao, 2001), thus consequently more desertification.

By other way, dune formation is a process mainly observed in the major sand deserts of the world. Desertification, understood as the progress of desert areas, is well pictured in these regions. Great dunes rapidly change overtime and disappear any productive land such as agriculture or small towns (Kusky, 2009). Another interaction relating wind and soil after land degradation is the generation of dust storms. Particles flying in the atmosphere reflect solar energy over a broadband, including the infrared and thus have a cooling effect (Hardy, 2003). In addition, dust storms cause several impacts. For example, dust storms in Asia have a great impact on the air quality of cities of China, Korea and Japan. The same is for African dust storms which effects reach European countries; in America, dust storms also disrupt social an economic activities. Several efforts of land management to minimize dust storms have done worldwide by planting trees and shrubs that endure long dry periods, and in some cases establishment of induced grasslands can be used to feed cattle and for improving quality of living of people in arid lands (Shao & Dong, 2006).

2.5 Overgrazing

Rangelands in semi-arid zones are especially vulnerable to desertification. Nowadays around 73% of this type of land has been degraded (Oldfield, 2005). Grass provides a complete cover and perennial protection of the ground, which result in minimal soil erosion due to the high probability of seed germination, and to a strong root network that gives more cohesion to soil particles. But, when cattle exert overgrazing, the phantom of desertification comes in. Overgrazing is widely regarded as one of the major causes of desertification in arid lands due to depletion of pasture and scrub and accelerated soil loss. When soil is trampled and compacted by livestock, it loses its ability to establish vegetation and moisture conservation, resulting in increased evaporation and runoff. Additionally, an overgrazing process can be detected as a plant cover decrease, the replacement of original species and the undeniable soil erosion footprint. Thus, overgrazing can exacerbate the impact of drought by modifying the microclimate of the soil, altering the water retention, exposing bare soil to erosion (Manzano & Návar, 2000; Santini et al., 2010; Amiraslani & Dragovich, 2011).

Viglizzo & Frank (2006) described a complete picture of interaction between ecosystem and climate caused by the mismanagement in the Argentine Pampas. Deforestation, overgrazing, intensive agriculture and improper harvesting technology with the interaction of extreme drought conditions between 1930 and 1940 caused severe sandstorms, mortality in livestock, crop failure, failure of bankers and migration. Subsequently, as weather improved they returned to agricultural activity. But during the period 1970-2002 recurrent episodes of flooding in the basin River Quinto happened, probably, by an excessive intensification of agricultural activities in the area.

2.6 Land degradation and agriculture

Soil degradation assessment in Mexico suggests that almost 1'200,000 ha are affected by some degree of salinization. The states with the highest incidence are Tamaulipas, Sonora, Baja California, Chihuahua, Coahuila, Colima (CONAZA, 1994; SEMARNAT & COLPOS, 2003). The high presence of salts degrades soil fertility and promotes abandonment of the land. Additionally, surface albedo increases after progressive salinization. Fujimaki et al. (2003) found that increasing the concentration of salt in the soil at a rate of 1 mg/cm^2 results in an increase in the albedo of 0.0002. By increasing the albedo, the temperature decreases because there is less absorption of solar energy.

Another feedback or interaction relating agriculture, degradation and climate, is the potential retention of carbon by the soil. It is estimated that the ratio between carbon stocks in soil and standing biomass is about 20/1 respectively in semiarid lands, compared with a typical moist forest with a 1/1 ratio (Grainger et al., 2000). When sustainable agriculture practices such as conservation tillage are used, carbon stocks in soil, root and aboveground biomass (284-306Mg/ha/year) can be comparable to those in forested lands. So, climate impact if soil carbon is released to atmosphere could be comparable to that caused by typical deforestation of aboveground forest carbon pools (Etchevers et al., 2009).

3. Influences of climate on desertification

Considering the climate variability reported in recent times, it inspires revisiting the potential effects of climate on land desertification following land degradation, but under a climate-change perspective.

3.1 Effects on pests and land cover change

Allen et al. (2010) conducted an extensive global review of the experiences reported from studies of tree mortality related to climate, particularly drought and heat. In the case of North America, they reported approximately 20 million ha affected from Alaska to Mexico (Table 1). This mortality was led jointly by climatic and biotic factors: 1) multi-year drought, 2) high temperatures preceded by very cold winters, and 3) dramatic incidence of forest pests such as bark beetles of the genera *Dendroctonus spp*, *Ips spp* and *Scolytus spp* or defoliators of the genus *Malacosoma sp*.

Place	Host	Area affected (ha)	Reference
Alaska	*Picea spp*	>1'000,000	Berg et al. (2006)
British Columbia, Canada	*Pinus contorta*	>10'000,000	Kurz et al. (2008)
Saskatchewan and Alberta, Canada	*Populus tremuloides*	1'000,000	Hogg et al. (2008)
Southwest United States	*Pinus edulis*	1'000,000	Breshears et al. (2005)
Missouri and South Carolina	*Quercus velutina* *Q. coccinea*	Not reported	Voelker et al. (2008) Clinton et al. (1993)
California, U.S. and Baja California, Mexico.	*Pinus jeffreyi y* *Abies concolor*	Not reported	Savage et al. (1997)

Table 1. Some cases of tree mortality in North America (Allen et al., 2010)

Low temperatures in winter had naturally controlled insect populations. When not enough cool years came, insects proliferated and then husked or defoliated millions of trees. This fact increased the vulnerability of forest to fires due to the amount of combustible material generated by tremendous mortality (Henson, 2008, Letcher, 2009). The devastation caused by tree mortality and poor restoration of these ecosystems could be the beginning of a long and painful process of desertification, even though such areas are not near major deserts of the world.

Several impacts of climate change on forestry are also expected. Although some researchers view positively the increase in global temperature, which could augment productivity in U.S. forests, they also accept that higher incidence of forest fires, pests and diseases may occur (Latta et al., 2010). Thus, forest spatial distribution and composition changes and their effects on biodiversity are the main forecasting. Changes in forest habitat distributions are also expected. Forest species will look for cooler areas to establish themselves, which probably will be at higher altitudes or latitudes. Interesting colonisations of broadleaved species in temperate ecosystems are expected because of warmer conditions to grow.

Other researchers have highlighted some others vulnerabilities to climate change relating water into vegetation. Conifers are more capable of endure drought than broadleaves because of their higher resistance to xylem cavitation processes (Maherali et al., 2004; Cochard, 2006). Thus, climate change could lead negative effects on world forestry because severe consequences on investing for intensive forest management and supply of forest products could be occur in a near future (Lal et al., 2011).

3.2 Effects on soil degradation

The possible negative effects of climate change on the soil could be that an increased temperature and low rainfall in arid zones would reduce the incorporation of organic matter up to 25% (Kemp et al., 2003). Organic matter acts as a bonding between soil particles and promotes soil unity and stability. When a soil lacks of organic matter can be more vulnerable to wind and water erosion. This can precede more dramatic situations. When increasing temperature, speed wind becomes higher and more aggressive causing sandstorms. These scenarios are common in the great deserts in the world which expand to higher latitudes, shifting regions that were once fertile (Goudie, 2008; Kusky, 2009).

Assessment of soil loss can be carried out by direct and indirect methods. Direct methods comprise *in situ* measurements that can be made by different ways. For example, Pando et al. (2002) used the method of nails and washers to determine the loss of soil in an arid basin located in Nuevo Leon, Mexico as an indicator of desertification. The values ranged from 15 ton/ha (0-8% slope, with zero and low vegetation) and 151.7 ton/ha (> 30% slope) for one year of measurement. Other researchers have made indirect estimates using databases and applying models to estimate water erosion by combining the rainfall erosivity, soil susceptibility to erosion, the degree and length of slope and vegetation cover integrated into an Universal Equation Soil Loss (Zhou et al., 2008, Beskow et al., 2009). Others have integrated to this information the predominant direction of winds and existing windbreaks using GIS to determine the risk of wind erosion (Podhrázscká & Novotný, 2007; Santini et al., 2010).

Temperate zones do not escape from degradation. The increase of storms, combined with the deforestation caused by illegal logging in some regions of Mexico has led to the outbreak of large-scale landslides. For example, during an unusual event of extreme precipitation attributed to climate change because occurred outside the period of rain (15 hours of continuous precipitation in January 2010), the town of Angangueo, Michoacan, Mexico was almost fully covered by liquefied soil, detached from the mountains by the extreme humidity. Navarrete et al. (2011) found that illegal logging explained 61% of total degradation in that area. This area has global ecological importance because its temperate forests serve as habitat for the monarch butterfly (*Danaus plexipus*) when migrating from northern areas such as Canada.

Desertification is a multifactor process. When it was started from bad conditions, high variability of climate, social instability and overexploitation of natural resources all factors will work together following a downward spiral of overall degradation with no stop. But, the opposite is also possible. Increased biological productivity consciously managed can lead to effectively conserve ecosystem and to foster political and economic stability. It is easier to move from right to left due to worldwide conditions (Figure 1). But, there are experiences of people who though different, and started from a degraded environment and slummy economic circumstances, improving their future with great courage and a great vision.

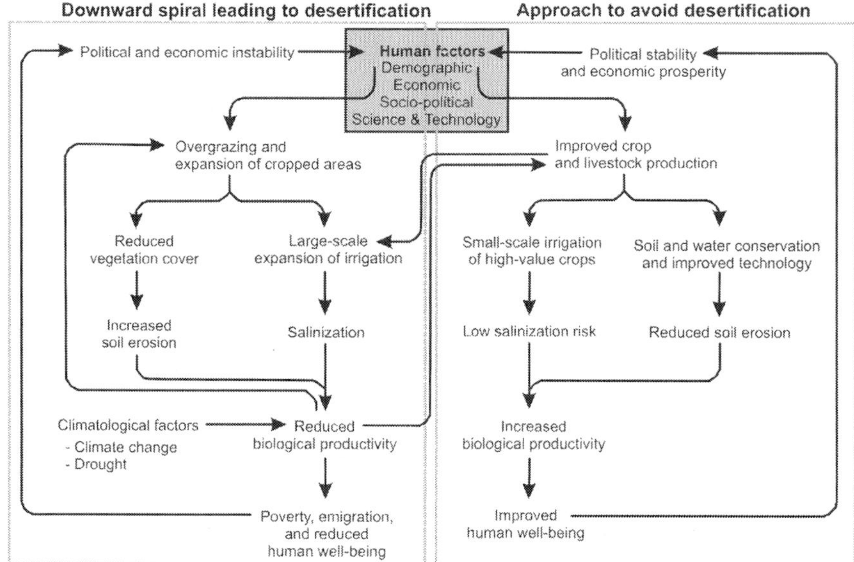

Fig. 1. Flowchart of desertification. (Adapted from Millennium Ecosystem Assessment, 2005)

4. Actions against desertification in Mexico

One of the main problems that have increased the country's vulnerability to climate change has been the rapid deforestation of the country. This can boost severe cases of desertification, an extreme situation of land degradation that is very difficult to reverse. In developing countries, like Mexico, deforestation is due to illegal logging and the expansion of the agriculture activities at the cost of land suitable for forestry, destroying fragile ecosystems as complex as both forests and woodlands. Payment for environmental services can be a good choice (SEMARNAT, 2009).

Currently the federal government of Mexico, through the Ministry of Environment and Natural Resources has created a council or committee called the National System for Combating Desertification and Degradation of Natural Resources (SINADES, as called in Spanish) whose primary objective is to coordinate efforts against desertification and degradation of natural resources by promoting programs that involve different levels of government and civil society.

For proper functioning of SINADES with relevance to the current scenario of Mexico, it was raised the need to formally update the Action Plan to Combat Desertification in Mexico made in 1994 by the Arid Zones National Commission (CONAZA, as called in Spanish). After wide participation of society, through a public consultation including 27 state workshops, 6 regional forums, one national workshop and an electronic query, it was made the document called National Strategy for Sustainable Land Management presented during the World Day to Combat Desertification and Drought (June 17, 2011) in Mexico. This document integrates eight key strategic lines: 1) to promote awareness throughout society and focus on Sustainable Land Management, 2) to promote the integrated planning of land use, 3) to strengthen institutional coordination and harmonization of policies, 4) to promote the generation and dissemination of information for Sustainable Land Management, 5) to promote co-responsible citizen participation, inclusion and gender and ethnic equity, 6) to strengthen research and transfer of good management practices, 7) to foster international cooperation, and 8) to design integrated financing strategies (SINADES, 2009).

The National Forestry Commission of Mexico (CONAFOR, as called in Spanish) through the ProArbol program provides incentives to forest owners who wish to perform actions for the protection, conservation, restoration and sustainable use of forest resources in Mexican ecosystems. CONAFOR has applied financial support for conservation and restoration in almost 494,000 ha from 2001-2007. Since 2005, CONAFOR has been the main partner of United Nations Convention to Combat Desertification in Mexico. One of the major projects implemented to combat desertification include the reforestation of degraded areas for soil conservation (Figure 2).

Fig. 2. Successful cases of ecosystem restoration. Ejido 16 de septiembre, Durango, Mexico. Source: CONAFOR (National Forestry Commission of Mexico)

Also CONAZA, through the program for the conservation and sustainable use of soil and water, provides incentives for rural residents to contribute to the sustainable use and management of soil resources, water and vegetation. This program make investments in projects oriented to soil and water conservation, improvement and rational use of land cover. Financial support can be agreed either through arrangements with state governments or as direct execution with rural people. Among the major projects implemented by CONAZA to combat desertification include grassland establishment. This approach is a good idea to stop desertification just when the soil is un-vegetated. But, elimination of native vegetation such as shrubland to establish new plantations could not be the best alternative. Franklin et al. (2006) studied the extensive conversion from desert vegetation to grassland (*Pennisetum ciliare*), and they found that net primary productivity decreased because original vegetation was composed by trees and shrubs. They stated that replacement of vegetation will tend to loss biodiversity, which is more valuable in ecological terms, than milk or meat production.

Poor people living in arid and semiarid areas are more vulnerable to the effects of desertification. Other federal agencies such as the Ministry of Social Development are working to provide economic support for infrastructure development and social housing to communities with high and very high marginalization that make up the priority focus areas. These actions alleviate underlying causes of desertification related to social drivers. For example, this Ministry gave away 82,307 firewood-saving stoves in 125 municipalities in the whole Mexico for people who are living in extreme poverty. Each firewood-saving stove avoids releasing 2.7 CO_2 tonnes per year to the atmosphere, meaning 222,229 tonnes less to global atmospheric concentrations of CO_2, as an alternative of mitigating climate change and its effects on environment (SEMARNAT & INE, 2009).

Other documents of equal importance on environmental issues are the Mexico's National Communications to the United Nations Framework Convention on Climate Change (1st: 1997 2nd: 2001, 3rd: 2006, 4th: 2009 and currently preparing the 5th: for the end of 2012) containing information for the Inventory of Greenhouse Gases updated for each period of analysis, as well as programs containing measures to mitigate and adapt to climate change (SEMARNAP, 1997; SEMARNAT & INE, 2001, 2006, 2009). For the specific case of desertification, the 4th National Communication of Mexico pinpoints the need to generate a national map updated on the vulnerability to desertification, which will serve as support for decision making. Likewise, each of the 32 Mexican states already have or are just working to develop their own programs of action to combat climate change through the state Ministries responsible for environment. These documents outline different alternatives and set up specific goals to achieve for mitigation and adaptation to climate change.

There are also non-governmental organizations widely involved in environmental issues. Such organizations serve as intermediaries between holders of forest resources and business sector who has the need to neutralize their emissions of greenhouse gases by investing in projects of environmental services such as carbon capture and other co-benefits. There are also individuals concerned on environment who are working to stop the line of degradation and desertification. A region of Oaxaca Mexico, characterized by semiarid climate, known as the Mixteca is proud to have a social organization that has successfully restored its ecosystem. This initiative is from a man named Jesús León Santos, who started to work in the early 1980's restoring the degraded lands in the Mixteca region. He is a leader of a democratic, farmer-led local environmental organization called the Center for Integral Small Farmer Development in the Mixteca (CEDICAM) who adopted several practices of sustainable agriculture and forest restoration in order to recover productivity in ecosystem. Nowadays, his organization involves 12 communities. They have planted more than one million trees and reforested more than 1000 ha. Main outcomes from these activities include 50% increase in agricultural production accompanied with soil and water conservation resulting in ecological, social and economic benefits. León was awarded the Goldman prize in 2008 honouring the titanic and prosperous effort to conserve and restore environmental capacities of semi-arid lands in Mexico. In addition, he has given various talks and visited environmental forums worldwide to share his experience as successful environmentalist.

Others believe that climate geoengineering could be an alternative to reverse the process of climate change. There are several alternatives for controlling climate and its effects on environment and societies, although much of these proposals are just in theory. In other words, these works just picture using environmental modelling to infer the effects of geoengineering alternatives in climate. For example, some visionary researchers have proposed changing leaf-albedo of crop plants through genetic manipulation to decrease global warming (Ridgwell et al., 2009). Other proposals include fertilization of the oceans with iron, because it was discovered that, large amounts of carbon can be fixed by phytoplankton (Keith et al., 2000, Güssow et al., 2010). Most of these alternatives of geoengineering have a great environmental risk, are very expensive and finally unaffordable. So, the most suitable approach to apply is the example of restoring forest ecosystems as described before. Semiarid ecosystems can be successfully restored and land degradation successfully reversed with appropriate practices directly on field.

5. Conclusions

Arid and semi-arid regions are characterized by short rainy seasons. When they experience dry periods, degradation of vegetation promotes erosion and thus desertification. Temperate regions with steep slopes are also vulnerable to loss of productive capacity because when vegetation that protects the soil is, naturally or induced, affected either storms or wind starts pouring the floor that lasted years to form complicating ecosystem restoration. At the same time, land use changes also have an impact on the climate; in fact, deforestation increases the evaporation of available water in the soil, minimizing transpiration by plants and reducing water storage. The increase of arid zones also has an impact on production and suspension of particles in the atmosphere and altering the climate mechanisms. Finally, the reduction of biomass and organic matter in degraded lands, reduces carbon storage in soil, contributing to global warming of the atmosphere, altering global atmospheric circulation and dramatically intensifying the frequency and impact of extreme events related to climate change.

Mexico, including government, non-governmental organizations and people, is actively fighting against degradation and desertification process in a variety of ways. Mexico's agenda related to the topic of environmental protection from the focus on mitigation of climate change is densely filled in all fields of action. These include environmental policy, financial issues and even direct action in the field to stop and reverse processes of degradation and desertification. Seeing the interactions of climate over ecosystems, and thus in people, was the only way, as the environmental issue becomes relevant in making decisions of governments and society, whose current major concerns are either mitigate or adapt to climate change effects. Mexico is playing correctly the key role of mitigating the effects of climate change. Perhaps the goal is still reachable, regaining the balance between society, ecosystem and climate.

6. References

Allen, C.; Macalady, A.; Chenchouni, H.; Bachelet, D.; McDowell, N.; Vennetier, M.; Kitzberger, T.; Rigling, A.; Breshears, D.; Hogg, E.; Gonzalez, P.; Fensham, R; Zhang, Z.; Castro, J.; Demidova, N.; Lim, J.; Allard, G.; Running, S.W.; Semerci, A. & Cobb, N. (2010). A global overview of drought and heat-induced tree mortality reveals emerging climate change risks for forests. *Forest Ecology and Management,* Vol.259, No.4, (February 2010), pp. 660–684, ISSN 0378-1127

Amiraslani, F. & Dragovich, D. (2011). Combating desertification in Iran over the last 50 years: an overview of changing approaches. *Journal of Environmental Management,* Vol.92, No.1, (September 2010), pp. 1-13, ISSN 0301-4797

Andréassian, V. (2004). Waters and forests: from historical controversy to scientific debate. *Journal of Hydrology,* Vol.291, No.1, (May 2004), pp. 1-27, ISSN 0022-1694

Arnalds, A. (2004). Carbon sequestration and the restoration of land health. Example of Iceland. *Climatic Change,* Vol.65, No.(3), (August 2004), pp. 333-346, ISSN 1573-1480

Barua, A.; Majumbder, M. & Das, R. (2010). Estimating spatial variation of river discharge in face of desertification induced uncertainty, In: *Impact of Climate Change on Natural Resource Management,* B.K. Jana & M. Majumder (Eds), 111-130, Springer, ISBN: 111-130

Berg, E.; Henry, J.; Fastie, C.; De Volder, A. & Matsuoka, S. (2006). Spruce beetle outbreaks on the Kenai Peninsula, Alaska, and Kluane National Park and Reserve, Yukon Territory: relationship to summer temperatures and regional differences in disturbance regimes. *Forest Ecology and Management,* Vol.227, No. 3, (June 2006), pp. 219–232, ISSN 0378-1127

Beskow, S.; Mello, C.; Norton, L.; Curi, N.; Viola, M. & Avanzi, J. (2009). Soil erosion prediction in the Grande River Basin, Brazil using distributed modeling. *Catena* Vol.79, No.1, (October 2009), pp. 49-59, ISSN 0341-8162

Bied-Charreton, M. (2008). Integrating the combat against desertification and land degradation into negotiations on climate change: a winning strategy, *French Scientific Committee on Desertification,* (November 2008)

Breshears, D.; Cobb, N.; Rich, P.; Price, K.; Allen, C.; Balice, R.; Romme, W.; Kastens, J.; Floyd, M.; Belnap, J.; Anderson, J.; Myers, O. & Meyer, C. (2005). Regional vegetation die-off in response to global-change-type drought. *Proceedings of the National Academy of Sciences of the United States of America,* Vol.102, No.42, (October 2005), pp. 15144–15148, ISSN 0027-8424

Burroughs, W. (2007). *Climate change. A Multidisciplinary approach.* 2d Ed. Cambridge University Press, ISBN 978-0-521-69033-1.

Clinton, B.; Boring, L. & Swank, W. (1993). Canopy gap characteristics and drought influences in oak forests of the Coweeta Basin. *Ecology,* Vol.74, No.5, (July 1993), pp. 1551-1558, ISSN 0012-9658

Cochard, H. (2006). Cavitation in trees. *Comptes Rendus Physique,* Vol.7, No.9, (November-December 2006), pp. 1018-1026, ISSN 1631-0705

CONAZA. (1994). Plan de acción para combatir la desertificación en México. Comisión Nacional de Zonas Áridas. Secretaría de Desarrollo Social. Saltillo, Coah. 160 p.

Davoudi, S.; Crawford, J. & Mehmood, A. (2009). *Planning for climate change. Strategies for mitigation and adaptation for spatial planners. Earthscan,* ISBN 978-1-84407-662-8

DOF (Diario Oficial de la Federación). (2011). Reglamento de la Ley de Desarrollo Rural Sustentable. Ultima modificación publicada en el Diario Oficial de la Federación el 26 de mayo de 2011. Mexico, D.F.

Etchevers, J.; Prat, C.; Balbontin, C.; Bravo, M. & Martínez, M. (2009). Influence of land use on carbon sequestration and erosion in México: a review, In: *Sustainable Agriculture,* E. Lichtfouse, M. Navarrete, P. Debaeke, V. Souchere & C. Alberola (Eds.), 87-96, Springer, ISBN 978-90-481-2665-1

Franklin, K.; Lyons, K.; Nagler, P.; Lampkin, D.; Glenn, E.; Molina, F.; Markow, T. & Huete, A. (2006). Buffelgrass (*Pennisetum ciliare*) land conversion and productivity in the plains of Sonora, Mexico. *Biological Conservation,* Vol.127, No.1, (January 2006), pp: 62-71, ISSN 0006-3207

Fujimaki, H.; Shiozawa, S. & Inoue, M. (2003). Effect of salty crust on soil albedo. *Agricultural and Forest Meteorology,* Vol. 118. No.1-2, (August 2003), pp. 125-135, ISSN: 0168-1923

Geist, H. & Lambin, E. (2004). Dynamic causal patterns of desertification. *Bioscience,* Vol.54, No.9, (September 2004), pp. 817-829, ISSN 0006-3568

Goudie, A. (2008). The history and nature of wind erosion in deserts. *Annual Review of Earth and Planetary Sciences,* Vol.36, (May 2008), pp. 97-119, ISSN 0084-6597

Grainger, A.; Stafford, M.; Squires, V. & Glenn, E. (2000). Desertification and climate change: the case for greater convergence. *Mitigation and Adaptation Strategies for Global Change,* Vol.5, No.4, (December 2000), pp. 361-377, ISSN 1381-2386

Günter, H.; Oswald, Ú.; Grin, J. & Czeslaw, M. (2009). *Facing global environmental change. Environmental, human, energy, food, health and water security concepts,* Springer, ISBN 978-3-540-68487-9

Güssow, K.; Proelss, A.; Oschlies, A.; Rehdanz, K & Rickels, W. (2010). Ocean iron fertilization: why further research is needed. *Marine Policy,* Vol.34, No.5, (September 2010), pp.911-918, ISSN 0308-597X

Hardy, J. (2003). *Climate change. Causes, Effects and Solutions.* John Wiley & Sons Ltd. ISBN 0-470-85018-3

Henson, R.H. (2008). *The rough guide to Climate Change. The symptomps. The science. The Solutions.* Rough Guides Ltd, ISBN 9-781-85828-105-5

Hogg, E.; Brandt, J.; & Michaellian, M. 2008. Impacts of a regional drought on the productivity, dieback, and biomass of western Canadian aspen forests. *Canadian Journal of Forest Research,* Vol.38, No.6, (April 2008), pp. 1373–1384.

Juárez, A. & Rodríguez, D. (2003). Efecto de los incendios forestales en la regeneración de Pinus oocarpa. var. ochoterenae. *Revista Chapingo. Serie Ciencias Forestales y del Ambiente,* Vol.9, No.2, (Diciembre 2003), pp. 125-130, ISSN 0186-3231

Keith, D. (2000). Geoengineering the Climate. History and Prospect. *Annual Review of Environment and Resources,* Vol.25, (November 2000), pp. 245-284, ISSN 1543-5938

Kemp, P.; Reynolds, J.; Virginia, R.A. & Withford, W. (2003). Decomposition of leaf and root litter of Chihuahuan desert shrubs: effects of three years of summer drought. *Journal of Arid Environments*, Vol.53, No.1, (January 2003), pp. 21-39, ISSN 0140-1963

Kurz, W.; Dymond, C.; Stinson, G.; Rampley, G.; Neilson, E.; Carroll, A.; Ebata, T; & Safranyik, L. (2008). Mountain pine beetle and forest carbon feedback to climate change. *Nature.* Vol.452, (January 2008), pp. 987–990, ISSN 0028-0836

Kusky, T. (2009). *The hazardous Earth. Climate Change. Shifting Glaciers, Deserts and Climate Belts.* Facts on File, Inc. Infobase Publishing. ISBN 978-0-8160-6466-3

Lal, P.; Alavalapati, J. & Mercer, E. (2011). Socio-economic impacts of climate change on rural United States. *Mitigation and Adaptation Strategies for Global Change*, Vol.7, No.7, (October 2011), pp. 819-844, ISSN 1381-2386

Latta, G.; Temesgen, H.; Adams, D. & Barrett, T. (2010). Analysis of potential impacts of climate change on forests of the United States Pacific Northwest. *Forest Ecology and Management*, Vol.259, No.4, (February 2010), pp. 720-729, ISSN 0378-1127

Letcher, T. (2009). *Climate Change: observed impacts on planet Earth.* Elsevier. ISBN 978-0-444-53301-2

Lu, H. & Shao, Y. (2001). Toward quantitative prediction of dust storms: an integrated wind erosion modelling system and its applications. *Environmental Modelling and Software*, Vol.16, No.3, (April 2001), pp. 233–249, ISSN 1364-8152

Maherali, H.; Pockman, W. & Jackson, R. (2004). Adaptive variation in the vulnerability of woody plants to xylem cavitation, *Ecology*, Vol.85, No.8, (August 2004), pp. 2184–2199, ISSN 0012-9658

Manzano, M. & Návar, J. (2000). Processes of desertification by goats overgrazing in the Tamaulipan thornscrub (matorral) in north-eastern Mexico. *Journal of Arid Environments*, Vol.44, No.1, (January 2000), pp. 1-17, ISSN 0140-1963

Millennium Ecosystem Assessment. (2005). Ecosystems and human well-being: Desertification Synthesis. World Resources Institute, Washington, D.C., USA. ISBN 1-56973-590-5

Nagaz, K.; Moncef, M. & Ben-Mechlia, N. (2011). Impacts of irrigation regimes with saline water on carrot productivity and soil salinity. *Journal of the Saudi Society of Agricultural Sciences*, In press, (Available online 15 June 2011), ISSN 1658-077X

Navarrete, J.; Ramírez, M. & Pérez D. (2011). Logging within protected areas: spatial evaluation of the monarch butterfly biosphere reserve, Mexico. *Forest Ecology and Management*, Vol.262, No.4, (August 2011), pp. 646-654, ISSN 0378-1127

Oldfield, F. (2005). *Environmental Change. Key issues and alternative approaches.* Cambridge University Press. ISSN 978-0-521-53633-2

Pando, M.; Gutierrez, M.; Maldonado, A.; & Jurado, E. (2002). Evaluación de los procesos de desertificación en una cuenca hidrológica del NE de México. *Ciencia UANL*, Vol.5, No.4, (Octubre-Diciembre 2002), pp. 519-524, ISSN 2007-1175

Perry, R. (1986). Desertification processes and impacts in irrigated regions. *Climatic Change*, Vol.9, No.1, (August 1986), pp. 43-47, ISSN 1573-1480

Podhrázscká, J. & Novotný, I. (2007). Evaluation of the wind erosion risks using GIS. *Soil and Water Research*, Vol.2, No.1, pp. 10-14, ISSN 1801-5395

Rafferty, J. (2011). *The living Earth. Climate and Climate Change.* Britannica Educational Publishing, ISBN 978-1-61530-388-5

Rappaport, A. & Hammond, S. (2007). *Degress that matter. Climate change.* The MIT Press. ISBN 978-0-262-18258-4

Rasmussen, K.; Fog, B. & Madsen, J. (2001). Desertification in reverse? Observations from northern Burkina Faso. *Global Environmental Change,* Vol.11, No.4, (December 2001) pp. 271-282, ISSN 0959-3780

Ridgwell, A.; Sigarayer, J.; Hetherington, A. & Valdes, P. 2009. Tackling regional climate change by leaf-albedo bioengineering. *Current Biology,* Vol. 19, No.2, (January 2009), pp. 146-150, ISSN 0960-9822

Runnström, M. (2000). Is northern China winning the battle against desertification?. *Ambio,* Vol.29, No.8, (April 2000), pp. 468-476, ISSN 0044-7447

Santini, M.; Caccamo, G.; Laurentu, A.; Noce, S. & Valentini, R. (2010). A Multi-component GIS framework for desertification risk assessment by an integrated index. *Applied Geography,* Vol.30, No.3, (July 2010), pp. 394-415, ISSN 0143-6228

Savage, M. (1997). The role of anthropogenic influences in a mixed-conifer forest mortality episode. *Journal of Vegetation Science,* Vol.8, No.1, (February 1997), pp. 95–104, ISSN 1100-9233

SEMARNAP. (1997). *México, 1ra Comunicación Nacional ante la Convención Marco de las Naciones Unidas sobre el Cambio Climático.* Secretaría del Medio Ambiente, Recursos Naturales y Pesca. México, D.F. 149 p.

SEMARNAT. (2009). *Consecuencias sociales del cambio climático en México. Análisis y propuestas.* Secretaría del Medio Ambiente y Recursos Naturales. ISBN 978-968-817-933-8

SEMARNAT & COLPOS. (2003). *Evaluación de la degradación del suelo inducida por el hombre, escala 1:250000.* México, Secretaría del Medio Ambiente y Recursos Naturales, Colegio de Postgraduados

SEMARNAT & INE. (2001). *México, 2da Comunicación Nacional ante la Convención Marco de las Naciones Unidas sobre el Cambio Climático.* Secretaría del Medio Ambiente y Recursos Naturales, Instituto Nacional de Ecología, ISBN 9968-817-494-7

SEMARNAT & INE. (2006). *México, 3ra Comunicación Nacional ante la Convención Marco de las Naciones Unidas sobre el Cambio Climático.* Secretaría del Medio Ambiente y Recursos Naturales, Instituto Nacional de Ecología, ISBN 968-817-811-X

SEMARNAT & INE. (2009). *México, 4ta Comunicación Nacional ante la Convención Marco de las Naciones Unidas sobre el Cambio Climático.* Secretaría del Medio Ambiente y Recursos Naturales, Instituto Nacional de Ecología, ISBN 978-607-7908-00-5

Shao, Y. & Dong, C. (2006). A review on East Asian dust storm climate, modelling and monitoring. *Global and Planeraty Change,* Vol.52, No.1, (July 2006), pp. 1-22, ISSN 0921-8181

SINADES. (2009). *Estrategia Nacional de Manejo Sustentable de Tierras. Sistema Nacional de Lucha contra la Desertificación y la Degradación de los Recursos Naturales (SINADES).* Secretaría del Medio Ambiente y Recursos Naturales. Dirección General del Sector Primario y Recursos Naturales Renovables, Mexico, ISBN 978-607-7908-42-5

Sivakumar, M. (2007). Interactions between climate and desertification. *Agricultural and Forest Meteorology,* Vol.142, No.2, (February 2007), pp. 143-155, ISSN: 0168-1923

UNCCD. (2004). *Preserving our common ground. UNCCD 10 years on.* United Nations Convention to Combat Desertification. Bonn, Germany. 20 p.

Ustin, S.; Palacios-Orueta, P.; Whiting, M.; Jacquemoud, S. & Li, L. (2009). Remote sensing based assessment of biophysical indicators for land degradation and desertification. In: *Recent Advances in Remote Sensing and Geoinformation Processing for Land Degradation Assessment*, A. Röder & J. Hill (Eds), CRC Press/Balkema/Taylor & Francis Group, pp. 15-44, ISBN 978-0-415-39769-8

Viglizzo, E. & Frank, F. (2006). Ecological interactions, feedbacks, thresholds and collapses in the Argentine Pampas in response to climate and farming during the last century. *Quaternary International,* Vol.158, No.1, (December 2006), pp. 122-126, ISSN 1040-6182

Voelker, S.; Muzika, R. & Guyette, R. (2008). Individual tree and stand level influences on the growth, vigor, and decline of Red Oaks in the Ozarks. *Forest Science,* Vol.54, No.1, (February 2008), pp. 8–20, ISSN 0015-749X

Zhou, P.; Luukkanen, O.; Tokola, T. & Nieminen, J. (2008). Effect of vegetation cover on soil erosion in a mountainous watershed. *Catena,* Vol.75, No.3, (November 2008), pp. 319-325, ISSN 0341-8162

9

Envisioning Ecosystems – Biodiversity, Infirmity and Affectivity

Diana Domingues, Cristiano Miosso,
Lourdes Brasil, Rafael Morgado and Adson Rocha
Universidade de Brasília, Faculdade UnB Gama (FGA/UnB)
Programa de Pós-Graduação em Engenharia Biomédica
LART – Laboratório de Pesquisa em Arte e TecnoCiência
Conselho Nacional de Desenvolvimento Científico e Tecnológico – CNPq
Coordenação de Aperfeiçoamento de Pessoal de Nível Superior – CAPES
Brazil

1. Introduction

Earth's biosphere, climate biodiversity crises and environmental issues are raising a profound level of awareness concerning the collective responsibility toward Earth's life and demanding the responsibility for promoting a healthy ecosystem. A new transdisciplinar group of Brazilian researchers at the University of Brasilia at Gama (FGA), working on the Biomedical Engineering Graduate Program, shares with the international community of Art and TechnoScience several themes related to the ecosystem biodiversity and the extremophile condition (Bec, 2007). We consider our responsibility and the urgent attention to life in our country's huge territory, while facing the effects of an endemic infection of tropical climates and working for the preservation of the Biomes in Amazon Forest, the lung of the planet. Our collaborative projects are concerned with infirmity of the territory and the human invasion and destruction of the ecosystem self-organizing defence. Consequently, we work on health care and affective geographies, in the sense postulated by the geographer/philosopher Milton Santos (Santos, 2008 & Santos, 2009) renewed by geolocated and ubiquitous condition and sentient technologies. Interventions in social networks allow people to exist in the sense of being here and there, consisting of a largely sense of place, to be co-located, being connected everywhere, with an awareness of place amplified by the power to take care of the ecosystem. Our project leads with landscapes - as a living organism - in a given geography, being socially engineered, from the conceptual subjective use of space to the advanced mobile telematic computing and ubiquitous data processing and data information visualization by applying the perspective of acting everywhere - and specially in extreme and hostile space.

Assuming the role pointed out by Louis Bec, we are engaged extremophiles (Bec, 2007), working in the direction of a cultural and anthropological paradigm, and, in case, concerned with the planet's health. We face the ecosystem as a living organism that claims for health care projects to build an inhabitable world for future generations. How to be silent when technologies provide us with sentient devices, satellites, geographical interfaces, mobile technologies, and, for the expansion of the limits of our bodies, by sensor systems? The presence of wireless devices and the possibilities of affective mobile computing (Picard, 1997) modified the traditional concepts of ecosystem, by allowing informational flows, biofeedback data and affectiveness and health (Rocha, 2008) that can be reached by wireless networks, in an ecolocated world.

Until when will the ecosystem's biodiversity crisis persist, given that researchers discuss environment in molecular levels, nanolevels, and satellites put human's eyes, ears, hands and bots in the sky, and global positioning transforms us into extremophiles? As extremophiles artists and scientists, may we collaborate to deal with the environmental and ecological challenges regarding life on the planet, the biological transformations, the environmental and cultural changes? Can we advance the technological apparatus adapted to the ecosystem in a cultural, ecological and healthy social life?

We agree with Malina's beliefs and historical contribution of New Leonardos (Malina, 2007) in collaborative levels of convergent theories and practices which conduct the creative mind of artists, scientist and humanists, from the given to the extremophiles biodiversity engineered world.

We conduct art and Techno Science researches in order to collect, visualize, and analyse data from biological systems and the environment. This work is based on innovative collaborative practices involving artists, engineers, physicists, geographers and humanists. Our main goal is to advance and explore the available technologies for measuring physical parameters and processing biomedical signals, thus allowing an advanced study of complex life systems and the redefinition of the interaction between humans and the environment.

1.1 Biocybrid systems and the engineered nature

For facing the challenges of the ecosystem and the blurring limits in natural and the engineered nature on creative technological levels regarding ecosystem, the "nature itself" and the "future and engineered nature", the Group of researchers at UnB Gama in the Biomedical Engineering Graduate Program and in the Laboratory of Research in Art and TechnoSciences - LART develops biocybrid systems (Domingues, 2009). We consider human existence is nowadays co-located in the continuum and symbiotic zone between body and flesh - cyberspace and data - and the hybrid properties of physical world. That continuum generates a biocybrid zone (Bio+cyber+hybrid) and life is reinvented (Domingues, 2011).

The results reaffirm ontological levels of creative reality and mutual influences with environment information, coincident to the James Gibson's ecological perception theory (Gibson, 1986). The ecosystem in its dynamical relations between human, animal, plants, landscapes, urban life and objects brings questions and challenges for artworks and the reengineering of life discussed in our artworks in art and technoscience that challenges the regarding of the ultimate nature of the planet's life.

Our biocybrid systems are founded on networked cyberinfrastructure and physic territory actions. These actions use local mobile technologies and advanced scientific methods for analysing and evaluating biodiversity research on continental-scale environmental issues. Also, they are based on accelerated research strategies, and on communication strategies for collaborative work on biodiversity, ecological and geographic informatics, social platforms, health process and learning experiences in the physical world and in the digital environment, in collaborations and reciprocity, by using technological common network protocols, for taking on 21st century environmental scientific challenges.

We will describe two of the types of research we conduct: SAPIO – biodiversity, infirmity and affective geography (Dengue infirmity and health care) and Frogs' signatures – Pantanal Bioma in Amazon Forest (preservation of ecosystem and biological community).

2. SAPIO: Biodiversity, infirmity and affective ecosystem

The System for the Acquisition and Processing of Ovitrampas Images (SAPIO) is a project developed, in its origin, by the University of Brasilia at Gama (FGA/UnB) and the Federal University of Pernambuco (UFPE). It aims at developing an automated tool for monitoring, studying, fighting, and preventing dengue. Before describing the project in more detail, we briefly comment on the nature of this disease, and how it affects the lives of millions of people in Brazil and neighbour countries. Our recent proposal expands SAPIO in Art and TechnoScience, through creative extremophile plans under the theme of the infirmity of landscapes, and techniques for data visualization of social information underlying information processing and communication technologies (Diamond, 2009). Data visualization design methods and affective aesthetics focused around the central theme of affective geographies information visualization deal with creative technologies and the research tools and computational technologies developed regarding the practice of analysing data patterns, making these invisible ecosystem structures and social patterns be visible. In case of social platforms, then data visualization process becomes interactive, and is transformed into learning methods for health care and dengue infirmity in our territory.

An endemic disease in the Southwest Asia and with several epidemics having occurred in Brazil and other tropical countries, the dengue is caused by an arbovirus transmitted by the female Aedes Aegypti mosquito. The initial symptoms may vary significantly from people to people, and include high fever, muscular pains, headaches and nausea. A haemorrhagic variety of the disease can occur, and is characterized, after the first stages, by mouth and nose bleeding, internal bleeding, and intravascular coagulation. Infection can also happen to different organs, and can lead to death. Between 1995 and 2001, for instance, more than two million cases of dengue in American countries were reported to the Pan American Health Organization (PAHO), including around 50 thousand cases of the haemorrhagic type and more than five hundred deaths (WHO, 2011).

The prevention of the dengue disease depends mainly on fighting the Aedes Aegypti mosquito, as dengue vaccines are still not commercially available. The population of risk areas is instructed to avoid accumulating water in recipients for several days in exposed areas, as the female mosquito can deposit the eggs in these recipients, and the water allows the larvae to develop. Specific, yet affordable, measures can help preventing the mosquito to reproduce, including the use of coffee grounds in the exposed areas, as coffee was found to kill the larvae.

Some other methods used to combat the Aedes Aegypti mosquito are based the ovitrampas, which are special traps that collect the mosquito's eggs and allow the study of infestation tendencies and the prompt definition of control actions. In this context, the SAPIO project is aimed at obtaining and analysing ovitrampas images, in order to automatically count the deposited eggs and to disseminate the collected information through the World Wide Web.

The SAPIO is divided into two main stages: the Ovitrampas Images Processing (PIO-SAPIO) and the Geographic Information System (SIG-SAPIO).

2.1 PIO-SAPIO

The PIO-SAPIO explores image processing and Artificial Neural Network (ANN) techniques, which it is one of the intelligence system techniques, in order to automatically estimate the number of deposited eggs. The images from the ovitrampas are filtered and binarized, and morphological operations allow the evidencing of present objects. Finally, an ANN classifies the depicted objects, thus allowing the eggs to be detected and counted (Elpídio, 2010).

The present proposal addresses the use of an ANN Multi Layer Perceptron (MLP) (Haykin, 1999) to automate the process of identifying Aedes Aegypti eggs in areas of Lamina of the Ovitraps. The algorithm developed to automatically identify the Aedes Aegypti eggs in digitalized images of ovitraps was elaborated with Matlab software, version 7.6. In this study, 5 samples of typical ovitraps, previously digitalized were used. The images measure, on average, 23000X6000 pixels and were submitted to a subdivision process to generate a bank of sub-samples of original images to test and validate the ANN. The sub-sampling of the original images has as an objective, to help the training of the ANN and later to compare the results obtained with the manual analysis completed by trained technicians.

In the process of segmentation the Regions of Interest (ROI) were extracted (Gonzalez, 2008), resulting in 217 sub-images, organized in 7 data banks, being that 6 were used in training and one for testing. This strategy resulted in 7 possibilities of system validation with the variation of the bank designed for the test. Figure 1 illustrates the process of sub-sampling and ROI extraction of the digitalized images of the ovitraps.

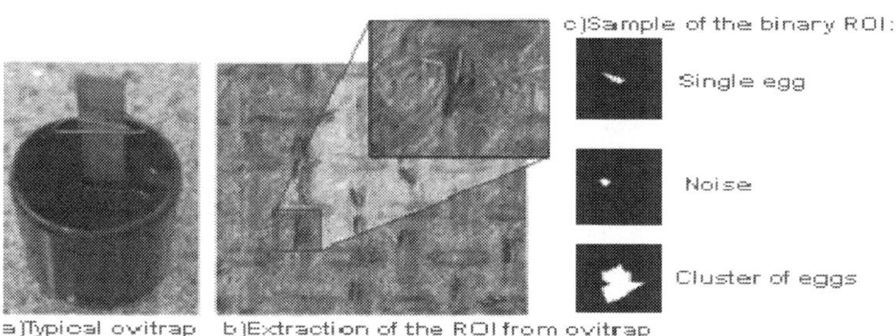

Fig. 1. Example of extraction and classification of ROI obtained from a typical ovitramp.

Before the classification process of the Aedes Aegypti eggs with the ANN MLP, each sub-image was submitted to a pre-processing process, where they were binarized by tonality of the ROI. An average filter was applied and the morphological operations of opening and closing to minimize noise effects (Gonzalez, 2008). These operations are useful in the identification and extraction of image characteristics, besides helping in the identification and minimization of noise that interferes in the form of objects contained in the image. In general, the application of the morphological operations simplifies the objects of interest and enhances essential characteristics of the same. Therefore, the objective of the interactive application of the opening and closing operations is the elimination of artefacts of the image that interfere in the identification of the ROI without distorting the global geometry of the characteristics that it intends to preserve.

Finally, the labelling of the objects present in the binary image, whose result was used as a desired value in the classification of ANN, was completed. The desired value was classified in: no labelled object, noise, single egg or cluster of eggs. The classification of the desired value considered factors such as average size of the egg, area of the labelled object, among others, to complete classification. The entrance of the ANN was extracted from the histogram of the binarized sub-images, originated after the process of segmentation.

The ANN MLP perfects the process of identification of Dengue mosquito eggs, on the digitalized image of the ovitraps, and acts as a deciding mechanism to facilitate classification of ROI. The ANN MLP used was tested with various configurations and the best results were obtained with the following architecture: a rate of learning from 0.95 and one term momentum with the value of 0.01, two hidden layers containing 16 and 50 neurons, respectively. The exit layer was structured with 4 neurons that correspond to the possibilities of classifications of desired value.

In tests developed with the ANN MLP, it was possible to achieve an average quadratic error (Haykin, 1999) of approximately 0.09 and about 94% of cases classified correctly. During the classification, the performance of the ANN was analysed with the hyperbolic functions tangent and sigmoid (Haykin, 1999). It was observed that the ANN MLP obtained better performance during the classification with the use of the Sigmoid function. Figure 2 presents the results of the training with both functions and the consequential performance of the ANN MLP.

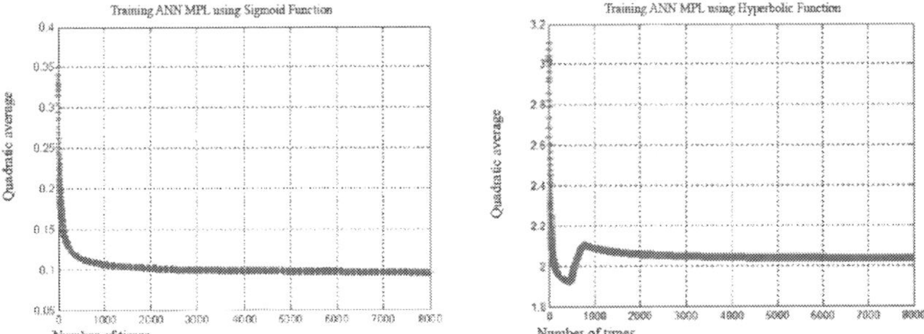

Fig. 2. Performance of the ANN MLP (variation of average quadratic error) according to the type of function.

Table 1 presents the set of data resulting in the classification of the ANN MLP according to the set of image database.

Image Database	Number of images submitted to testing	Average Quadratic Error	Correct Classifications	Percentage of correct classifications
1	23	0,0414	21	91,30%
2	41	0,0475	37	90,24%
3	23	0,0497	23	100 %
4	33	0,0425	30	90,91%
5	35	0,0387	33	94,29%
6	35	0,042	33	94,29%
7	27	0,046	26	96,3%

Table 1. Results of the ANN MLP Classification.

It was found that the major part of the ANN classification resulted in the cluster of *Aedes Aegypti* mosquito eggs. Even though the quantifying of egg clusters has already been considered in earlier work (Elpídio, 2010), as the objective of this approach is to provide a new strategies to validate the algorithm developed, besides making improvements in its performance, currently the efforts made are being invested in broadening knowledge of the ANN MLP to identify the limit between eggs in clusters and making the automated counting of eggs in ovitraps viable.

2.2 SIG-SAPIO

The main objective of the SIG-SAPIO, which the geographic information systems are considered one of the intelligence systems, on the other hand, is to develop a World Wide Web platform that describes the geographical proliferation of the mosquito, based on information periodically extracted from the ovitrampas and automatically uploaded to an online database. It represents an important effort in keeping the local populations informed about the mosquito and the risk areas, thus helping to maintain prevention measures. Our developed tool is ready to be used in large-scale, to provide dynamic and easy-to-read information in an illustrated and didactic manner. The version first of the SIG-SAPIO was presented in the (Amvame Nze, 2011). At following, we present the stages of the new version of this system.

One of the SIG-SAPIO stages is the trapping technique that uses plastic buckets are filled of stained water with a wood slide on the edge, Figure 3. This bucket and slide (ovitrap) is currently used as a trap for the mosquito. A smell from the wood slide is used to attract the Aedes Aegypti into the trap and lay the larva on it: this is the ovitrap technique.

A human health expert collects the wood slides and manually counts the eggs deposited by the mosquito. The first part of this project, automates using a special algorithm the eggs counting by image processing techniques, and compares it with the manual technique, Figure 3.

The resulting data collection obtained from the Acquisition and Image Processing group (SAPIO's first stage) is filtered by a script then directed into a free relational database being part of the Web-based group (SAPIO's second stage). This database is dynamically maintained as data comes from the first stage.

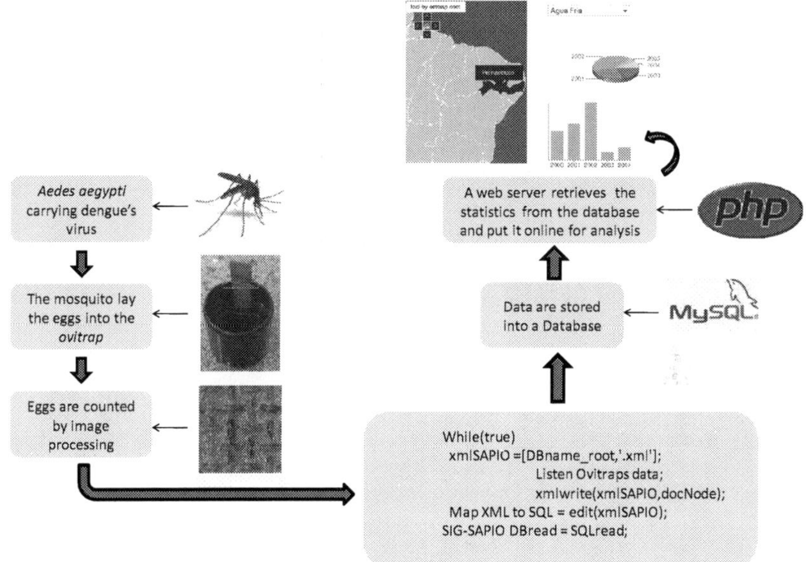

Fig. 3. Diagram showing the methodology leading from egg counting to web-based monitoring.

After data acquisition, a web-based is use to show mapping and statistic information collected from the first stage as shown in Figure 3. Google Maps API is used to support current implementation and phpMyAdmin is used for database acquisition. The prototype demonstrates how dengue monitoring would be displayed for web users. A general map is presented for the user to have direct statistic information from any state, and in need of more detailed information from that state, a link calls Google Maps to provide accurate positioning of the dengue proliferation.

This work has shown the development of an automated web-based application that can provide the monitoring and analysis of dengue proliferation. Its database is filled with data coming from an automated Aedes Aegypti image processing egg counting from ovitraps. This platform should be used for monitoring and prevention of the disease as an alert for citizens and government officials in real time.

Moreover, we have built a PIO-SAPIO/SIG-SAPIO with a human/environment/net, natural/artificial, remote/local and rural/urban structure in mutual contamination, with dengue ecological information analogous to the principle of the ouroborus mythic serpent, and self-regenerating emergent narratives about health care and dengue.

At the convergence of the scientific methods, the efficiency of social biocybrid platforms provides the awareness of the territory's infirmity and of people's behaviour in the physical environment and the cyberspace. This is made possible by the interaction with large amounts of data, especially those provided by geographic interfaces such as GPS systems, Google Maps, Google Earth, SMS, MMS and the tweeter, Facebook, Messenger or other networks. These interfaces create an affective geography, concerned with the infirmity – locus, and inform about the health conditions. The individuals are co-located in the virtual environment and the physical environment, and they live in a social platform where their share skills, affectivity and knowledge about dengue. The individuals are positioned in the virtual and physical environments, and they share an existence in the social platform, where they learn and teach. Their experiences and knowledge are shared, in favour of the group´s health and against dengue.

The project uses virtual environment to reengineer reality and blurs people's life and social behaviour in cyberspace and in daily life always looking for a healthier physical environment. Data collected by interface devices generate a locus by processes fundamentally different of photography, television or other media information that, until now, are dedicated to dengue public campaigns. The social platforms allow access to different sources of data signals coming from cyberconnections and from mobile devices providing a networked process of learning about the dengue epidemic during collaborations and reciprocities, by sharing behaviour and affectivity regarding the territory and their experiences and knowledge in favour of the group's health and against the dengue epidemic. The existence is an endless social process of learning and teaching, in mutual and intertwined influences referred to the ecosystem and the infirmity of the landscape.

Data visualization scientific techniques and their information processing of biosignals of the ecosystem, and communication technologies information are visualized and transform data to representations intelligible on a human code which culminate into icono(geo)graphies. In the art history, the artists revigorate the landscape language and methods, enhancing the abstracionism domain, and in close relation to the *land art* and *environmental art* and participatory forms of landscape in social models. Consequently, information processing and scientific forms in data signals processing of natural phenomena transform the artistic landscape in data visualization landscapes as new forms of landscapes in Contemporary Arts. The participation of communities in communication technologies and data mining systems reveal the affectivity to the territory in social behaviours and the data visualization of *graphos* coming from search engines are transformed in emergent maps of the affection and the care of the landscape.

This project is outstanding and should cover the immediate needs of the dengue configuration in the national territory and the increasingly expansion of the disease in several states, mainly reaching the Distrito Federal which is affected since 1991, and producing autochthons cases in 1997 as affirms Catan, Fontes et al. *"Its distribution in the Distrito Federal is uneven, presenting higher rates in places the perverse urbanization in the peripheral countries and the decline of the public health system in these countries. In Brazil this disease reemerge in the 1980's, after almost 60 years without any case in without infrastructure."*

Thus, the aim of our work is to improve the Dengue extinction in Brazil and dealing with biomedical research topics and Art and Technoscience contribution in data visualization landscapes and collaborative affective platforms regarding a healthier territory.

3. Biocybrid system: Frogs' signatures in the Pantanal Biome

A second ecosystem-related line of research developed at LART investigates the frogs' populations in the Brazilian Pantanal area, and explores the richness of information conveyed by the frogs' vocalizations. A set of microphones will be strategically distributed over an ecological sanctuary, and the acquired audio signals will be transmitted to a remote cave (Domingues, 2003) for data visualization and analysis. In this stage is possible to use an ANN, which it is one of the intelligence system techniques, to classify the frogs' vocalizations.

Once the audio signals are received, the visualization system will then prefilter these signals, in order to extract the frogs' callings. Finally, it will extract and interpret different properties of interest, which we briefly describe below, and it will project dynamic images that change, over time, according to these properties.

Different types of physical and behavioural properties, as well as environmental parameters, can be extracted from the frogs calling activity, as shown in recent research in bioacustics. For instance, the emitted sounds are an important characteristic of each species, and have thus been used as a tool for species differentiation (Glaw, 1991), as well as for the discovery of new species (Vences, 2010). Furthermore, they provide valuable behavioural information and reflect reactions to the environmental conditions (Vielliard, 2010).

We will explore this aspect to visualize the acquired sounds and different properties extracted from them. The system will be developed in four different stages, which we briefly discuss below.

3.1 First stage: Dynamic spectrograms

A spectrogram is a graphical representation of the time-frequency distribution of a signal. Typically, in the bidimensional case, the horizontal axis shows different time slots, whereas the vertical axis corresponds to the frequency components; in this case, each point in the plane correspond to an specific time and a single frequency component, and the colour or gray level associated to this point indicates the magnitude of that particular frequency found in the signal at the indicated time slot. In a tridimensional representation, the third axis indicates the magnitudes of the components.

Note that in order to build the spectrogram corresponding to a particular signal, it is necessary to decompose it in the frequency domain, for each time interval considered. Different time-frequency techniques, such as the short-time Fourier transform and filter bank decompositions, are found in the literature (Cohen, 1995 & Cohen, 1998), each with its advantages and disadvantages and with different levels of compromise between time and frequency resolution. In this project, we will compare different techniques as applied to the frogs' calls.

After the prefiltering stage to isolate the frogs' sounds from the environmental noise, we will divide them into time intervals by applying appropriate windows, and build a spectrogram for each considered interval. In fact, since we want a dynamic visualization that changes as the audio signals evolve, spectrograms will be built for time segments of a few seconds, with the represented times in each spectrogram being the subdivisions of those segments. The main objective will be to improve immersion by providing a visual perception of the frogs' vocalization. Using 3D spectrograms evolving with time, this will show how the sound amplitudes of each frequency increase or decrease over time in the observed regions.

In order to illustrate some of the data visualization techniques we wish to apply, we show, in Fig. 4, an example of a segment of frog vocalization, as a function of time. In Fig. 5 and Fig. 6, we show the corresponding spectrogram, respectively in its 2D and 3D forms.

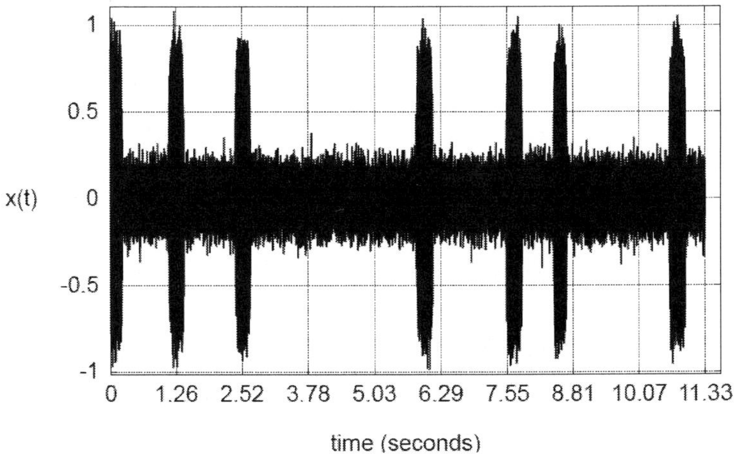

Fig. 4. A segment of a frog vocalization, as a function of time.

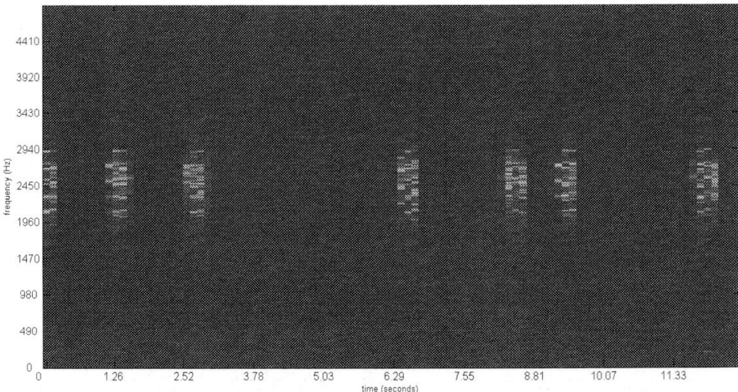

Fig. 5. The 2D spectrogram corresponding to the frog vocalization in Fig. 4.

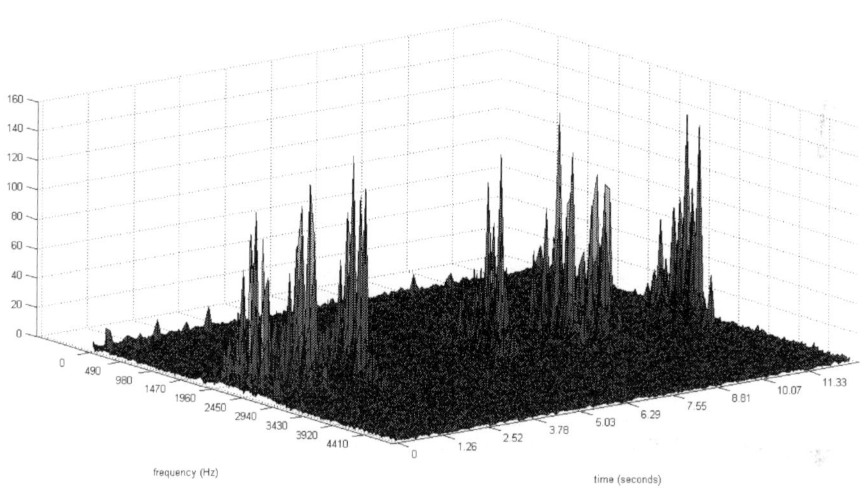

Fig. 6. The 3D spectrogram corresponding to the frog vocalization in Fig. 4.

3.2 Second stage: Species classification and localization

There are around 20 frog species inside the analysed ecological sanctuary. As the vocal activity of frogs has very specific characteristics for each of the species, it is possible to differentiate them from the collected sounds (Glaw, 1991); actually, a description of the vocalizations is almost mandatory when characterizing newly found species of amphibians (Vielliard, 2010).

In the second stage of the visualization development, we will implement a classifier that will allow us to differentiate the acquired sounds in terms of the species (neural networks and statistical classifiers will be compared). As we dynamically project the different classes inside the cave, this will illustrate the behaviour of the species, in terms of geographical distribution. The dynamic images will also show how they respond to each other's movements and possibly to human presence.

3.3 Third stage: Individuals classification and counting

The sounds emitted by the frogs can also be used to distinguish different individuals in each of the species present. Hence, after the sounds from each of the species are separated, it is possible to estimate the number of individuals, in each considered region.

Inside the cave, we will project a graphical representation, by region, of the estimated number of individuals. This will provide a more in-depth visualization of their behaviours. Additionally, the technique can be used in a different context, over longer periods of time, to estimate the growing or shrinking of the populations, in an indirect, noninvasive manner.

3.4 Fourth stage: Dynamic abstract images related to physical properties

In the last planned stage of the visualization development, we will build abstract images that show physical properties like temperature, humidity and light conditions, which are extractable from the sounds. In these images, the positions in the plane will map the different areas from where the sounds are collected, and different colours will correspond to different levels of the analysed variables.

The dynamic images show the evolution of such parameters over time, as well as, combined with the visualizations from the first three stages, the individuals' and species' responses to the environmental conditions.

3.5 Study of the local biology

Besides improving the immersion experience using a visualization tool, this development will represent a contribution to the study of the local biology, by providing information about the species localization and population sizes, as well as their behaviour and reactions to the environment. Other potential applications include the measurement of the impact, to several biological groups, of cities overlapping their natural habitats.

Also, in developing the biocybrid system we provide the human proximity to the datalandscape and the manipulation of visual and sonic information, during enactions of human body dialogues with the distant Pantanal Biome. People wearing rings, bracelets, T-shirts, shoes and glasses equipped with biosensors are users working close to the environment. Metaphorically, we propose the frogs' signatures and the human behaviour dealing with laws and phenomena of the cosmos, by influencing life of nature inside the world as a living organism exchanging electrical potentials, heats, sounds and vibrations and the sense of presence being advanced by the technological apparatus.

4. Life systems and pattern formation

Note that, in developing research on complex life systems, we emphasize the study of pattern formation. The scientific community has recently devoted significant attention to the study of this process. In fact, pattern formation occurs in several natural phenomena, and thus constitutes the object of study in different scientific fields (Cross, 1993). Some of these phenomena are well known, such as Rayleigh-Bénard convection (Bénard, 1900; Rayleigh, 1916 & Getling, 1991) and the Belousov-Zhabotinsky reactions (Petrov, 1997). The essence of biology itself is the self-organization and the subsequent pattern formation with the raising of emergent properties. Some examples include: morphogenesis, which comprises a loss of symmetry, as the spatial differentiation rises from a homogeneous environment (Ingham, 1998); the non-homogeneous distribution of plankton in the oceans (Vilar, 2003; Lehman, 1982 & Mackas, 1985) and the distribution of other species in ecological systems (Levin, 1992).

One can argue that all those systems seem to violate the second law of thermodynamics, when in fact other principles are violated, like mixing, ergodicity, and in some cases the fluctuation-dissipation theorem (Costa, 2003). In our research, we note that, in most cases, the rise of emergent properties, such as spatio-temporal synchronization (Morgado, 2007) and pattern formation (da Cunha, 2009), are consequences of some kind of memory.

In (Morgado, 2007), we show that the synchronization phenomena in systems described by the logistic equation with stochastic noise, an equation that is widely used for population models, are accomplished when we introduce in the equation a memory term, which describes the temporal interaction between generations. We find that synchronization only occurs when the memory has sufficient intensity and/or range, indicating that both parameters are important. In fact, we find a subset in this parameter space where synchronization is present.

In (da Cunha, 2009), we study the problem of bacteria growth, using the Fisher Equation. This equation extends the logistic equation and adds a diffusive term, while retaining the growth and competition terms. But in these, the interactions are local, which does not correspond to the reality in most cases. For example, the spread of chemical toxic substances as a result of metabolic activity is a non-local interaction that affects the growth rate and the competition for resources. To address these problems, we propose a generalization of the traditional Fisher equation, in order to include non-local growth and competition terms, by means of memory terms. These memories extend the growth and competition over a controlled range in the growth domain. We find that the pattern formation only occurs in a subset of the parameter space, just like the synchronization phenomena discussed before. More than just a coincidence, these results indicate that memory plays a significant role in the development of emergent properties.

5. Conclusion

Cyberculture, scientific, artistic, technological and social aims in self-organizing complex biocybrid systems favour the human responsibility, reciprocity and generosity, and contributes to the ecosystem health.

Transdisciplinar methods and blurred boundaries of disciplines allow researchers to incorporate their subjects' concerns and questions as Ecosystem, biodiversity, affective aesthetics, art of landscape and social behaviour into the research process. Wireless networks have all created an ideal environment for an explosion of human capacity of reinvention of life and the desire of reengineering nature in data visualization; mobile devices add several levels for being ecolocated. Biocybrid systems and the layering capacity of GPS, Bluetooth, geo-tagging, SIG, augmented reality, neural networks, our eyes, ears and hands in the sky by the extrusion in the satellites and sensors, provide localization and personalization capacities in new forms of life and the responsibility to the ecosystem. The innovations in scientific data visualization , and social data visualization information and communication enable a healthier future for the planet in ways that were unpredictable, not visible, intensely giving us the sense to take care the health and to deal with the biodiversity challenges of the earth's biosphere and control biodiversity crises and environmental issues, thus responding to our profound level of awareness concerning the collective responsibility toward Earth's life and providing us with some extremophile initiatives to the responsibility for promoting a healthy ecosystem.

In Malina's words, *"a new Renaissance is under way, and at no time in human history have the challenges of a sustainable civilization on our planet been more daunting. It is surely through the work of these new Leonardos that we will change our ideas about the world and how to act in the world."*

6. References

Amvame Nze, G. D; Brasil, L. M. Sistema WEB para Monitoração e Análise de Proliferação do Aedes Aegypti. In: Pan American Health Care Exchanges (PAHCE); PAHCE 2011, March 28 – April 1, Rio de Janeiro, Brazil.

Bec, L. We are extremophiles, MetaMorphosis: Challenging Arts and Sciences, International conference, key note, November 8-10, 2007. Prague, Czech Republic.

Bénard, H. Les tourbillons cellulaires dans une nappe liquide transportent de la chaleur par convection en regime permanent, Ann. Chim. Phys. 7, 62 (1900).

Catão, R. de C.; Guimarães, R. F.; Júnior, O. A. de C. and Gomes, R. A. T. Análise da distribuição do Dengue no Distrito Federal. , In: Espaço & Geografia, Vol.12, No 1 (2009), 81:103. Available at http://www.red.unb.br/index.php/ geografia/article/viewPDFInterstitial/2815/2428.

Cohen, L. (1995). Time-frequency analysis. Englewood Cliffs, N.J, Prentice Hall PTR.

Cohen, L. and P. Loughlin (1998). Recent developments in time-frequency analysis. Boston, Kluwer Academic Publishers.

Costa, I.V.L.; Morgado, R.; Lima, M. V. B. and Oliveira, F.A. The Fluctuaction- dissipation theorem fails for fast superdiffusion, Europhysics Lett. 63, 173 (2003).

Cross, M. C. and Hohenberg, P. C. Pattern formation outside of equilibrium, Review of Modern Physics 65, 851 (1993).

da Cunha, J.A.R.; Penna, A.L.A.; Vainstein, M.H.; Morgado, R.; Oliveira, F.A. Self-organization analysis for a nonlocal convective Fisher equation, Physics Letters A 661–667 (2009).

Diamond, Sara. A Tool for Online Collaboration: CodeZebraOS. PhD thesis. University of East London. Available at: http://www.codezebraosphd.com/ CodeZebraOSPhD_Thesis.pdf. 2009.

Domingues, D. G.; Rocha, A. F.; Handam, C. A.; Augusto, L. and Miosso, C. J. Biocybrid Systems and the Reengineering of Life. Electronic Imaging - Imaging, Interaction, and Measurement. IS&T/SPIE Session: The Engineering Reality of Virtual Reality 2011. Ian E. McDowall, Fakespace Labs, Inc.; Margaret Dolinsky, Indiana University. 78641a-1-78641a-9, 2011.

Domingues, D. G. Research Project: Art and TechnoScience: Expanded Interactions and Cybrid Condition in Software Art, CNPq Minister of Science and Technology 2008, CAPES 2009.

Domingues, D. G. The immersive Poetics of Artificial Worlds, in Hybrid Reality: Art Technology and Human Factor, H. Twaites, Ed. Montreal, Canada: Hexagram Institute, 2003.

Elpídio, F. G. G. et al. Automatic Identification of Aedes Aegypti Eggs Deposited in Ovitraps Slides Using Image Processing Techniques. In: XXII Brazilian Congress of Biomedical Engineering, 2010.

Getling. Formation of spatial structures in Rayleigh-Bénard convection, Usp. Fiz. Nauk. 161, 1 (1991).

Gibson, James. J. Ecological Approach to Visual Perception.: Lawrence Erlbaum. Associates Publishers, Hillsdate, 1986.

Glaw, F. and Vences, M. Bioacoustic differentiation in painted frogs (Discoglossus). Amphibia-Reptilia, 12:385–394, 1991.

Gonzalez, R. C. and Woods, R. Digital Image Processing. Upper Saddle River, New Jersey: Prentice-Hall, Edinburgh, 2008.

Haykin, S. Neural Networks: A Comprehensive Foundation, 2nd. Ed., USA, Prentice Hall, 1999.

Ingham, P. W. The molecular genetics of embryonic pattern-formation in drosophila, Nature 335, 25 (1988).

Lehman, J. T. and Scavia, D. "Microscale patchiness of nutrients in plankton communities", Science 216, 729 (1982).

Levin, S. A. "The problem of pattern and scale in ecology", Ecology 73, 1943 (1992).

Mackas, D. L.; Denman, K. L. and Abbott, M. R. "Plankton patchiness: Biology in the Physical Vernacular", Bulletin of Marine Science 37, 652 (1985).

Malina, R. MetaMorphosis: Challenging Arts and Sciences, International conference, key note, November 8-10, 2007. Prague, Czech Republic.

Morgado, R., Ciesla, M.; Longa, L.;Oliveira, F.A. Synchronization in the presence of memory, Europhys. Lett. 79 10002 (2007).

Petrov, V.; Ouyang, Q. and Swinney, H. L. Resonant pattern formation in a chemical system, Nature 388, 655 (1997).

Picard, R. W. *Affective computing*, Cambridge, Mass.: MIT Press, 1997.

Rayleigh. On the dynamics of revolving fluids Proc. R. Soc. London A 93, 148 (1916).

Rocha, A. F. As redes de sensores e o monitoramento da saúde humana, in Informática em Saúde, L. Brasil, Ed.: Brasília: Universa, 2008, pp. 489-510.

Santos, Milton. Da totalidade ao lugar. São Paulo: Edusp, 2008.

Santos, Milton. A natureza do Espaço – técnica e tempo, razão e emoção. São Paulo: Edusp, 2009.

Vences, M.; Glaw, F. et al. Molecular phylogeny, morphology and bioacoustics reveal five additional species of arboreal microhylid frogs of the genus Anodonthyla from Madagascar. Contributions to Zoology, 79(1):1–32, 2010.

Vielliard, J. M. E and Silva, M. L. A bioacústica como ferramenta de pesquisa em comportamento animal. In Regina Brito e William Lee Martin Grauben Assis, editor, Estudos do Comportamento II (in press). Editora da UFPA, 2010.

Vilar, J. M. G.; Sole, R. V. and Rubi, J. M. "On the origin of plankton patchiness", Physica A 317, 239 (2003).

WHO. Dengue. Available at http://www.who.int/denguecontrol/en/index.html, 2011. Accessed on the 10th of October, 2011.

10

Ecological Research of Arctic Restricted Exchange Environments (Kandalaksha Bay, White Sea, Russian Arctic)

Sofia Koukina, Alexander Vetrov and Nikolai Belyaev
P.P. Shirshov Institute of Oceanology of Russian Academy of Sciences
Russia

1. Introduction

Estuaries and continental shelf areas comprise 5.2% of the earth surface, and only 2% of the oceans' volume. However, they carry a disproportionate human load (Wolanski, 2007). Hypoxia also occurs in shallow coastal seas and estuaries, where their occurrence and severity appear to be increasing (Rabalais, 2010).

The Arctic has come under intense scrutiny by the scientific community in recent years. This is due to the key role played by the region in the variability of the world ocean and, hence, the Earth's climate, its vast and still not fully realized natural resources, and rapidly increasing technogenous load on the environment (Shapovalov, 2010). The White Sea of Russian Arctic (the only Russian inland sea) is characterized by extreme diversity of enclosed estuarine systems that are inhabited by unique biota (Krasnova, 2008, Lisitzin et al, 2010). These environments, which were wildernesses not long ago, are often sites of anthropogenic impacts that expose them to risk (Peresypkin & Romankevich, 2010). In the recent decades significant research efforts have been undertaken to evaluate concentration of potential contaminants (such as trace metals, oil pollutants, and chemically resistant anthropogenic substances) in sea waters, sediments, and food chains of the White Sea with special attention devoted to shallow areas and enclosed estuarine systems (Koukina et al., 2001, 2010, Savenko, 2006). Most of these areas were found to be characterized by very active fronts and biogeochemical barriers creating small scale variability in hydrodynamic, lithological and geochemical patterns that significantly complicates the differentiation of sites of natural element accumulation from sites of anthropogenic contamination (Koukina et al, 2003, Pantiulin, 2001).

The major geological feature of the coastal zone of the White Sea is modern endogenous crustal uplift of 4 mm per year an average (Oliunina & Romanenko, 2007). Due to this process numerous small bays, lagoons and straits are isolating from the sea with further transformation into brackish and fresh water lakes. These unique ecosystems named separating basins (in a wider sense, restricted exchange environments) were not sufficiently studied until recent time due to the tiny (to oceanographic ranges) sizes and sea-isolated setting. Characteristic for such reservoirs specific hydrological and hydrochemical regime was earlier observed in the relict Arctic lake Mogil'noe located on the Kil'din Island of the Barents Sea (Sapozhnikov et al, 2001). The study of two separating lakes of Kandalaksha Bay

(Kislosladkoe Lake and Cape Zeleny Lake) first revealed the specific features of separating process within Karelian coastaline of Kandalaksha Bay of the White Sea (Shaporenko et al, 2005). These reservoirs being separating basins of different isolation phase were shown to be characterized by sharp water column stratification, extreme values of hydrological and hydrochemical parameters and ultra-contrast oxidizing conditions complicated by subsurface oxygen maximums (O_2>20 mg/l) and anoxic hydrosulfuric (H_2S>90 mg/l) zone occurrence in the central bottom depressions. The complex of environmental changes occurring under separation, of which the anoxia and its consequences are of most hazarding, often leads to the disturbance within the ecosystem. Thus, in the separating basins under investigation the mass extinction of marine biota and simultaneous microbial community extension were observed (Peresypkin & Belyaev, 2009, Vetrov & Peresypkin, 2009). The further study of such restricted exchange environments is essential for comprehensive environmental assessment of the region as well as for objective anoxia prognosis in Arctic ecosystems.

Sediments in separating basins are unique traps for elements entering these aquatic environments. The upper sediment column in estuarine and coastal environments can be regarded as a slowly stirred reactor to which substrates are added at the top and mixed downward (Martin, 2010). The comparative biogeochemical study of Karelian shore coastal basins showed the pronounced enrichment of sediments from separating bays in clay fraction, organic matter and in nutrients and metals in comparison to the neighboring open shallow sea (Koukina et al, 2003, 2006). Further study of the region revealed the extreme high total Fe (160000 µg/g) as well as enhanced content of labile (acid soluble) Fe, Mn and Cr in sediments from central parts of two separating lagoons of Chernorechenskaya Bay (Koukina et al, 2010). Since all previously made biogeochemical assessments of the Karelian shore showed no significant contamination, the enrichments of sediments from separating basins (sometimes reaching the artifact concentrations) were preliminary related to the natural accumulation of organic matter, Fe and trace metals and, especially, their labile forms within the solid sediment phase, and this may occur due to the specific conditions of separating basin being: (1) permanent (or seasonal) inputs of terrestrial material that supply organic matter, Fe and Mn oxides and elements associated with, (2) pronounced trapping effect under restricted water exchange with an open sea, (3) contrast oxidizing conditions within the water column and upper sediments, that influence the element cycling and speciation and, hence, may increase potential contaminant bioavailability and toxity. Therefore, it is essential to know more about the carriers and transport modes of potential contaminants such as trace heavy metals and their response to the specific biogeochemical conditions of separating basins. The objective environmental assessment of such ecosystems requires comprehensive biogeochemical examination, while the complex characterization of sediments from such unstable environments can serve as a basis of ecological monitoring of the region and allows some deductions as to carriers, transport modes and potential bioavailability of major and trace elements entering and within these different systems.

The present chapter continues and summarizes the series of ecological researches of the White Sea restricted exchange environments. Hence, in this work the TOC, *n*-alkanes and metal (Fe, Mn, Cu, Zn, Cr and Pb) forms distribution in surface sediment samples from representative small exchange environments of the Karelian shore being Kislosladkoe Lake, Cape Zeleny Lake, Chernorechenskaya Bay and Porchalische Lagoon was determined and discussed in relation to the hydrological and hydrochemical features of representative basins. Since major and trace element speciation is essential in processes such as the toxity and bioavailability of pollutants in natural systems, the special goal of the study is the comparative study of ecologically significant metal forms in sediments of separating basins.

2. Materials and methods

2.1 Study area and sampling

2.1.1 Environmental setting

The White Sea is a sub-Arctic inland sea (Fig.1). It is the only Arctic sea with the major part of its basin south to the Polar Circle. The climate of the region is transitional from oceanic to continental, with long severe winters and cool summers. The period of ice cover lasts, on average, from November until May. The total area of the White Sea is approximately 90000 km², with a water volume of some 6000 km³. The average bottom depth is 67 m, with a maximum depth of 350 m (Glukhovsky, 1991). Despite its small area, the White Sea is characterized by a relatively large discharge of river water – around 180 km³ per year (Dobrovolsky & Zalogin, 1982). This discharge is spread between four major bays: Mezensky Zaliv (Mezen Bay), Dvinskoy Zaliv (Dvina Bay), Onegsky Zaliv (Onega Bay) and Kandalakshsky Zaliv (Kandalaksha Bay). The first three bays are located at the mouths of the major rivers of the White Sea: the Mezen, Severnaya Dvina and Onega, respectively. The deepest bay, Kandalaksha Bay, is a wide estuary with evenly distributed fresh-water discharge (Pantiulin, 2001). A pronounced tidal asymmetry with an ebb tide lasting twice as long as the flood tide is characteristic of the area. The Karelian coastline of Kandalaksha bay forms a continuum of small bays of 1-35 km in length. Due to the endogenous crustal uplift (4 mm per year an average), this bay contains a continuum of shallow environments, ranging from estuaries of different types to separating basins where water exchange is

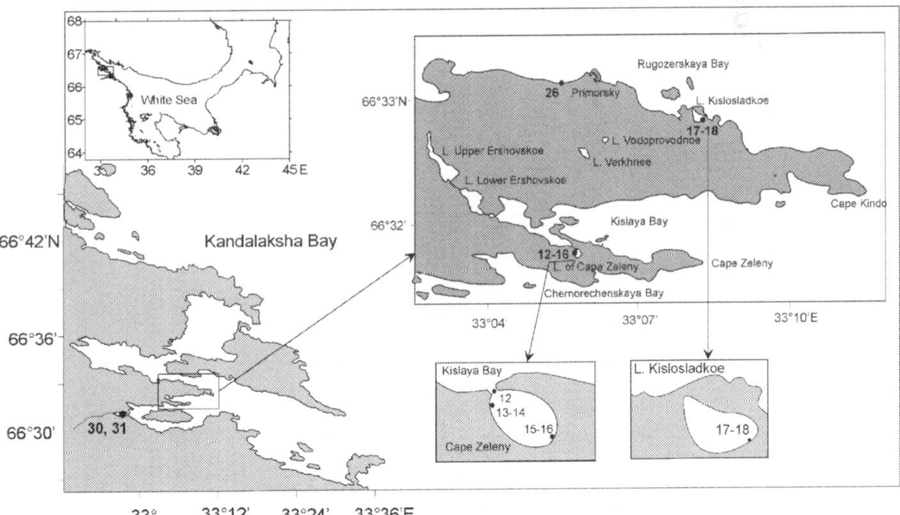

Fig. 1. Map of Kandalaksha Bay of the White Sea showing the location of the sampling stations.

severely restricted. The present study focuses on four small basins along the Karelian shore of Kandalaksha Bay, in the vicinity of the White Sea Biological Station of Moscow state Lomonosov University: Chernorechenskaya Bay (the inner part), Kislosladkoe Lake, Cape Zeleny Lake, and Porchalische lagoon (Fig. 1).

Chernorechenskaya Bay, located at the mouth of the Chernaya river, is 2 km long and consists of upper and lower basins with maximum depths of 5 and 10 m, respectively. The basins are separated from each other by shallow bars. It is a typical small estuary characterized by permanent fresh-water input throughout the year, resulting in significant vertical and horizontal salinity gradients.

Kislosladkoe Lake, located 2 km to the east from Primorsky village, is oval shaped with 180 m length and 100 m width. It is a typical separating basin isolated from Velikaya Salma strait by an uplift of two underwater rock bars. There is an insignificant surface water exchange between the Kandalaksha Bay and the lake. The swampy southern shore is characterized by insignificant, but constant, fresh water input provided by a small stream.

The separating lake at Cape Zeleny is located 6 km to the south from Primorsky village. It is round shaped with 200 m length and 120 m width. The former lagoon was isolated from Kislaya Bay by a rock bar. The small strait still provides a water exchange between lake and bay. At the highest flood tide sea water enters the lake, while during the rest of the time the lake water is discharged into the bay. There is no permanent fresh water input, while seasonal fresh water outflow from the southern shore may occur.

Porchalische lagoon is a littoral pool located 300 m to the north from Primorsky village at the marsh runoff. During the flood tide the pool is encroached by sea water, during the ebb tide – fully filled with fresh water.

2.1.2 Sampling

Ten surface sediment samples were collected between 28.08-10.09.2008 in the littoral zone of Karelian shore of Kandalaksha Bay in small exchange environments Kislosladkoe Lake, Cape Zeleny Lake, Chernorechenskaya Bay and Porchalische lagoon (Fig. 1). Sediment samples were obtained during the ebb tide, with a 0.2 l plastic box-corer designed at the P.P. Shirshov Institute of Oceanology. The upper layer (0-6 cm) of the samples was retrieved with stainless steel spatulas, transferred into pre-cleaned polyethylene containers and frozen until analyses. Where present, surface oxic yellowish layer (0-3 cm) was retrieved and collected separately. Sampling, sample transportation and preparation procedures were carried out using standard clean techniques described elsewhere (Koukina et al, 1999, Loring & Rantala, 1992).

2.2 Methods

2.2.1 Total organic carbon (TOC) and carbonates analyses

From each station, 100-150 g of sediment sample was dried at 60 °C, ground and homogenized.

Total organic carbon (TOC) and carbonates contents were determined in dry sediment samples on Shimadzu analyzer TOC Vcph (Shimudzu Co.) in the Ocean Chemistry Laboratory of P.P. Shirshov Institute of Oceanology of RAS.

2.2.2 *n*-alkanes analyses

The *n*-alkanes were isolated from dry powdered sample by 45 min hexane extraction in Branson 1210 ultrasonic bath (Branson Ultrasonic B.V. Co) preceded by 15 min degassing. The extract obtained was filtered through GF/F glass fiber filters under vacuum. The solvent was evaporated at 35 °C in Yamato RE-52 vacuum rotor vaporizer. To remove dissolved sulfur, the extract obtained was passed through an activated metallic copper column, evaporated in a nitrogen flow and stored in a refrigerator at +5°C until instrumental analysis. Further determination of *n*-alkanes contents in the extracts was carried out on Shimadzu GCMS-QP 5050 chromatographic mass spectrometer (Shimudzu Co.) in the Ocean Chemistry Laboratory of P.P. Shirshov Institute of Oceanology of RAS.

2.2.3 Metal analyses

In order to determine acid soluble (labile), organically bound and total contents of Fe, Mn, Cr, Pb, Zn and Cu, samples were parallel treated by 25% acetic acid (CH_3COOH), 0.1 M sodium pyrophosphate ($Na_4P_2O_7 \cdot 10H_2O$) and aqua regia (HNO_3:HCl =1:3), respectively. The difference between amounts of acid soluble and total metal contents was considered a measure of mineral incorporated (acid insoluble) metal form.

To isolate the acid soluble metals, 15 ml of 25% acetic acid was added to 1.1 g of dry sample in a polypropylene vial and shake in a mechanical shaker for 6 hours. The extract with sediment was then filtrated into a 25 ml glass volumetric flask. The sediment on the filter was washed with 10 ml of distilled water, the wash water added to the flask.

To isolate the organically bound metals, 15 ml of 0.1 M sodium pyrophosphate was added to 1.1 g of dry sample, shake for 15 min in a mechanical shaker and left for 24 hours. The extract with sediment was then filtrated into a 25 ml glass volumetric flask. The filter was washed with 10 ml of 0.1 M sodium pyrophosphate, the wash liquid added to the flask.

For total metal content analysis, 0.5 g of dried powdered sample was kept in the oven at 300 °C until constant weight. Afterwards, the sample was objected to a triple successive treatment by aqua regia (HNO_3:HCl =1:3) with evaporating. After the last evaporating, residual sediment was objected to dissolution in 1M HNO_3 with heating.

Further determination of metal contents in the extracts was carried by an atomic absorption spectrometer (AAS) Hitachi 180-8 in the Analytical Centre of Moscow State Lomonosov University. The relative accuracy for AAS determinations was within the standard deviations of the certified sediment reference material SDO-1 (Berkovitz & Lukashin, 1984).

3. Results and discussion

3.1 Water column

The water column structure discussion is based on the research of Cape Zeleny Lake and Kislosladkoe Lake by Shaporenko S.I. and co-authors (Shaporenko et al, 2005). The specific conditions of shallow separating basins being restricted water exchange with an open sea, permanent fresh water input and bottom relief complicated by depressions lead to extreme values of hydrological and hydrochemical parameters within the water column during the summer period (Fig. 2). The hydrological and hydrochemical stratification of the water column strongly differs from the open coastal White Sea. The most specific features are: thin surface fresh water layer, sharp halocline, high temperature and extreme high oxygen concentrations (more than 20 mg/l that makes up to 300% of saturation capacity) in subsurface water layer, while low temperature and extreme high H_2S concentrations (more than 90 mg/l) in the bottom depressions. The active water reaction (pH) in the upper aerobic layers of the lakes was slightly alkaline (8-9), near the oxic-anoxic layers boundary and in the anaerobic near bottom layer – close to neutral (7-7.5). The water alkalinity in the lakes varied within the ranges of 40 to 70 mg $CaCO_3$ per 1 liter, while its distribution followed that of salinity. The CO_2 concentration in the lakes waters increased from surface (9-11 mg/l) to the bottom layer (up to 400 mg/l). The phosphorus forms contents in the waters of the lakes did not exceed the respective averaged values for Kandalaksha Bay. However, the lakes waters were enriched in all nitrogen forms in comparison to the open White Sea, while NH_4^+ concentration was especially enhanced probably due to the anaerobic zones occurrence. The total dissolved iron varied within the ranges 0.01-0.03 mg/l and was found twice orders higher the background values.

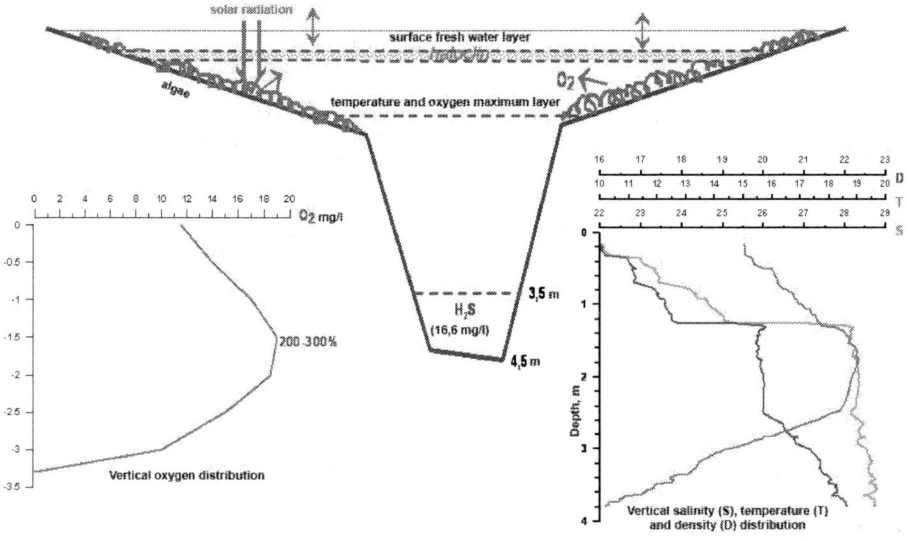

Fig. 2. Water column structure of typical separating basin (Kislo-Sladkoe lake, summer 2001; Shaporenko et al, 2005)

The thin fresh water layer is separated from the temperature and oxygen maximum layer by the sharp halocline. The halocline provides the vertical sustainability of the water stratification and preserves the lower layers from mixing. In shallow waters, under sufficient light and heat, the active growth of thread algae occurs. The algae may cover up to 80-90% of shallow lake bottom. The intensive photosynthesis at the algae extent causes the permanent input of O_2 into the water column. Since the salinity and density cline preserves the water layers from mixing, the oxygen is not released to the surface water layer contacting the atmosphere – and is therefore accumulated within the subsurface waters forming the oxygen maximum layer. In the bottom depressions, on the contrary, the poor water mixing and intensive organic matter accumulation and decomposition cause the extensive sulfate reduction and elevated H_2S concentration in the near bottom waters. Thus, within the water column the sharp change from oxic to redox conditions is observed. The absence of sharp density gradient between oxygen and hydrosulfuric zones is a specific feature of the Karelian shore separating basins. Hence, the upper aerobic waters here are relatively enriched in NH_4^+ and CO_2, and other biogenous elements.

During the seasons with low photosynthesis, the hydrosulfuric zone may spread towards the water surface up to the halocline. Moreover, in winters, under the ice cover, the anaerobic conditions may occur in the whole water column. Due to this, the resident species within the ecosystem are stressed and the regular mass biota extinctions occur.

3.2 Sediment distribution and biogeochemical characteristics

3.2.1 Sediment type distribution

Karelian shore sediments can be divided in to two main lithological facies according to sedimentary conditions in the White Sea: the facies of coastal zone and the facies of bays formed by the rugged cost (Lisitzin et al, 2010). In the coastal zone, tidal and wind generated currents form coarse sediments, of which well sorted fine-grained sands are of most importance. In the small bays of the rugged coast, the hydrodynamic and thus sedimentary conditions are very calm, leading to accumulation of fine-grained silty material (Koukina et al, 2003). These sediments are mainly comprised of aleurite silts, which often have a high content of coarse fragments transported by ice. The heavy mineral fraction of quartzo-feldspathic sand-size material is mainly comprised of pyroxene, epidote, garnet, and hornblende. The major clay minerals of the fine-grained fraction are illite (50-80%) and chlorite (20-30%), with an admixture of montmorillonite and kaolinite (Pavlidis, 1995). Most of the source material was originally derived from the Archean igneous (widely ranged from ultrabasic to persilic and from normal to alkaline row) and metamorphic (mostly gneisses, crystalline schists, amphibolites) rocks that form the White Sea geological block of the Baltic Shield (Milanovsky, 1987). In addition, a few appearances of Paleozoic ultrabasic alkaline dike were found at the Rugozerskaya site in the vicinity of the White Sea Biological Station of Moscow State University (Shurkin, 1984). Sediments derive from the submarine erosion of rock outcrops, the reworking of old glacial deposites during post-glacial sea-level rises, glacial outwash, and ice-rafting.

Textural studies indicate that sediment distribution along the studied part of the Karelian shore is a patchwork of brownish silty sands, brownish and grayish sandy aleurite silts, and fine-grained brownish and grayish aleurite silt deposits. Most of sediments are characterized by a yellowish oxic surface layer 1-3 cm thick. The most fine-grained organic-rich and sometimes hydrosulfuric sediments are found in separating basins (Koukina et al, 2010). The evolution of sediments here is caused by specific depositional conditions, which are strongly affected by small-scale hydrological, hydrodynamic and hydrochemical processes unique for each particular area, of which the hypoxia is of most importance (Koukina et al, 2003).

The physical and chemical environments of estuarine and coastal sediments and within the Karelian coastline, in particular, vary over wide ranges. Sediments underlying oxic bottom waters, have layered structure. The upper centimeters of such usually muddy impermeable sediments have an oxic "cap" just below the sediment-water interface, where Fe and Mn oxides precipitate; below the oxic layer is a region of Fe and S reduction, but without buildup of dissolved H_2S. Below this layer – if the supply of organic matter supports extensive sulfate reduction – is a layer of elevated dissolved H_2S (Martin, 2010). Many trace metals both form insoluble sulfides and tend to sorb on to precipitating Fe and Mn oxides. Hence, they are released to pore waters when Fe and Mn oxide reduction occurs, but tend to be immobilized by precipitation as sulfides. In sediments, underlying permanently or seasonally anoxic bottom water, hydrosulfuric layer is spread to the sediment-water interface providing the free upward transport of reduced Fe, Mn and S from sediments to near bottom waters. Thus, the dissolved H_2S, Fe and Mn exhibited sharp content maximums in the near bottom anaerobic water layers of lakes Kislosladkoe and Cape Zeleny (Shaporenko et al, 2005). The other important coastal sediment type is permeable sandy sediment, in which the aerobic conditions usually prevail but element accumulation hardly occurs due to the low adsorption capacity of sand material. Most of the sediments studied from Kislosladkoe, Cape Zeleny and Chernorechenskaya underlie the oxic waters and, therefore, have the oxic cap on top 1-3 cm. In these muddy samples, the oxic cap and the underlying sediment were sampled and analyzed separately (st. 13-14, 15-16, 17-18, 30-31). Hydrosulfuric sediment from Porchalische lagoon (st. 26) and silty sand from the stream bed of Cape Zeleny Lake (st. 12) had no visible layers.

3.2.2 Total organic carbon and carbonates distribution

Sedimentary organic content showed great variability, ranging from 0.3% to 18.7% for TOC (an average 4.5%) (Table 1). The most found values of organic carbon are within the range of average marine and Arctic Ocean sediments (Romankevich, 1984).

Study area	Station	Layer (cm)	Sediment type	$CaCO_3$, %	TOC, %	n-alkanes $\mu g\ g^{-1}$	i-C_{19}/ i-C_{20}	CPI	$\sum C_{12}+C_{22}/ \sum C_{23}+C_{40}$
Cape Zeleny Lake	12	0-1	silty sand	50.7	2.10	0.52	0.71	3.48	1.10
	13	0-1	silt	0.38	3.40	3.82	1.11	7.82	0.16
	14	1-6	silt	0.33	1.60	2.11	1.37	6.40	0.19
	15	0-1	silt	0.30	4.90	5.90	0.42	6.59	0.24
	16	1-6	silt	0.58	3.90	7.37	0.61	6.69	0.15
Kislosladkoe Lake	17	0-3	silt	1.97	8.30	5.24	1.15	9.57	0.13
	18	3-6	silty sand	0.30	0.25	0.24	1.30	2.74	0.59
Porchalische lagoon	26	0-1	silt	0.83	18.7	5.87	0.22	11.9	0.10
Cherno-rechenskaya Bay	30	0-1	silt	2.07	0.40	1.11	0.53	4.70	1.06
	31	1-3	silt	0.32	1.00	0.89	1.48	3.87	0.78
	Mean	-	-	5.77	4.46	3.31	0.89	6.38	0.45

Table 1. TOC, carbonates and n-alkanes distribution in surface sediments.

Among sites studied, the lowest TOC sediment content is related to the mouth of Chernaya river in Chernorechenskaya Bay (Fig. 1). At this site the loose surface yellowish layer showed twice lower TOC content of 0.4 % in comparison to the underlying grayish silty sediment (1%) (st. 30, 31). This might be due to an enhanced destruction (oxidation) of newly deposited organic matter that is characteristic for surface oxic sediment layer (Romankevich & Vetrov, 2001). At the same time, in high-energy areas (such as estuarine water mixing zones) the effects of sediment resuspension can dominate sediment-water-column exchange (Postma, 1980). Thus, estuarine currents may transport the loose mud within the water mixing zone, so that the surface muddy layer may not be necessarily bound to the subsurface sediment.

Within Cape Zeleny Lake, the strait bed sediment sample exhibited the moderate TOC content of 2.1% (st. 12). In the western shallow part of the lake, sediment comprised of grayish silt with TOC of 1.6% was overlied by an oxic layer with the twice higher TOC content of 3.4% (st.13, 14). Sediments from the southern part of the lake, influenced by seasonal fresh water input and hence terrigenous organic matter supply, were mostly enriched in TOC (4.9% and 3.9% for oxic and underlying layer, respectively (st.15, 16)).

In Kislosladkoe lake, the sediment collected in the eastern part of the lake near the rocky bar strait (st. 17, 18), exhibited a thick (3 cm) organic rich oxic layer (8.3% of TOC) underlying by a poor in organic matter dark colored silty sand (0.3% of TOC). The surface sediment is most probably newly formed under the calm hydrodynamic of separating basin, while the subsurface sand originates from the times when the basin had a free exchange with Velikaya Salma Strait.

The TOC maximum content (18.7%) is related to Porchalische lagoon. The fine-grained silt from Porchalische was characterized by a noticeable hydrosulfuric smell and the absence of the yellowish oxic layer that is typical of sediments underlying the anoxic bottom waters and experiencing the high flux of organic matter. The marine algae brought into the lagoon by storms cannot survive under the stress from regular tidal sea-fresh water change (Vetrov & Peresypkin, 2009). Further deposition and decomposition of the algal mass might contribute to an increased accumulation rate of organic matter and hence to formation of anoxic sediments.

Sediment inorganic carbon, presumably supplied by carbonates (0.3-2%), was uniformly low in the samples studied (Table 1). The extreme high (for the region studied) carbonate content of 50% was found in the strait connecting Cape Zeleny and Kislaya Bay in the vicinity of a large boulder (st. 12). The strait bed is comprised of silty rubble and silty sand with a high content of shell residues coming out of mussel colony inhabiting the boulders.

3.2.3 *n*-alkanes distribution

The total *n*-alkanes content ranged from 0.2 μg g^{-1} to 7.4 μg g^{-1} (mean 3.3 μg g^{-1}) (Table 1). These values are considered to be relatively low for the area studied (Peresypkin & Belyaev, 2009). Spearman correlation coefficient revealed strong positive correlation (r_s=0,88) between TOC and *n*-alkanes content for the sediments studied.

The *n*-alkanes distribution (Fig.3) relative to hydrocarbon chain length showed the maximums in C_{23}-C_{35} zone in the most of samples studied marking the major input of organic matter from terrestrial plant remains. The terrigenous hydrocarbons make up to 80% of total n-alkanes content. The mean values of the low-molecular to high-molecular homologues ratio ($\sum C_{12}$-C_{22}/$\sum C_{23}$-C_{35}) of 0.45 (ranging from 0.1 to 1.1) and CPI (Carbon Preference Index) of 6.38 (ranging from 2.74 to 11.9) also contribute to the permanent terrigenous organic matter input into the sediments within the area studied (Table 1). The mean i-C_{19}/i-C_{20} value of 0.89 (ranging from 0.22 to 1.48) may indicate that transformation of the original organic matter in sediments studied often occurs under redox conditions, which are especially pronounced in the Porchalische lagoon (st. 26).

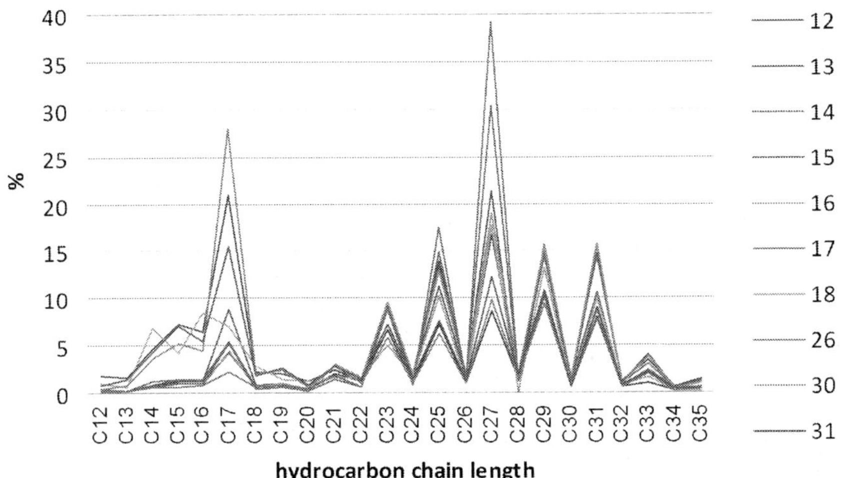

Fig. 3. The distribution of n-alkanes in surface sediments.

In some samples (st. 12, 30, 31), the elevated content of low-molecular hydrocarbons C_{14}-C_{17} with clear maximum at C_{17} marks an autochthonous microbial source of organic matter. At these sites, the total n-alkanes content in sediments was low (0.5-1 µg g^{-1}), and the sum of low molecular homologues ($\sum C_{12}$-C_{22}) made up to 45-60%. The found n-alkanes distribution is typical for the sediments covered by bacterial mats (e.g. purpur sulfur and green cyanobacteria species) (Peresypkin & Belyaev, 2009, Vetrov & Peresypkin, 2009). The highest C_{17} peak (27 %) is related to the upper oxic sediment layer in Chernaya river estuary (st. 30), where the lowest TOC content of 0.4% was determined.

The sample from Porchalische lagoon (st. 26) is characterized by highest TOC (18.7%), $\sum C_{23}$-C_{35} (>80%) and C_{27} peak (up to 40 %), while relatively low total n-alkanes (5.87 µg g^{-1}) content and lowest i-C_{19}/i-C_{20} index (0.22) among sites studied. This indicates the significant terrigenous organic matter input from terrestrial plant remains (despite the algal mass supply characteristic for this site), and their decomposition under redox conditions. Hence, the significant part of organic matter might be comprised of organic substances other than n-alkanes which require special investigation.

Thus, the found TOC and n-alkanes distribution showed that sediments from studied basins tend to be terrigenous with major input of organic matter from terrestrial plant remains. The transformation of the original organic matter in sediments studied often occurs under redox conditions. The minor presence of autochthonous microbial sources may indicate the microbial community extension.

3.3 Metals in sediments

3.3.1 Abundance and distribution of metals in surface sediments

Trace heavy metals enter coastal environments with a terrigenous fresh water discharge from the drainage area, with the precipitates and aerosols. Within the water column, dissolved metal forms are actively adsorbed on to suspended particulate matter (SPM). Further sedimentation of suspended particles contributes to trace heavy metal accumulation within the sediment phase (Foerstner & Wittmann, 1981). Within the sea-fresh waters barrier zone, flocculation and coagulation of the dissolved (colloidal) and fine suspended matter (such as clay particles, oxyhydroxides and organic substances) initiated at the salinity gradient lead to metal enrichment of the deposited material, since the dissolved and particulate metals are actively incorporated in (sorbing on) newly formed aggregates (Morris, 1986). Although, some dissolved metal entering the aquatic environments may be directly fixed by bottom sediments "escaping" the SPM phase (Dauvalter, 1998a). Therefore, bottom sediments being a long-term natural element "traps" may be used as preferential indicators of the trace heavy metals environmental pollution degree.

Anthropogenic materials (e.g. trace heavy metals) can both be present in the sediments from deposition in the past and can arrive with current deposits, while sediment resuspension and subsequent reaction and transport in the water column can be an important mechanism for redistribution of trace contaminant metal. Within the Karelian coastline, the small-scale geochemical patterns and disparities complicate the assessment of natural geochemical background of the region under investigation (Koukina et al, 2003). Therefore, the found mean trace heavy metal content values were compared with those found in previous studies for surface sediments from the other White Sea regions, where the anthropogenic trace heavy metal contamination was not detected. These systems are: the Karelian shore small bays (Ermolinskaya, Griaznaya, Poyakonda, Podvolochie, Chernorechenskaya), Chupa estuary (Chupa Bay of Kandalaksha Bay) and Dvina Bay of the White Sea (Table 2). To assess the ecological status of the Karelian shore sediments, the found content values were also related to the available sediment quality guidelines for trace heavy metals.

The mean contents of Pb, Cu, Zn and Cr in Karelian shore separating basins are comparable or lower the respective values from other Kandalaksha Bay regions and Dvina Bay of the White Sea. According to effects range-low (ERL) and effects range-median (ERM) sediment quality guidelines for trace metals, all found contents for Karelian shore separating basins were below both the ERL and ERM levels generally suggested to have potential adverse biological effects of 20-30% and 60-90% of incidence, respectively (Long et al, 1995). Hence, the sediments studied were related to uncontaminated by trace heavy metals and by Pb, Cu, Zn and Cr, in particular.

	n	Pb	Cu	Zn	Cr	Source
Karelian shore separating basins	10	4	16	118	41	Present study
	SD	2	12	70	18	
Karelian shore small bays	18	16	16	141	169	(Koukina et al 2003)
	SD	8	11	26	46	
Chupa Bay*	11	21	18	85	nd	(Millward et al 1999**)
	Range	12-25	15-22	57-110	nd	
Dvina Bay	13	20	21	76	84	(Koukina et al 2001)
	SD	14	8	27	26	
ERL guideline value	Incidence of adverse effect 20-30%	47	34	150	81	(Long et al, 1995)
ERM guideline value	Incidence of adverse effect 60-90%	218	270	410	370	(Long et al, 1995)

Table 2. Mean metal contents ($\mu g\ g^{-1}$) in the surficial sediments from the Karelian shore separating basins, Karelian shore small bays, Chupa estuary, Dvina Bay, effects range-low (ERL) and effects range-median (ERM) sediment quality guidelines values for trace metals; n - number of samples, SD – standard deviation to nearest $\mu g\ g^{-1}$, * - range of contents for <63 μm sediment fraction, ** - as cited in Koukina et al, 2003.

St.	Pb_{tot}	Cu_{ac}	Cu_{alk}	Cu_{tot}	Zn_{ac}	Zn_{alk}	Zn_{tot}	Cr_{ac}	Cr_{alk}	Cr_{tot}	Mn_{ac}	Mn_{alk}	Mn_{tot}	Fe_{ac}	Fe_{alk}	Fe_{tot}
12	2	1	1	3	3	4	159	<0.5	<0.5	12	7	2	26	286	252	3560
13	6	2	2	18	5	5	63	2	1	68	4	1	150	756	665	22710
14	1	2	1	13	5	4	151	1	<0.5	50	4	<0.5	115	594	509	17400
15	6	1	1	8	6	4	31	3	2	49	5	2	88	1504	1595	19240
16	6	1	<0.5	13	7	4	61	2	1	52	4	1	106	1900	1710	20500
17	3	1	3	14	8	7	32	1	1	21	9	5	54	330	1145	8180
18	3	1	1	5	3	2	241	1	<0.5	15	3	1	44	146	47	5560
26	1	2	3	59	12	19	190	<0.5	<0.5	43	4	1	79	357	630	12890
30	5	1	1	12	5	4	128	1	<0.5	45	5	1	80	775	418	16100
31	5	1	1	12	5	3	128	2	1	51	25	19	102	3030	2390	20500
m	4	1	1	16	6	6	118	1	1	41	7	3	85	968	936	14664

Table 3. Acid soluble (labile, Me_{ac}), alkali soluble (organically bound, Me_{alk}) and total metal (Me_{tot}) distribution in surface sediments ($\mu g\ g^{-1}$).

Trace metal distribution in coastal sediments depends on the sediment type and major elements – the carriers (TOC, Fe, Mn) abundance (Koukina et al, 2002, 2010). Table 3 shows the total metal and metal forms contents in the Karelian shore sediments, while table 4 shows the Spearman correlations between elements studied. Among elements studied, Fe is major element and sediment indicator of Fe-rich aluminosilicates, Fe-rich heavy minerals, and hydrous Fe oxides. The total Fe in Karelian shore sediments was relatively low and varied within the range of 0.4-2.3% in the samples studied. Total Mn ranged from 26 to 150 $\mu g\ g^{-1}$, and total Cr ranged from 11 to 68 $\mu g\ g^{-1}$, while Mn and Cr distribution was similar to that of Fe (Spearman correlation coefficient r_s was 0.92 and 0.98, respectively, Table 4). The elevated contents of Fe, and thus of Mn and Cr, were found in the fine-grained silts from

Cape Zeleny Lake and Chernorechenskaya Bay (Table 3, st. 13-16, 30-31). These sites experience a significant input of terrestrial material from permanent and/or seasonal fresh water discharge that supply Fe and Mn oxides and Fe-Mn bearing minerals mobilized from adjacent soils, stream and river beds, and ground waters. Total Pb, Cu and Zn varied in the ranges 1-6, 3-60, and 30-240 µg g^{-1}, respectively. Total Pb content was uniformly low and significantly correlated with total Fe (r_s=0.69). Total Cu positively correlated with TOC (r_s=0.57) and had a clear content maximum in the organic rich sediment from Porchalische lagoon (59 µg g^{-1}). Total Zn showed no pronounced affinity to major and trace elements studied but exhibited sporadic sharp enrichments (160-240 µg g-1) in some sediments of Cape Zeleny Lake (st.12, 14, 16), Kislosladkoe Lake (st. 18) and organic rich sediment of the Porchalische lagoon (st. 26).

The vertical metal distribution within the upper sediment column is influenced by the oxidation-reduction reactions and authigenic mineral formation (Dauvalter, 1997, Moiseenko et al, 2009a). At most of the sites, where the muddy sediments underlie the oxic bottom water, the top oxic layer was enriched in Fe, Mn, Cr and, in a lesser extent, in Cu and Pb. Increased contents of Fe, Mn and elements associated with in the top oxic sediment layer are due to the precipitation and accumulation of Fe and Mn oxides at the sediment-water interface. Fe and Mn may be also supplied to the top layer in the reduced dissolved forms, which diffuse from the lower layers of sediment column where Fe and SO$_4$ reduction under anaerobic conditions occurs. Cr, being a siderophile element, may be both associated with Fe-Mn bearing minerals and precipitated with Fe and Mn oxides that can lead to enrichments in fine-grained solids of surface sediments. However, at some sites, the subsurface sediments were slightly enriched in Fe, Mn and Cr comparing to top oxic layer (st. 15-16, 30-31). In the layer of Fe and S reduction and below, sulfates in sediments are reduced to form of hydrogen sulfide. The newly formed hydrogen sulfide is bound by mobile forms of ferrous oxide to monosulfide and, in the same time, experiences bacterial transformation into free sulfur. In this form, sulfur and the lower metal oxides form such minerals as pyrite, chalcopyrite, pentlandite, covellite etc (Dauvalter, 1998b, Dauvalter, 2000, Moiseenko et al, 2009b). Thus, the post-depositional diagenetic effects of Fe-Mn cycling both at and near the sediment-water interface may lead to subsurface metal accumulation (Loring et al, 1998).

In the most of the samples with an oxic cap, subsurface sediments exhibited at least twice higher Zn content comparing to the surface oxic layer. These are sediments from Cape Zeleny Lake (st. 13-14, Zn content 60-150 µg g^{-1} , st. 15-16, Zn 30-60 µg g^{-1}) and especially Kislosladkoe Lake (st. 17-18, Zn 30-240 µg g^{-1}). According to the vertical structure of these sediments, below the oxic cap (in the anaerobic layer) FeS and other metal insoluble sulfides may be formed due to the precipitation of Fe (III) and SO$_4{}^{2-}$ reduction products. The presence of sulfides is considered to be one of the major factors controlling the immobilization of metals in sediments, since most of divalent metals form insoluble sulfides or coprecipitate and adsorb on to iron sulfides. Thus, divalent Zn is shown to be bound by reduced S forming sphalerite ZnS in the flooded soils and lake sediments. Moreover, under anaerobic conditions Zn2+ may be coordinated by reduced S-containing functional groups of humic substances (Vodianitzky, 2008, 2009). Thus, under anaerobic conditions Zn may form insoluble sulfides and complexes with organic substances that may cause Zn subsurface content maximums. The abundance and sporadic enrichments in Zn may be also attributed to mineralogical composition of the sediment and hence anomalous concentrations of detrital heavy minerals containing the element. Heavy minerals, enriched in microelements, are always present in the dust fraction of soils and terrigenous sediments that originate from the bedrocks of respective biogeochemical province (Motuzova, 2009). Thus, Zn as a chalcophile element may preferentially concentrate in some of the rock-building minerals, characteristic for Karelian shore, for example, ferromagnesium silicates. Indeed, some hornblendes and pyroxenes of alkaline rocks characteristic for the area may contain up to 600 g tonne $^{-1}$ of Zn (Ivanov et al, 1973, Ivanov, 1997).

Therefore, the Spearman correlation analysis and metal distribution in relation to oxic-anoxic boundary within the upper centimeters of sediment column revealed following element associations. Fe, Mn, Cr and, in a lesser extent, Pb preferentially accumulate within the top oxic layer due to the precipitation with Fe-oxyhydroxides. Cu also tends to accumulate in surface sediments, but its sedimentation is to a higher degree bound to organic matter sedimentation processes than for other metals studied (Koukina et al, 1999, 2001, 2010). Zn enrichments in the subsurface zone are most likely controlled by a sulfide-associated phase.

	Cu_{ac}	Cu_{alk}	Cu_{tot}	Zn_{ac}	Zn_{alk}	Zn_{tot}	Pb_{tot}	Cr_{ac}	Cr_{alk}	Cr_{tot}	Mn_{tot}	Fe_{ac}	Fe_{alk}	Fe_{tot}	TOC
Cu_{ac}	1,00	**0,68**	**0,70**	0,36	**0,80**	-0,07	-0,25	-0,03	-0,01	0,23	0,40	-0,09	-0,11	0,17	0,35
Cu_{alk}	**0,68**	1,00	0,54	0,47	**0,85**	-0,10	-0,23	-0,26	-0,09	-0,23	-0,18	-0,30	-0,08	-0,19	0,62
Cu_{tot}	**0,70**	0,54	1,00	**0,73**	**0,74**	-0,16	-0,05	0,03	0,17	0,44	0,48	0,11	0,32	0,38	0,57
Zn_{ac}	0,36	0,47	**0,73**	1,00	0,57	-0,45	0,02	0,06	0,42	0,21	0,20	0,20	0,46	0,21	**0,82**
Zn_{alk}	**0,80**	**0,85**	**0,74**	0,57	1,00	-0,04	-0,31	-0,36	-0,11	-0,07	0,01	-0,28	-0,08	-0,12	**0,63**
Zn_{tot}	-0,07	-0,10	-0,16	-0,45	-0,04	1,00	-0,60	**-0,75**	**-0,96**	-0,47	-0,41	-0,57	**-0,72**	-0,53	-0,44
Pb_{tot}	-0,25	-0,23	-0,05	0,02	-0,31	-0,60	1,00	**0,68**	**0,70**	0,62	0,46	0,60	0,49	**0,69**	-0,04
Cr_{ac}	-0,03	-0,26	0,03	0,06	-0,36	**-0,75**	0,68	1,00	**0,86**	**0,76**	**0,72**	**0,74**	**0,72**	**0,82**	0,04
Cr_{alk}	-0,01	-0,09	0,17	0,42	-0,11	**-0,96**	0,70	0,86	1,00	**0,64**	0,57	0,67	**0,78**	**0,69**	0,37
Cr_{tot}	0,23	-0,23	0,44	0,21	-0,07	-0,47	0,62	**0,76**	**0,64**	1,00	**0,96**	**0,79**	**0,66**	**0,98**	0,05
Mn_{tot}	0,40	-0,18	0,48	0,20	0,01	-0,41	0,46	**0,72**	0,57	**0,96**	1,00	**0,68**	0,53	**0,92**	0,03
Fe_{ac}	-0,09	-0,30	0,11	0,20	-0,28	-0,57	0,60	**0,74**	**0,67**	**0,79**	**0,68**	1,00	**0,79**	**0,84**	0,03
Fe_{alk}	-0,11	-0,08	0,32	0,46	-0,08	**-0,72**	0,49	**0,72**	**0,78**	**0,66**	0,53	**0,79**	1,00	**0,72**	0,44
Fe_{tot}	0,17	-0,19	0,38	0,21	-0,12	-0,53	**0,69**	**0,82**	**0,69**	**0,98**	**0,92**	**0,84**	**0,72**	1,00	0,08
TOC	0,35	0,62	0,57	**0,82**	**0,63**	-0,44	-0,04	0,04	0,37	0,05	0,03	0,03	0,44	0,08	1,00

Table 4. Spearman rank order correlation matrix for acid soluble (Me_{ac}), alkali soluble (Me_{alk}) and total (Me_{tot}) metals and TOC in surface sediments, (n=10, marked correlations are significant at p<0.05).

3.3.2 Metal forms in surface sediments

Trace metals are introduced into coastal sediments as constituents of, or in association with, solid inorganic and organic particles supplied from natural and anthropogenic sources or derived from solution (Loring et al, 1998). The most important metal carriers within these sediments are aluminosilicate clay minerals, Fe-Mn oxides, marine and terrestrial organic materials, carbonates, and detrital heavy minerals (Postma, 1980, Foerstner & Wittmann, 1981). Thus, because the metal is distributed over different phases, a simple measurement of total concentrations of the metal is inadequate to assess its bioavailability (Gerson et al, 2008, Krishnamurti, 2008). It is necessary to measure the metal in the various phases and this is usually done using selective or/and sequential extraction (Perin et al, 1997, Riedel & Sanders, 1998, Szefer et al, 1995). The previous sequential extraction studies of Karelian shore sediments showed that metals occur mainly in a biogeochemically stable mineral-incorporated form, which comprises 77-99% of total metal content. (Koukina et al, 2001, 2010). These are metals incorporated into aluminosilicate lattices, bound to resistant iron and manganese minerals or organic compounds, integrated with detrital heavy minerals or discrete hydroxides, carbonates and sulfides. Metals in mineral form cannot be easily recycled into the water column due to a higher vulnerability of mineral components to post-depositional diagenetic processes.

In this work, the two selective extractions were performed in order to assess potentially most bioavailable metal fractions in sediments being labile (acid soluble) and organically bound (alkali soluble). The labile (weakly bound) part of total metal content was extracted using the 25% acetic acid. The acetic acid removes metals held in ion exchange positions, easily soluble amorphous compounds of iron and manganese, carbonates and those metals weakly held in organic matter (Loring & Rantala, 1992). The proportion of the total metal removed by the extraction is defined as acid soluble (non-detrital) metal fraction in sediments. The 0.1M sodium pyrophosphate (pH 10) extract was supposed to remove the organically bound metals from sediments (Beckett, 1989). Organic matter could play a critical role in concentrating metals via ligand binding (Gerson et al, 2008, Naidu et al, 1997). Organic substances of different origin are first subjected to numerous biotic and abiotic diagenetic transformations in the surficial sediments, while associated elements may become more bioavailable. 0.1M sodium pyrophosphate is known to mobilize also some part of easily soluble amorphous Fe-oxides and hence the elements associated with (Vorobyova, 2006).

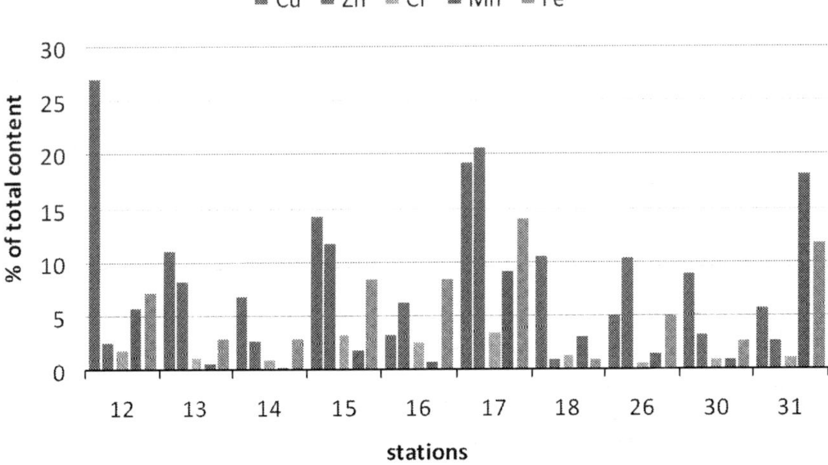

Fig. 4. The distribution of acid soluble (labile) metals in surface sediments.

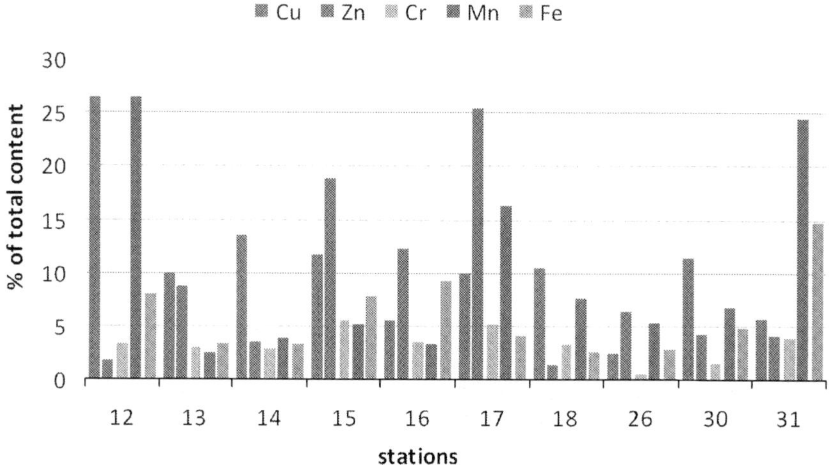

Fig. 5. The distribution of alkali soluble (organically bound) metals in surface sediments.

The amounts of metals extracted by 25% acetic acid and 0.1M sodium pyrophosphate are given in Table 2. For Pb, contents of both labile acid soluble and organically bound alkali soluble forms were negligible or below the detection limit. The amounts of labile acid soluble form varied within the ranges of 146-3030 µg g^{-1} for Fe, 3-25 µg g^{-1} for Mn, 1-3 µg g^{-1} for Cr, 1-2 µg g^{-1} for Cu, and 3-12 µg g^{-1} for Zn. This form made up 3-15% (an average 6.1%) of total contents for Fe, 2-27% (10.2%) for Mn, 0.5-5.5% (3.3%) for Cr, 3-27% (10.7%) for Cu and 1-25% (8.6%) for Zn (Fig. 3). The amounts of organically bound form varied within the ranges of 47-2390 µgg^{-1} for Fe, 1-19 µg g^{-1} for Mn, 1-2 µg g^{-1} for Cr, 1-3 µg g^{-1} for Cu, and 2-19 µg g^{-1} for Zn. This form made up 1-14% (an average 6.4%) of total contents for Fe, 0.5-18% (4.1%) for Mn, 0.5-3% (1.6%) for Cr, 3-2% (11.7%) for Cu and 1-21 (6.9%) for Zn (Fig. 4). Assuming that the determined amounts (parts) of acid soluble and alkali soluble forms could be a measure of potential metal bioavailability in sediments, the elements studied may be arranged in the following decreasing sequence: Cu>Zn>Mn>Fe>Cr.

For Fe, the distribution of both labile and organically bound forms in the samples studied is to a great extent a function of total Fe content (r_s=0.71-0.83, Table 4), while relative enrichments in labile and organically bound Fe were related to the sediments from Cape Zeleny Lake (st. 12, 15, 16) and especially to the subsurface sediments from Chernaya river estuary (st. 31). Labile, organically bound and total Cr was significantly positively correlated with respective forms of Fe (r_s=0.67-0.98). Labile and organically bound Mn distribution did not reveal any correlations with other elements studied (therefore these correlations were excluded from table 4). However, enrichments in organically bound and, to a higher extent, in labile Mn were related to Cape Zeleny Lake strait bed (st. 12), organic rich surface sediment from Kislosladkoe Lake (st. 17), and subsurface sediment from Chernaya river estuary (st. 31). Strong positive correlation is revealed between organically bound forms of Zn and Cu (r_s=0.85), which were both positively correlated with organic carbon (r_s=0.62-0.63), while labile Zn exhibited even stronger correlation with TOC (r_s=0.82). The relative percent content of labile acid soluble form (in relation to total metal content) significantly exceeded these of organically bound alkali soluble form at all sites for Mn, and at most of the sites for Cr and Zn (Fig. 3 and 4). Mn is known to be precipitated from solution at pH>8, while the 0.1M sodium pyrophosphate has pH 10, at which the labile Mn and associated elements might be immobilized (Vodianitzky, 2009). For Fe and Cu, acid soluble form slightly exceeded the alkali soluble form, except sites where sediments were mostly enriched in TOC (st.17 and 26, 8.3% and 18.7% of TOC, respectively). In these organic rich sediments, the content and the part of organically bound Fe, Cu, and sometimes Zn (st. 26) were at least twice higher than these of labile metals. Therefore, in sediments studied the most bioavailable metal fraction is mostly comprised of metals held in ion exchange positions, weakly bound to organic matter and, in a greater extent, associated with easily soluble amorphous Fe-oxides, abundant in terrigenous sediments. Organic matter starts to play a critical role in concentration metals via ligand binding at the sufficient TOC content of organic rich sediments. Among elements studied the most bioavailable part of Zn and Cu is probably bound to organic substances, while bioavailable Cr and Mn are in a greater extent controlled by Fe-oxyhydroxide formation.

Among sites studied, the elevated content of both bioavailable forms of most metals studied is related to samples 12, 17, 26, 31. The surface silts from Kislosladkoe Lake (st. 17) and Porchalische lagoon (st. 26) experience the significant supply of terrestrial organic matter and are mostly enriched in TOC. The surface silty sand collected from the bed of the strait connecting Cape Zeleny Lake and Kislaya Bay (st. 12) experiences the regular sea-fresh water change. The subsurface silt collected in the vicinity of Chernaya river mouth (st. 31) is located within the estuarine water mixing zone. In both samples (12 and 31), the clear imprint of the microbial organic matter was detected by the *n*-alkanes distribution study. Hence, the role of microbial community in element speciation in sediments needs special study. The elevated contents of bioavailable metal forms are related to sediments enriched in organic matter and/or located within the sea-fresh water barrier zones.

Taking into account the individual metal affinity to organically bound or labile fraction, an average, the acetic acid and sodium pyrophosphate seem to release comparable amounts of metals from the sediments (3-11% and 2-12 % of total metal content, respectively). This might be to some extent due to a crossing selectivity of these reagents. Thus, both acetic acid and sodium pyrophosphate are able to remove metals weakly held in organic matter and associated with easily soluble amorphous Fe-oxides. However, the found amounts of labile and organically bound metal forms in studied separating basins exceeded the respective values of labile metal fraction for sediments from previously studied open small bays of the Karelian shore (that made up an average only 2-5% of total metal content) and corresponds to the enhanced labile metal (6-8 % for Fe, Mn, Zn, Cr) contents found in separating lagoons of Chernorechenskaya Bay (Koukina et al, 2010). Major and trace elements entering the separating basins with fresh water discharge are scavenged by Fe-Mn oxides and bound to coagulating organic substances in the course of sedimentation processes, with further burial in surface sediments in labile forms. The anaerobic conditions in the bottom depressions and their winter extension within the water column cause the release of labile metals to pore and bottom waters. While some trace metals may be immobilized by precipitation as sulfides at sufficient H_2S concentration, regular change of oxidizing conditions at permanent element input from fresh water supply contributes to relative accumulation of bioavailable metal forms in sediments. Hence, at the found low or close to background trace heavy metal contents, in separating basins the relative part of labile bioavailable metals is still enhanced in relation to the neighboring open coastal sea.

4. Conclusion

Restricted exchange environments studied varied from typical small estuary Chernorechenskaya, complicated by shallow bars, to separating basins of different isolation stage being Cape Zeleny Lake and Kislosladkoe Lake, and to the isolated littoral pool - Porchalische lagoon. The specific conditions of separating basins under investigation being restricted water exchange with open sea, permanent or seasonal fresh water input and bottom relief complicated by depressions lead to extreme values of hydrochemical parameters during the summer period, while the most specific features are contrast oxidizing conditions within the water column with anoxia zones occurrence in the bottom depressions that may be spread on to the whole water body in winters. The evolution of sediments here is strongly affected by small-scale variability in oxidizing conditions.

The found TOC and n-alkanes distribution showed that sediments from studied basins tend to be terrigenous with major input of organic matter from terrestrial plant remains. The transformation of the original organic matter in sediments studied often occurs under redox conditions which are especially pronounced in the Porchalische lagoon. The revealed presence of autochthonous microbial sources of the organic matter in sediments may indicate the microbial community extension within the ecosystems.

The total trace heavy metal contents in sediments from separating basins studied corresponded to the respective values for other White Sea regions uncontaminated by trace heavy metals and were below the threshold levels according to sediment quality guidelines. Hence, in the restricted exchange environments studied no significant contamination by trace heavy metals (Pb, Cu, Zn and Cr, in particular) was detected. Spearman correlation analysis and total metal distribution in relation to oxic-anoxic boundary within the upper centimeters of sediment column revealed element affinity to different sediment components. Thus, Fe, Mn, Cr and, in a lesser extent, Pb preferentially accumulate within the top oxic layer due to the precipitation with Fe-oxyhydroxides. Total Cu also tends to accumulate in surface sediments, but its sedimentation is to a higher degree bound to organic matter sedimentation processes than for other metals studied. Total Zn enrichments in the subsurface zone are most likely controlled by a sulfide-associated phase and/or bound to Zn-rich detrital minerals.

The comparative study of two most bioavailable metal forms being labile (acid soluble) and organically bound (alkali soluble) showed that acetic acid and sodium pyrophosphate release comparable amounts of metals from the sediments being 3-11% and 2-12 % of total metal content, respectively. Therefore, the most bioavailable part of metals is comprised of metals held in ion exchange positions, weakly bound to organic matter and, in a greater extent, associated with easily soluble amorphous Fe-oxides, abundant in terrigenous sediments. Organic matter starts to play a critical role in concentration of metals at the sufficient TOC content of organic rich sediments. The found amounts of labile and organically bound metal forms exceeded the respective values for previously studied small bays of the Karelian shore. Hence, in separating basins the relative part of labile bioavailable metals is enhanced in relation to the neighboring open coastal sea. In separating basins, the regular change of oxic to anoxic conditions in sediments and waters by permanent element input from fresh water supply contributes to relative accumulation of bioavailable metal forms in sediments.

Among sites studied, the elevated contents of bioavailable metal forms are related to sediments enriched in organic matter and/or located within the sea-fresh water barrier zones. Major and trace elements entering the separating basins with fresh water discharge are scavenged by newly forming Fe-Mn oxides and bound to coagulating organic substances in the course of sedimentation processes, with further burial in surface sediments in potentially bioavailable labile forms that can be further recycled to water column. Among elements studied the most bioavailable part of Zn and Cu is probably bound to organic substances, while bioavailable Cr and Mn are in a greater extent controlled by Fe-oxyhydroxide formation. According to their averaged potential bioavailability, the elements studied may be arranged in the following decreasing sequence: Cu>Zn>Mn>Fe>Cr.

The present study is essential for the prognosis of Arctic ecosystems vulnerability and, in particular, anoxia occurrences in Arctic coastal environments. The most vivid example is the Porchalische lagoon, where at highest (among sites studied) organic matter accumulation rate the anoxic sediments exhibited the enhanced contents of potentially bioavailable trace heavy metals. Such ecologically unfavorable situation might be typical for central bottom depressions of separating basins in summer and may be aggravated in winter periods.

5. Acknowledgment

The study was supported by the Russian Foundation of Basic Research (09-05-00011), the Presidium of Russian Academy of Sciences (Programs 21 and 23).

6. References

Beckett P.H.T. (1989) The use of extractants in studies on trace metals in soils, sewage sludges and sludge-treated soils, *Advances in Soil Science*, Vol. 9, pp. 143-176.

Berkovitz L.A., Lukashin V.N. (1984). Three sediment reference samples: SDO-1, SDO-2, SDO-3, *Geostandard Newsletters*, Vol. 8, No 1, pp. 761-795.

Dauvalter V.A. (1997). Metal concentrations in sediments in acidifying lakes in Finnish Lapland. *Boreal Environmental Research*, Vol. 2, pp. 369-379, ISSN 1239-6095.

Dauvalter V.A. (1998a). Concentration of metals in bottom sediments of acid lakes. *Water resources*, Vol. 25, No. 3, pp. 385-335, ISSN 0097-8078.

Dauvalter V.A. (1998b). Heavy metals in the bottom sediments of the Inari-Pasvik lake-river system. *Water resources*, Vol. 25, No. 4, pp. 451-457, ISSN 0097-8078.

Dauvalter V.A., Moiseenko T.I., Kudryavtseva L.P., and Sandimirov S.S. (2000). Accumulation of heavy metals in Lake Imandra because if its pollution with Industrial Waste. *Water resources*, Vol. 27, No. 4, pp. 451-457, ISSN 0097-8078.

Dobrovolsky A.D., Zalogin B.A. (1982). *The Seas of the USSR*, MSU, Moscow, Russia, 192 pp.

Foerstner U., Wittmann G.T.W. (1981). *Metal pollution in the aquatic environment*, Springer-Verlag, Berlin, Germany, 486 pp.

Gerson A.R., Anastasio C., Crowe S., Fowle D., Guo B., Kennedy I., Lombi E., Nico P.S., Marcus M.A., Martin R.R., Naftel S.J., Nelson A.J., Paktunc D., Roberts J.A., Weisener C.G. and Werner M.L. (2008). Frontiers in assessing the role of chemical speciation and natural attenuation on the bioavailability of contaminants in the terrestrial environment, in: *Chemical Bioavailability in Terrestrial Environments*, 1, Naidu R. (Ed.), Elsevier, Oxford, UK, pp. 99-136, ISBN 978-0-444-52169-9.

Glukhovsky B.N. (Ed.) (1991). *Hydrometeorology and hydrology of the seas of the USSR,* Vol. 2, White Sea, Issue 1, Hydrometeoizdat, Leningrad, USSR, 240 pp.

Ivanov V.V. and 16 others (1973). *Mean contents of admixture elements in minerals*, Nedra, Moscow, USSR, 208 pp.

Ivanov V.V. (1995). *Ecological element geochemistry*. Vol. 4, Ecologiya, Moscow, 416 pp., ISBN 5-7120-0647-2.

Koukina S.E., Calafat-Frau A., Hummel H., and Palerud R. (2001). Trace metals in suspended particulate matter and sediments from the Severnaya Dvina estuary, Russian Arctic, *Polar Record*, Vol. 37, pp. 249-256, ISSN 0032-2474.

Koukina S.E, Korneeva G.A, Ametistova L.E, and Bek T.A. (2003). A comparative biogeochemical study of sediments from Kandalaksha bay, White Sea, Russian Arctic, *Polar Record*, Vol. 39, pp. 357-356, ISSN 0032-2474.

Koukina S.E., Korneeva G.A., and Bek T.A. (2010). Forms of metals in littoral sediments in Kandalaksha Bay of the White Sea in the Russian Arctic, *Oceanology International*, Vol. 50, No. 6, pp. 926-932, ISSN 0001-4370.

Koukina S.E., Korneeva G.A., Kolbasov G.A, Ametistova L.E., Gantzevich M.M., Bek T.A. and Romankevich E.A. (2006). Sediment biogeochemistry and macrobenthos community structure in small coastal environments of Kandalaksha bay, in: *Proceedings of White Sea Biological Station,* Vol. 10, Grif&K, Tula, Russia, pp. 204-215.

Koukina S.E., Sadovnikova L.K., Calafat-Frau A., Palerud R., and Hummel H. (1999). Forms of metals in bottom sediments from some estuaries of the basins of the White and Barents seas, *Geochemistry International*, Vol. 37, No. 12, pp. 1197-1202, ISSN 0016-7029.

Krasnova E.D. (2008). *Travels over the Kindo-Mis*, Grif & K, Tula, Russia, 144 pp., ISBN 978-5-8125-1120-3.

Krishnamurti G.S.R. (2008). Chemical methods for assessing contaminant bioavailability in soils, in: *Chemical Bioavailability in Terrestrial Environments*, 1, Naidu R. (Ed.), Elsevier, Oxford, UK, pp. 495-520, ISBN 978-0-444-52169-9.

Lisitzin A.P., Nemirovskaya I.A., Shevchenko V.P. (Eds.) (2010). *Natural environment of the catchment area of the White Sea*. The White Sea System, Vol. 1, Scientific World, Moscow, Russia, 480 pp., ISBN 978-5-91522-194-8.

Long E.R., MacDonald D.D., Smith S.L., and Calder F.D. (1995). Incidence of adverse biological effects within ranges of chemical concentrations in marine and estuarine sediments, *Environmental Management,* Vol. 19, pp. 81-97, ISSN 0364-152X.

Loring D.H., Rantala R.T.T (1992): Manual for geochemical analyses of marine sediments and suspended particulate matter, *Earth-Science Reviews*, Vol. 32, pp. 235-283, ISSN 0012-8252.

Loring D.H., Dahle S., Naes K., Dos Santos J., Skei J.M., and Matishov G.G.(1998): Arsenic and other trace metals in sediments from the Pechora Sea, Russia, *Aquatic Geochemistry*, Vol. 4, pp. 233-252, ISSN 1380-6165.

Martin, A. J., McNee, J. J., and Pedersen, T. F. (2001): The reactivity of sediments impacted by metal mining in Lago Junin, Peru, J. *Journal of Geochemical Exploration,* Vol. 74, No. 1-3, pp. 175-187, ISSN 0375-6742.

Martin W.R. (2010): Chemical processes in estuarine sediments, in: *Marine Chemistry and Geochemistry*, Steele J.H., Thorpe S.A and Turekian K.K. (Eds.), Academic Press, Elsevier Ltd, London, UK, pp. 445-456, ISBN 978-0-08-096483-6, London, UK.

Milanovsky E.E. (1987). *Geology of the USSR*. Part 1. Moscow State University, Moscow, USSR, 415 pp., ISBN 5-1646475.

Moiseenko T.I., Gashkina N.A., Sharov A.N., Vandysh O.I., and Kudryavtseva L.P. (2009). Anthropogenic transformation of the Arctic Ecosystem of Lake Imandra: tendencies for recovery after long period of pollution. *Water resources*, Vol. 36, No. 3, pp. 296-309, ISSN 0097-8078.

Moiseenko T. I. A. N. Sharov, O. I.Vandish, L. P. Kudryavtseva, N. A. Gashkina (2009). Long-term modification of arctic lake ecosystem: reference condition, degradation and recovery. *Limnologica*, V. 39, No. 1, pp. 1-13, ISSN 0075-9511.

Morris A.W. (1986). Removal of trace metals in the very low salinity region of the Tamar estuary, England. *Science of the total Environment*, Vol. 49, pp. 297-304, ISSN 0048-9697 *Water Resources*, Vol. 36, No.3, pp. 321-325, ISSN 0097-8078.

Motuzova G.V (2009). *Microelement compounds in soils: Systematic organization, ecological significance, monitoring*, LIBROCOM, Moscow, Russia, 168 pp., ISBN 978-5-397-00140-3.

Naidu A.S., Blanchard A., Kelley J.J., Goering G.G., Hameedi M.J. and Baskaran M. (1997). Heavy metals in Chukchi Sea sediments as compared to selected circum-Arctic shelves, *Marine Pollution Bulletin*, Vol. 35, pp. 260-269, ISSN 0025-326X.

Oliunina O.S., Romanenko F.A. (2007). Uplift of the Karelian shore of the White Sea based on results of the peatlands study, in: Fundamental problems of the Quarter: study results and basic directions for future investigations, *Proceedings of the 5th All-Russian meeting on the Quarternary period study*, GEOS, Moscow, Russia, pp. 312-315

Pantiulin A.N. (2001). White Sea as a large-scale estuarine system, *Geophysical Research Letters*, Vol. 3, pp. 5-21, ISSN 0094-8276.

Pavlidis Y.A., Scherbakov F.A. and Scevchenko A.Y. (1995) Clay minerals of Cuba and White Sea shelves: geology and climate – a comparison. *Oceanology International*, Vol. 35, No. 1, pp. 121-127, ISSN 0001-4370.

Peresypkin V.I. and Belyaev N.A. (2009). The composition of *n*-alkanes in the sediments and bacterial matte from separated lakes of Kandalaksha Bay of the White Sea, in: *Proceedings of the 18th International School/Conference on Marine Geology*, Moscow, Russia, 16-20 November 2009, Part 4, pp. 125-128, ISBN 978-5-89118-479-4.

Peresypkin V.I., Romankevich E.A. (2010) *Lignin Biogeochemistry*, GEOS, Moscow, Russia, 340 pp., ISBN 978-5-89118-484-8.

Perin G., Fabris R., Manente S., Rebello Wagener A., Hamacher C., Scotto S. (1997). A five-year study on the heavy metal pollution of Guanabra Bay sediments (Rio de Janeriro, Brazil), and evaluation of the metal bioavailability by means of geochemical speciation, *Water Research*, Vol. 31, No. 12, pp. 3017-3028, ISSN 0043-1354.

Postma H. (1980). *Chemistry and biogeochemistry of estuaries*, Wiley, Chichester, UK, 452 pp.

Rabalais N.N. (2010). Hypoxia, in: *Marine Chemistry and Geochemistry*, 2, Steele J.H., Thorpe S.A and Turekian K.K. (Eds.), Academic Press, Elsevier Ltd, London, UK, pp. 306-314, ISBN 978-0-08-096483-6.

Riedel G.F. and Sanders J.G. (1998). Trace element speciation and behavior in the tidal Delaware River, *Estuaries*, Vol. 21, pp. 78-98.

Romankevich E.A. (1984). *Geochemistry of organic matter in the ocean*, Springer-Verlag, Berlin, Heideberg, N.Y., Tokyo, 334 pp., ISBN 3-540-12754-2

Romankevich E.A., Vetrov A.A. (2001). *Cycle of carbon in the Russian Arctic Seas*, Nauka, Moscow, Russia, 302 pp., ISBN 5-02-004358.

Sapozhnikov V.V., Arzhanova N.V., Titov O.V., Torgunova N.I., and Rusanov I.I. (2001). Hydrochemical and microbiological features of Lake Mogil'noe. *Water Resources*, Vol. 28, No.1, pp. 54-62, ISSN 0097-8078.

Savenko V.S. (2006). *Chemical composition of world river's suspended matter*, GEOS, Moscow, Russia, 175 pp., ISBN 5-89118-345-5.

Shaporenko S.I., Korneeva G.A., Pantiulin A.N., and Pertzova H.M. (2005). Characteristics of the ecosystems of water bodies separating from Kandalaksha Bay of the White Sea. *Water Resources*, Vol. 32, No. 5, pp. 469-483, ISSN 0097-8078.

Shapovalov S.M. (2010). *Physical, geological and biological researches of oceans and seas*, Scientific World, Moscow, Russia, 630 pp., ISBN 978-5-91522-184-9.

Shurkin K.A. (1984). *Precambrian of the White Sea region*, Nauka, Leningrad, USSR, 341 pp.

Szefer P., Glasby G.P., Pempkowiak J., and Kaliszan R. (1995). Extraction studies of heavy-metal pollutants in surficial sediments from the southern Baltic Sea off Poland, *Chemical Geololy*, Vol. 120, pp. 111-126, ISSN 0009-2541.

Vetrov A.A. and Peresypkin V.I. (2009). Organic carbon in separated reservoirs of Kandalaksha Bay of the White Sea, in: *Proceedings of the 18th International School/Conference on Marine Geology*, Moscow, Russia, 16-20 November 2009 Part 4, pp. 29-33, ISBN 978-5-89118-479-4.

Vodianitzky Yu.N. (2008). *Heavy metals and metalloids in soils*, Dokuchaev Soil Science Institute, Moscow, Russia, 164 pp, ISBN 978-5-85941-267-9.

Vodianitzky Yu.N. (2009). *Heavy metals and metalloids in contaminated soils*, Dokuchaev Soil Science Institute, Moscow, Russia, 184 pp, ISBN 978-5-85941-327-0.

Vorobyova L.F. (Ed.) (2006). *Theory and Practice of Chemical Analysis of Soils*, GEOS, Moscow, Russia, 400 pp., ISBN 5-89118-344-7.

Wolanski E. (2007). *Estuarine Ecohydrology*, Elsevier, Amsterdam, Netherlands, 157 pp., ISBN 978-0-444-53066-0.

11

Two Species and Three Species Ecological Modeling - Homotopy Analysis

Venkata Sundaranand Putcha[*]
*Center for Mathematical Sciences - DST, C.R. Rao Advanced Institute of Mathematics,
Statistics and Computer Science (AIMSCS), UoH Campus,
Gachibowli, Hyderabad, A.P
India*

1. Introduction

It is observed that big trees are older than little trees and there are more little trees than big trees since not every little tree grows up to be a big tree - most die young. But seeds produced and shed by the bigger trees will naturally form little trees. It can also be noticed that there are some dead needles and leaves on the ground and some standing dead trees which will eventually fall to the soil, the result of the deaths of those young trees and plant parts. It is natural that the live trees have roots in the soil formed partly from those dead leaves and logs and surmise that the trees obtain some nutrients from them Mangroves are the forests positioned at the convergence of land and Sea in inter tidal zones of the world. Mangrove forests are architecturally simple when compared to rainforests, often lacking under storey of leafs and shrubberies and are generally less species rich than other tropical forests. Mangroves have been heavily used traditionally for food, timber, fuel and medicine, and presently occupy about $181000\,km^2$ of tropical and subtropical coastline. Mangroves are a precious ecological and economic resource, being vital nursery grounds and breeding sites for birds, fish, crustaceans, shellfish, reptiles,Polychaete, Crabs, Prawn, Zooplankton and mammals; a renewable source of wood; accrual sites for sediment, contaminants, carbon and nutrients; and offer Fortification against coastal erosion. Major reasons for obliteration of mangroves are urban development, aquaculture, mining and over exploitation for timber, fish, crustaceans and shellfish.

Ecological modeling does not deal directly with natural objects, it deals with the mathematical objects and operations which are offered as analogs of nature and natural processes. These mathematical models do not contain all information about nature that we may know, but only what we think are the most pertinent for the problem at hand. Ecological modeling helps us understand the logic of our thinking about nature to help us avoid making plausible arguments that may not be true or only true under certain restrictions. It helps us avoid wishful thinking about how we would like nature to be in favor of rigorous thinking about how nature might actually work.

It has been observed by many ecologist that Living organisms and their non living environment are inseparable, interrelated and interacts up on each other.When species interact the population dynamics of each species is affected. We consider here systems involving

---

[*]Dedicated to my grandmother Balatripura Sundari Putcha

two-species systems and three-species systems. There are three main types of interaction, (i) If the growth rate of one population is decreased and the other increased the populations are in a predator-prey situation, (ii) If the growth rate of each population is decreased then it is competition, (iii) If each population's growth rate is enhanced then it is called mutualism or symbiosis.

Eco system comprising organisms and a biotic environment is a functional unit of ecology and non linear differential equations of first order are used to represent complex ecological phenomena. In 1925 Volterra [20] first proposed a simple model for the predation of one species by another to explain the oscillatory levels of certain fish catches in the Adriatic with the following assumptions (i) The prey in the absence of any predation grows unboundedly in a Malthusian way (ii) The effect of the predation is to reduce the prey's per capita growth rate by a term proportional to the prey and predator populations (iii) In the absence of any prey for sustenance the predator's death rate results in exponential decay (iv) The prey's contribution to the predators' growth rate is proportional to the available prey as well as to the size of the predator population. Ayala etal [21] studied the theoretical models through experimental tests. Later Albrecht etal [22], Gopalaswamy [23], Brown etal [24] studied the stability of two species systems.

The systems of non linear differential equations are perfect and an abstract representation of the real problem and attracted the attention of mathematicians [3], [4], [7] and [9] and ecologists. May [25] studied the nonlinear aspects of competition between three species. Schuster etal [26] and Zhang Jinyan [27] studied three species systems.So it is of great interest to formulate a mathematical models on Two Species and Three Species Ecological systems and find their approximate analytical solutions. In general construction of analytic approximations of nonlinear problems with strong nonlinearity is not easy and constructed Solutions are mainly determined by the type of nonlinearity and the corresponding technique. The region of convergence of obtained solutions depend on physical parameters rather than the analytical technique.

Though numerical techniques can be used to solve the nonlinear system of differential equations the functional form of reasonably good solution, which is of vital use for detailed study of the system, can be constructed by Homotopy methods. In general it is difficult to obtain analytic approximations of nonlinear problems with strong nonlinearity. Traditionally, solution expressions of a nonlinear problem are mainly determined by the type of nonlinear equations and the employed analytic techniques, and the convergence regions of solution series are strongly dependent of physical parameters.

The Homotopy analysis method HAM, initially proposed by Liao in his Ph.D. thesis [15], is a powerful analytic method for nonlinear problems. A systematic and clear exposition on HAM is given in [16]. In recent years, this method has been successfully employed by many authors Sami etal [32], Francisco M. Fernandez [33], Faghidian [34], Shijun Liao [35], Zoua L [36], Rafei M [37], Cheng-shi Liu etal [38], Vipul K. Baranwal etal [39] to solve many types of nonlinear problems in science and engineering.

The references cited may not be exhaustive since much literature is available on methods for solving nonlinear systems by approximate analytical solutions in general and Homotopy methods in particular. This chapter is organized as follows. In section 2, outline of Perturbation method, Adomian decomposition method for system, Homotopy perturbation method, Homotopy analysis method and Homotopy analysis method for system to find approximate analytical solutions of non linear differential equations is presented. Section

3 is concerned with the two species systems and present approximate analytical solutions obtained by using perturbation method, Homotopy perturbation method and Homotopy analysis method. In section 4, Three species ecological systems are considered and are solved by using perturbation method, Adomian decomposition method and Homotopy analysis method. While remembering the famous words "In learning Science examples are useful than rules" of Isaac Newton numerical examples are presented to illustrate the simplicity and accuracy of the methods in section 5. Section 6 presents the summary and conclusions.

2. Methods

2.1 Perturbation method

Perturbation method is a widely used method to solve non linear differential equations.Consider a system of nonlinear differential equations of the type $\dot{x}_1(t) = f_1(x_1, x_2, x_3, t) + \varepsilon g_1(x_1, x_2, x_3, t), \dot{x}_2(t) = f_2(x_1, x_2, x_3, t) + \varepsilon g_2(x_1, x_2, x_3, t), \dot{x}_3(t) = f_3(x_1, x_2, x_3, t) + \varepsilon g_3(x_1, x_2, x_3, t)$ where the functions f_1, f_2, f_3 are linear and the functions g_1, g_2, g_3 are nonlinear and ε is a small parameter, satisfying the initial conditions $x_1(0) = a, x_2(0) = b, x_3(0) = c$. The solutions $x_1(t), x_2(t), x_3(t)$ are expressed as a power series of ε as $x_1(t) = x_{1,0}(t) + \varepsilon x_{1,1}(t) + \varepsilon^2 x_{1,2}(t) + ..., x_2(t) = x_{2,0}(t) + \varepsilon x_{2,1}(t) + \varepsilon^2 x_{2,2}(t) + ..., x_3(t) = x_{3,0}(t) + \varepsilon x_{3,1}(t) + \varepsilon^2 x_{3,2}(t) + ...$. These series will converge rapidly if ε is very small. The terms $x_{1,0}(t), x_{2,0}(t), x_{3,0}(t)$ are the solutions of the corresponding linear equations $\dot{x}_1(t) = f_1(x_1, x_2, x_3, t), \dot{x}_2(t) = f_2(x_1, x_2, x_3, t), \dot{x}_3(t) = f_3(x_1, x_2, x_3, t)$ and are known as the generating solutions. The terms $x_{1,1}(t), x_{2,1}(t), x_{3,1}(t), x_{1,2}(t), x_{2,2}(t), x_{3,2}(t)$ are called correction terms. It is an established fact that analytic approximations of nonlinear problems often fail for strong nonlinearities and the perturbation approximations work only for weak nonlinearities.

2.2 Adomian decomposition method for system of differential equations

The decomposition method was developed by G. Adomian [5] to linear and non-linear differential and partial differential equations. With the assumptions like weak non-linearity and small perturbation, the solution of the simpler mathematical problem may not be a good approximation to the solution of the original problem. The advantage of the decomposition method is that it provides analytical approximation to a wide class of non-linear equations without linearization as in perturbation, closure approximation or discritization methods. In this section, we describe the Adomain's decomposition method for non-linear matrix differential equation. Consider a matrix differential equation $FX(t) = G(t)$, where F represents a general non-linear ordinary differential operator involving both linear and non-linear terms, $X(t)$ and $G(t)$ are square matrices. Decompose the nonlinear operator into $M + N + R$, where M is easily invertible operator and R is the remainder of the linear operator and N represents non-linear operator. Thus the equation

$$FX(t) = G(t) \tag{1}$$

may be written as

$$MX(t) + RX(t) + NX(t) = G(t) \tag{2}$$

$$i.e., \quad MX = G - RX - NX$$

$$i.e., M^{-1}MX = M^{-1}G - M^{-1}RX - M^{-1}NX \tag{3}$$

since M is easily invertible operator. If this corresponds to an initial value problem, the integral operator M^{-1} may be a definite integral from t_0 to t. Solving (3) for X yields,

$$X = E_1 + E_2 t + M^{-1}G - M^{-1}RX - M^{-1}NX \tag{4}$$

The non linear term NX will be equated to $\sum_{n=0}^{\infty} A_n$, where the A_n's are special polynomials, which can be defined further by (7), and X will be decomposed into $\sum_{n=0}^{\infty} X_n$, with X_0 the initial approximate matrix identified as $E_1 + E_2 + M^{-1}G$. Thus (4) can be expressed as

$$\sum_{n=0}^{\infty} X_n = X_0 - M^{-1}R \sum_{n=0}^{\infty} X_n - M^{-1} \sum_{n=0}^{\infty} A_n \tag{5}$$

Consequently we can write, the matrix approximations,

$$X_1 = -M^{-1}RX_0 - M^{-1}A_0, X_2 = -M^{-1}RX_1 - M^{-1}A_1, ..., X_{n+1} = -M^{-1}RX_n - M^{-1}A_n \tag{6}$$

The matrix polynomials A_n are generated for each non-linearity so that A_0 depends only on X_0, A_1 depends only on X_o and X_1, A_2 depends on X_0, X_1 and X_2 and so on. All of the matrix components X_n are calculable, and thus $X = \sum_{n=0}^{\infty} X_n$. If the series converges, then the nth partial sum $\Phi_n = \sum_{i=0}^{n-1} X_i$ will be the solution since $\Phi_n \longrightarrow \sum_{i=0}^{n-1} X_i = X$. To calculate the matrix polynomials A_n's consider an equation for which $X(t)$ is the solution, containing a non-linear term $NX = Q(X) = \sum_{n=0}^{\infty} A_n$. These A_n matrix polynomials are defined as

$$A_0 = Q(X_0), A_1 = (e^{X_1 d/dx_0} - 1)Q(X_0)$$
$$A_2 = (e^{X_2 d/dx_0} - 1)Q(X_0) + (e^{X_1 d/dx_0} - 1)(e^{X_2 d/dx_0} - 1)Q(X_0) \tag{7}$$

Thus $\sum_{i=0}^{n} A_i = Q(X_0) + (X - X_0)dQ/dX_0 + ...$ That is, the partial sums consist of the essential terms of a Taylor expansion about the function $X_0(t)$ rather than about a point. Thus, addition of the first (n+1) terms of the $A_0, A_1,...,A_n$ approaches $(e^{Xd/dx_0} - 1)Q(X_0)$. With the product terms, products of n! occur in the denominator which also can be ignored after some n. Thus A_n can be written as

$$A_n = B_n Q(X_0), \text{where } B_n = C_n \sum_{j=0}^{n} B_j, \ n \geq 1, \ B_0 = 1, \ C_n = (e^{X_n d/dx_0} - 1) \tag{8}$$

2.3 Homotopy perturbation method

Various perturbation methods applied to solve nonlinear problems are based on the strict assumption of the existence of a small parameter which is not satisfied in many non linear problems. To overcome this difficulty in 1999, J H He [14] proposed a method called Homotopy perturbation method which is a combination of the classical perturbation technique and Homotopy technique. To explain the basic idea of the Homotopy perturbation method for solving nonlinear differential equations, we consider the following nonlinear differential equation

$$A(u) - f(r) = 0, r \in \Omega \tag{9}$$

satisfying the boundary condition

$$B(u, \frac{\partial u}{\partial t}) = 0, r \in \Gamma \tag{10}$$

where A is a general differential operator , B is a boundary operator, $f(r)$ is a known analytical function, Γ is the boundary of domain Ω and $\frac{\partial}{\partial t}$ denotes differentiation along the normal drawn outwards from Ω. The operator A can be divided into a linear part L and a nonlinear part N as

$$L(v) + N(v) - f(r) = 0 \tag{11}$$

Construct a Homotopy $v(r,p) : \Omega \times [0,1] \longrightarrow \mathbb{R}$ where the embedding parameter $p \in [0,1], r \in \Omega$, the initial guess u_0 satisfying

$$H(v,p) = (1-p)[L(v) - L(u_0)] + p[A(v) - f(r)] = 0, \tag{12}$$

The above equation is equivalent to $H(v,p) = L(v) - L(u_0) + pL(u_0) + p[N(v) - f(r)] = 0$. As p changes from 0 to 1 the initial guess $u_0(r)$ of $v(r,p)$ goes to $u(r)$. In this context $L(u) - L(u_0)$ and $A(v) - f(r)$ are homotopic in the topological sense. Now the solution of (12) is of the form $v = v_0 + pv_1 + p^2 v_2 + ...$ The approximate analytical solution of (9) is given by $u = \lim_{p \longrightarrow 1} v = \sum_{i=0}^{\infty} v_i$. This series is convergent for most cases, however, the convergent rate depends upon the nonlinear operator $A(v)$ and 1. From [12] the series converges if (i) the second derivative of $N(v)$ with respect to v must be small, because the parameter p may be relatively large,(i.e. $p \longrightarrow 1$) and (ii) the norm of $L^{-1}\frac{\partial N}{\partial v}$ must be smaller than one. perturbation techniques use perturbation quantities to transfer a nonlinear problem into an infinite number of linear sub-problems and then approximate it by the sum of solutions of the first several sub-problems. The existence of perturbation quantities is obviously a cornerstone of perturbation techniques, however, it is the perturbation quantity that brings perturbation techniques some serious restrictions.It is not necessary for every nonlinear problem to contain such a perturbation quantity. This is a major restriction of perturbation method.

2.4 Homotopy analysis method

From the literature it is observed that if a nonlinear problem has a unique solution, there may exist an infinite number of different solution expressions whose convergence region and rate are dependent on an auxiliary parameter. Homotopy analysis method was first proposed by Liao [15] based on Homotopy, a fundamental concept in topology and differential geometry. The HAM is based on construction of Homotopy which continuously deforms an initial guess approximation to the exact solution of the given problem. An auxiliary linear operator is chosen to construct the Homotopy and an auxiliary linear parameter is used to control and adjust the region of the convergence region and rate of the solution series, which is not possible in the other methods like perturbation techniques, Homotopy perturbation methods, decomposition methods. The Homotopy analysis method provides the greater flexibility in choosing initial approximations and auxiliary linear operators. Moreover, unlike all previous analytic techniques, the Homotopy analysis method provides great freedom to use different base functions to express solutions of a nonlinear problem so that one can approximate a nonlinear problem more efficiently by means of better base functions. Thus, this method is valid for nonlinear problems with strong nonlinearity. Furthermore, the Homotopy analysis method logically contains some previous techniques such as perturbation method, AdomianŠs decomposition method, LyapunovŠs artificial small parameter method, and the δ-expansion method. Thus, it can be regarded as a unified or generalized theory of these previous methods. Consider the following non linear problem

$$N(u(x,t)) = 0, t > 0 \tag{13}$$

where N is a nonlinear operator and $u(x,t)$ is unknown function of the independent variables x,t. The zero order deformation equation

$$(1-q)L[\phi(x,t,q) - u_0(x,t)] = qhH(x,t)N(\phi(x,t,q)) \tag{14}$$

where $q \in [0.1]$ is the Homotopy or embedding parameter, $h \neq 0$ an auxiliary parameter, $H(x,t) \neq 0$ is an auxiliary function, L is an auxiliary linear operator, $u_0(x,t)$ an initial guess of $u(x,t)$ and $\phi(x,t,q)$ is an unknown function. By substituting $q=0$ and $q=1$ in (14) it can easily be observed that $\phi(x,t,q)$ deforms continuously from the initial guess $u_0(x,t)$ to the exact solution $u(x,t)$ as the embedding parameter q increases from 0 to 1. By expanding $\phi(x,t,q)$ in a Taylor series, we get

$$\phi(x,t,q) = u_0(x,t) + \sum_{m=1}^{\infty} u_m(x,t)q^m \tag{15}$$

$$\text{where } u_m(x,t) = \frac{1}{m!}\frac{\partial^m \phi(x,t,q)}{\partial q^m} \tag{16}$$

The convergence of the series (15) is controlled by h . Assume that the auxiliary parameter h the auxiliary function H, the initial approximation $u_0(x,t)$ and the auxiliary linear operator L are so properly chosen that the series (15) converges at $q=1$. Then, exact solution of (15) is given by

$$u(x,t) = u_0(x,t) + \sum_{m=1}^{\infty} u_m(x,t)q^m \tag{17}$$

Now the functions $u_m(x,t)$ for $m=1,2,3,...$ are determined by differentiating the zero order deformation equation (14) m times with respect to the embedding parameter q, dividing by $m!$ and then setting $q=0$ we get the mth order deformation equation

$$L[u_m(x,t) - \chi_m u_{m-1}(x,t) = hH(t)R_m(u_{m-1}(x,t)) \tag{18}$$

$$\text{where} \qquad R_m(u_{m-1}(x,t)) = \frac{1}{(m-1)!}\frac{\partial^{m-1}\phi(x,t,q)}{\partial q^{m-1}} \tag{19}$$

$$u_m = \{u_0(x,t), u_1(x,t), u_3(x,t), ..., u_m(x,t)\} \tag{20}$$

$$\chi_m = \begin{cases} 0, m \leq 1 \\ 1, otherwise \end{cases} \tag{21}$$

For any given operators L and N we get the mth order deformation Equation (18) and solving it we get $u_m(x,t)$ for different m. The solution of problem (13) is obtained by substituting the obtained $u_m(x,t)$Šs in (17) and choosing a suitable value of h for the convergence of the series.

Homotopy analysis method is based on the following assumptions: (i) There exists the solution of the $zero^{th}$ order deformation equation in the whole region of the embedding parameter $q \in [0,1]$. (ii) All the higher order deformation equations have solutions. (iii) All Taylor series expanded in the embedding parameter q converge at $q=1$. Till now, there are no rigorous theories to direct us to choose the initial approximations, auxiliary linear operators, auxiliary functions, and auxiliary parameter are available in the literature though some fundamental rules based on practical problems such as the rule of solution expression, the rule of coefficient ergodicity, and the rule of solution existence, which play important roles within the Homotopy analysis method are available.

2.5 Homotopy analysis method for system of differential equations

Consider the system of differential equations

$$N_i(x_i(t))) = y_i(t), i = 1, 2, 3, ..., n \tag{22}$$

where N_i are nonlinear operators, t denotes the independent variable, $x_i(t)$ are unknown functions, and $y_i(t)$ are known analytic functions representing the nonhomogeneous terms. Equation (22) becomes a homogeneous equation if $y_i(t) = 0$. The zeroth-order deformation equation is constructed as

$$(1-q)L[\phi_i(t,q) - x_{i,0}(t)] = qh\{N_i(\phi_i(t,q)) - y_i(t)\} \tag{23}$$

where $q \in [0,1]$ is an embedding parameter, h is a nonzero auxiliary function, L is an auxiliary linear operator, $x_{i,0}(t)$ are the initial guesses of $y_i(t)$, and $\phi_i(t,q)$ are unknown functions. It is important to note that one has great freedom to choose the auxiliary objects such as h and L in HAM. Obviously, when $q = 0$ and $q = 1$, both

$$\phi_i(t,0) = x_{i,0}(t); \phi_i(t,1) = x_i(t) \tag{24}$$

hold. Thus, as q increases from 0 to 1, the solutions $\phi_i(t,q)$ vary from the initial guesses $x_{i,0}(t)$ to the solutions $x_i(t)$. By expanding $\phi_i(t,q)$ in Taylor series with respect to q, we get

$$\phi_i(t,q) = x_{i,0}(t) + \sum_{m=1}^{\infty} x_{i,m}(t)q^m \tag{25}$$

$$\text{where} \qquad x_{i,m}(t) = \frac{1}{m!} \frac{\partial^m \phi_i(t,q)}{\partial q^m} \Big|_{q=0} \tag{26}$$

If the auxiliary linear operator L, the initial guesses $x_{i,0}(t)$, the auxiliary parameters h, and the auxiliary functions are chosen properly such that the series (25)converges at $q = 1$ and

$$\phi_i(t,1) = x_i(t) = x_{i,0}(t) + \sum_{m=1}^{\infty} x_{i,m}(t) \tag{27}$$

will be one of the solution of the original nonlinear equation. For the choice of $h = -1$, (23) becomes

$$(1-q)L[\phi_i(t,q) - x_{i,0}(t)] + qh\{N_i(\phi_i(t,q)) - y_i(t)\} = 0 \tag{28}$$

Define the vector

$$\vec{x}_{i,n}(t) = [x_{i,0}(t), x_{i,1}(t), ..., x_{i,n}(t)] \tag{29}$$

The m^{th}order deformation equation is obtained by differentiating the equation (23) with respect to the embedding parameter q, m times , then setting $q = 0$, and finally dividing them by $m!$ and is given by

$$L[x_(i,m) - \chi_m x_{i,m-1}(t)] = hH(t)R_{i,m}(\vec{x}_{i,m-1}) \tag{30}$$

$$\text{where} \quad R_{i,m}(\vec{x}_{i,m-1}) = \frac{1}{(m-1)!} \frac{\partial^{m-1}\{N_i[\phi_i(t,q)] - y_i(t)\}}{\partial q^{m-1}} \Big| q = 0 \tag{31}$$

$$\text{and} \qquad \chi_m = \begin{cases} 0, m \le 1 \\ 1, otherwise \end{cases} \tag{32}$$

It is evident that $x_{i,m-1}(t)$ $(m \ge 1)$ is governed by the linear equation (30) with the linear initial or boundary conditions that come from the problem of consideration.

3. Two species ecological systems

Consider a two species ecological system represented by Lokta Volterra equations of the form

$$\dot{N_1}(t) = N_1(t)[a_1 - b_{11}N_1(t) - b_{12}N_2(t) \tag{33}$$

$$\dot{N_2}(t) = N_2(t)[a_2 - b_{21}N_1(t) - b_{22}N_2(t) \tag{34}$$

where $N_1(t), N_2(t)$ denote the densities of the two species , $a_1, b_{11}, a_2,\ b_{22}$ are the logistic parameters for the first and second species and b_{12}, b_{21} arise from from the inhibiting effect on the growth of the first species due to the presence of the second and from the effect on the growth of the second species due to the presence of the first species. Assume that the per capita growth rate of each population at any instant is a linear function of the densities of the two competing populations at that instant. Each population would grow logistically in the absence of the other. In 1977 [17] obtained the exact solutions of the Lokta Volterra equations $\dot{n_1}(t) = n_1(t)[a_1 - c_1 n_2(t)]$, $\dot{n_2}(t) = n_2(t)[-a_2 - c_2 n_1(t)]$ when $a_1 = a_2$. The form of exact solution of this system when all the coefficients $a_1, a_2, c_1 and c_2$ time varying and $a_1(t) = a_2(t), c_1(t) = c_2(t)$ Wilson [18]. With another assumption $[a_1(t) - a_2(t)]c_1(t)c_2(t) = c_2(t)\dot{c_1}(t) - c_1(t)\dot{c_2}(t)$ Burnside [19] gave the exact solution. Murty and Rao [3] constructed approximate analytical solution of $\dot{n_1}(t) = n_1(t)[a_1 - b_{11}n_1(t) - b_{12}n_2(t)]$, $\dot{n_2}(t) = n_2(t)[-a_2 - b_{21}n_1(t)]$ with the assumption that the prey population $n_1(t)$ interfere with one another but not the predators $n_2(t)$. As all these assumptions may not depict the real world phenomena it is important to find the approximate analytical solutions of the system of equations (33-34) because of their non linear nature.

3.1 Two species system - perturbation method

The equilibrium point of (33) is given by

$$N_1^* = [a_1 b_{22} - a_2 b_{12}]/[b_{11}b_{22} - b_{21}b_{12}] \tag{35}$$

$$N_2^* = [a_2 b_{11} - a_1 b_{21}]/[b_{11}b_{22} - b_{21}b_{12}] \tag{36}$$

Define new variables by

$$X(t) = N_1(t) - N_1^* \tag{37}$$

$$Y(t) = N_2(t) - N_2^* \tag{38}$$

i.e., the variables X and Y represent densities of the two populations deviating from the equilibrium point. Now the equations (33) become

$$\dot{X}(t) = -b_{11}N_1^* X(t) - b_{12}N_1^* Y(t) - b_{11}X^2(t) - b_{12}X(t)Y(t) \tag{39}$$

$$\dot{Y}(t) = -b_{21}N_2^* X(t) - b_{22}N_1^* Y(t) - b_{22}Y^2(t) - b_{21}X(t)Y(t) \tag{40}$$

By introducing the parameter μ in to non linear terms of the above equation we get

$$\dot{X}(t) = -b_{11}N_1^* X(t) - b_{12}N_1^* Y(t) - \mu b_{11}X^2(t) - \mu b_{12}X(t)Y(t) \tag{41}$$

$$\dot{Y}(t) = -b_{21}N_2^* X(t) - b_{22}N_1^* Y(t) - \mu b_{22}Y^2(t) - \mu b_{21}X(t)Y(t) \tag{42}$$

We seek solutions of the (33) in the form

$$X(t) = X_0 + \mu X_1(t) + \mu^2 X_2(t) + \ldots \tag{43}$$

$$Y(t) = Y_0 + \mu Y_1(t) + \mu^2 Y_2(t) + \ldots \tag{44}$$

By equating like powers of μ after substituting (41),(42),(43),(44) in (37), (38),(39), (40) we get

$$\dot{X}_0(t) = -b_{11}N_1^* X_0(t) - b_{12}N_1^* Y_0(t) \tag{45}$$

$$\dot{Y}_0(t) = -b_{21}N_2^* X_0(t) - b_{22}N_1^* Y_0(t) \tag{46}$$

$$\dot{X}_1(t) = -b_{11}N_1^* X_1(t) - b_{12}N_1^* Y_1(t) - b_{11}X_0^2(t) - b_{12}X_0(t)Y_0(t) \tag{47}$$

$$\dot{Y}_1(t) = -b_{21}N_2^* X_1(t) - b_{22}N_1^* Y_1(t) - b_{22}Y_0^2(t) - b_{21}X_0(t)Y_0(t) \tag{48}$$

By applying the initial conditions $X(0) = p_0, Y(0) = q_0$, zero initial conditions for correction terms and by using Laplace transforms the solutions of the above equations are given by

$$X_0(t) = \frac{1}{\alpha - \beta}[p_0(\alpha e^{\alpha t} - \beta e^{\beta t}) + A_1(e^{\alpha t} - e^{\beta t})] \tag{49}$$

$$Y_0(t) = \frac{1}{\alpha - \beta}[q_0(\alpha e^{\alpha t} - \beta e^{\beta t}) + A_2(e^{\alpha t} - e^{\beta t})] \tag{50}$$

where $A_1 = b_{22}N_2^* p_0 - b_{12}N_1^* q_0$, $A_2 = b_{11}N_1^* q_0 - b_{21}N_2^* p_0$, $\alpha = -B_1 + \sqrt{B_1^2 - 4C_1}/2$,

$\beta = -B_1 - \sqrt{B_1^2 - 4C_1}/2$, $B_1 = (b_{11}N_1^* + b_{22}N_2^*$, $C_1 = (b_{11}b_{22} - b_{21}b_{12})N_1^* N_2^*$

Now $X_1(t)$ and $X_2(t)$ are given by

$$X_1(t) = P_1\alpha e^{\alpha t} + P_2 e^{\beta t} + P_3 e^{2\alpha t} + P_4 e^{2\beta t} + P_5 e^{(\alpha+\beta)t} \tag{51}$$

$$Y_1(t) = Q_1\alpha e^{\alpha t} + Q_2 e^{\beta t} + Q_3 e^{2\alpha t} + Q_4 e^{2\beta t} + Q_5 e^{(\alpha+\beta)t} \tag{52}$$

where the constants $P_i, Q_i (i = 1, 2, 3, 4, 5)$ are given by

$$P_1 = \frac{D_1\alpha + L_1}{\alpha(\beta - \alpha)} + \frac{D_2\alpha + M_1}{(\alpha - \beta)(\alpha - 2\beta)} + \frac{D_3\alpha + N_1}{\beta(\beta - \alpha)}$$

$$P_2 = \frac{D_1\beta + L_1}{(\beta - \alpha)(\beta - 2\alpha)} + \frac{D_2\beta + M_1}{\beta(\alpha - \beta)} + \frac{D_3\beta + N_1}{\alpha(\alpha - \beta)}, \quad P_3 = \frac{2D_1\alpha + L_1}{\alpha(2\alpha - \beta)}, \quad P_4 = \frac{2D_2\beta + M_1}{\beta(2\beta - \alpha)},$$

$$P_5 = \frac{D_3(\alpha + \beta) + N_1}{\alpha\beta}, \quad Q_1 = \frac{E_1\alpha + L_2}{\alpha(\beta - \alpha)} + \frac{E_2\alpha + M_2}{(\alpha - \beta)(\alpha - 2\beta)} + \frac{E_3\alpha + N_2}{\beta(\beta - \alpha)}$$

$$Q_2 = \frac{E_1\beta + L_2}{(\beta - \alpha)(\beta - 2\alpha)} + \frac{E_2\beta + M_2}{\beta(\alpha - \beta)} + \frac{E_3\beta + N_2}{\alpha(\alpha - \beta)}, \quad Q_3 = \frac{2E_1\alpha + L_2}{\alpha(2\alpha - \beta)}, \quad Q_4 = \frac{2E_2\beta + M_2}{\beta(2\beta - \alpha)},$$

$$Q_5 = \frac{E_3(\alpha + \beta) + N_2}{\alpha\beta} \quad \text{and} L_1 = b_{22}N_2^* D_1 - b_{12}N_1^* E_1, \quad M_1 = b_{22}N_2^* D_2 - b_{12}N_1^* E_2$$

$$N_1 = b_{22}N_2^*D_3 - b_{12}N_1^*E_3, \quad L_2 = b_{11}N_1^*E_1 - b_{21}N_2^*D_1, \; M_2 = b_{11}N_1^*E_2 - b_{21}N_2^*D_2,$$

$$N_2 = b_{11}N_1^*E_3 - b_{21}N_2^*D_3 \text{ where } D_1 = -b_{11}B_2^2 - b_{12}B_2C_2, \quad D_2 = -b_{11}B_3^2 - b_{12}B_3C_3$$

$$D_3 = 2b_{11}B_2B_3 + b_{12}(B_2C_3 + B_3C_2), \quad E_1 = -b_{22}C_2^2 - b_{12}B_2C_2, \; E_2 = -b_{22}C_3^2 - b_{21}B_3C_3,$$

$$E_3 = 2b_{22}C_2C_3 + b_{21}(B_2C_3 + B_3C_2)$$

$$B_2 = \frac{\alpha p_0 + A_1}{\alpha - \beta}, \; B_3 = \frac{\beta p_0 + A_1}{\alpha - \beta}, C_2 = \frac{\alpha q_0 + A_2}{\alpha - \beta}, \; C_3 = \frac{\beta q_0 + A_2}{\alpha - \beta}$$

Now the second order approximation is given by $X(t) \simeq X_0(t) + X_1(t)$, $Y(t) \simeq Y_0(t) + Y_1(t)$. Now the approximate analytical solution in terms of N_1 and N_2 is given by

$$N_1(t) \simeq \frac{a_1 b_{22} - a_2 b_{12}}{b_{11}b_{22} - b_{21}b_{12}} + [P_1 + \frac{p_0\alpha + A_1}{(\alpha - \beta)}]e^{\alpha t} + [P_2 - \frac{p_0\beta + A_1}{(\alpha - \beta)}]e^{\beta t}$$

$$+ P_3 e^{2\alpha t} + P_4 e^{2\beta t} + P_5 e^{(\alpha + \beta)t}$$

$$N_2(t) \simeq \frac{a_2 b_{11} - a_1 b_{21}}{b_{11}b_{22} - b_{21}b_{12}} + [Q_1 + \frac{q_0\alpha + A_2}{(\alpha + \beta)}]e^{\alpha t} + [Q_2 - \frac{q_0\beta + A_2}{(\alpha - \beta)}]e^{\beta t}$$

$$+ Q_3 e^{2\alpha t} + Q_4 e^{2\beta t} + Q_5 e^{(\alpha + \beta)t}$$

The accuracy of the solution can be improved by finding higher order approximations.

3.2 Two species system - Homotopy perturbation method

Now we consider Lokta Volterra system (33) with the conditions that $b_{11} = 0$ and $b_{22} = 0$

$$\dot{N}_1(t) = N_1(t)[a_1 - b_{12}N_2(t) \tag{53}$$

$$\dot{N}_2(t) = N_2(t)[a_2 - b_{21}N_1(t)] \tag{54}$$

and apply the Homotopy Perturbation Method to construct a Homotopy of the above resultant system (53),(54) as follows

$$(1 - p)(\dot{v}_1 - \dot{N}_{1_0}) + p(\dot{v}_1 - v_1(t)[a_1 - b_{12}v_2(t)]) = 0 \tag{55}$$

$$(1 - p)(\dot{v}_2 - \dot{N}_{2_0}) + p(\dot{v}_2 - v_2(t)[a_2 - b_{21}v_1(t)]) = 0 \tag{56}$$

(. denotes differentiation with respect t) the initial approximations are given by

$$v_{1,0}(t) = N_{1_0}(t) = N_1(0) \tag{57}$$

$$v_{2,0}(t) = N_{2_0}(t) = N_2(0) \tag{58}$$

and

$$v_1 = v_{1,0} + pv_{1,1} + p^2 v_{1,2} + p^3 v_{1,3} + +p^4 v_{1,4} + \dots \tag{59}$$

$$v_2 = v_{2,0} + pv_{2,1} + p^2 v_{2,2} + p^3 v_{2,3} + +p^4 v_{2,4} + \dots \tag{60}$$

where the functions v_{ij} $(i = 1, 2, j = 1, 2, 3)$ are to be determined. By using equations (57),(58),(59),(60) in equations (55), (56) we get

$$(1 - p)[(\dot{v}_{1,0} + p\dot{v}_{1,1} + P^2\dot{v}_{1,2} + P^3\dot{v}_{1,3} + ...) - \dot{N}_1(0)] + p[(\dot{v}_{1,0} + p\dot{v}_{1,1} + P^2\dot{v}_{1,2} + P^3\dot{v}_{1,3} + ...)$$

$$- (v_{1,0} + pv_{1,1} + p^2v_{1,2} + p^3v_{1,3} + ...)(a_1 - b_{12}v_{2,0} + pv_{2,1} + p^2v_{2,2} + p^3v_{2,3} + ...) = 0 \qquad (61)$$

$$(1 - p)[(\dot{v}_{2,0} + p\dot{v}_{2,1} + P^2\dot{v}_{2,2} + P^3\dot{v}_{2,3} + ...) - \dot{N}_2(0)] + p[(\dot{v}_{2,0} + p\dot{v}_{2,1} + P^2\dot{v}_{2,2} + P^3\dot{v}_{2,3} + ...)$$

$$- (v_{2,0} + pv_{2,1} + p^2v_{2,2} + p^3v_{2,3} + ...)(a_2 - b_{21}v_{1,0} + pv_{1,1} + p^2v_{1,2} + p^3v_{1,3} + ...) = 0 \qquad (62)$$

i.e., $(\dot{v}_{1,1} - a_1 N_{1_0} + b_{12}N_{1_0}N_{2_0})p + (\dot{v}_{1,2} - a_1 v_{1_1} + b_{12}N_{1_0}v_{2_1} + b_{12}N_{2_0}v_{1_1})p^2 +$

$$(\dot{v}_{1,3} - a_1 v_{1_2} + b_{12}N_{1_0}v_{2_2} + b_{12}N_{2_0}v_{1,2} + b_{12}v_{1,1}v_{2,1})P^3 = 0 \qquad (63)$$

$(\dot{v}_{2,1} - a_2 N_{1_0} + b_{21}N_{1_0}N_{2_0})p + (\dot{v}_{2,2} + a_2 v_{2_1} - b_{21}N_{1_0}v_{2_1} - b_{21}N_{2_0}v_{1_1})p^2 +$

$$(\dot{v}_{2,3} - N_{2_0}b_{21}v_{1,2} - b_{21}v_{1,1}v_{2,1} + v_{2,2}(a_2 - b_{21}N_{1_0}))p^3 = 0 \qquad (64)$$

By considering the initial conditions $v_{ij} = 0$, $i = 1, 2, j = 1, 2, 3$

$$\dot{v}_{1,1} - a_1 N_{1_0} + b_{12}N_{1_0}N_{2_0} = 0 \dot{v}_{1,2} - a_1 v_{1_1} + b_{12}N_{1_0}v_{2_1} + b_{12}N_{2_0}v_{1_1} = 0$$

$$\dot{v}_{2,3} - N_{2_0}b_{21}v_{1,2} - b_{21}v_{1,1}v_{2,1} + v_{2,2}(a_2 - b_{21}N_{1_0}) = 0 \qquad (65)$$

$$\dot{v}_{2,1} - a_2 N_{1_0} + b_{21}N_{1_0}N_{2_0} = 0, \dot{v}_{2,2} + a_2 v_{2_1} - b_{21}N_{1_0}v_{2_1} - b_{21}N_{2_0}v_{1_1} = 0$$

$$\dot{v}_{2,3} - N_{2_0}b_{21}v_{1,2} - b_{21}v_{1,1}v_{2,1} + v_{2,2}(a_2 - b_{21}N_{1_0}) = 0$$

From the above equations we get

$$v_{1,1} = a_1 N_{1_0} t - b_{12}N_{1_0}N_{2_0}t$$

$$v_{1,2} = \frac{1}{2}N_{1_0}[-2a_1 b_{12}N_{2_0} + a_1^2 + b_{12}a_2 N_{2_0} - b_{12}b_{21}N_{1_0}N_{2_0} + b_{12}^2 N_{2_0}^2]t^2$$

$$v_{1,3} = -\frac{1}{6}N_{1_0}[4a_1 b_{12}b_{21}N_{1_0}N_{2_0} - 3a_1 b_{12}^2 N_{2_0}^2 - 3a_1 b_{12}a_2 N_{2_0} - a_1^3 +$$
$$b_{12}a_2^2 N_{2_0} + b_{12}^3 N_{2_0}^3 - 4b_{12}^2 b_{21}N_{1_0}N_{2_0}^2]t^3$$

$$v_{2,1} = -a_2 N_{2_0}t - b_{21}N_{1_0}N_{2_0}t$$

$$v_{2,2} = \frac{1}{2}N_{2_0}[2a_2 b_{21}N_{1_0} - a_2^2 - b_{21}a_1 N_{1_0} + b_{12}b_{21}N_{1_0}N_{2_0} + b_{21}^2 N_{1_0}^2]t^2$$

$$v_{2,3} = \frac{1}{6}N_{2_0}[4a_2 b_{12}b_{21}N_{1_0}N_{2_0} - 4b_{12}b_{21}^2 N_{1_0}^2 N_{2_0} - 3a_1 a_2 b_{21}N_{1_0} - 3a_1 b_{21}^2 N_{1_0} - 2a_1 b_{12}b_{21}N_{1_0}N_{2_0}$$

$$+ b_{12}^2 b_{21}N_{1_0}N_{2_0}^2 + a_1^2 b_{21}N_{1_0} + 3a_2^2 b_{21}N_{1_0} - 3a_2 b_{21}^2 N_{1_0}^2 + b_{21}^3 N_{1_0}^3 - a_2^3]t^3$$

Now the approximate analytical solution containing four terms by using Homotopy Perturbation method is given by

$$N_1(t) = \lim_{p \to 1} v_1(t) = v_{1,1}(t) + v_{1,2}(t) + v_{1,3}(t) + v_{1,4}(t)$$

$$N_2(t) = \lim_{p \to 1} v_2(t) = v_{2,1}(t) + v_{2,2}(t) + v_{2,3}(t) + v_{2,4}(t)$$

The accuracy of the solution can be improved by considering more terms in the above solution expressions.

3.3 Two species system - homotopy analysis method

Consider the system of differential equations representing a two species ecological system (33),(34) satisfying the initial conditions

$$N_1(0) = 4, N_2(0) = 10 \tag{66}$$

According to Homotopy Analysis Method, the initial approximations of (118) are satisfying the initial conditions

$$N_{1,0}(0) = 1, N_{2,0}(0) = 2 \tag{67}$$

and the auxiliary linear operators $\mathcal{L}$ for $i = 1,2$ are

$$\mathcal{L}[\phi_i(t,q)] = \frac{\partial \phi_i(t,q))}{\partial t}, \, i = 1,2, \text{ with } \mathcal{L}([C_i] = 0. \tag{68}$$

and the nonlinear operators

$$\mathcal{N}_1[\phi_i(t,q)] = \frac{\partial \phi_1(t,q))}{\partial t} - \phi_1(t,q)[a_1 - b_{11}\phi_1(t,q) - b_{12}\phi_2(t,q)] = g_1(t) \tag{69}$$

$$\mathcal{N}_2[\phi_i(t,q)] = \frac{\partial \phi_2(t,q))}{\partial t} - \phi_2(t,q)[a_2 - b_{21}\phi_1(t,q) - b_{22}\phi_3(t,q)] = g_2(t) \tag{70}$$

The zeroth-order deformation equation is constructed as

$$(1-q)\mathcal{L}[\phi_1(t,q) - L_0(t)] = qh\{\mathcal{N}_1(\phi_1(t,q)) - g_1(t)\} \tag{71}$$

$$(1-q)\mathcal{L}[\phi_2(t,q) - D_0(t)] = qh\{\mathcal{N}_2(\phi_2(t,q)) - g_2(t)\} \tag{72}$$

where $q \in [0,1]$ is an embedding parameter, h is a nonzero auxiliary function, $\mathcal{L}$ is an auxiliary linear operator, $N_{1,0}(t), N_{2,0}(t)$ are the initial guesses of $N_1(t), N_2(t)$, and $\phi_i(t,q)$ are unknown functions.

3.3.1 Solution as polynomial functions

Now, by taking polynomials as base functions the mth order deformation equations are given by

$$\mathcal{L}[N_{1,m}(t) - N_{1,m-1}(t)] = qhR_{1,m}(\overrightarrow{N}_{1,m-1}) \tag{73}$$

$$\mathcal{L}[N_{2,m}(t) - N_{2,m-1}(t)] = qhR_{2,m}(\overrightarrow{N}_{2,m-1}) \tag{74}$$

As q increases from 0 to 1, the solutions $\phi_1(t,q), \phi_2(t,q)$ vary from the initial guesses $N_{1,0}(t), N_{2,0}(t)$ to the solutions $N_1(t), N_2(t)$. where

$$R_{1,m}(\overrightarrow{N}_{1,m}) = \dot{N}_{1,m-1} - a_1 N_{1,m-1}(t) +$$

$$b_{11} \sum_{i=0}^{m-1} N_{1,i}(t) N_{1,m-1-i}(t) + b_{12} \sum_{i=0}^{m-1} N_{1,i}(t) N_{2,m-1-i}(t) \tag{75}$$

$$R_{2,m}(\overrightarrow{N}_{2,m}) = \dot{N}_{2,m-1} - a_2 N_{2,m-1}(t) +$$

$$b_{21} \sum_{i=0}^{m-1} N_{2,i}(t) N_{1,m-1-i}(t) + b_{22} \sum_{i=0}^{m-1} N_{2,i}(t) N_{2,m-1-i}(t) \tag{76}$$

subject to $N_{1,m} = 0, N_{2,m} = 0$

where the dot denotes differentiation with respect to the variable t. The solution of the mth order deformation equation for $m \geq 1$ is given by

$$N_{1,m}(t) = \chi_2 N_{1,m-1}(t) + h\mathcal{L}^{-1}[R_{1,m}(\overrightarrow{N}_{1,m-1})] \tag{77}$$

$$N_{2,m}(t) = \chi_2 N_{2,m-1}(t) + h\mathcal{L}^{-1}[R_{2,m}(\overrightarrow{N}_{2,m-1})] \tag{78}$$

$$\text{and } \chi_m = \begin{cases} 0, m \leq 1 \\ 1, \text{otherwise} \end{cases} \tag{79}$$

The integration constants c_i are determined by the initial condition. Now the successive approximations are given by

$$N_{1,1}(t) = [-a_1 + b_{11} + 2b_{12}]ht \tag{80}$$

$$N_{2,1}(t) = [-2a_2 + 2b_{21} + 4b_{22}]ht \tag{81}$$

$$N_{1,2}(t) = [-a_1 + b_{11} + 2b_{12}]ht + [-a_1 + b_{11} + 2b_{12}]h^2 t + \{[-a_1[-a_1 + b_{11} + 2b_{12}] +$$

$$2[-a_1 + b_{11} + 2b_{12}]b_{11} + b_{12}[2[-a_2 + b_{21} + 2b_{22}] + 2[-a_1 + b_{11} + 2b_{12}]]\}[(h^2 t^2/2) \tag{82}$$

$$N_{2,2}(t) = [-2a_2 + 2b_{21} + 4b_{22}]ht + [-2a_2 + 2b_{21} + 4b_{22}]h^2 t + \{[-a_2[-2a_2 + 2b_{21} + 4b_{22}] +$$

$$[-8a_2 + 8b_{21} + 16b_{22}]b_{12} + b_{21}[2[-a_1 + b_{11} + 2b_{12}] + [-2a_2 + 2b_{21} + 4b_{22}]]\}[(h^2 t^2/2) \tag{83}$$

Now the analytic solution via the polynomial base functions of the system has the general form

$$N_1(t) = \sum_{m=1}^{\infty} a_{i,m}(h)t^m, \quad N_2(t) = \sum_{m=1}^{\infty} a'_{i,m}(h)t^m$$

The above expressions represent a family of solution expressions in the auxiliary parameter h. By using h-curves [16] valid regions of a convergent solution series can be determined.

3.3.2 Solution as exponential functions

Now, by taking exponential functions as base functions, the initial approximations of (33),(34) satisfying the initial conditions

$$N_1(0) = 1, N_2(0) = 2 \tag{84}$$

and the auxiliary linear operators $\mathcal{L}$ for $i = 1, 2, 3$ are

$$\mathcal{L}[\phi_i(t,q)] = \frac{\partial \phi_i(t,q))}{\partial t} + \phi_i(t,q), \; i = 1, 2, \text{ with } \mathcal{L}([C_i e^{-t}] = 0. \tag{85}$$

The solution of the mth order deformation equation for $m \geq 1$ is given by

$$N_{1,m}(t) = \chi_2 N_{1,m-1}(t) + h e^{-t} \int_0^t e^{\tau} H(\tau) R_m(\overrightarrow{N}_{1,m-1}) d\tau + C_1 e^{-t} \tag{86}$$

$$N_{2,m}(t) = \chi_2 N_{2,m-1}(t) + h e^{-t} \int_0^t e^{\tau} H(\tau) R_m(\overrightarrow{N}_{2,m-1}) d\tau + C_2 e^{-t} \tag{87}$$

$$\text{and } \chi_m = \begin{cases} 0, m \leq 1 \\ 1, otherwise \end{cases} \tag{88}$$

The integration constants c_i are determined by the initial condition. Now the successive approximations obtained by taking $H(\tau) = 1$ in the above equations are given by

$$N_{1,1}(t) = [-a_1 + b_{11} + 2b_{12}]h(1 - e^{-t}) \tag{89}$$

$$N_{2,1}(t) = [-2a_2 + 2b_{21} + 4b_{22}]h(1 - e^{-t}) \tag{90}$$

$$N_{1,2}(t) = [-a_1 + b_{11} + 2b_{12}]h(1 - e^{-t}) + [-a_1 + b_{11} + 2b_{12}]h^2 t e^{-t}$$
$$\{-a_1[-a_1 + b_{11} + 2b_{12}] + 2b_{11}[-a_1 + b_{11} + 2b_{12} + b_{12}[[-2a_2 + 2b_{21}$$
$$+ 4b_{22}] + 2[-a_1 + b_{11} + 2b_{12}]]]\}[h^2(e^{-t} + t - 1)] \tag{91}$$

$$N_{2,2}(t) = [-2a_2 + 2b_{21} + 4b_{22}]h(1 - e^{-t}) + [-2a_2 + 2b_{21} + 4b_{22}]h^2 t e^{-t}$$
$$\{-a_2[-2a_2 + 2b_{21} + 4b_{22}] + 4b_{12}[-2a_2 + 2b_{21} + 4b_{22}] + b_{21}[[-2a_2 + 2b_{21}$$
$$+ 4b_{22}] + 2[-a_1 + b_{11} + 2b_{12}]]]\}[h^2(e^{-t} + t - 1)] \tag{92}$$

Now the analytic solution via the polynomial base functions of the system has the general form

$$N_1(t) = \sum_{m=1}^{\infty} N_{1,m}(t), N_2(t) = \sum_{m=1}^{\infty} N_{2,m}(t) \tag{93}$$

By using h-curves [16] valid regions of a convergent solution series can be determined.

4. Three species ecological system

We consider a three species ecological system

$$\dot{L}(t) = L(t)[P_1 - \gamma D(t) + \beta W(t)]$$
$$\dot{D}(t) = D(t)[P_2 + \gamma L(t) - \alpha W(t)] \tag{94}$$
$$\dot{W}(t) = W(t)[P_3 - \beta L(t) + \alpha D(t)]$$

where $L(t)$, $D(t)$ and $W(t)$ denote the amount of litter, detritus and predators in the sea at time t in the mangrove area respectively, $\alpha, \beta, \gamma, P_1, P_2, P_3$ denote the parameters.

4.1 Three species system - perturbation method

We making the following assumptions (i) W(t) preys on D(t), D(t) preys on L(t)and L(t) preys on W(t) (ii)W(t) preys on D(t)and L(t), in the absence of litter and detritus population of coastal organisms decrease. D(t) preys on L(t), L(t) has sufficient bio chemicals, so in the absence of detritus and predators Litter increases without limit.ie., (i) $\alpha, \beta, \gamma > 0$ (ii) $\alpha, \gamma > 0, \beta < 0, P_1 > 0, P_3 > 0$ and

$$P_1\alpha + P_2\beta + P_3\gamma = 0 \tag{95}$$

The equilibrium point for the system (94) is calculated by using the concept of generalized inverse [2] and is given by solving the following system

$$P_1 = \gamma D^*(t) - \beta W^*(t), P_2 = -\gamma L^*(t) + \alpha W^*(t), P_3 = \beta L^*(t) - \alpha D^*(t) \tag{96}$$

By solving the above system using algorithm [2] for the generalized inverse the equilibrium point of (94) is given by

$$L^* = \frac{P_3\beta - P_2\gamma}{\alpha^2 + \beta^2 + \gamma^2}, \quad D^* = \frac{P_1\gamma - P_3\alpha}{\alpha^2 + \beta^2 + \gamma^2}, \quad W^* = \frac{P_2\alpha - P_1\beta}{\alpha^2 + \beta^2 + \gamma^2} \tag{97}$$

Now we define the new densities deviating from the equilibrium point as

$$L_1(t) = L(t) - L^*, D_1(t) = D(t) - D^*, W_1(t) = W(t) - W^* \tag{98}$$

Now the system of equations (94)become

$$\begin{aligned}
\dot{L}_1(t) &= -\gamma L_1^* D_1(t) + \beta L_1^* W_1(t) - \gamma L_1(t)D_1(t) + \beta L_1(t)W_1(t) \\
\dot{D}_1(t) &= \gamma D^* L_1(t) - \alpha D^* W_1(t) + \gamma L_1(t)D_1(t) - \alpha D_1(t)W_1(t) \\
\dot{W}(t)_1 &= -\beta W^* L_1(t) + \alpha W^* D_1(t) - \beta L_1(t)W_1(t) + \alpha D_1(t)W_1(t)
\end{aligned} \tag{99}$$

We introduce the parameter μ in to the non linear terms of the above system we get the following

$$\begin{aligned}
\dot{L}_1(t) &= -\gamma L^* D_1(t) + \beta L^* W_1(t) - \mu\gamma L_1(t)D_1(t) + \mu\beta L_1(t)W_1(t) \\
\dot{D}_1(t) &= \gamma D^* L_1(t) - \alpha D^* W_1(t) + \mu\gamma L_1(t)D_1(t) - \mu\alpha D_1(t)W_1(t) \\
\dot{W}_1(t) &= -\beta W^* L_1(t) + \alpha W^* D_1(t) - \mu\beta L_1(t)W_1(t) + \mu\alpha D_1(t)W_1(t)
\end{aligned} \tag{100}$$

We seek solutions of the (100) in the form

$$\begin{aligned}
L_1(t) &= L_{1_0}(t) + \mu L_{1_1}(t) + \mu^2 L_{1_2}(t) + \dots \\
D_1(t) &= D_{1_0}(t) + \mu D_{1_1}(t) + \mu^2 D_{1_2}(t) + \dots \\
W_1(t) &= W_{1_0}(t) + \mu W_{1_1}(t) + \mu^2 W_{1_2}(t) + \dots
\end{aligned} \tag{101}$$

By equating like powers of μ after substituting (101) in (100) we get

$$L_{1_0}(t) = -\gamma L^* D_{1_0}(t) + \beta L^* W_{1_0}(t)$$

$$D_{1_0}(t) = \gamma L D^* L_{1_0}(t) - \alpha D^* W_{1_0}(t) \tag{102}$$

$$W_{1_0}(t) = -\beta W^* L_{1_0}(t) + \alpha W^* D_{1_0}(t)$$

$$L_{1_1}(t) = -\gamma L^* D_{1_1}(t) + \beta L^* W_{1_1}(t) - \gamma L_{1_0} D_{1_0}(t) + \beta L_{1_0} W_{1_0}(t)$$

$$D_{1_1}(t) = \gamma L D^* L_{1_1}(t) - \alpha D^* W_{1_1}(t) + \gamma L_{1_0} D_{1_0}(t) - \alpha D_{1_0} W_{1_0}(t) \qquad (103)$$

$$W_{1_1}(t) = -\beta W^* L_{1_1}(t) + \alpha W^* D_{1_1}(t) - \beta L_{1_0} W_{1_0}(t) + \alpha D_{1_0} W_{1_0}(t)$$

The solution of the equation (102) with the initial condition $L_1(0) = C_1, D_1(0) = C_2, W_1(0) = C_3$ by using Laplace transformation and the generating functions $L_1(t), D_1(t), W_1(t)$ is given by

$$L_{1_0}(t) = (C_1 - \frac{Q_1}{K_1^2})Cos(K_1 t) + (\frac{C_3 L^* \beta - C_2 L^* \gamma}{k_1})Sin(k_1 t) + \frac{Q_1}{k_1^2}$$

$$D_{1_0}(t) = (C_2 - \frac{Q_2}{K_1^2})Cos(K_1 t) + (\frac{C_1 D^* \gamma - C_3 D^* \alpha}{k_1})Sin(k_1 t) + \frac{Q_2}{k_1^2} \qquad (104)$$

$$W_{1_0}(t) = (C_3 - \frac{Q_3}{K_1^2})Cos(K_1 t) + (\frac{C_2 W^* \alpha - C_1 W^* \beta}{k_1})Sin(k_1 t) + \frac{Q_3}{k_1^2}$$

where
$k_1^2 = L^* D^* \gamma^2 + L^* W^* \beta^2 + D^* W^* \alpha^2$, $Q_1 = C_1 D^* W^* \alpha^2 + C_2 L^* W^* \alpha \beta$
$+ C_3 L^* D^* \alpha \gamma$, $Q_2 = C_1 D^* W^* \alpha \beta + C_2 L^* W^* \beta^2 + C_3 L^* D^* \beta \gamma$, $Q_3 = C_1 D^* W^* \alpha \gamma + C_2 L^* W^* \beta \gamma + C_3 L^* D^* \gamma^2$; C_1, C_2 and C_3 are the initial values of $L_{1_0}(t), D_{1_0}(t)$ and $W_{1_0}(t)$ respectively. Now the solutions of (103) with the initial condition $L_1(0) = 0, D_1(0) = 0, W_1(0) = 0$ are given by

$$L_{1_1}(t) = \frac{A_1}{2K_1}Sin(2K_1 t) + A_2 Cos(2k_1 t) + \{\frac{A_3}{k_1} + \frac{A_5}{2k_1^3} + \frac{A_6 t}{2k_1}\}Sin(k_1 t) + \{A_4 + \frac{A_5 t}{2k_1^2}\}$$
$$Cos(k_1 t) + A_7$$

$$D_{1_1}(t) = \frac{B_1}{2K_1}Sin(2K_1 t) + B_2 Cos(2k_1 t) + \{\frac{B_3}{k_1} + \frac{B_5}{2k_1^3} + \frac{B_6 t}{2k_1}\}Sin(k_1 t) + \{B_4 + \frac{B_5 t}{2k_1^2}\}$$
$$Cos(k_1 t) + B_7 \qquad (105)$$

$$W_{1_1}(t) = \frac{D_1}{2K_1}Sin(2K_1 t) + D_2 Cos(2k_1 t) + \{\frac{D_3}{k_1} + \frac{D_5}{2k_1^3} + \frac{D_6 t}{2k_1}\}Sin(k_1 t) + \{D_4 + \frac{D_5 t}{2k_1^2}\}$$
$$Cos(k_1 t) + D_7$$

where $A_1 = \frac{4R_1}{3} - \frac{X_2}{3k_1^2} + \frac{X_3}{12k_1^4}$, $A_2 = -\frac{X_1}{3k_1^2}$, $A_3 = R_3 - \frac{R_1}{3} + \frac{X_2}{3k_1^2} - \frac{X_3}{3k_1^4}$, $A_4 = \frac{X_1}{3k_1^2} - \frac{X_6}{k_1^4}$, $A_5 = X_5 - R_3 k_1^2$, $A_6 = X_4 - \frac{X_6}{k_1^2}$, $A_7 = \frac{X_3}{4k_1^4} + \frac{X_6}{k_1^4}$,

$B_1 = \frac{4S_1}{3} - \frac{Y_2}{3k_1^2} + \frac{Y_3}{12k_1^4}$, $B_2 = -\frac{Y_1}{3k_1^2}$, $B_3 = S_3 - \frac{S_1}{3} + \frac{Y_2}{3k_1^2} - \frac{Y_3}{3k_1^4}$, $B_4 = \frac{Y_1}{3k_1^2} - \frac{Y_6}{k_1^4}$, $B_5 = Y_5 - S_3 k_1^2 B_6 = Y_4 - \frac{Y_6}{k_1^2}$, $B_7 = \frac{Y_3}{4k_1^4} + \frac{Y_6}{k_1^4}$,

$D_1 = \frac{4T_1}{3} - \frac{Z_2}{3k_1^2} + \frac{Z_3}{12k_1^4}$, $D_2 = -\frac{Z_1}{3k_1^2}$, $D_3 = T_3 - \frac{T_1}{3} + \frac{Z_2}{3k_1^2} - \frac{Z_3}{3k_1^4}$, $D_4 = \frac{Z_1}{3k_1^2} - \frac{Z_6}{k_1^4}$, $D_5 = Z_5 - T_3 k_1^2$, $D_6 = Z_4 - \frac{Z_6}{k_1^2}$, $D_7 = \frac{Z_3}{4k_1^4} + \frac{Z_6}{k_1^4}$

where $X_1 = 2K_1 R_2 - S_1 \gamma L^* - T_1 L^* \beta$, $X_2 = \alpha^2 D^* W^* R_1 - 2k_1 S_2 \gamma L^* + T_1 \alpha \gamma L^* D^* +$

$2k_1 T_2 \beta L^*$, $X_3 = 2k_1 R_2 \alpha^2 D^* W^* + 2k_1 T_2 \alpha \gamma L^* D^* + 2k_1 S_2 \alpha \beta L^* W^*$, $X_4 = k_1 R_4 - S_3 \gamma L^* + T_3 \beta L^*$, $X_5 = R_3 \alpha^2 D^* W^* - k_1 S_4 \gamma L^* + T_3 \alpha \gamma L^* W^* + S_3 \alpha \beta L^* W^* + k_1 T_4 \beta L^*$, $X_6 = k_1 R_4 \alpha^2 D^* W^* + k_1 T_4 \alpha \gamma L^* D^* + k_1 S_4 \alpha \beta L^* W^*$,

$Y_1 = 2K_1 S_2 - T_1 \alpha D^* + R_1 D^* \gamma$, $Y_2 = 2k_1 R_2 \gamma D^* - 2k_1 T_2 \alpha D^* + R_1 \alpha \beta D^* W^* + T_1 \beta \gamma L^* D^* + S_1 \beta^2 L^* W^*$, $Y_3 = 2k_1 R_2 \alpha \beta D^* W^* + 2k_1 T_2 \beta \gamma L^* D^* + 2k_1 S_2 \beta^2 L^* W^*$, $Y_4 = k_1 S_4 - T_3 \alpha D^* + R_3 \gamma D^*$, $Y_5 = R_4 k_1 \gamma D^* + R_3 \alpha \beta D^* W^* - T_4 k_1 \alpha D^* + T_3 \beta \gamma L^* D^* + S_3 \beta^2 L^* W^*$, $Y_6 = k_1 R_4 \alpha \beta D^* W^* + k_1 T_4 \beta \gamma L^* D^* + k_1 S_4 \beta^2 L^* W^*$,

$Z_1 = 2K_1 T_2 + S_1 \alpha W^* - R_1 W^* \beta$, $Z_2 = 2k_1 S_2 \alpha W^* + T_1 \gamma^2 L^* D^* + S_1 \gamma \beta L^* W^* + R_1 \alpha \gamma W^* D^* - 2k_1 R_2 \beta W^*$, $Z_3 = 2k_1 T_2 \gamma^2 D^* L^* + 2k_1 S_2 \beta \gamma L^* W^* + 2k_1 R_2 \alpha \gamma D^* W^*$, $Z_4 = k_1 T_4 + S_3 \alpha W^* - R_3 \beta W^*$, $Z_5 = S_4 k_1 \alpha W^* + T_3 \gamma^2 D^* L^* + S_3 k_1 \gamma \beta L^* W^* + R_3 \alpha \gamma W^* D^* - k_1 R_4 \beta W^*$, $Z_6 = k_1 T_4 \gamma^2 D^* L^* + k_1 S_4 \beta \gamma L^* W^* + k_1 R_4 \alpha \gamma D^* W^*$,

$R_1 = (C_1 - \frac{Q_1}{k_1^2})(C_3 \beta - C_2 \gamma)$, $R_2 = \frac{L^{*2} k_1^2 (C_3 \beta - C_2 \gamma)^2 - (C_1 k_1^2 - Q_1)^2}{2L^* k_1^3}$, $R_3 = \frac{Q_1}{k_1^2}(C_3 \beta - C_2 \gamma)$, $R_4 = \frac{Q_1}{L^* k_1}(\frac{Q_1}{k_1^2} - C_1)$, $S_1 = (C_2 - \frac{Q_2}{k_1^2})(C_1 \gamma - C_3 \alpha)$, $S_2 = \frac{D^{*2} k_1^2 (C_1 \gamma - C_3 \alpha)^2 - (C_2 k_1^2 - Q_2)^2}{2D^* k_1^3}$, $S_3 = \frac{Q_2}{k_1^2}(C_1 \gamma - C_3 \alpha)$, $S_4 = \frac{Q_2}{D^* k_1}(\frac{Q_2}{k_1^2} - C_2)$, $T_1 = (C_3 - \frac{Q_3}{k_1^2})(C_2 \alpha - C_1 \beta)$, $T_2 = \frac{W^{*2} k_1^2 (C_2 \alpha - C_1 \beta)^2 - (C_3 k_1^2 - Q_3)^2}{2W^* k_1^3}$, $T_3 = \frac{Q_3}{k_1^2}(C_2 \alpha - C_1 \beta)$, $T_4 = \frac{Q_3}{W^* k_1}(\frac{Q_3}{k_1^2} - C_3)$.

Now the solution of (99) is given by

$$L_1(t) \simeq L_{1_0}(t) + L_{1_1}(t), \quad D_1(t) \simeq D_{1_0}(t) + D_{1_1}(t), \quad W_1(t) \simeq W_{1_0}(t) + W_{1_1}(t)$$

Now the solution of the three species ecological system (94) is given by

$$L = \frac{P_3 \beta - P_2 \gamma}{\alpha^2 + \beta^2 + \gamma^2} + (C_1 - \frac{Q_1}{K_1^2} + A_4 + \frac{A_5 t}{2k_1^2}) Cos(K_1 t) + (\frac{C_3 L^* \beta - C_2 L^* \gamma}{k_1} + \frac{A_3}{k_1} + \frac{A_5}{2k_1^3} + \frac{A_6 t}{2k_1}) Sin(k_1 t) + \frac{A_1}{2K_1} Sin(2K_1 t) + A_2 Cos(2k_1 t) + \frac{Q_1}{k_1^2} + A_7,$$

$$D = \frac{P_1 \gamma - P_3 \alpha}{\alpha^2 + \beta^2 + \gamma^2} + (C_2 - \frac{Q_2}{K_1^2} + B_4 + \frac{B_5 t}{2k_1^2}) Cos(K_1 t) + (\frac{C_1 D^* \gamma - C_3 D^* \alpha}{k_1} + \frac{B_3}{k_1} + \frac{B_5}{2k_1^3} + \frac{B_6 t}{2k_1}) Sin(k_1 t) + \frac{B_1}{2K_1} Sin(2K_1 t) + B_2 Cos(2k_1 t) + \frac{Q_2}{k_1^2} + B_7,$$

$$W = \frac{P_2 \alpha - P_1 \beta}{\alpha^2 + \beta^2 + \gamma^2} + C_3 - \frac{Q_3}{K_1^2} + D_4 + \frac{D_5 t}{2k_1^2}) Cos(K_1 t) + (\frac{C_2 W^* \alpha - C_1 W^* \beta}{k_1} + \frac{D_3}{k_1} + \frac{D_5}{2k_1^3} + \frac{D_6 t}{2k_1}) Sin(k_1 t) + \frac{D_1}{2K_1} Sin(2K_1 t) + D_2 Cos(2k_1 t) + \frac{Q_3}{k_1^2} + D_7.$$

These solutions are valid for larger deviations from the equilibrium point since the additional correction terms improves accuracy. By taking the transformation

$$L = L_1 + L^*, \; D = D_1 + D^*, \; W = W_1 + W^* \tag{106}$$

where L^*, D^* and W^* are the equilibrium points given by (97), the three species ecological system becomes

$$\begin{aligned}
\dot{L}_1 &= [L^* + L_1][-\gamma D_1 + \beta W_1] \\
\dot{D}_1 &= [D^* + D_1][-\alpha W_1 + \gamma L_1] \\
\dot{W}_1 &= [W^* + W_1][-\beta L_1 + \alpha D_1]
\end{aligned} \tag{107}$$

By defining a Liapunov function $V(t)$, positive definite function as

$$\begin{aligned}
V(t) = {} & d_1[L_1 - L^* log(1 + L_1/L^*)] + d_2[D_1 - D^* log(1 + D_1/D^*)] + \\
& d_3[W_1 - W^* log(1 + W_1/W^*)]
\end{aligned}$$

with d_1, d_2, d_3 as positive constants, $[d_2 - d_1]\gamma \leq 0$, $[d_3 - d_2]\alpha \leq 0$ and $[d_1 - d_3]\beta \leq 0$ we can observe that the three species ecological system is asymptotically stable since

$$\dot{V}(t) = d_1 L_1[-\gamma D_1 + \beta W_1] + d_2 D_1[-\alpha W_1 + \gamma L_1] + d_3 W_1[-\beta L_1 + \alpha D_1] \leq 0$$

4.2 Three species system - Adomian decomposition method

It is a well recognized fact that the enormous production of mangrove detritus provides a necessary energy to drive the biological machinery of mangrove estuarine ecosystem besides supporting to certain extent the coastal population. In the process of litter decomposition, if the food material is not ingested by other animals, the decomposition process of litter completes and this liberates nutrients into water and soil. Part of the detritus produced in mangrove estuarine ecosystem may be carried to the adjacent coastal waters through tidal fluxes. In fact here detritus serves a couple of purposes. (i) Regenerates nutrients by undergoing complete decomposition process and (ii) Serves as a food source to estuarine and coastal organisms like Polychaete, Crabs, Prawn, Fish and Zooplankton. Here after we call all these consumers as predators. Now we are in a position to formulate the mathematical model. If $L(t)$, $D(t)$ and $W(t)$ denote the amount of litter, detritus and predators in the sea at time t in the mangrove area respectively then the following assumptions are reasonable. (1) If there is no litter from the mangroves the detritus formation decreases in a way proportional to litter and if there is no detritus formation, the amount of litter increases exponentially within a limited time. (2) If there are no bacteria, fungi and protozoa, the formation of detritus decreases and the amount of litter increases in a limited time and (3) In The absence of consumers like lower Carnivores and higher Carnivores, fish etc... the amount of detritus increases and in the absence of detritus in the sea, the growth rate of above predators decreases. These assumptions give rise to the following system of non linear first order differential equations,

$$\dot{L} = \alpha_1 L - \beta_1 LD, \dot{D} = \alpha_2 D + \beta_2 LD - \gamma_2 DW, \dot{W} = -\alpha_3 W + \gamma_3 DW \tag{108}$$

where $\alpha_i (i = 1, 2, 3), \beta_j (j = 1, 2)$ and $\gamma_k \, (k = 2, 3)$ are all positive constants. The presence of both litter and detritus is beneficial to the growth of predators like fish and carnivores in the sea. More specifically the predator species increases and the detritus decreases at rates proportional to the product of the two. If they also satisfy the condition,

$$\alpha_1 \gamma_3 - \alpha_3 \beta_1 = 0 \tag{109}$$

It amounts to saying that the growth of predators is directly proportional to the formation of the detritus. The interior equilibrium point for the system (108) is calculated by pseudo inverse concept [2] and on solving the system of equations

$$\beta_1 D = \alpha_1, \quad -\beta_2 L + \gamma_2 W = \alpha_2, \quad \gamma_3 D = \alpha_3 \tag{110}$$

The possible equilibria are (i) Trivial equilibrium point $L = 0$, $D = 0$ and $W = 0$. (ii) Equilibrium in the absence of predators $L_w = -\alpha_2/\beta_2$, $D_W = \alpha_1/\beta_1$ (iii) Equilibrium in the absence of Detritus $L_D = 0$, $W_D = 0$(iv) Equilibrium in the absence of Litter $D_L = \alpha_3/\gamma_3$, $W_L = \alpha_2/\gamma_2$ and (v)The interior equilibrium

$$L = -(\alpha_2 \beta_2)/(\beta_2^2 + \gamma_2^2), D = (\alpha_1 \beta_1 + \alpha_3 \gamma_3)/(\beta_1^2 + \gamma_3^2), W = (\alpha_2 \gamma_2)/(\beta_2^2 + \gamma_2^2) \tag{111}$$

Equilibrium (ii) is the case when the entire litter is decomposed in to detritus which is usual prey predator equilibrium, whereas (iii) is the equilibrium with no detritus and equilibrium (iv) is the one with no predators. Neither all these three equilibria (ii), (iii) and (iv) nor the trivial equilibrium are of much interest to us. Define densities of the three species deviating from the interior equilibrium values given in (111) as

$$L_1 = L - a, D_1 = D - b, W_1 = W - c \tag{112}$$

where $a = -(\alpha_2\beta_2)/(\beta_2^2 + \gamma_2^2), b = (\alpha_1\beta_1 + \alpha_3\gamma_3)/(\beta_1^2 + \gamma_3^2) and$
$c = (\alpha_2\gamma_2)/(\beta_2^2 + \gamma_2^2)$. From equation (108) we get

$$\begin{aligned}
\dot{L}_1 &= -a\beta_1 D_1 - \beta_1 L_1 D_1 \\
\dot{D}_1 &= b\beta_2 L_1 - b\gamma_2 W_1 + \beta_2 L_1 D_1 - \gamma_2 D_1 W_1 \\
\dot{W}_1 &= c\gamma_3 D_1 + \gamma_3 D_1 W_1
\end{aligned} \tag{113}$$

The above equations can be represented by the system

$$\dot{X} = E_1 XB + E_2 XC + E_1 XF_1 X + E_2 XF_2 X \tag{114}$$

Where $X = \begin{bmatrix} L_1 & 0 & 0 \\ 0 & D_1 & 0 \\ 0 & 0 & W_1 \end{bmatrix}, E_1 = \begin{bmatrix} 0 & 1 & 0 \\ 0 & 0 & 1 \\ 1 & 0 & 0 \end{bmatrix}, E_2 = \begin{bmatrix} 0 & 0 & 1 \\ 1 & 0 & 0 \\ 0 & 1 & 0 \end{bmatrix}, B = \begin{bmatrix} 0 & 0 & 0 \\ -a\beta_1 & 0 & 0 \\ 0 & -b\gamma_2 & 0 \end{bmatrix},$

$C = \begin{bmatrix} 0 & b\beta_2 & 0 \\ 0 & 0 & c\gamma_3 \\ 0 & 0 & 0 \end{bmatrix}, F_1 = \begin{bmatrix} 0 & 0 & 0 \\ -\beta_1 & 0 & 0 \\ 0 & -\gamma_2 & 0 \end{bmatrix}, F_2 = \begin{bmatrix} 0 & \beta_2 & 0 \\ 0 & 0 & \gamma_3 \\ 0 & 0 & 0 \end{bmatrix}$. The initial conditions $L(0) = p_1$,

$D(0) = p_2$, $W(0) = p_3$ now become $L_1(0) = p_1 - a$, $D_1(0) = p_2 - b$, $W_1(0) = p_3 - c$ and these can be written in the matrix notation as

$$\begin{bmatrix} L_1(0) & 0 & 0 \\ 0 & D_1(0) & 0 \\ 0 & 0 & W_1(0) \end{bmatrix} = \begin{bmatrix} p_1 - a & 0 & 0 \\ 0 & p_2 - b & 0 \\ 0 & 0 & p_3 - c \end{bmatrix} \tag{115}$$

Now, applying Adomain's decomposition method to the system (114), we get $MX = E_1 XB + E_2 XC + E_1 XF_1 X + E_2 XF_2 X = RX + NX$, where $RX = E_1 XB + E_2 XC$, the linear term and $NX = E_1 XF_1 X + E_2 XF_2 X$ is the non-linear term of the system, and M denotes the differential operator. Therefore,

$$X = M^{-1}[RX] + M^{-1}[NX] \tag{116}$$

Where $X = \sum_{n=0}^{\infty} X_n$ and $NX = \sum_{n=0}^{\infty} A_n$. Thus (116) will become

$$\sum_{n=0}^{\infty} X_n = M^{-1}\{E_1 \sum_{n=0}^{\infty} X_n B + E_2 \sum_{n=0}^{\infty} X_n C\} + M^{-1} \sum_{n=0}^{\infty} A_n$$

Now the components of X_n are therefore easily identified as,

$$X_0 = diag\left[p_1 - a, p_2 - b, p_3 - c \right] \tag{117}$$

$$\begin{aligned}
X_1 &= M^{-1}[E_1 X_0 B + E_2 X_0 C] + M^{-1} A_0 \\
X_2 &= M^{-1}[E_1 X_1 B + E_2 X_1 C] + M^{-1} A_1 \quad \cdots \\
X_n &= M^{-1}[E_1 X_{n-1} B + E_2 X_{n-1} C] + M^{-1} A_{n-1}
\end{aligned}$$

Now the polynomials A_n for the above non-linear equations are calculated by using (7) and are given by

$A_0 = E_1 X_0 F_1 X_0 + E_2 X_0 F_2 X_0$

$A_1 = X_1[E_1 F_1 X_0 + E_1 X_0 F_1 + E_2 F_2 X_0 + E_2 X_0 F_2] + (X_1^2/2!)[2E_1 F_1 + 2E_2 F_2]$

$A_2 = X_2[E_1 F_1 X_0 + E_1 X_0 F_1 + E_2 F_2 X_0 + E_2 X_0 F_2] + (X_2^2/2!)[2E_1 F_1 + 2E_2 F_2] + X_1 X_2[2E_1 F_1 + 2E_2 F_2]$.

Thus, the initial approximation matrix X_0 is given by (117) and special polynomial $A_0 = diag[Q_1, Q_2, Q_3]$ where $Q_1 = -\beta_1(p_1 - a)(p_2 - b)$, $Q_2 = (p_2 - b)[\beta_2(p_1 - a) - \gamma_2(p_3 - c)]$, $Q_3 = \gamma_3(p_2 - b)(p_3 - c)$.

The second approximation X_1 is given by

$X_1 = M^{-1}[E_1 X_0 B + E_2 X_0 C] + M^{-1} A_0 = diag[(Q_1 + Q_4)t, (Q_2 + Q_5)t, (Q_3 + Q_6)t]$ where $Q_4 = -a\beta_1(p_2 - b), Q_5 = b[\beta_2(p_1 - a) - \gamma_2(p_3 - c)], Q_6 = c\gamma_3(p_2 - b)$.

The third approximation X_2 is given by $X_2 = M^{-1}[E_1 X_1 B + E_2 X_1 C] + M^{-1} A_1$.

The special polynomial A_1 is given by

$$A_1 = diag\begin{bmatrix} (Q_1 + Q_4)Q_7 t & (Q_2 + Q_5)Q_8 t+ & (Q_3 + Q_6)Q_9 t+ \\ -\beta_1(Q_1 + Q_4)^2 t^2, & ((\beta_2 - \gamma_2)(Q_2 + Q_5)^2 t^2), & \gamma_3(Q_3 + Q_5)^2 t^2 \end{bmatrix}$$ Where $Q_7 = -\beta_1(p_1 - a + p_2 - b), Q_8 = (p_2 - b)(\beta_2 - \gamma_2) - \gamma_2(p_3 - c) - \beta_2(p_1 - a), Q_9 = \gamma_3(p_2 - b + p_3 - c))$.

Therefore $X_2 = diag[(Q_{10}t^2 + Q_{13}t^3), (Q_{11}t^2 + Q_{14}t^3), (Q_{12}t^2 + Q_{15}t^3)]$, where $Q_{10} = [(Q_1 + Q_4)Q_7 - a\beta_1(Q_2 + Q_5)]/2$, $Q_{11} = [(Q_2 + Q_5)Q_8 + b(\beta_2(Q_1 + Q_4) - \gamma_2(Q_3 + Q_6)]/2$, $Q_{12} = [(Q_3 + Q_6)Q_9 + \gamma_3 c(Q_2 + Q_5)]/2$, $Q_{13} = [-\beta_1(Q_1 + Q_4)^2]/3$, $Q_{14} = [(\beta_2 - \gamma_2)(Q_2 + Q_5)^2]/3$, $Q_{15} = [\gamma_3(Q_3 + Q_6)^2]/3$.

The fourth approximation X_3 is given by $X_3 = M^{-1}[E_1 X_2 B + E_2 X_2 C] + M^{-1} A_2$. Where the special polynomial A_2 is given by $A_2 = diag[q_1, q_2, q_3]$, Where

$$q_1 = Q_7 Q_{13} t^2 + [Q_7 Q_{13} - 2\beta_2(Q_1 + Q_4)Q_{10}]t^3 -$$
$$[\beta_2 Q_{10}^2 + 2\beta_2(Q_1 + Q_4)Q_{13})]t^4 - 2\beta_2 Q_{10}Q_{13}t^5 - \beta_2 Q_{13}^2 t^6$$

$$q_2 = Q_8 Q_{11} t^2 + [Q_8 Q_{14} + 2(\beta_2 - \gamma_2)(Q_2 + Q_5)Q_{11}]t^3 + 2[(\beta_2 - \gamma_2)Q_{11}^2 + (\beta_2 - \gamma_2)$$
$$2(Q_2 + Q_5)Q_{14}]t^4 + (\beta_2 - \gamma_2)(Q_2 + Q_5)Q_{11}Q_{14}t^5 + (\beta_2 - \gamma_2)Q_{14}^2 t^6,$$

$$q_3 = Q_9 Q_{12} t^2 + [Q_9 Q_{15} + 2\gamma_3(Q_3 + Q_6)Q_{12}]t^3 +$$
$$[\gamma_3 Q_{12}^2 + 2\gamma_3 Q_{15}(Q_3 + Q_6)]t^4 + 2\gamma_3 Q_{12}Q_{15}t^5 + \gamma_3 Q_{15}^2 t^6$$

then the fourth approximation X_3 is calculated and is given by

$$X_3 = diag\begin{bmatrix} Q_{16}t^3 + Q_{17}t^4 & Q_{21}t^3 + Q_{22}t^4 & Q_{26}t^3 + Q_{27}t^4 \\ +Q_{18}t^5 + Q_{19}t^6, & +Q_{23}t^5 + Q_{24}t^6, & +Q_{28}t^5 + Q_{29}t^6 \\ +Q_{20}t^7 & +Q_{25}t^7 & +Q_{30}t^7 \end{bmatrix}$$

Where

$Q_{16} = [Q_7 Q_{10} - a\beta_1 Q_{11}]/3$, $Q_{17} = [Q_7 Q_{13} - 2\beta_2(Q_1 + Q_4)Q_{10} - a\beta_1 Q_{14}]/4$, $Q_{18} = [-\beta_2 Q_{10}^2 - 2\beta_2(Q_1 + Q_4)Q_{13}]/5$, $Q_{19} = [-2\beta_2 Q_{10}Q_{13}]/6$, $Q_{20} = [-\beta_2 Q_{13}^2]/7$, $Q_{21} = [Q_8 Q_{14} + b\beta_2 Q_{10} - b\gamma_2 Q_{12}]/3$, $Q_{22} = [Q_8 Q_{14} + 2(\beta_2 - \gamma_2)(Q_2 + Q_5)Q_{11} + b\beta_2 Q_{13} + b(\beta_2 Q_{13} - \gamma_2 Q_{15})]/4$, $Q_{23} = [(\beta_2 - \gamma_2)Q_{11}^2 + 2(\beta_2 - \gamma_2)(Q_2 + Q_5)Q_{14}]/5$, $Q_{24} = [2(\beta_2 - \gamma_2)Q_{11}Q_{14}]/6$, $Q_{25} = [(\beta_2 - \gamma_2)Q_{14}^2]/7$, $Q_{26} = [\gamma_3 c Q_{11} + Q_9 Q_{12}]/3$, $Q_{27} =$

$[Q_9 Q_{15} + 2\gamma_3 Q_{12}(Q_3 + Q_6) + \gamma_3 c Q_{14}]/4$, $Q_{28} = [\gamma_3 Q_{12}^2 + 2\gamma_3 Q_{12}(Q_3 + Q_6)]/5$, $Q_{29} = [2\gamma_3 Q_{12} Q_{15}]/6$ and $Q_{30} = [\gamma_3 Q_{15}^2]/7$

Thus the approximate analytical solution of (114) is given by

$X \approx X_0 + X_1 + X_2 + X_3$ i.e.,

$$L_1 = (p_1 - a) + (Q_1 + Q_4)t + Q_{10}t^2 + (Q_{13} + Q_{16})t^3 + Q_{17}t^4 + Q_{18}t^5 + Q_{19}t^6 + Q_{20}t^7$$

$$D_1 = (p_2 - b) + (Q_2 + Q_5)t + Q_{11}t^2 + (Q_{14} + Q_{21})t^3 + Q_{22}t^4 + Q_{23}t^5 + Q_{24}t^6 + Q_{25}t^7$$

$$W_1 = (p_3 - c) + (Q_3 + Q_6)t + Q_{12}t^2 + (Q_{15} + Q_{26})t^3 + Q_{27}t^4 + Q_{28}t^5 + Q_{29}t^6 + Q_{30}t^7$$

In terms of original densities of the three species L(t), D(t) and W(t) the approximate solutions are given by

$$L(t) = p_1 + (Q_1 + Q_4)t + Q_{10}t^2 + (Q_{13} + Q_{16})t^3 + Q_{17}t^4 + Q_{18}t^5 + Q_{19}t^6 + Q_{20}t^7$$

$$D(t) = p_2 + (Q_2 + Q_5)t + Q_{11}t^2 + (Q_{14} + Q_{21})t^3 + Q_{22}t^4 + Q_{23}t^5 + Q_{24}t^6 + Q_{25}t^7$$

$$W(t) = p_3 + (Q_3 + Q_6)t + Q_{12}t^2 + (Q_{15} + Q_{26})t^3 + Q_{27}t^4 + Q_{28}t^5 + Q_{29}t^6 + Q_{30}t^7$$

The accuracy of the solution is increased by finding the higher iterations. In the absence of predators, the transformation $L = L_1 - \alpha_2/\beta_2$ and $D = D_1 + \alpha_1/\beta_1$ in (108) yields the following system of equations

$$\dot{L}_1 = \alpha_1(L_1 - \alpha_2/\beta_2) - \beta_1(L_1 - \alpha_2/\beta_2)(D_1 + \alpha_1/\beta_1)$$

$$\dot{D}_1 = \alpha_2(D_1 + \alpha_1/\beta_1) + \beta_2(L_1 - \alpha_2/\beta_2)(D_1 + \alpha_1/\beta_1).$$

By defining a positive definite function $V_1(t)$ as

$$V_1(t) = d_1[(-\alpha_2/\beta_2)\log(1 - \beta_2 L_1/\alpha_2) - L_1] + d_2[(\alpha_1/\beta_1)\log(1 + \beta_1 D_1/\alpha_1) - D_1]$$

In the absence of predators system (108) is asymptotically stable if there exists positive constants d_1 and d_2 suchthat $[\beta_1 d_1 - \beta_2 d_2] \leq 0$ since

$$\dot{V}_1(t) = -d_1[\dot{L}_1 L_1/(L_1 - \alpha_2/\beta_2)] - d_2[\dot{D}_1 D_1/(D_1 + \alpha_1/\beta_1)] = L_1 D_1(\beta_1 d_1 - \beta_2 d_2) \leq 0$$

Similarly in the absence of litter, by defining a Liapunov function $V_2(t)$, a positive definite function as

$$V_2(t) = d_2[(\alpha_3/\gamma_3)\log(1 + \gamma_3 D_1/\alpha_3) - D_1] + d_3[(\alpha_2/\gamma_2)\log(1 + \gamma_2 D_1/\alpha_2) - W_1]$$

it is observed that the system (108) is asymptotically stable if there exists positive constants d_2 and d_3 suchthat $[\gamma_2 d_2 - \gamma_3 d_3] \leq 0$. By defining a Liapunov function $V_3(t)$, a positive definite function as

$$V_3(t) = d_1[a\log(1 + L_1/a) - L_1] + d_2[b\log(1 + D_1/b) - D_1] + d_3[c\log(1 + W_1/c) - W_1]$$

It is observed that the system (108) is asymptotically stable if there exists positive constants d_1, d_2 and d_3 such that $[\beta_1 d_1 - \beta_2 d_2] \leq 0$, $[\gamma_2 d_2 - \gamma_3 d_3] \leq 0$.

4.3 Three species system - Homotopy Analysis Method

Consider the system of differential equations representing a three species ecological system

$$\dot{L}(t) = L(t)[P_1 - \gamma D(t) + \beta W(t)]$$
$$\dot{D}(t) = D(t)[P_2 + \gamma L(t) - \alpha W(t)] \tag{118}$$
$$\dot{W}(t) = W(t)[P_3 - \beta L(t) + \alpha D(t)]$$

satisfying the initial conditions

$$L(0) = 0.2, D(0) = 0.3, W(0) = 0.5 \tag{119}$$

According to Homotopy Analysis Method, the initial approximations of (118) are satisfying the initial conditions

$$L_0(0) = 0.2, D_0(0) = 0.3, W_0(0) = 0.5 \tag{120}$$

and the auxiliary linear operators $\mathcal{L}$ for $i = 1, 2, 3$ are

$$\mathcal{L}[\phi_i(t, q)] = \frac{\partial \phi_i(t, q))}{\partial t}, i = 1, 2, 3, \text{with } \mathcal{L}([C_i] = 0. \tag{121}$$

and the nonlinear operators

$$N_1[\phi_i(t, q)] = \frac{\partial \phi_1(t, q))}{\partial t} - \phi_1(t, q)[P_1 - \gamma\phi_2(t, q) + \beta\phi_2(t, q)] = g_1(t) \tag{122}$$

$$N_2[\phi_i(t, q)] = \frac{\partial \phi_2(t, q))}{\partial t} - \phi_2(t, q)[P_2 + \gamma\phi_1(t, q) - \alpha\phi_3(t, q)] = g_2(t) \tag{123}$$

$$N_3[\phi_i(t, q)] = \frac{\partial \phi_3(t, q))}{\partial t} - \phi_3(t, q)[P_3 - \beta\phi_1(t, q) - \alpha\phi_2(t, q)] = g_3(t) \tag{124}$$

The zeroth-order deformation equation is constructed as

$$(1 - q)\mathcal{L}[\phi_1(t, q) - L_0(t)] = qh\{N_1(\phi_1(t, q)) - g_1(t)\} \tag{125}$$

$$(1 - q)\mathcal{L}[\phi_2(t, q) - D_0(t)] = qh\{N_2(\phi_2(t, q)) - g_2(t)\} \tag{126}$$

$$(1 - q)\mathcal{L}[\phi_3(t, q) - W_0(t)] = qh\{N_3(\phi_3(t, q)) - g_3(t)\} \tag{127}$$

where $q \in [0, 1]$ is an embedding parameter, h is a nonzero auxiliary function, $\mathcal{L}$ is an auxiliary linear operator, $L_0(t), D_0(t), W_0(t)$ are the initial guesses of $L(t), D(t), W(t)$, and $\phi_i(t, q)$ are unknown functions.

4.3.1 Solution as polynomial functions

Now, by taking polynomials as base functions the mth order deformation equations are given by

$$\mathcal{L}[L_m(t) - L_{m-1}(t)] = qhR_{1,m}(\overrightarrow{L}_{m-1}) \tag{128}$$

$$\mathcal{L}[D_m(t) - D_{m-1}(t)] = qhR_{2,m}(\overrightarrow{D}_{m-1}) \tag{129}$$

$$\mathcal{L}[W_m(t) - W_{m-1}(t)] = qhR_{3,m}(\overrightarrow{W}_{m-1}) \tag{130}$$

As q increases from 0 to 1, the solutions $\phi_1(t,q), \phi_2(t,q), \phi_3(t,q)$ vary from the initial guesses $L_0(t), D_0(t), W_0(t)$ to the solutions $L(t), D(t), W(t)$. where

$$R_{1,m}(\overrightarrow{L}_m) = \dot{L}_{m-1} - P_1 L_{m-1}(t) + \gamma \sum_{i=0}^{m-1} L_i(t) D_{m-1-i}(t) - \beta \sum_{i=0}^{m-1} L_i(t) W_{m-1-i}(t) \tag{131}$$

$$R_{2,m}(\overrightarrow{L}_m) = \dot{D}_{m-1} - P_2 D_{m-1}(t) + \gamma \sum_{i=0}^{m-1} D_i(t) L_{m-1-i}(t) + \alpha \sum_{i=0}^{m-1} D_i(t) W_{m-1-i}(t) \tag{132}$$

$$R_{3,m}(\overrightarrow{W}_m) = \dot{W}_{m-1} - P_3 W_{m-1}(t) + \beta \sum_{i=0}^{m-1} W_i(t) L_{m-1-i}(t) - \alpha \sum_{i=0}^{m-1} W_i(t) D_{m-1-i}(t) \tag{133}$$

subject to $L_{1,m} = 0, D_{1,m} = 0, W_{1,m} = 0$ \hfill (134)

where the dot denotes differentiation with respect to the variable t. The solution of the mth order deformation equation for $m \geq 1$ is given by

$$L_m(t) = \chi_2 L_{m-1}(t) + h\mathcal{L}^{-1}[R_{1,m}(\overrightarrow{L}_{m-1})] \tag{135}$$

$$D_m(t) = \chi_2 D_{m-1}(t) + h\mathcal{L}^{-1}[R_{2,m}(\overrightarrow{D}_{m-1})] \tag{136}$$

$$W_m(t) = \chi_2 W_{m-1}(t) + h\mathcal{L}^{-1}[R_{3,m}(\overrightarrow{W}_{m-1})] \tag{137}$$

$$\text{and } \chi_m = \begin{cases} 0, m \leq 1 \\ 1, otherwise \end{cases} \tag{138}$$

The integration constants c_i are determined by the initial condition. Now the successive approximations are given by

$$L_1(t) = [-0.2P_1 h + 0.06\gamma h - 0.1\beta h]t \tag{139}$$
$$D_1(t) = [-0.3P_2 h - 0.06\gamma h + 0.15\alpha h]t \tag{140}$$
$$W_1(t) = [-0.5P_3 h + 0.1\beta h - 0.15\alpha h]t \tag{141}$$

$$L_2(t) = [-0.2P_1 + 0.06\gamma - 0.1\beta](ht) + [-0.2P_1 + 0.06\gamma - 0.1\beta](h^2 t) + [0.2P_1^2 - 0.12P_1\gamma +$$
$$0.2P_1\beta - 0.06P_2\gamma + 0.006\gamma^2 + 0.03\alpha\gamma - 0.06\beta\gamma + 0.1P_3\beta + 0.03\beta^2 + 0.03\alpha\beta](h^2 t^2/2) \tag{142}$$

$$D_2(t) = [-0.3P_2 - 0.06\gamma + 0.15\alpha](ht) + [-0.3P_2 - 0.06\gamma + 0.15\alpha](h^2 t) + [0.3P_2^2 + 0.12P_2\gamma -$$
$$0.3P_2\alpha + 0.06P_1\gamma - 0.006\gamma^2 + 0.03\beta\gamma - 0.06\alpha\gamma - 0.15P_3\alpha + 0.03\alpha\beta + 0.03\alpha^2](h^2 t^2/2) \tag{143}$$

$$W_2(t) = [-0.5P_3 - 0.1\beta - 0.15\alpha](ht) + [-0.5P_3 + 0.1\beta - 0.15\alpha](h^2 t) + [0.5P_3^2 - 0.2P_3\beta +$$
$$0.3P_3\alpha - 0.1P_1\beta + 0.03\beta\gamma - 0.03\beta^2 - 0.06\alpha\beta + 0.15P_2\alpha + 0.03\alpha\gamma - 0.03\alpha^2](h^2 t^2/2) \tag{144}$$

Now the analytic solution via the polynomial base functions of the system has the general form

$$L(t) = \sum_{m=1}^{\infty} a_{i,m}(h)t^m, \quad D(t) = \sum_{m=1}^{\infty} a'_{i,m}(h)t^m, \quad W(t) = \sum_{m=1}^{\infty} a''_{i,m}(h)t^m$$

The above expressions represent a family of solution expressions in the auxiliary parameter h. By using h-curves [16] valid regions of a convergent solution series can be determined. Accuracy can easily be obtained by finding more terms by using symbolic computation softwares such as Maple and Mathematica.

4.3.2 Solution as exponential functions

Now, by taking exponential functions as base functions, the initial approximations of (118) satisfying the initial conditions

$$L_0(0) = 0.2, D_0(0) = 0.3, W_0(0) = 0.5 \tag{145}$$

and the auxiliary linear operators $\mathcal{L}$ for $i = 1, 2, 3$ are

$$\mathcal{L}[\phi_i(t,q)] = \frac{\partial \phi_i(t,q))}{\partial t} + \phi_i(t,q), \, i = 1,2,3, \text{with } \mathcal{L}([C_i e^{-t}] = 0. \tag{146}$$

The solution of the mth order deformation equation for $m \geq 1$ is given by

$$L_m(t) = \chi_2 L_{m-1}(t) + he^{-t} \int_0^t e^{\tau} H(\tau) R_m(\overrightarrow{L}_{m-1}) d\tau + C_1 e^{-t} \tag{147}$$

$$D_m(t) = \chi_2 D_{m-1}(t) + he^{-t} \int_0^t e^{\tau} H(\tau) R_m(\overrightarrow{D}_{m-1}) d\tau + C_2 e^{-t} \tag{148}$$

$$W_m(t) = \chi_2 W_{m-1}(t) + he^{-t} \int_0^t e^{\tau} H(\tau) R_m(\overrightarrow{W}_{m-1}) d\tau + C_3 e^{-t} \tag{149}$$

$$\text{and } \chi_m = \begin{cases} 0, m \leq 1 \\ 1, \, otherwise \end{cases} \tag{150}$$

The integration constants c_i are determined by the initial condition. Now the successive approximations obtained by taking $H(\tau) = 1$ in the above equations are given by

$$L_1(t) = [-0.2P_1 + 0.06\gamma - 0.1\beta]h(1 - e^{-t}) \tag{151}$$

$$D_1(t) = [-0.3P_2 - 0.06\gamma + 0.15\alpha]h(1 - e^{-t}) \tag{152}$$

$$W_1(t) = [-0.5P_3 + 0.1\beta - 0.15\alpha]h(1 - e^{-t}) \tag{153}$$

$$L_2(t) = [-0.2P_1 + 0.06\gamma - 0.1\beta]h(1 - e^{-t}) + [-0.2P_1 + 0.06\gamma - 0.1\beta](h^2 te^{-t}) +$$
$$[0.2P_1^2 - 0.12P_1\gamma + 0.2P_1\beta - 0.06P_2\gamma + 0.006\gamma^2 + 0.03\alpha\gamma - 0.06\beta\gamma + 0.1P_3\beta +$$
$$0.03\beta^2 + 0.03\alpha\beta][h^2(1 - e^{-t} - te^{-t})] \tag{154}$$

$$D_2(t) = [-0.3P_2 - 0.06\gamma + 0.15\alpha]h(1 - e^{-t}) + [-0.3P_2 - 0.06\gamma + 0.15\alpha](h^2 te^{-t}) +$$
$$[0.3P_2^2 + 0.12P_2\gamma - 0.3P_2\alpha + 0.06P_1\gamma - 0.006\gamma^2 + 0.03\beta\gamma - 0.06\alpha\gamma - 0.15P_3\alpha +$$
$$0.03\alpha\beta + 0.03\alpha^2][h^2(1 - e^{-t} - te^{-t})] \tag{155}$$

$$W_2(t) = [-0.5P_3 - 0.1\beta - 0.15\alpha]h(1 - e^{-t}) + [-0.5P_3 + 0.1\beta - 0.15\alpha](h^2 te^{-t}) +$$
$$[0.5P_3^2 - 0.2P_3\beta + 0.3P_3\alpha - 0.1P_1\beta + 0.03\beta\gamma - 0.03\beta^2 - 0.06\alpha\beta + 0.15P_2\alpha +$$
$$0.03\alpha\gamma - 0.03\alpha^2][h^2(1 - e^{-t} - te^{-t})] \tag{156}$$

Now the analytic solution via the polynomial base functions of the system has the general form

$$L(t) = \sum_{m=1}^{\infty} L_m(t), D(t) = \sum_{m=1}^{\infty} D_m(t), W(t) = \sum_{m=1}^{\infty} W_m(t) \tag{157}$$

By using h-curves [16] valid regions of a convergent solution series can be determined.Accuracy can easily be obtained by finding more terms by using symbolic computation softwares such as Maple and Mathematica.

5. Examples

Example 1. *Consider a homogeneous system of equations*

$$\dot{x}_1 = -102x_1 + 88x_2, \quad \dot{x}_2 = 88x_1 + 102x_2 \tag{158}$$

where . represents derivative with respect to t with the initial conditions

$$x_1(0) = 1, x_2(0) = 3 \tag{159}$$

Now apply the Homotopy Analysis Method to solve the equations (158) by expressing the solutions $x_1(t)$ and $x_2(t)$ by a set of base functions $\{t^n / n = 0, 1, 2...\}$ as $x_1(t) = \sum_{n=0}^{\infty} a_n t^n$, $x_2(t) = \sum_{n=0}^{\infty} b_n t^n$ where a_n and b_n are coefficients which are to be evaluated. We choose the initial approximations as $x_{1,0}(t) = x_1(0) = 1$, $x_{2,0}(t) = x_2(0) = 3$ and the auxiliary linear operator as $L[\phi_i(t;q)] = \frac{\partial \phi_i(t;q)}{\partial t}, i = 1, 2$ with the property $L[C_i] = 0$, where $C_i (i = 1, 2)$ are integral constants. Define a system of nonlinear operators as $\quad N_1[\phi_i(t;q)] = \frac{\partial \phi_i(t;q)}{\partial t} + 102\phi_1(t;q) - 88\phi_2(t;q)$, $N_2[\phi_i(t;q)] = \frac{\partial \phi_i(t;q)}{\partial t} - 88\phi_1(t;q) - 102\phi_2(t;q)$. Now the zeroth order deformation equation as

$$(1 - q)L[\phi_i(t;q) - x_{i,0}(t)] = qh_i N_i[\phi_i(t;q)], i = 1, 2 \tag{160}$$

From the definition of Homotopy we get $\phi_i(t, 0) = x_{i,0}(t), \phi_i(t, 1) = x_i(t)$. It can be understood that as q increases from 0 to 1 initial approximation $\phi_i(t;q)$ approaches the solution $x_i(t)$ for $i = 1, 2$. Now $\phi_i(t;q) = x_{i,0}(t) + \sum_{m=1}^{\infty} x_{i,m}(t)q^m$ where

$$x_{i,m}(t) = \frac{1}{m!} \frac{\partial \phi_i(t;q)}{\partial q^m} \Big|_{q=0} \tag{161}$$

Define the vector $\vec{x}_{i,n} = \{x_{i,0}(t), x_{i,1}(t), ..., x_{i,n}(t)\}$. The mth deformation equation is given by

$$L[x_{i,m}(t) - \chi_m x_{i,m-1}(t)] = h_i R_{i,m} \vec{x}_{i,m} \tag{162}$$

subject to the initial condition $x_{1,0} = 0, x_{2,0} = 0$ where

$$R_{1,m} \vec{x}_{i,m} = \dot{x}_{1,m-1} + 102x_{1,m-1} - 88x_{2,m-1}$$
$$R_{2,m} \vec{x}_{i,m} = \dot{x}_{2,m-1} - 88x_{1,m-1} - 102x_{2,m-1}$$

Now the solution of the mth order deformation equation 162 for $m \geq 1$ is

$$x_{i,m}(t) = \chi_m x_{i,m-1}(t) + h_i L^{-1}[R_{i,m}(\vec{x}_{i,m-1})] \tag{163}$$

Now get the successive approximations as

$$x_{1,1}(t) = -120ht, x_{2,1}(t) = 280ht, \tag{164}$$

$$x_{1,2}(t) = -120ht - 120h^2t - 18400h^2t^2, \quad x_{2,1}(t) = 2800ht + 280h^2t + 21600h^2t^2 \tag{165}$$

Now by taking $h = -1$ (without plotting the h-curve)the solution of (158)is given by

$$x_1(t) = \sum_{m=0}^{\infty} x_{1,m}(t) = -e^{-200t} + 2e^{-40t}, \quad x_2(t) = \sum_{m=0}^{\infty} x_{2,m}(t) = e^{-200t} + 2e^{-40t}$$

Example 2. *Consider the system of nonlinear ordinary differential equations*

$$\dot{x}_1 = -1002x_1 + 1000x_2^2, \quad \dot{x}_2 = x_1 - x_2 - x_2^2 \tag{166}$$

with the initial conditions

$$x_1(0) = 1, x_2(0) = 1 \tag{167}$$

Now we apply Homotopy analysis method and take the initial approximation of the solution of the above system as

$$x_{1,0}(t) = x_1(0) = 1, \ x_{2,0}(t) = x_2(0) = 1$$

and the auxiliary linear operator as $L[\phi_i(t;q)] = \frac{\partial \phi_i(t;q)}{\partial t}, i = 1,2.$ *Define nonlinear operators as*

$$N_1[\phi_i(t;q)] = \frac{\partial \phi_i(t;q)}{\partial t} + 1002\phi_1(t;q) - 1000\phi_2(t;q)$$

$$N_2[\phi_i(t;q)] = \frac{\partial \phi_i(t;q)}{\partial t} - \phi_1(t;q) + \phi_2(t;q) + \phi_2(t;q)^2$$

Now the zeroth order deformation equation as

$$(1-q)L[\phi_i(t;q) - x_{i,0}(t)] = h_i R_{i,m}(\overrightarrow{x}_{i,m-1}) \qquad i = 1,2 \tag{168}$$

where

$$R_{1,m}(\overrightarrow{x}_{i,m-1}) = \dot{x}_{1,m-1} + 1002x_{1,m-1} - 1000 \sum_{j=0}^{m-1} x_{2,m-1j}(t)$$

$$R_{2,m}(\overrightarrow{x}_{i,m-1}) = \dot{x}_{2,m-1} - x_{1,m-1} - \sum_{j=0}^{m-1} x_{2,m-1-j}(t)$$

where the "." denotes differentiation with respect t. Now the solution of the mth order deformation is given by

$$x_{i,m}(t) = \chi_m x_{i,m-1}(t) + h \int R_{i,m}(\overrightarrow{x}_{i,m-1}) dt + C_i, \qquad i = 1,2$$

the integration constants Ci (i = 1,2) can be determined by the initial conditions.

$$x_{1,1}(t) = 2ht, x_{1,2}(t) = 2ht + 2h^2 t + 2h^2 t^2 + \dots$$
$$x_{2,1}(t) = ht, x_{1,2}(t) = ht + h^2 t + h^2 t^2 + \dots$$

Now by taking $h = -1$ the successive approximations are computed as (The proper value of h can be chosen by plotting h-curve of $\dot{x}_i(0)$ and $\dot{x}_i(0)$ as discussed in [16]

$$x_{1,1}(t) = -2t, x_{1,2}(t) = 2t^2, x_{1,3}(t) = -\frac{4}{3}t^3 \dots$$

$$x_{2,1}(t) = -t, x_{1,2}(t) = \frac{1}{2}t^2, x_{1,3}(t) = -\frac{1}{6}t^3 \dots$$

Thus the solution of (166) is given by

$$x_1(t) = \sum_{m=0}^{\infty} x_{1,m}(t) = \sum_{m=0}^{\infty} \frac{(-2t)^m}{m!} = e^{-2t} \tag{169}$$

$$x_2(t) = \sum_{m=0}^{\infty} x_{2,m}(t) = \sum_{m=0}^{\infty} \frac{(-t)^m}{m!} = e^{-t} \tag{170}$$

The solutions (169),(170) obtained by Homotopy analysis are coinciding with the exact solutions of (166) satisfying the initial conditions (167).

Example 3. *Consider the system of nonlinear ordinary differential equations*

$$\dot{x}_1 = x_2, \ \dot{x}_2 = x_3, \ \dot{x}_3 = -6x_1 - 2.92x_2 - x_3 + x_1^2 \tag{171}$$

satisfying the initial conditions

$$x_1(0) = 0.2, x_2(0) = -0.3, x_3(0) = 0.1 \tag{172}$$

Now we apply Homotopy Analysis method expressing the solutions $x_1(t)$, $x_2(t)$ and $x_3(t)$ by a set of base functions $\{e^{(-nt)}/n \geq 0\}$ and take the initial approximation of the solution of the above system as

$x_{1,0}(t) = x_1(0) = 0.2, \ x_{2,0}(t) = x_2(0) = -0.3, \ x_{3,0}(t) = x_3(0) = 0.1$

and the auxiliary linear operator as $L[\phi_i(t;q)] = \frac{\partial \phi_i(t;q)}{\partial t} + \phi_i(t,q), \quad i = 1,2,3$ with the property $L[C_i e^{-t} = 0], i = 1,2,3$ where $C_i(i = 1,2,3)$ are integral constants. Now the first approximations of solutions are given by
$x_{1,1}(t) = 0.3h - 0.3he^{-t}, \ x_{2,1}(t) = -0.1h + 0.1he^{-t}, \ x_{3,1}(t) = 0.404h - 0.404he^{-t}.$
Second approximations of solutions are given by
$x_{1,2}(t) = 0.3h - 0.3he^{-t} + 0.1h^2 - 0.1h^2 e^{-t} + 0.2h^2 te^{-t},$
$x_{2,1}(t) = -0.1h + 0.1he^{-t} - 0.404h^2 + 0.404h^2 e^{-t} + 0.304h^2 te^{-t},$
$x_{3,1}(t) = 0.404h - 0.404he^{-t} + 1.8728h^2 - 1.8728h^2 e^{-t} - 1.4688h^2 te^{-t}.$

Now the solution of the system (171) can be expressed as $v_i(t) = \sum_{m=0}^{\infty} \sum_{n=0}^{\infty} d_{i,m,n}(h)t^m e^{-nt}, \quad i = 1,2,3$ where the coefficients $d_{i,m,n}$ depend on h and $v_i(t) = x_i(t), i = 1,2,3$. By plotting $\dot{v}(t)$ and $\ddot{v}(t)$ as discussed in [16] we can choose $h = -1$ for x_1, $h = -0.86$ for x_2, $h = -1.1$ for x_3. We can easily observe that higher order Homotopy Analysis Method solutions are very nearer to the solutions obtained by RK-method and Adams bash forth predictor corrector method.

Example 4. *Consider differential equation*

$$\dot{x} = -x^2 \tag{173}$$

satisfying the initial conditions

$$x(0) = 1 \tag{174}$$

By applying Homotopy Perturbation method we construct Homotopy as

$$(1 - p)(\dot{\phi} - \dot{x}_0) + p(\dot{\phi} + \phi^2) = 0 \tag{175}$$

Now the initial approximation is taken as $x_0 = 1$. Let the solution of (175) is of the form

$$\phi = \phi_0 + p\phi_1 + p^2\phi_2 + \cdots \tag{176}$$

By equating the like powers of p after using (176) in (175), we get
$\dot{\phi}_0 = \dot{x}_0$, $\dot{\phi}_1 + \dot{x}_0 + \phi_0^2 = 0$, $\phi_1(0) = 0$, $\dot{\phi}_2 + \dot{x}_0 + 2\phi_0\phi_1 = 0$, $\phi_2(0) = 0$. *If we take* $\phi_0 = x_0 = 1$
then $\phi_1 = -t$ *and* $\phi_2 = t^2$. *Thus the second order approximation of the solution of (175) obtained by using Homotopy Perturbation method is given by* $x(t) = \phi_0 + p\phi_1 + p^2\phi_2 = 1 - t + t^2$.

Example 5. *Consider the system of nonlinear ordinary differential equations*

$$\dot{x}_1 = x_1 - x_1x_2, \quad \dot{x}_2 = -0.1x_2 + x_1x_2 \tag{177}$$

with the initial conditions $x_1(0) = 14, x_2(0) = 18$. *By applying Homotopy perturbation method we get the third order approximation to the solution as* $x_1 = 14 - 238t + 271.6t^2 + 20191.32t^3$, $x_2 = 18 + 250.2t - 403.11t^2 - 20087.33t^3$. *We can observe that higher order approximations agree with the solutins obtained by RK-method and Adams bash forth predictor corrector method.*

6. Summary and conclusions

In this chapter we studied the dynamics of Two species and Three species ecological systems. Since Two species systems represent the various prey predator interactions and Three species systems represent the behavior of the litter, detritus and predators in the mangrove areas their respective dynamics are modeled by nonlinear differential equations. Heuristic procedures for Perturbation method, Adomian decomposition method for system, Homotopy perturbation method and Homotopy analysis method are presented so as to familiarize the reader with the application of these important methods. Two species and Three species models are solved by using the above stated methods. The Homotopy perturbation method which was used to solve the nonlinear system of differential equations, governing the prey and predator problem, is very easy and accurate to employ with reliable results. There is less computation needed in comparison with the Adomian decomposition and power series methods.

Homotopy analysis method was applied to solve both the Two species and Three species ecological systems by considering both Polynomial base functions and exponential base functions. The accuracy of the continuous solution obtained by using Homotopy analysis method ia almost same as that of a numerical method. This is convenient for practical applications with minimum requirements on calculation and computation and validity of the Homotopy analysis method series solutions can be enhanced by finding more terms and/or using the Pade technique. The functional form of the solution would be useful in the study of the stability of the system. Owing to the generality, the considered two species and Three species models exhibit very rich dynamics. From the illustrative examples it can be observed that Homotopy analysis method is a powerful method for nonlinear problems and provides us with a convenient way of controlling the convergence of approximation series, which is a major advantage when compared with other methods.

7. Acknowledgements

This work is supported by DST-CMS project Lr.No.SR/S4/MS:516/07, Dt.21-04-2008 and the support is gratefully acknowledged.

8. References

[1] Pielou, E.C, *An Introduction to Mathematical Ecology*, Wiley-Interscience, New York (1969)

[2] Rao C.R and Mitra S.K, *Generalized Inverse of Matrices and its Applications*, John Wiley and Sons, New York (1971)

[3] Murty K.N and Rao D.V.G, Approximate analytical solution of general Lotka-Volterra equations, *J. Math. Anal. Appl.* 122(1987), pp. 582-588

[4] Murty K.N, Srinivas M.A.S and Prasad K.R, Certain mathematical models for biological systems and their approximate analytical solutions, In: *Proc. Int. Conf. on Non-linear analysis and applications to Bio-mathematics*, Andhra Univ., Visakhapatnam, 1987

[5] Adomian G, A review of the decomposition method in applied mathematics, *J. Math. Anal. App.* 135(1988), pp. 501–544

[6] Venkata Sundaranand Putcha, Ramu Malladi., A Mathematical Model on Detritus In Mangrove Estuarine Ecosystem, International Journal of Pure and Applied Mathematics Volume 63 No. 2 2010, 169-181

[7] Murty K.N, Srinivas M.A.S and Prasad K.R, Approximate analytical solutions to the three-species ecological system, *J. Math. Anal. Appl.* 145(1990), pp. 89-99

[8] Nayfeh A.H, Perturbation Methods, Wiley Classics Library, John Wiley & Sons, New York, NY, USA, 2000.

[9] Srinivasu P.D.N and Gayatri I.L, Influence of prey reserve capacity on predator-prey dynamics, *Ecological Modelling* 181(2005), pp. 191-202

[10] Murray J. D, *Mathematical Biology I An Introduction*, 3rd ed, Springer, Berlin, 2002

[11] Freedman H.I and Waltman P, Persistence in models of three interacting predator-prey populations, *Mathematical Biosciences* 68(1984), No. 2, 213Ű231

[12] He J.H, Homotopy perturbation technique, Computer Methods in Applied Mechanics and Engineering 178 (3Ű4) (1999) 257Ű262.

[13] He J.H, Homotopy perturbation technique:A new nonlinear analytical technique, Applied Mathematics and Computation 135(2003) 73Ű79.

[14] He J.H, Asymptotology by Homotopy perturbation method, Applied Mathematics and Computation 156(2004) 591Ű596.

[15] Liao S.J., The proposed Homotopy analysis techniques for the solution of nonlinear problems, Ph.D. dissertation, Shanghai Jiao Tong University, Shanghai, China, 1992.

[16] Liao S.J, Beyond perturbation, Introduction to the Homotopy analysis method, CRC Press, Boca Raton, Chapman and Hall, 2003

[17] Varma V.S, Exact Solutions for a special Prey Predator Competing species system, Bulltin Mathematical Biology, 39 (1977) 619-622

[18] Wilson, A J, On Varma's Prey Predaor Poblem, 42 (1980) 599-600

[19] Burnside R.R, A note on exact Solutions of Prey Predator equations, Bulltin Mathematical Biology, 44 (1982) 893-897

[20] Lotka A.J, Elements of Physical Biology, Williams & Wilkins, Baltimore, Md, USA, 1925.

[21] Ayala F.J, Gilpin M.E and Ehrenfeld J.G, Competition between species: Theoretical Models and experimental tests, Theoret. Population Biology,4 (1973), 331-356

[22] Albrecht F, Gatzke H, Haddad A and Wax N, The dynamics of two interacting Populations, Journal of Mathematical Analysis and Applications,46 (1974), 658-670

[23] Gopalaswamy K, Time lags and global Stability in two species competition, Bulletin Mathematical Biology,42 (1980), 728-737

[24] Brown G.C, Stability in an insect pathogen model incorporating age-dependent immunity and seasonal host reproduction, Bulletin Mathematical Biology,46 (1984), 139-153

[25] May R.M and Leonard W.J, Nonlinear aspects of competition between three species, SIAM Journal of Applied Mathematics,29,2 (1975), 243-253

[26] Schuster P , Sigmund K and Wolff R, on w-limits for competition between three species, SIAM Journal of Applied Mathematics, 29,1 (1979), 49-54

[27] Zhangjinyan, Three dimensional Gradiant conjugate system and applications, Proceedings of the International Conference on Non-linear analysis and applications , USA, 109 (1986) 613-621.

[28] Venkata Sundaranand Putcha and Ramu Malladi, A Mathematical Model on Detritus In Mangrove Estuarine Ecosystem, International Journal of Pure and Applied Mathematics, 63, 2 (2010), 169-181

[29] Venkata Sundaranand Putcha, Non-Linear Differential and Difference Equations Existence, Uniqueness, Stability, Observability and Controllability, Ph.D Dissertation, Andhra University, Visakhapatnam, India, 1995

[30] Mantripragada Ananta Satya Srinivas, Contributions to Certain Nonlinear Biomathematical Models, Ph.D Thesis, Andhra University, Visakhapatnam, India, 1989

[31] John Pastor, Mathematical Ecology of Populations and Ecosystems, Wiley-Blackwell , A John Wiley & Sons, Ltd., Publication, 2008

[32] Sami Bataineh A, Noorani M.S.M, Hashim I, Solving systems of ODEs by Homotopy analysis method, Communications in Nonlinear Science and Numerical Simulation 13 (2008) 2060-2070

[33] Francisco M. Fernt'andez, On some approximate methods for nonlinear Models, arXiv:0904.4044v1 [math-ph] 26 Apr 2009

[34] Faghidian Sa, Moghimi Zand, M, Farjami Y, Farrahi G.H, Application of Homotopy-Pade Technique To The Volterra's Prey And Predator Problem, Appl. Comput. Math., V.10, N.2, 2011, pp.262-270

[35] Shijun Liao, Comparison between the Homotopy analysis method and Homotopy perturbation method, Applied Mathematics and Computation 169 (2005) 1186-1194

[36] Zoua L, Zonga Z, Dongb G.H, Generalizing Homotopy analysis method to solve Lotka-Volterra equationI, Computers and Mathematics with Applications 56 (2008) 2289-2293

[37] Rafei M, Daniali H, Ganji D.D, Pashaei H, Solution of the prey and predator problem by Homotopy perturbation method, Applied Mathematics and Computation 188 (2007) 1419-1425

[38] Cheng-shi Liu, The essence of the Homotopy analysis method, arXiv:1105.6183v1 [nlin.SI] 31 May 2011

[39] Vipul K Baranwal, Ram K Pandey, Manoj P Tripathi, Om P Singh, Analytic algorithms for Some Models of Nonlinear Age-Structured Population Dynamics and Epidemiology, Journal of Modern Physics, 2011, 2, 236-247

12

Stable Isotope Research in Southern African Birds

Craig T. Symes
School of Animal, Plant and Environmental Sciences
University of the Witwatersrand
South Africa

1. Introduction

The use of stable isotopes in ornithological research has risen exponentially in the past few decades, mainly due to strides in technology that permit the processing of large numbers of samples at lower and lower cost. However, most likely because of the availability of resources in "developed countries", studies are concentrated in predominantly Palaearctic and Nearctic countries. Africa is largely neglected but offers unique opportunities for both collaborative research on a global scale, and novel studies in unchartered fields. This chapter provides a brief summary of, 1) stable isotopes and some of the applications thereof in ornithology, 2) examples of the type of work currently completed and underway in Africa and how it might be applicable in a global perspective, and, 3) prospects for future studies in the use of stable isotopes for understanding African ornithology. Given the high diversity of birds in southern Africa (at least 950 species), associated with at least seven terrestrial biomes, e.g. fynbos, succulent Karoo, Nama-Karoo, grassland, savanna, forest, and thicket, numerous opportunities exist for addressing research questions of biological interest and conservation concern.

2. Stable isotopes

2.1 What are stable isotopes?

The atom of an element is defined by the number of protons within the nucleus which, together with the neutrons it contains, define its mass (Hoefs, 2009). For a particular element, additional neutrons within the nucleus will add mass to an atom but will not change its chemistry (Hoefs, 2009). These atoms of different mass are called isotopes, stable if they are not radioactive, and may behave differently during chemical reactions because of this difference in mass (Hobson and Wassenaar, 2008; Hoefs, 2009). Only 21 elements occur as pure elements, the rest are mixtures of at least two isotopes (Bigeleisen, 1965; Hoefs, 2009). However, a handful of elements, particularly light isotopes that are important in key life processes, e.g. hydrogen, carbon, nitrogen, oxygen and sulphur, are of interest to biologists (West et al., 2006; Hobson and Wassenaar, 2008).

Isotope fractionation, or the partial separation of light and heavier stable isotopes, may occur during physical processes, e.g. diffusion and evaporation, or biological processes, e.g.

photosynthesis and metabolism. The proportions of these isotopes in the environment will subsequently become disparate and it is the understanding of these differences, and how, when and why they occur, that are valuable for scientists in understanding complex interactions and processes in the world in which we live (Ehlringer, 1991; West et al., 2006).

Isotope ratios are expressed relative to arbitrary element-specific standards using delta-value (δ) notation (Bigeleisen, 1965; Ehleringer and Osmond, 1989; McKechnie, 2004; Hoefs, 2009). For numerous reasons absolute measurements are not reliable and it is more common to calculate these ratios relative to a standard reference material (Gonfiantini, 1978; Ehleringer and Osmond, 1989), e.g. standard mean ocean water (SMOW), or more recently Vienna standard mean ocean water (V-SMOW), for hydrogen (D/H) and oxygen ($^{18}O/^{16}O$) in water; *Belemnitella americana* from the Cretaceous Peedee formation, South Carolina (PDB or Peedee Belemnite), for carbon ($^{13}C/^{12}C$) and oxygen ($^{18}O/^{16}O$) in carbonates and organic material; atmospheric air (N_2) for nitrogen (N^{15}/N^{14}); and troilite (FeS) from the Canyon Diablo iron meteorite (Canyon Diablo meteorite, CD) for sulphur (Gonfiantini, 1978; Ehleringer and Osmond, 1989; Lee-Thorp and Thalma, 2000; McKechnie, 2004; Hoefs, 2009). The isotopic compositions of natural materials can be measured with great accuracy with a mass spectrometer (Peterson and Fry, 1987) and the accepted isotope ratio δ-value, expressed in parts per thousand (per mil, ‰), is calculated by:

$$\delta_{sample} (‰) = [(R_{sample}/R_{standard}) - 1] \times 1000 \qquad (1)$$

where R represents the isotopic ratio (heavier to lighter isotope, e.g. $^{13}C/^{12}C$, $^{18}O/^{16}O$) of the sample and standard respectively (Peterson and Fry, 1987; Hobson and Wassenaar, 2008; Hoefs, 2009).

Thus, for example, the isotope ratio of carbon would be represented as follows;

$$\delta_{sample} (‰) = [(^{13}C/^{12}C)_{sample}/(^{13}C/^{12}C)_{standard} - 1] \times 1000 \qquad (2)$$

Isotope standards, obtained from the IAEA (International Atomic Energy Agency), are often used to calibrate other laboratory standards that in turn are run with samples. This is because some of these original standards are long exhausted. Therefore, when expressing results, the standard used is always noted. This also facilitates conversions that relate to other standards, e.g. $\delta^{18}O_{SMOW} = 1.03091.\delta^{18}O_{PDB} + 30.91$, when comparing different studies (Friedman and O'Neill, 1977).

Stable isotopes have been used as a tool in palaeontological research for decades (Ehleringer and Osmond, 1989; Lee-Thorp and Thalma, 2000; Hoefs, 2009). Botanists have also long studied the patterns of ^{13}C distribution in plants resulting from climatic, altitudinal, latitudinal, and photosynthetic factors (Park and Epstein, 1960, 1961; Smith and Epstein, 1971; O'Leary, 1981; Dawson et al., 2002) and it is only recently that the technique has been applied in the zoological arena (Peterson and Fry, 1987), with a plethora of research publications appearing in recent decades (e.g. Lee-Thorp et al., 1989; Hobson & Clark, 1992a,b; Cerling and Harris 1999; Hobson et al. 2001; McKechnie, 2004; Herrera et al., 2003, 2006; West et al., 2006). The ratios of stable isotope signatures unique to particular environments, and the changes and variation in these ratios due to physical and biological processes, has gained popularity as a research tool for biologist's worldwide (Bigeleisen, 1965; Ehleringer and Osmond, 1989; Lee-Thorp and Thalma, 2000; McKechnie, 2004). The reason for this is twofold; a reduction in the cost of analysing samples and improved technology that allows for the rapid processing of samples.

2.2 Carbon

Isotope compositions change in predictable ways as elements are cycled through ecosystems (Bigeleisen, 1965; Peterson and Fry, 1987). In incomplete reactions more of the lighter isotope is used up and the reaction products become depleted (i.e. they contain more of the lighter isotope). The unreacted material thus becomes enriched (i.e. containing more of the heavier isotope). In photosynthesis, the degree of carbon isotope fractionation is established during two rate-controlling processes; the diffusion of CO_2 into the chloroplasts, and the carboxylation process itself (Ehleringer and Monson, 1993; Hoefs, 2009). In photosynthesis there is thus a depletion of ^{12}C in the remaining CO_2 because the light ^{12}C is concentrated in the synthesized organic material (Hoefs, 2009). In C_3 plants (trees, shrubs and herbs, and temperate or shade grasses), where the initial product of photosynthesis is a three-carbon molecule, the CO_2 fixing enzyme Rubisco (ribulose-1,5-bisphosphate carboxylase/ oxygenase) discriminates against ^{13}C more strongly than another CO_2 fixing enzyme, PEP (phosphoenolpyruvate) carboxylase, that occurs in C_4 (mainly tropical grasses) and CAM (Crassulacean acid metabolism) plants (Park and Epstein, 1960, 1961; Smith and Epstein, 1971; Ehleringer, 1991). Plants exhibiting C_3 photosynthesis thus become more depleted in ^{13}C relative to C_4 and CAM plants, with typical $\delta^{13}C$ values for these different groups of plants as follows (Vogel et al., 1978; Ehleringer and Osmond, 1989; Ehleringer, 1991; Dawson et al., 2002);

$$-26.5\text{‰} \; (-37 \text{ to } -24\text{‰}) \text{ for } C_3$$

$$-12.5\text{‰ for } (-16 \text{ to } -9\text{‰}) \text{ for } C_4$$

$$-17\text{‰} \; (-19 \text{ to } -9\text{‰}) \text{ for CAM}$$

Implicit in the application of isotope techniques for biologists is the assumption that the isotope ratio of a consumers' tissue is related in some way to its diet (De Niro and Epstein, 1981; Ehleringer and Osmond, 1989; Hobson and Clark, 1992a,b). Typically the whole body of an animal is enriched in ^{13}C relative to its diet by 1‰, although this fractionation can vary under different conditions (see later; De Niro and Epstein, 1978; Hobson and Clark, 1992). Because of this it may be necessary to establish levels of fractionation in the laboratory under controlled conditions (Gannes et al., 1997).

The use of stable isotopes has provided insight into the use of food resources by animal's not otherwise possible (Ehleringer and Osmond, 1989; Kelly, 2000; Dalerum and Angerbjörn, 2005). Whereas diet analyses have usually considered ingested material, gut contents or excreted matter to infer diets for animals, stable isotopes are able to provide an interpretation of assimilated matter using non-destructive means (Hobson and Clark, 1993; Phillips and Gregg, 2003).

The differences in the ratios of stable isotopes in different plants and environments are therefore useful in understanding more complex processes related to, for example, food chains and food webs, and sources of carbon. Carbon stable isotopes are thus used to reconstruct and quantify the proportions of isotopically distinct diets in different animals.

These dietary reconstructions rely on linear mixing models, with the proportions of two food sources in an animal's diet calculated by:

$$\delta X_{tissues} = p\delta X_A + (1 - p)\delta X_B + \Delta \tag{3}$$

where $\delta X_{tissues}$ is the isotope ratio in the animal's tissues, δX_A and δX_B are the isotope ratios of the respective food sources and p is the proportion of food A in the diet. The discrimination factor between the diet and the food source is represented by Δ (McKechnie, 2004; Hoefs, 2009). However, the dietary input for an animal is more often than not made up of more than two isotopic endpoints. Therefore, more isotopes and more sources can be used in multiple mixing models, e.g. IsoSource, MixSIR, ISOCONC1_01.xls, to quantify dietary proportions.

2.3 Nitrogen

Nitrogen isotopes are a useful tool in testing hypotheses concerning trophic levels and food web structure, and energy and nutrient transfer, because they are fractionated to a greater degree than carbon isotopes (De Niro and Epstein, 1981; Minagawa and Wada, 1984; Peterson and Fry, 1987; Mizutani et al., 1992; Hobson et al., 1994; Hobson and Wassenaar, 1999; McKechnie, 2004). Isotopic enrichment along a food chain is typically greater for ^{15}N than for ^{13}C, so the $\delta^{15}N$ values of a consumer's tissues becomes enriched by 3-5‰ relative to its diet (De Niro and Epstein, 1981; Minagawa and Wada, 1984; Fry, 1988; Mizutani et al., 1992; Hobson et al., 1994; Hobson and Wassenaar, 1999; McCutchan et al., 2003). This fractionation results from the differences between nitrogen assimilation and nitrogen excretion. An animals' trophic position can thus be estimated by:

$$\textbf{Trophic level} = \lambda + [(\delta^{15}N_{secondary\ consumer} - \delta^{15}N_{base}) / \Delta_n] \tag{4}$$

where λ is the trophic position of the organism (λ = 1 for primary producers) used to estimate the $\delta^{15}N$ base, $\delta^{15}N_{secondary\ consumer}$ refers to the tissues of the consumer of interest, $\delta^{15}N_{base}$ is the corresponding value at the base of the food web, and Δ_n is the ^{15}N enrichment per trophic level (i.e. fractionation between diet and tissue).

The accuracy in determining the trophic position of an organism thus relies on, 1) a clear understanding of the food base on which a particular animal relies, and 2) the ^{15}N enrichment per trophic level (Hobson and Wassenaar, 2008). Also, for a clearer understanding of trophic positions reliable estimates of $\delta^{15}N$ at the base, and changing diet-tissue fractionation factors related to diet switches, particularly over a long period of time, are required (Post, 2002). This is one reason for the call for more controlled laboratory experiments in which stable isotopes under particular conditions can be better understood (Gannes et al., 1997; Martínez del Rio et al., 2009).

2.4 Research applications of carbon and nitrogen stable isotopes for birds

Stable isotope analysis has been used in the field of ornithology since the 1980s and has grown exponentially in the past few decades (Hobson, 2011). The biologically important stable isotopes of C, N, H, O and S have subsequently been measured in a variety of avian tissues to greater understand bird biology in mainly three realms; 1) diet and trophic relationships, 2) tracing the relative contribution of endogenous and exogenous nutrient inputs into reproduction, and 3) determining the origin of migratory populations or individuals (Peterson and Fry, 1987; Kelly, 2000; Inger and Bearhop, 2008; Hobson, 2011). For other reasons including those stated above, and besides the fact that they are often measured concurrently on a mass spectrometer, combined use of carbon and nitrogen are useful in understanding carbon sources and interpreting trophic level interactions (Hobson et al., 1994). In particular they may be useful in understanding resource use and niche partitioning in diverse bird communities (Kelly, 2000; Herrera et al., 2003, 2006; Symes and Woodborne, 2009).

2.5 Hydrogen

2.5.1 Global fingerprint

The global distribution of hydrogen isotopes in precipitation (δDp) has been used as a useful tool in determining the migratory connectivity of bird species in the Americas and Europe (e.g. Chamberlain et al., 1997; Hobson and Wassenaar, 1997; Chamberlain et al., 2000; Meehan et al., 2001; Wassenaar and Hobson, 2001; Hobson et al., 2004a,b; Clark et al. 2006). This is because of a significant and predictable change in δDp across the globe, with a general depletion of δD (more negative δ values) in precipitation with an increase in latitude (Hobson, 1999; Bowen and Revenaugh, 2003; Lott and Smith, 2006; Inger and Bearhop, 2008; Hobson, 2011). A relatively robust model of hydrogen in precipitation across the globe, which also incorporates other variables such as distance from the sea, elevation and precipitation, has been developed (Bowen and Revenaugh, 2003). However, when making links between the hydrogen isotope values in feathers and the origins of migrating populations, it is important to have a sound understanding of moult and feather growth in the study species (Hobson and Wassenaar, 2008). This is because it is important to know where a feather is grown in its annual cycle (Clark et al. 2006). Also important is a sound understanding of fractionation processes from precipitation to feathers; for example most studies on raptor species show a depletion in δD of feathers from precipitation of 37–52‰ (Lott et al., 2003, Meehan et al., 2003; Hobson et al., 2009). However, it has been demonstrated that under different environmental conditions fractionation values may vary (McKechnie et al., 2004). Once again, for clearer interpretations regarding our understanding of bird migrations more laboratory and fieldwork experiments are required (Gannes et al., 1997; Martínez del Rio et al., 2009). Possible confounding factors that will obscure results of δD analysis include, for example, 1) animals consuming drinking water of very different isotopic composition when compared to food, 2) the enrichment of δD due to water stress in arid regions or during reproduction, 3) the change in δD patterns due to climate change, and 4) food webs relying on groundwater or lakes and wetlands (Hobson, 2005). Also, there must be clarity on the hydrogen component, i.e. exchangeable or non-exchangeable hydrogen, that is analysed. Feather samples must therefore be handled in a manner that ensures that hydrogen exchange between keratin and the ambient water is accounted for (Chamberlain, 1997; Hobson, 1999).

In Africa the hydrogen isotope gradient on either side of the equator is low so movements of birds within the continent are more difficult to study using stable isotopes (but see Yohannes et al., 2005, 2007, 2009, 2011; Wakelin et al., 2011). However, these challenges have not prevented research into African bird migrations using stable isotopes (see later).

2.5.2 Water turnover

Analysing hydrogen stable isotopes in tissues can provide valuable information in understanding the sources of water for organisms and water turnover, particularly in arid environments. To date, there has been no work conducted on understanding water sources of birds in southern Africa. However, in the Americas there have been investigations into the use of water sources in arid environments (Wolf and Martínez del Rio, 2003; McKechnie et al., 2004). For example, ratios of deuterium were used to determine the contribution of saguaro plant resources to water requirements of desert White-winged Doves *Zenaida asiatica mearsnii* (Wolf and Martínez del Rio, 2000; Wolf et al., 2002; Wolf and Martínez del Rio, 2003).

2.6 Tissue turnover

Important in interpretations of diet, water use, trophic level delineation etc., are assumptions regarding tissue turnover. Although it is not the task of this chapter to comprehensively review this topic, important points related to our understanding of avian biology and stable isotopes will be addressed.

The isotopic composition of a tissue that remains metabolically inert after formation should reflect that of the food and water consumed during synthesis (Chamberlain et al., 1997; Hobson, 1999; Hobson, 2005). Feathers are grown at least once a year and remain metabolically inert, so when moulted retain information of the previous location of moult. Claws of birds are metabolically inert but continuously grown, and can be useful to infer dietary and habitat information over a period of time from weeks to many months (Bearhop et al., 2003). Tissues that are metabolically active will reflect the average diet of the individual during a certain period into the past, according to the tissue's turnover rate, ranging from days (liver and blood plasma), to weeks (muscle and whole blood), to months, years or a lifetime (bone collagen) (Hobson and Clark, 1992a; Hobson, 1999; Hobson, 2005; Podlesak et al., 2005; Inger and Bearhop, 2008; Larson and Hobson, 2009; Hobson, 2011). Breath, reflecting immediately metabolised nutrients, and faeces representing the isotopic signature of a recent meal, may also be sampled non-destructively (Podlesak et al., 2005; Carleton et al., 2006; Voigt et al., 2008; Symes et al., 2011). Therefore, by analysing different tissues with different turnover rates temporal and/or spatial trophic shifts may be assessed.

Interpretations of temporal changes in $\delta^{15}N$ values, and associated trophic level shifts, can be complicated by insufficient knowledge on diet-tissue discrimination factors, dietary changes in different species and temporal changes in $\delta^{15}N$ of food sources, i.e. vegetation and insects. Numerous other factors can affect diet-tissue fractionation and studies to date have addressed these issues for a range of animal species, mostly under controlled conditions in captivity (see Post, 2002; Vanderklift and Ponsard, 2003; Robbins et al., 2005). For example, fractionation factors may be affected by, i) prey type, ii) prey quality, iii) temperature, iv) form of nitrogen excretion, v) habitat type, vi) water stress, and vii) nutritional status (including differences in dietary C:N ratios) (Ambrose, 1991; Hobson and Clark, 1992a,b; Hobson et al., 1993; Hobson et al., 1994.; Pinnegar and Polunin, 1999; Perkins and Speakman, 2001; Bearhop et al., 2002; Vanderklift and Ponsard, 2003; Pearson et al., 2003; Evans Ogden et al., 2004; Cherel et al., 2005; Robbins et al., 2005; Podlesak and McWilliams, 2006). In cases where we do not know more about discrimination factors related to these factors we may be unable to make clear interpretations concerning trophic positions for particular species or animal communities.

Unfed animals have been found to show higher (enriched) $\delta^{15}N$ values due to recycling of endogenous nitrogen as body mass is lost without replacement of preferentially excreted ^{14}N (Hobson et al., 1993). Some studies may present contradictory results; in studies where animals have high C:N ratio diets (nitrogen becomes limiting at higher C:N ratios) the diet-tissue discrimination factor is shown to be higher (Adams and Sterner, 2000; McCutchan et al., 2003; Vanderklift and Ponsard, 2003; Robbins et al., 2005; Tsahar et al., 2008), although Hobson and Bairlein (2003) found no significant difference in discrimination factor for different C:N diets for Garden Warblers *Sylvia borin*.

When Garden Warblers were fed a predominantly insect-based diet the diet-whole blood discrimination factor for carbon was 2.5‰ (Hobson and Bairlein, 2003), and House Sparrows *Passer domesticus* under different temperature conditions diet-whole blood discrimination factor was 1.5-1.8‰ (Carleton and Martínez del Rio, 2005). Fractionation factors for $\delta^{13}C$ and $\delta^{15}N$ for four tissues (plasma, breast muscle, cellular fraction of blood, whole blood, kidney, liver) of dunlin *Calidris alpina pacifica* fed controlled diets were 0.5-1.9‰ and 2.9-4.0‰ respectively (Evans Ogden et al., 2004). In calculating dietary proportions in tropical dry forest birds, Herrera et al. (2006) used derived values from other studies of birds; for carbon this was 1.4‰. It has commonly been accepted that diet-tissue discrimination factors for ^{13}C are lower than ^{15}N, in the order of 1-2‰ (Fry et al., 1978; Mizutani et al., 1992; Hobson and Wassenaar, 1999). In mammals controlled diet experiments on *Rattus rattus*, diet-whole blood discrimination factors for carbon ranged from -8.79‰ to 0.64‰ (Caut et al., 2008). Recent works therefore suggest that caution be applied in the use of isotope models because of uncertainty in diet-tissue fractionation factors for a wide variety of organisms and conditions (Caut et al., 2008b). Some have considered nectar feeding species and breath samples of broad-tailed hummingbirds *Selaphorus platycercus*, in a study that controlled diet, the diet-breath fractionation factor was calculated at -2.3‰ to -1.6‰ (Carleton et al., 2006).

To my knowledge no stable isotope laboratory studies have been done on African birds, and more specifically sunbirds. For wild caught birds $\delta^{15}N$ values for feathers of nectar feeding species were greater than expected; Southern Double-collared Sunbird *Cinnyris chalybeus* measured 7.5±1.2‰ (n = 5; Symes and Woodborne, 2009) and Malachite Sunbird *Nectarinia famosa* 11.3‰ (n = 1; Symes et al., 2011). In 15 Northern Double-collared Sunbirds *Cinnyris reichenowi* the $\delta^{15}N$ values of feather samples ranged from 8.3-10.3‰ (Procházka et al., 2010); the range being less than that in a single White-bellied Sunbird *Cinnyris talatala* where samples from primary feathers (n = 10) ranged from 7.4-10.1‰ (mean value=8.6±0.8‰; Symes and Woodborne, 2011). Sunbirds have a diet that is typically low in nitrogen and to meet nitrogen requirements supplement their diet with insects (Skead, 1967; Daniels, 1987; Maclean, 1990, 1993; Markman et al., 1999, 2004; Gartrell, 2000; Roxburgh and Pinshow 2002). However, morphological and physiological features of nectarivores may be directed at utilizing a food source typically low in nitrogen. The reabsorption of nitrogen, and subsequent loss of lighter nitrogen isotopes, may thus explain the observed enrichment (Tsahar et al., 2005). Alternatively, the diet of arthropods may contain a high proportion of spiders, organisms with possibly high $\delta^{15}N$ values. For this reason diet-tissue fractionation rates were investigated in two sunbird species. White-bellied Sunbird and Amethyst

Sunbirds *Chalcomitra amethystina* (Nectariniidae), two common and broadly sympatric species on the South African highveld (Hockey et al., 2005), were captured and held under controlled laboratory conditions during a separate study investigating gastro-intestinal sucrase activity (Napier et al., 2008). They were fed a diet of sucrose [20% (w/w) sucrose] and protein supplement (2%; Ensure®, Abbott Laboratories, Johannesburg, South Africa) with a $\delta^{13}C$ value of -13.1±0.1‰ and $\delta^{15}N$ value of 1.1±0.8‰. After 0, 5, 11 and 300 days 1-2 individuals of each species were euthanized and the carbon and nitrogen stable isotope signatures of blood, fat (dissected from a layer deposited on the pectoral muscle), feather, liver and muscle (pectoral) determined. In both species the values were greater than that reported for other bird species in the literature for South African birds (e.g. Symes and Woodborne, 2009, 2011). Although only one individual of each species was sampled on day 300, where tissue values had more than sufficient time to equilibrate with diet, the values are remarkable, given the known diet-tissue fractionation factors for other bird species (Table 1).

Tissue	Species	Carbon (‰)	Nitrogen (‰)
Liver	Amethyst Sunbird	3.4	5.5
	White-bellied Sunbird	4.1	8.3
Muscle	Amethyst Sunbird	4.1	6.7
	White-bellied Sunbird	4.5	8.0
Blood	Amethyst Sunbird	5.4	10.3
	White-bellied Sunbird	3.4	9.5

Table 1. Diet-tissue fractionation values for different tissues of White-bellied Sunbird *Cinnyris talatala* and Amethyst Sunbirds *Chalcomitra amethystina* (n = 1 for each value) determined in controlled laboratory experiments and euthanized at 300 days in captivity (Symes and Woodborne, 2011; Symes and Woodborne unpubl. data).

3. Links with the north – Migrant birds in southern Africa

3.1 Before stable isotopes

Bird ringing has been a valuable tool in tracking bird movements in the past (Nichols and Kaiser, 1999; Bairlein, 2003; Hartley, 2003). However, understanding movement patterns requires that ringed birds are recovered and when they are only the start and end points of movements can be determined. More recently, radar has been used to quantify migration in space and time, to determine the altitude of migration and to understand the influence of weather conditions on migration (Alerstam and Hedenström, 1998; Bairlein, 2003). For years the tracking of migrating species has proved difficult for all but conspicuous species (Chamberlain et al., 1997; Wassenaar and Hobson, 1998; Hobson, 1999; Wassenaar and Hobson, 2000; Hobson, 2002; Kelly et al., 2002; Rubenstein and Hobson, 2004; Hobson, 2005). Satellite telemetry, or radiotracking by satellite, allows for collection of migration data of greater detail both spatially and temporally (Meyberg et al., 1995; Nichols and Kaiser, 1999; Meyberg et al., 2001; Bairlein, 2003; Hartley, 2003). It also allows for tracking of species even when recoveries of birds are low, and aids in the identification of important stopover and refuelling sites (Bairlein, 1985, 2003). However, it was previously only be applied to relatively large species due to the weight of the transmitters (Bairlein, 2003). Birds can also be tagged with geolocation devices (which use real-time measurement of light intensity) and global positioning system devices (which use satellite data) to establish the geographic co-ordinates of a bird as it travels, with or without recapture (Bairlein, 2003). Tracking birds in these ways is often far more efficient and informative than ringing, and can be used on small bird species. However, these less traditional methods can be time-consuming and expensive, and track only the routes of individual birds that in turn are used to infer patterns for a species.

Recently established chemical and molecular markers appear to be extremely useful in the study of bird migration. DNA analysis can be used to determine different populations and track their movements (Hobson, 1999; Bairlein, 2003; Clegg et al., 2003; Hartley, 2003; Rubenstein and Hobson, 2004).

The use of stable isotopes (hydrogen, carbon, nitrogen, strontium) in deciphering the movements of a number of species has proved successful and efficient (Hobson and Wassenaar, 1999; Hobson et al., 2001; Kelly et al., 2002; Graves et al., 2002). It has the advantage over marked-recapture techniques in which the same individual does not need to be recaptured again later. In addition it can include the study of smaller organisms such as migratory insects (Wassenaar and Hobson, 1998; Hobson, 2002). Stable isotopes may be used to pinpoint the movements of birds as they travel between distant locations with distinct isotopic composition and incorporate isotopic signatures into their tissues (Hobson, 1999; Bairlein, 2003; Szép et al., 2003; Inger and Bearhop, 2008). This is based on the principle that foodwebs from different regions differ in their isotopic signatures due to several biogeochemical processes, and birds that move from one foodweb to another isotopically distinct foodweb will retain the isotopic signature of their previous location (Hobson, 1999; Hobson, 2005; Podlesak et al., 2005; Larson and Hobson, 2009; Hobson, 2011). The use of stable isotope analysis requires only one sampling of an individual, is less time-consuming and will not influence the behaviour of the individual like an extrinsic marker (Hobson, 1999; Hobson, 2011). Retrospective studies on museum or archived samples are also possible (Hobson, 2011) and may even be used to study long term changes in bird diets brought about by anthropogenic changes (Chamberlain et al., 2005). Ringing cannot be completely replaced by stable isotope analysis (Bairlein, 2003), but the use of stable isotopes will no doubt enhance our understanding of the patterns and processes of migration in many species where very little is known.

Many migrant species are known to be currently undergoing population declines (Holmes, 2007). For birds that migrate, the sites of origin, stopover and destination are important during different stages of the annual period of migrants (Hahn et al., 2009). Thus studies into the largely unknown details of migration, e.g. differences in migration of age classes, sexes and populations; location of key stopover and wintering areas; connections between migration, moulting and breeding; migratory connectivity between populations of birds; carry-over effects between winter habitat occupancy and breeding success; and external factors regulating migration are required for use in conservation planning (Chamberlain et al., 1997; Hobson, 1999; Bairlein, 2003; Holmes, 2007).

Stable isotopes allowed Kelly et al. (2002) to report the leapfrog pattern of Wilson's Warbler *Wilsonia pusilla*. They analysed hydrogen stable isotope ratios (δD) in breeding, migrating and wintering warblers and established that northerly breeding populations migrated earliest in autumn, flying over southerly breeding birds (Kelly et al., 2002). These northern breeders migrated to the southern edge of the Wilson's Warbler's range in south Central America (Kelly et al., 2002), a phenomenon that would have been difficult to detect through other known research methods. Other species in which migratory connectivity have been revealed include Cooper's Hawks *Accipiter cooperii* (Meehan et al., 2001), Ecuadorean hummingbirds (Hobson et al., 2003) and the Aquatic Warbler *Acrocephalus paludicola* (Pain et al., 2004), although these studies are confined to the Americas.

3.2 Intra-African migrants

Very few studies have used stable isotope analysis to interpret intra-African avian migratory patterns. This is probably due to two reasons. Firstly, limited resources in many African countries limit studies and secondly, and possibly more importantly, the isotopic gradient of the distribution of deuterium is not as great as that in the Americas and Palaearctic. However, despite these challenges some detailed studies have highlighted important findings. For example, Wakelin et al. (2011), used hydrogen, carbon and nitrogen stable isotopes to investigate migratory connectivity of the Blue Swallow *Hirundo atrocaerulea*, a threatened intra-African migrant with breeding populations in three geographically disjunct regions in southern Africa (Spottiswood, 2005). Their results indicated that there is overlap in the wintering ranges of at least two of the three major breeding populations (Wakelin et al., 2011). Because deuterium may not be a useful isotope to use in Africa, a multiple isotope approach may be more efficient (Wakelin et al., 2011).

3.3 Palaearctic migrants

Several fairly recent studies in Europe have dealt with Palaearctic-African migrants, i.e. those avian species that migrate between the Palaearctic (Europe, Africa north of the Sahara, and most of Asia north of the Himalayas) and Africa, south of the Sahara (Chamberlain et al., 2000; Symes and Woodborne, 2010). Examples of these studies are discussed below.

3.3.1 Amur falcon

In a study of Amur Falcon *Falco amurensis*, a small (138-160g, Schäfer, 2003) insectivorous raptor that undergoes a one-way migration between southern Africa and the eastern Palaearctic of ~13,000 km (Figure 1a; Ferguson-Lees and Christie, 2001; Mendelsohn, 1997; Jenkins, 2005), Symes and Woodborne (2010) demonstrated that South African populations do not show strong site fidelity but move widely across the subregion, feeding on a broad range of arthropods that become seasonally abundant during the austral summer. They have a wide breeding range in the eastern Palaearctic; through Mongolia, Siberia and northern China with a distribution range at least eight times that of the southern African range (Cheng, 1987, Mendelsohn, 1997; Ferguson-Lees and Christie, 2001; Greenberg and Marra, 2005; Global Raptor Information Network, 2008; Symes and Woodborne, 2010). They arrive late in southern Africa, compared to other migrants, during November to early-December, and roost in colonies that number thousands of individuals (Figure 1b & c; Benson, 1951; Cade, 1982; Tarboton and Allan, 1984). For a number of months they are a common sight on the eastern South African Highveld, before departing north to their breeding grounds in April to May (Mendelsohn, 1997; Jenkins, 2005). Although details on migration routes and feeding grounds were limited regarding stable isotope analyses, results did provide valuable information on the biology of Amur Falcons. The recent tracking of individual birds using satellite tracking devices may further contribute to a greater understanding of the movements and migration patterns of this species.

Fig. 1. a. Female Amur Falcons *Falco amurensis*; b & c. communal roost of Amur Falcons (Heidelberg, Gauteng, South Africa; 29 December; 26°30'11"S, 28°21'31"E, ~1,600m a.s.l.). Photographs: Craig Symes.

3.3.2 Willow Warblers

Hedenström and Pettersson (1987) noticed two different migratory directions within Willow Warbler *Phylloscopus trochilus* populations, and endorsed C and N stable isotope analysis as a method to determine wintering ranges of migrants in place of more traditional methods. It also emphasised the need for regional isotopic maps of feather keratin for migratory species in their African wintering grounds in order to pinpoint the exact localities of these areas.

Subsequently, a study by Chamberlain et al. (2000) used stable isotopes to investigate the migratory divide of Willow Warblers in Scandinavia. The breeding ranges of two subspecies of Willow Warbler occur adjacent to one another and overlap in a migratory divide. *Phylloscopus t. trochilus* is known to occur south of 61°N and *P. t. acredula* north of 63°N. The migratory divide should occur between these two ranges. Additionally, according to ringing recoveries, *P. t. trochilus* is thought to migrate to West Africa and *P. t. acredula* to East Africa. In this study $\delta^{15}N$ and $\delta^{13}C$ isotope signatures of feathers collected from males of both subspecies in Scandinavia were determined. The study found more enriched values of $\delta^{15}N$ and $\delta^{13}C$ in *P. t. acredula*, suggesting that they do indeed winter in different regions of Africa. This is because East Africa, where *P. t. acredula* winters, has a more arid climate and more C4 plants than West Africa, where *P. t. trochilus* winters. The study also found intermediate $\delta^{15}N$ ratios in the birds of the contact zone, suggesting that they probably represent both subspecies.

A later study conducted stable isotope analysis in feathers of two Willow Warbler subspecies' in Africa, and attempted to map the subspecies wintering ranges (Bensch et al., 2006). However, many of the isotopic signatures documented in the Scandinavian study by Chamberlin et al. (2000) did not match those of the same subspecies in Africa as documented by Bensch et al. (2006). This prompts the need for more information before Willow Warbler wintering ranges can be effectively mapped (see proposed study later).

3.3.3 Aquatic Warblers

The globally threatened Aquatic Warbler *Acrocephalus paludicola* is threatened by anthropogenic habitat destruction and has a fragmented breeding population across the western Palaearctic with 90% of the population restricted to Belarus, Poland and Ukraine (Pain et al., 2004). Little is known about its wintering quarters which are assumed to be located in sub-Saharan Africa. Conservation of the species necessitates knowing where the wintering range occurs. Pain et al. (2004) sampled flight feathers of Aquatic Warblers in Europe and analysed δD, $\delta^{13}C$ and $\delta^{15}N$ ratios to determine whether different breeding subpopulations formed a single mixed population on the African wintering grounds. Neither $\delta^{15}N$ nor δD varied between subpopulations. Although this suggests that subpopulations do not differ in trophic level in their wintering ranges, it does not necessarily suggest that subpopulations winter in the same place, as similar δD ratios are found at many different latitudes in Africa. However, significant differences in mean $\delta^{13}C$ ratios between subpopulations suggest different wintering latitudes. Birds that breed further north or west in Europe possibly winter further north in Africa.

Subsequently, a study by Oppel et al. (2011) again collected feathers of the species across Europe to investigate $\delta^{15}N$, $\delta^{13}C$, δD ratios and attempted to determine the species' wintering regions in Africa. Feathers were collected at the only known wintering site in Senegal. The feathers sampled in Africa showed similar degrees of variation in $\delta^{15}N$, $\delta^{13}C$

and δD ratios to those sampled in Europe, little evidence to suggest that subpopulations wintered in different regions. Nevertheless, 20% of the samples in Europe fell outside of the range of isotopic values recorded in Senegal, suggesting an over-wintering region elsewhere in Africa. It was concluded that the large variability of feather isotopic signatures within sites of specialist species like the Aquatic Warbler, due to the large variation in plant isotope values at the base, makes it difficult to identify wintering sites using stable isotope analysis. According to Flade et al. (2011), ringing and molecular studies have also not provided conclusive evidence about African wintering regions, yet geolocators recently attached to Aquatic Warblers in the Ukraine could be promising.

3.3.4 Reed and Great Reed Warblers

Procházka et al. (2008) attempted to link the breeding and wintering grounds of the Reed Warbler *Acrocephalus scirpaceus* by analysing all relevant sub-Saharan ringing recoveries and conducting stable carbon and nitrogen isotope analysis on moulted feathers collected in Africa. This species is thought to experience a migratory divide in Central Europe, and birds from populations either side of the divide will either travel southeast or southwest during migration to different wintering quarters in Africa. Ringing recoveries suggested strong connection between breeding and wintering grounds in Africa. Higher $\delta^{15}N$ values of southeast migrating birds suggested that they occupied more arid biome types than southwest migrating birds, as was expected. More work is necessary before the precise wintering grounds can be identified, but the combination of ringing and stable isotope analysis was successful in understanding more about migration patterns of this species.

Similarly, Yohannes et al. (2008) used δD, $\delta^{13}C$ and $\delta^{15}N$ isotope ratios of feathers of the Great Reed Warbler *Acrocephalus arundinaceus* to test whether isotope signatures of individuals remained consistent between years, and thus whether birds exhibited philopatry i.e. returned to the same location in successive years. The findings supported this behaviour, as similar isotopic signatures were obtained during different years. However, it could be that birds did not return from the same location as previous years, but rather that they returned from habitats with similar isotopic signatures. This study highlights the benefits of using multiple isotopes in stable isotope analysis of migration, and again highlights the effectiveness of using ringing data in combination with stable isotopes when attempting to understand bird movements.

4. Diet and trophic structure of southern African birds determined using stable isotopes

Numerous studies have demonstrated the complexities of links and trophic levels in avian food webs (Herrera et al., 2003, 2006; Symes et al., 2009). Usually this is done with tissues that can be sampled non-destructively such as feathers, blood, toe-nail and breath (Hobson and Wassenaar, 2008). However, in the case of feathers, interpretations using isotope values may be confounded by intra-individual variation (Symes and Woodborne, 2011). Recognising temporal dietary variation as represented in feathers grown at different times, e.g. moulting flight feathers, is thus important and needs to be critically assessed in any study of avian diets using stable isotope analysis. It may, however, be used advantageously where separate flight feathers are individually sampled to infer diet on a temporal scale (e.g. Ramos et al., 2009).

4.1 The forest-grassland matrix, and savannas and grasslands

The forest (C3 plant dominant) and grassland (C4 plant dominant) matrix in savanna and forest-grassland ecosystems of South Africa offers a suitable opportunity to understand carbon sourcing from two different vegetation types (Acocks, 1975). In each vegetation type the vegetation mosaic is different; in C3 forests the C4 grass environment usually surrounds the forest whilst in savanna the C4 grassy component occurs beneath the trees (Figure 2). However, grasslands are not entirely C4 and consist of an important C3 component of woody and herbaceous plants. Also, not all grasses in this biome are C4 and a gradient of increasing proportion of C3 grasses occurs with an increase in latitude south and an increase in altitude (Vogel et al., 1978). Never-the-less, in a montane forest the contribution of C3 (predominantly trees) and C4 (predominantly grasses, although few C4 grasses occur in forest) carbon to the diets of forest birds was investigated using stable isotope analysis of feathers (Symes and Woodborne, 2010). In addition, nitrogen isotopes were measured to investigate trophic partitioning. The forest bird community generally met predictions regarding known diets of different species, and overall very little isotopic partitioning of resources, as inferred from stable isotopes, was recognised. Birds, that are able to utilise the

(a)

(b)

(c)

(d)

Fig. 2. a. Mistbelt mixed *Podocarpus* forest in the KwaZulu-Natal midlands, South Africa, occurring naturally in fragments on south facing slopes, surrounded by grassland, note the well defined grassland-forest ecotone, b. forest interior with few grasses (usually in gaps) occurring in understorey, late-summer (March; 29°27'50"S, 29°52'43"E, ~1,530 m a.s.l.), c. savanna near Nelspruit, South Africa, showing extensive grassy understorey, early-spring (September), d. pale form of *Erythrina lysistemon*, an important nectar source for birds, in savanna near Nelspruit, early-spring (September; 25°34'22"S, 31°10'53"E, ~ 800 m a.s.l.). Photographs: Craig Symes.

environment in different ways to non-flying animals, may not need to partition resources in an "isoscape" (Bowen & West, 2006; West et al., 2010); partitioning resources may be more effectively recognised in space and time.

Current work on South African forest birds is now focussing on the suitability of stable isotopes in understanding bird diets. By sampling multiple "tissues", i.e. blood, faeces and feather, during different seasons, i.e. winter and summer, and comparing these data with comprehensive feeding observations in the forest, a clearer understanding of dietary acquisition and resource partitioning in a forest bird community can be obtained (Scott and

Symes unpubl. data). Preliminary results, that also analyse bird community census data, suggest that although there is very little change in avian diversity between seasons there is a significant seasonal turnover in the functional nature of the bird community between seasons (Scott and Symes unpubl. data). This turnover relates to the change in food resources available to birds between seasons (Scott and Symes unpubl. data).

4.2 Wetlands and marine environments

Stable isotopes provide a valuable tool in understanding the link between marine and terrestrial systems and numerous examples, from Alaskan Brown Bears *Ursus arctos* that derive marine nutrients through the consumption of migrating Pacific salmon (*Onchorhynchus gorbuscha, O. keta* and *O. kisutch*) (e.g. Ben-David et al., 2004) to numerous "maritime mammals" that consume intertidal energy resources and transfer these resources to the land (Carlton & Hodder, 2003). The ability of birds to fly enables them to transfer nutrients between terrestrial or marine systems and depositions of guano are testimony to the large amounts of nutrients transferred from sea to land. In marine feeding Great Black Cormorants *Phalacrocorax carbo* stable isotopes in feathers were used to show that birds had been feeding on an almost entirely freshwater prey source (Bearhop et al., 1999). Similarly, on the northern Chilean coast, two closely related songbirds (*Cincloides nigrofumosus* and *C. oustaleti*) were shown, using stable isotopes (δ^{13}C), to utilise coastal and freshwater environments quite differently; and the reliance on marine sources was accompanied by adjustments in the osmoregulatory mechanism (Sabat et al., 2006). As terrestrial productivity increased with an increase in latitude south so too did the incorporation of terrestrial carbon increase (Sabat et al., 2006). However *C. nigrofumosus* that lives exclusively on coastal environments, had a greater contribution of marine input and less variable δ^{13}C values than *C. oustaleti* which shifts seasonally from coastal to freshwater environments (Sabat et al., 2006). These studies are possible because of the distinct differences in marine and terrestrial food webs with marine organisms typically having more enriched isotope values for D, δ^{13}C, δ^{15}N, δ^{18}O and δ^{34}S (Hobson, 1999). Southern Africa's extensive coastline offers suitable working examples to assess the marine contribution to terrestrial birds. For example, pelicans (*Pelecanus onocrotalus* and *P. rufescens*) occur in freshwater and marine systems across the sub-region but are thought not to be involved in significant movements (Crawford, 2005; Ryan, 2005). Much like the study of Klaassen et al. (2001) that demonstrated that capital resources are not important for breeding arctic waders (10 species) so too might stable isotopes be used to identify the importance of marine resources, at varying distance from the coast, for many birds, e.g. pelicans. Many studies have considered nutrient transfers between marine and terrestrial systems whilst others have used stable isotopes successfully to track the movements of seabirds within a marine environment (e.g. Barbraud and Chastel, 1998; Ramos et al., 2009). Unique stable isotope signatures of different feeding grounds (related to latitude, species composition, sea temperature, etc), together with other tracking techniques, may provide valuable information of the use of oceanic resources by seabirds.

The Mangrove Kingfisher *Halcyon senegaloides* has an almost exclusively coastal distribution along the east coast of Africa where, as its name implies, has a strong association with mangroves (Figure 3; Fry et al., 1988, 1992). Because of this association it is suggested that it has a strong reliance on marine resources. However, it has recently been discovered breeding in arboreal termitaria in woodland in central Mozambique (Davies et al., 2012). It is not yet clear whether this population is migratory and if they are what role resources obtained from possible "overwintering" sites on the coast play in contributing to reproductive output and success (Davies et al., 2012). Stable isotopes might be a useful tool in answering this question. Additional species that might also rely on seasonal utilisation of marine and terrestrial resources, depending on their location and how they move across the landscape include, for example Pied Kingfisher *Ceryle rudis*, Reed Cormorant *Phalacrocorax africanus* and Greyheaded Gull *Larus cirrocephalus* (Hobson, 1987; Hockey et al., 2005).

(a)

(b)

Fig. 3. a. Mangrove swamp on KwaZulu-Natal north coast (28°58'51"S, 31°44'09"E, ~5m a.s.l.), site of Mangrove Kingfisher *Halcyon senegaloides*, b. arboreal termitarium on stem of *Sterculia* sp. (see arrow), used as nesting site by Mangrove Kingfisher in central Mozambique (18°09'17"S, 35°07'29"E, ~170m a.s.l.). Photographs: Craig Symes.

4.3 Specialist diets

Large tree-like CAM aloes such as *Aloe marlothii* and *A. ferox* are a conspicuous feature of the South African landscape (Figure 4), not unlike the saguaro cactus *Carnegeia gigantea* of the arid regions of the Americas (Fleming et al., 1996). Studies using stable isotopes in the Sonoran Desert have provided insight into the use of saguaro fruit for bird communities (Wolf and Martínez del Rio, 2000; Wolf et al., 2002; Wolf and Martínez del Rio, 2003). The fruit of saguaro cacti has a unique CAM stable isotope composition (Wolf et al., 2002). Using this feature, temporal analyses of $\delta^{13}C$ and δD revealed the importance of fruit as a source of nutrients and water for White-winged Doves *Zenaida asiatica mearsnii* (Wolf et al., 2002). These results were confirmed by a positive correlation between the $\delta^{13}C$ in dove liver tissues and percent saguaro fruit in the crop contents (Wolf and Martínez del Rio, 2000). Further investigations revealed that two different desert doves relied on saguaro fruit for different reasons; White-winged

(a)

(b)

Fig. 4. a. Flowering *Aloe marlothii* at Suikerbosrand Nature Reserve, Gauteng, South Africa, late-winter (August, 26°31'54"S, 28°10'11"E, ~1,630 m a.s.l.), b. Wattled Starling *Creatophora cinerea* feeding on *A. marlothii* nectar. Photographs: Craig Symes.

Doves predominantly perched atop saguaros and fed on flowers or fruit, whilst Mourning Doves *Zenaida macroura* fed exclusively on the ground (Wolf et al., 2002). These differences in feeding mode affected the importance of saguaro fruit for these two dove species. Saguaro provided an important source of nutrients and water for White-winged Doves, whereas for Mourning doves saguaro provided nutrients and less importantly water (Wolf et al., 2002). In South Africa, the importance of sugar in the dilute nectars of *A. marlothii* has been demonstrated as an important carbohydrate source for numerous opportunistic avian nectarivores (Symes et al., 2008; Symes, 2010; Symes et al., 2011). In particular the analysis of breath samples has indicated that ingested sugars are used as an income resource for a wide range of opportunistic nectar feeding species (Symes et al., 2011).

In southern Africa, specialised nectarivores are from the family Nectariniidae (sunbirds; 12 species) and Promeropidae (sugarbirds; 2 species) (Skead, 1967). However, additional opportunistic nectar feeders include white-eyes (Zosteropidae), weavers (Ploceidae), bulbuls (Pycnonotidae), barbets (Lybiidae), mousebirds (Coliidae) and starlings (Sturniidae) (Marloth, 1915; Oatley, 1964; Skead, 1967; Oatley and Skead, 1972; Jacot Guillarmod et al., 1979; Craig and Hulley, 1994; Oatley, 2001; Symes, 2010). Numerous studies have indicated the importance of nectar sources for bird communities in southern Africa (Frost and Frost, 1981; Daniels, 1987; Tree, 1990; Craig and Hulley, 1994; Symes et al., 2001; Symes, 2010) but only recently has this importance been quantified through the use of stable isotopes (Symes et al., 2011). The high diversity of aloes in southern Africa provides a valuable nectar source for many birds and although flowering occurs throughout the year there is a peak in the number of flowering species during the drier winter months (Figure 5).

The provision of fruit by saguaro cacti can be compared with the provision of nectar by aloes in southern Africa. *Aloe marlothii* is a CAM photosynthesizer (Eller et al., 1993), as is *A. arborescens* (Kluge et al., 1979); they display typical $\delta^{13}C$ values like many other aloes (*A. marlothii* $\delta^{13}C$ = -14.93 and -12.09‰; *A. greatheadii* var. *davyana* = -14.41‰; *A. arborescens* = -16.92‰; Figure 6). However, the $\delta^{13}C$ values of aloes vary for samples collected throughout Africa, ranging from -28.48‰ to -11.51‰ (mean±SD = 17.26±4.94‰; Figure 6), so the application of isotope techniques, in assessing the value of nectar for opportunistic nectar feeders, needs to be conducted with a sound understanding of the isotopic environment.

Important in dietary studies determining the proportion of resources from different food items is an understanding of the isotope ratios of the food base on which birds are feeding, and the context of these food resources. In the study of the importance of *A. marlothii* nectar for birds Symes et al. (2011) were able to detect more enriched changes in the isotope signatures of the whole blood and breath of species that normally assimilated carbon from C3 sources outside of the flowering period (Wolf and Hatch, 2011). However, for seed eating species (granivores), such as Southern Masked Weaver *Ploceus velatus* that feed mostly on C4 plants (C4 grasses $\delta^{13}C$ = 14.7±2.5‰) the shift in diet to nectar ($\delta^{13}C$ = 12.6±0.5‰) was less obvious and any changes in diet were difficult to detect in tissues during the pre-flowering and flowering months (Symes et al., 2011). In this study nectars are comprised mostly of hexose sugars with very little protein and the emphasis of the study was to identify the routing of nectar carbohydrates in the system (Symes et al., 2011). In other circumstances the routing of protein using nitrogen stable isotopes may be studied (Herrera et al., 2006). For many nectar feeders the acquisition of nitrogen may be obtained from animal protein. Sunbirds (Family Nectariniidae) may regularly be observed feeding on insects and it is important that they feed nestlings this rich protein source (Hockey et al.,

2005; Markman et al., 1999). In Painted Honeyeaters *Grantiella picta* (Family Meliphagidae) and Mistletoebirds *Dicaeum hirundinaceum* (Family Dicaeidae) the isotopically distinct values of Grey Mistletoe *Amyema quandang* (Loranthaceae) fruit (mean $\delta^{15}N$ = 4.4‰) and arthropods (mean $\delta^{15}N$ = 7.1‰) allowed Barea and Herrera (2009) to determine that approximately half of their nitrogen was obtained from mistletoe fruit. Similarly, the opportunity may exist in Africa for stable isotope analyses to determine the importance of mistletoes for mistletoe feeding specialists such as tinkerbirds (Family Lybiidae).

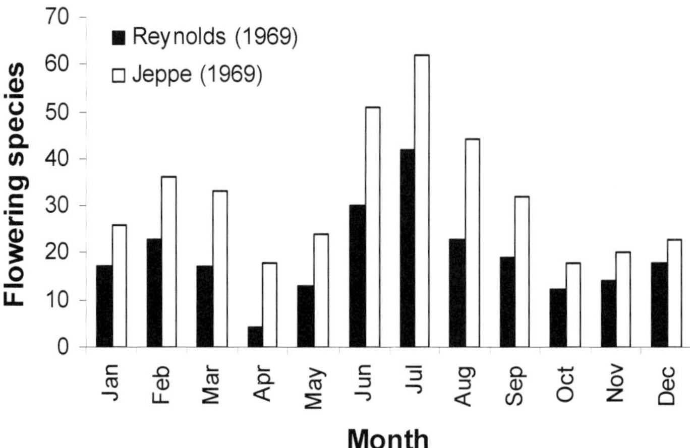

Fig. 5. Number of South African (after Jeppe, 1969) and southern African (after Reynolds, 1969, *n* = 133) *Aloe* species flowering in different months of the year. The different references may have included each aloe species flowering in more than one month.

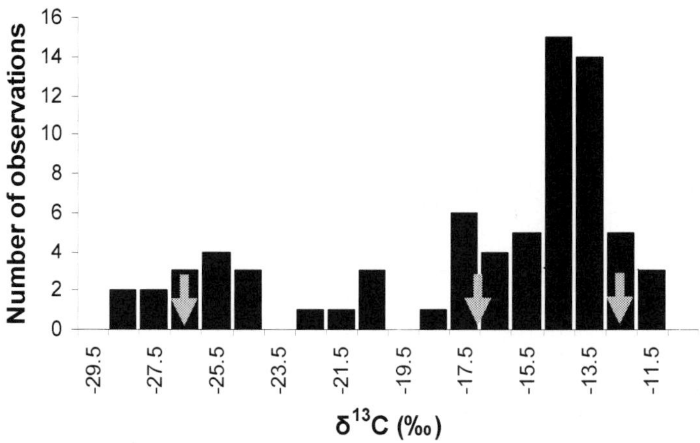

Fig. 6. Frequency distribution of $\delta^{13}C$ (‰) for 64 *Aloe* species (*n* = 72) (J. Vogel and S. Talma unpubl. data). Arrows indicate means for C_3 (-26.5‰), CAM (-17‰) and C_4 (-12.5‰) plants (Vogel et al., 1978; Ehleringer and Osmond, 1989; Ehleringer, 1991; Dawson et al., 2002).

5. Future research

5.1 Forging links

Scope for developing links between northern and southern hemisphere researchers and institutions is wide and collaborative efforts may strongly enhance our understanding of avian migration patterns. For any scientist the limiting factors in any research are a combination of resource availability and time. By thinking and working globally, rather than locally, broad questions regarding bird movements and behaviour can be addressed. In particular, questions relating to global change and the effects changes in the modern world are having on biodiversity can be addressed. Below I address in detail and give a number of possible studies on species with unique migration characteristics or patterns.

5.2 Deuterium and bee-eaters

Although there is a lack of a strong gradient in hydrogen isotopes distributed across Africa opportunities still remain for understanding the complexities of bird movements across the continent. The Palaearctic extends across a vast longitudinal range and from that wide region migrant species are funnelled down the African continent, many reaching South Africa (Chamberlain et al., 2000; Symes and Woodborne, 2010; Table 2).

Although there may be little conservation concern for certain species, rapid changes brought about by humans to the environment may significantly affect the long-term prospects for many of the worlds birds, particularly those that migrate and that rely on habitats in two different hemispheres during different times of the year. It is of added importance to study these patterns in the context of a world affected by climate change (Drent et al., 2003).

The European Bee-eater *Merops apiaster* has a widespread distribution across the western Palaearctic and migrates to Africa during the austral summer (Cramp and Perrins, 1993; Snow and Perrins, 1998; Fry, 2001). In South Africa there are two morphologically indistinct populations (Underhill, 1997; Barnes, 2005). Palaearctic-breeding birds migrating to South Africa (where they undergo primary moult) are thought to originate from the east of their Palaearctic range; birds from the western Palaearctic migrate to West Africa (Barnes, 2005). On the other hand southern Africa breeding birds arrive from central Africa (where they have undergone a primary moult) (Brooke and Herroelen, 1988). The funnelling effect of different members of each sub-population into southern Africa therefore offers a unique opportunity to understand how these populations interact. With an understanding of moult patterns (Holmgren and Hedenström, 1995) and analysing stable isotopes of primary feathers of individuals at different roosts (breeding and non-breeding) across South Africa inferences on migratory origins and use of resources can be made.

Region	Number of species
West Africa	175
Sudan	198
Eastern Democratic Republic of Congo, Rwanda, Burundi	127
Kenya	147
Zimbabwe	75
South Africa (Western Cape, Eastern Cape, Northern Cape)	58

Table 2. Approximate numbers of species of Palaearctic migrants to different regions of Africa (From Maclean, 1990, Table 6.2).

5.3 Warblers and flycatchers

Understanding the migratory connectivity between breeding and non-breeding sites of migratory species, particularly species that may be difficult to study because they are small and inconspicuous, is of ecological and conservation importance (Pain et al., 2004). The Spotted Flycatcher *Muscicapa striata* and Willow Warbler *Phylloscopus trochilus* are two of southern Africa's most common Palaearctic migrants (Herremans, 1997a,b; Dean, 2005; Johnson, 2005). Both have similar distributions, arriving during the non-breeding season (austral summer) in most of equatorial and southern Africa (Salewski et al., 2002, 2004; Herremans, 1997a,b).

In southern Africa the Willow Warbler is widespread in woodland and Karoo habitats, although is thinly distributed in the south-western Kalahari basin and western Karoo, and absent from the Namib Desert (Herremans, 1997a). There is a decreasing abundance in southern Africa from north to south and from east to west (Underhill et al., 1992b). It is one of the earliest Palaearctic migrants to arrive where, in the north of the sub-region it arrives from mid-September, with mid-arrival dates in mid-October in the north and December in the south (Herremans, 1997a). Arrival is rapid and annually consistent (Herremans, 1994) with arrivals in the Eastern Cape in early-December (Underhill et al., 1992b). Departure is synchronised in the whole region with mid-departure in mid-April (Underhill et al., 1992b); males precede females on northward migration by 10 days (Underhill et al., 1992a) and few are present in early-May (Underhill et al., 1992b). This may explain why moulting females in southern Finland began moulting 10-15 days later than males (Tiainen, 1981). Recoveries in southern Africa include five birds in Finland and two in Sweden (Hedenström and Pettersson, 1987; Dowsett et al., 1988; Oatley, 1995). However, no South African ringed birds have been caught north of the equator. Of 35 ringed and recaptured, 12 were caught at the same site the following summer, with 10 at the same site >1 yr later (Safring, unpubl. data 2010). There has been a decline in numbers across Europe since 1980 and because of this there is cause for its conservation (Baillie et al., 2010).

Three subspecies of Willow Warbler are currently recognised (Dean, 2005) although they are difficult to identify (Table 3; Figure 7a; Svensson, 1992). *Phylloscopus t. yakutensis* undergoes the longest migratory journey for a passerine (Curry-Lindahl, 1981) (to southern Africa from eastern Asia) and comprises 10% of the population in the sub-region (Clancey, 1970; Irwin, 1981). *P.t. acredula* is the most common, being slightly larger and paler than nominate, *P.t. trochilus* which occurs at intermediate abundance (Clancey, 1980; Irwin, 1981; Hopcroft, 1984). Blaber (1986) reports *P.t. trochilus* as most common in KwaZulu-Natal although Urban et al. (1997) note that the nominate race is unlikely in southern Africa.

The Spotted Flycatcher is a more specialized insectivorous feeder than the Willow Warbler, being the only flycatcher in the region that specializes in long upward hawking (Herremans, 1997b). In northern southern Africa arrivals occur from early-October (Irwin, 1981; Tarboton et al., 1987; Herremans, 1994), with most arrivals in late-October and November (Herremans, 1997b). Peak arrival in the south is early-December (Herremans, 1997b) and in Bloemfontein, arrivals are mostly during 5-30 November (median 21 November) (Kok et al., 1991). In Johannesburg the earliest arrivals were the 2 November (Bunning, 1977) and in Queenstown, Eastern Cape, the 14 November (Anon, 1953). Departures are quick and simultaneous, occurring in late-March and April, although are sometimes recorded in late-April (Irwin, 1981). In Botswana the latest departures were 5 May (Herremans, 1994) and in Johannesburg the latest departures were 25 March (Bunning, 1977). In Zimbabwe departures were 7-9 April (Borret, 1968; Tree, 1989, 1995). There is strong evidence of fidelity for non-breeding sites (Loske and Lederer, 1988; Maclean, 1993; Safring, unpubl. data 2010). Recoveries include four in Finland (Hunter and Hunter, 1970; Reed and Wells, 1957); one in Sweden (Underhill, 1994) and one in Britain (Tarboton et al., 1987; Underhill and Oatley, 1994).

Subspecies	Distribution	Breeding
P.t. yakutensis	To s Africa from c Asia (Curry-Lindahl, 1981); comprises 14.1% of the population in the sub-region, with likely concentration in e portion of South Africa (Clancey, 1970; Irwin, 1981)	Breeds e Siberia, from Taimyr Peninsula e to Anadyrland; from Ireland e & n through Europe to Russia & Ukraine (Snow & Perrins, 1998)
P.t. acredula	Winters e & s Africa, w to Cameroon and w to DRC; most common (54.7%)	Norway, n & s Sweden, ne Europe & Russia, e to Yenesei R
P.t. trochilus	Winters in w Africa, w to Cameroon (Urban et al., 1997); occurs at intermediate abundance (31.2%); Blaber (1986) reports *P.t. trochilus* as most common in Natal although Urban et al. (1997) note that it is unlikely in s Africa	Europe n to Sweden, e to s Poland & Romania (Urban et al., 1997)

Table 3. Distribution and breeding range of three subspecies of Willow Warbler *Phylloscopus trochilus* that are currently recognised (Dean, 2005) although difficult to identify (Svensson, 1992). All three are reported to winter in southern Africa (Clancey, 1970).

Five sub-species of Spotted Flycatcher are presently recognised (Table 4; Figure 7b; Clancey, 1980; Clancey et al., 1987; Johnson, 2005). In southern Africa *M.s. striata* is widespread and in Zimbabwe *M.s. neumanni* is most abundant (Benson et al., 1971). *M.s. balearica* is recorded in Damaraland, Namibia (Clancey, 1980), *M.s. sarudnyi* likely occurs in most of eastern southern Africa, and *M.s. inexpecta* is known in southern Africa from one moulting bird collected at Montclair, KwaZulu-Natal (31 Jan) (Clancey et al., 1987).

Subspecies	Distribution	Breeding
M.s. striata	Widespread in s Africa	Breeds w Europe e to w Siberia
M. s. neumanni	Massailand, n Tanzania, widespread in s Africa, most abundant subsp. in Zimbabwe (Benson et al., 1971)	Breeds e Mediterranean to Iran & w Siberia
M. s. balearica	Mallorca, Balearic Is, recorded Damaraland, Namibia (Clancey, 1980)	Breeds Balearic Island, w. Mediterranean
M. s. sarudnyi	e Iran & Transcapia; prob most in e of s Africa	Breeds Kazakhstan, s to e Iran
M.s. inexpecta	Tamak, Crimea, s Europe & Russia, known in s Africa from 1 moulting bird, collected 31 Jan (Montclair, KwaZulu-Natal; Clancey et al., 1987)	Breeds Crimean Peninsula

Table 4. Distribution and breeding range of the five subspecies of Spotted Flycatcher *Muscicapa striata* that are currently recognized (Clancey, 1980; Clancey et al., 1987).

(a)

(b)

(c)

Fig. 7. a. Willow Warbler *Phylloscopus trochilus* demonstrating new feathers moulted prior to migration north (Pretoria, 01 April; 25°46'33"S, 28°11'54"E, ~1,250m a.s.l.), b. Spotted Flycatcher *Muscicapa striata* (Pretoria, 01 April), c. Convergent moult in primary feathers of Spotted Flycatcher; arrow indicates primary feather moult direction, note three newly grown feathers (Colenso, KwaZulu-Natal, South Africa, 09 January; 28°41'21"S, 29°44'59"E, ~1,180m a.s.l.). Photographs: a. Graham Grieve; b & c. Craig Symes.

Although neither species are reported under threat in southern Africa (Herremans, 1997a,b) the Spotted Flycatcher has declined rapidly and consistently in the UK since the, 1960's (Baillie et al., 2005) with a decrease by 81% over the past 25 years (Baillie et al., 2008). Decreasing numbers may have been caused by deteriorating habitats, or by conditions on wintering grounds or along migration routes (Vanhinsbergh et al., 2003). Very little is known on the migratory origins of these species although low recovery rates of ringed birds have added to our knowledge of movements. In southern Africa there is still debate concerning the origins of many migrating birds and although individuals are known to show strong site fidelity little is known of population movements. Therefore, the opportunity exists to use stable isotopes to understand aspects of the migration biology, from an austral perspective, of two species in which the moult cycle is fairly well known. A growing network of bird ringers in South Africa have the potential to contribute to a comprehensive study. The Spotted Flycatcher moults body plumage on its breeding grounds during July-September (Craig, 1983) so old body plumage feathers can be collected for analysis. Spotted Flycatchers have a convergent primary moult (Williamson, 1972) that is sometimes irregularly started before migration south, but suspended during migration (Figure 7c; Cramp and Perrins, 1993). Most birds, however, have a complete flight feather moult in Africa (Ginn and Mellville, 1983). An old flight feather can therefore also be collected for comparative purposes with body plumage feathers. The Willow Warbler, unlike many other migrating passerines, undergoes a complete moult twice a year, one in Africa and one in Europe (Ginn and Melville, 1983; Jenni et al., 1994; Hedenström et al., 1995). A primary flight feather can thus be collected from Willow Warblers caught across southern Africa. Old feathers, grown in Europe, will likely represent the isotopic environment from which they originated. Collecting a primary feather that is moulted later in the season (just prior to migrating) is likely to be more representative since the bird will have spent longer at that particular site and other body tissues (e.g. endogenous reserves that may contribute to feather production) will have had longer to equilibrate isotopically with the environment.

These three species accounts provide detailed information on the potential for migratory studies using stable isotopes, and offer insight into further research on many other species, for example African Hoopoe *Upupa africana*, Yellow-billed Kite *Milvus aegyptius*, Steppe Buzzard *Buteo vulpinus* and Red-backed Shrike *Lanius collurio*, to name a few.

5.4 The way forward – An armoury of research tools

Stable isotopes can transform the way we do science. Ultimately, multiple isotope approaches may be more powerful (Hobson, 1999; Hobson, 2011) and statistical methods may aid interpretation of complex and confusing stable isotope datasets (Inger and Bearhop, 2008). Elements other than C, N, O, and H which are known to show isotopic patterns can be incorporated into stable isotope analysis (Hobson, 2005). Furthermore, isotopic base maps can be combined with additional non-isotopic layers e.g. biome coverage maps that are relevant to the species under consideration (Hobson, 2005). The use of stable isotopes in ornithology, and in general, is often criticised. However, the difficulties and problems that stable isotope analysis may pose can be overcome by dealing with the inherent uncertainty, applying it only to situations where it is useful, using it in addition to other more traditional methods like observation and faecal analysis, performing more controlled experiments to lay the foundation for interpretations, furthering the development of theory, and encouraging a more trans-disciplinary approach to ornithology that incorporates biology, chemistry and earth sciences (Kimura et al., 2002; Rubenstein and Hobson, 2004; Inger and Bearhop, 2008; Hobson, 2011).

The use of δD, $\delta^{13}C$, $\delta^{15}N$, $\delta^{18}O$ and other stable isotopes in ornithology is in its relative infancy, and many more comprehensive studies and laboratory experiments are needed to fully realise the potential in this field of science. The use of a method that is as far reaching as birds travel can be applied reliably and non-destructively at a relatively low cost to contribute to the survival of birds in a rapidly changing world (Simmons et al., 2004; Gordo, 2007; Both and Malverde, 2007).

6. Acknowledgments

Katie Roller is thanked for her input in the preparation of this chapter, and Sam Scott for editorial assistance. Graham Grieve is thanked for the use of his Willow Warbler photograph. Dr John Vogel and Dr Siep Talma (Council for Scientific and Industrial Research, Pretoria) are thanked for providing raw data on stable isotopes in aloes. Numerous colleagues and friends are thanked for their support and academic input over the few years in which I have been involved in stable isotope research; including Dr Stephan Woodborne (CSIR, Pretoria), Prof. Andrew McKechnie (University of Pretoria) and Prof. Sue Nicolson (University of Pretoria).

7. References

Acocks, J.P.H. (1975). *Veld types of South Africa. Memoirs of the Botanical Survey of South Africa No. 40.* pp. 1-128.

Adams, T.S. & Sterner, R.W. (2000). The effect of dietary nitrogen content on trophic level ^{15}N enrichment. *Limnology and Oceanography* 45(3): 601-607.

Alerstam, T. & Hedenström, A. (1998). The development of bird migration theory. *Journal of Avian Biology* 29: 343-369.

Ambrose, S.A. (1991). Effects of diet, climate and physiology on nitrogen isotope abundances in terrestrial foodwebs. *Journal of Archaeological Science* 18: 293-317.

Anon. (1953). Points from letters. *Ostrich* 24: 55-56.

Baillie, S.R., Marchant, J.H., Crick, H.Q.P., Noble, D.G., Balmer, D.E., Beaven, L.P., Coombes, R.H., Downie, I.S., Freeman, S.N., Joys, A.C., Leech, D.I., Raven, M.J., Robinson, R.A. & Thewlis, R.M. (2005). *Breeding birds in the wider countryside: their conservation status 2004.* BTO Research Report No. 385. BTO, Thetford.

Baillie, S.R., Marchant, J.H., Leech, D.I., Joys, A.C., Noble, D.G., Barimore, C., Grantham, M.J., Risely, K. & Robinson, R.A. (2009). *Breeding birds in the wider countryside: their conservation status 2008.* BTO Research Report No. 516. BTO, Thetford.

Baillie, S.R., Marchant, J.H., Leech, D.I., Joys, A.C., Noble, D.G., Barimore, C., Downie, I.S., Grantham, M.J., Risely, K. & Robinson, R.A. (2010). *Breeding birds in the wider countryside: their conservation status 2009.* BTO Research Report No. 541. BTO, Thetford.

Bairlein, F. (2003). The study of bird migrations – some future perspectives. *Bird Study* 50(3):243-253.

Bairlein, F. (1985). Body weights and fat deposition of Palaearctic passerine migrants in the central Sahara. *Oecologia* 66(1): 141-146.

Barbraud, C. & Chastel, O. (1998). Southern fulmars molt their primary feathers while incubating. *Condor* 100: 563–566

Barea. L.P. & Herrera, M.L.G. (2009). Sources of protein in two semi-arid zone mistletoe specialists: Insights from stable isotopes. *Austral Ecology* 34: 821-828.

Barnes, K.N. (2005). European Bee-eater *Merops apiaster*. pp. 194-195. In: *Roberts' birds of southern Africa*. (7th Edition). P.A.R. Hockey, W.R.J. Dean & P.G. Ryan, (Eds.). The Trustees of the John Voelcker Bird Book Fund, Cape Town.

Bearhop, S., Furness, R. W., Hilton, G. M., Votier, S. C. & Waldron, S. (2003). A forensic approach to understanding diet and habitat use from stable isotope analysis of (avian) claw material. *Functional Ecology* 17(2): 270-275.

Bearhop, S., Thompson, D.R., Waldron, S., Russel, I.C., Alexander, G. & Furness, R.W. (1999). Stable isotopes indicate the extent of freshwater feeding by cormorants *Phalacrocorax carbo* shot at inland fisheries in England. *Journal of Applied Ecology* 36: 75-84.

Bearhop, S., Waldron, S., Votier, S.C. & Furness, R.W. (2002). Factors that influence assimilation rates and fractionation of nitrogen and carbon stable isotopes in avian blood and feathers. *Physiological and Biochemical Zoology* 75(5): 451-458.

Ben-David, M., Titus, K. & Beier, L.R. (2004). Consumption of salmon by Alaskan brown bears: a trade-off between nutritional requirements and the risk of infanticide? *Oecologia* 138: 465–474.

Bensch, S., Bengtsson, G. & Åkesson, S. (2006). Patterns of stable isotope signatures in willow warbler *Phylloscopus trochilus* feathers collected in Africa. *Journal of Avian Biology* 37: 323-330.

Benson, C.W. (1951). A roosting site of the Eastern Red-footed Falcon, *Falco amurensis*. *Ibis* 93: 467-468.

Benson, C.W., Brooke, R.K., Dowsett, R.J. & Irwin, M.P.S. (1971). *The birds of Zambia*. Collins, London.

Bigeleisen, J. (1965). Chemistry of isotopes. *Science* 147: 463-471.

Blaber, S.J.M. (1986). Willow Warblers at Darvill, Pietermaritzburg. *Safring News* 15: 11-14 .

Borret, R. (1968). Migrant movements. *Honeyguide* 55: 10-11.

Both, B. & Marvelde, L. (2007). Climate change and timing of avian breeding and migration throughout Europe. *Climate Research* 35: 93-105.

Bowen G.J. & Revenaugh J. (2003). Interpolating the isotopic composition of modern meteoric precipitation. *Water Resources Research* 39(10): 1299. doi: 10.1029/2003WR002086

Bowen, G.J., Wassenaar, L.I. & Hobson, K.A. (2005). Global application of stable hydrogen and oxygen isotopes to wildlife forensics. *Oecologia* 143: 337-348.

Bowen, G.J. & West, J.B. (2008). Isotope landscapes for terrestrial migration research. pp. 79-105. In: *Tracking animal migration with stable isotopes*. K.A. Hobson & L.I. Wassenaar (Eds). Elsevier, Amsterdam, The Netherlands.

Brooke, R.K. & Herroelen, P. (1988). The nonbreeding range of southern African bred European Bee-eaters *Merops apiaster*. *Ostrich* 59: 63–66.

Bunning, L.J. (1977). Birds of the Melville Koppies Nature Reserve. *Southern Birds* 3: 1-58.

Cade, T.J. (1982). *Falcons of the world*. Cornell University Press, Ithaca, NY.

Carlton, J.T. & Hodder, J. (2003). Maritime mammals: terrestrial mammals as consumers in marine intertidal communities. *Marine Ecology Progress Series* 256: 271-286.

Carleton, S.A., Hartmann Bakken, B. & Martínez del Rio, C. (2006). Metabolic substrate use and the turnover of endogenous energy reserves in broad-tailed hummingbirds (*Selaphorus platycercus*). *Journal of Experimental Biology* 209: 2622-2627.

Carleton, S.A. & Martínez del Rio, C. (2005). The effect of cold-induced increased metabolic rate on the rate of ^{13}C and ^{15}N incorporation in house sparrows (*Passer domesticus*). *Oecologia* 144: 226-232.

Caut, S., Angulo, E. & Courchamp, F. (2008). Discrimination factors ($\Delta^{15}N$ and $\Delta^{13}C$) in an omnivorous consumer: effect of diet isotopic ratio. *Functional Ecology* 22(2): 255-263.

Cerling, T.E. & Harris, J.M. (1999). Carbon isotope fractionation between diet and bioapatite in ungulate mammals and implications for ecological and paleoecological studies. *Oecologia* 120: 347-363.

Chamberlain, C.P., Waldbauer, J.R., Fox-Dobbs, K., Newsome, S.D., Koch, P.L., Smith, D.R., Church, M.E., Chamberlain, S.D., Sorenson, K.J. & Risebrough, R. (2005). Pleistocene to recent dietary shifts in California condors. *PNAS* 102(46): 16707–16711.

Chamberlain, C.P., Blum, J.D., Holmes, R.T., Feng, X., Sherry, T.W. & Graves, G.R. (1997). The use of isotope tracers for identifying populations of migratory birds. *Oecologia* 109: 132-141.

Chamberlain, C.P., Bensch, S., Feng, X., Åkesson, S. & Andersson, T. (2000). Stable isotopes examined across a migratory divide in Scandinavian willow warblers (*Phylloscopus trochilus trochilus* and *Phylloscopus trochilus acredula*) reflect their African winter quarters. *Proc. R. Soc. London B* 267: 43-48.

Cheng, Tso-hsin. (1987). *A synopsis of the avifauna of China*. Beijing, Science Press.

Cherel, Y., Hobson, K.A. & Hassani, S. (2005). Isotopic discrimination between food and blood and feathers of captive penguins: implications for dietary studies in the wild. *Physiological and Biochemical Zoology* 78: 106-115

Clancey, P.A. (1970). Miscellaneous taxonomic notes on African birds XXVIII On the races of *Phylloscopus trochilus* (Linnaeus) occurring in southern Africa, with special reference to the status of *P. t. yakutensis* Ticehurst, 1935. *Durb. Mus. Novit.* 8: 331-334.

Clancey, P.A. (1980). *SAOS checklist of southern African birds*. Southern African Ornithological Society, Pretoria.

Clancey, P.A., Brooke, R.K., Crowe, T.M. & Mendelsohn, J.M. (1987). *SAOS checklist of southern African birds: first updating report*. Southern African Ornithological Society, Johannesburg.

Clark, R.G., Hobson, K.A. & Wassenaar, L.I. (2006). Geographic variation in the isotopic (δD, $\delta^{13}C$, $\delta^{15}N$, $\delta^{34}S$) composition of feathers and claws from Lesser Scaup and Northern Pintail: implications for studies of migratory connectivity. *Can. J. Zool.* 84: 1395-1401.

Clegg, S.M., Kelly, J.F., Kimura, M. & Smith, T.B. (2003). Combining genetic markers and stable isotopes to reveal population connectivity and migration patterns in a

Neotropical migrant, Wilson's warbler (*Wilsonia pusilla*). *Molecular Ecology* 12: 819-830.

Craig, A.J.F.K. (1983). Moult in southern African passerine birds: a review. *Ostrich* 54: 220-237.

Craig, A.J.F.K. & Hulley, P.E. (1994). Sunbird movements: a review with possible models *Ostrich* 65: 106-110.

Cramp, S. & Perrins, C.M. (1993). *The birds of the western Palaearctic. Vol. 7.* Oxford University Press, Oxford.

Crawford, R.J.M. (2005). Great White Pelican *Pelecanus onocrotalus*. pp. 614-615. In: *Roberts' birds of southern Africa.* (7th Edition). P.A.R. Hockey, W.R.J. Dean & P.G. Ryan, (Eds.), The Trustees of the John Voelcker Bird Book Fund, Cape Town.

Curry-Lindahl, K. (1981). *Bird migration in Africa. Vol. 1.* Academic Press, London.

Dalerum, F. & Angerbjörn, A. (2005). Resolving temporal variation in vertebrate diets using naturally occurring stable isotopes. *Oecologia* 144:647-658

Daniels, C.L. (1987). *The feeding ecology of nectarivorous birds in the Natal Drakensberg.* Unpublished. M.Sc. Thesis. University of Natal. Pietermaritzburg, South Africa.

Davies, G.B.P., Symes, C.T., Chittenden, H.N. & Peek, J.R. (2012). Mangrove Kingfishers (*Halcyon senegaloides*) nesting in arboreal termitaria in central Mozambique. *Annals of the Ditsong National Museum of Natural History* (Continues *Annals of the Transvaal Museum*) 2:146-152.

Dawson, T.E., Mambelli, S., Plamboeck, A.H., Templer, P.H. & Tu, K.P. (2002). Stable isotopes in plant ecology. *Annual Review of Ecology and Systematics* 33: 507-559.

De Niro, M.J. & Epstein, S. (1981). Influence of diet on the distribution of nitrogen isotopes in animals. *Geochimica et Cosmochimica Acta* 45: 341-351.

Dean, W.R.J. (2005). Willow Warbler *Phylloscopus trochilus*. pp. 807-808. In: *Roberts' birds of southern Africa.* (7th Edition). P.A.R. Hockey, W.R.J. Dean & P.G. Ryan, (Eds.), The Trustees of the John Voelcker Bird Book Fund, Cape Town.

Dowsett, R.J., Backhurst, G.C. & Oatley, T.B. (1988). Afrotropical ringing recoveries of Palaearctic migrants. *Tauraco* 1: 29-63.

Drent, R., Both, C., Green, M., Madsen, J. & Piersma, T. (2003). Pay-offs and penalties of competing migratory schedules. *Oikos* 103 (2): 274-292.

Ehleringer, J.R. (1991). $^{13}C/^{12}C$ fractionation and its utility in terrestrial plant studies. pp. 1-16. In: Rundel. P.W., Rundel, J.R. & K.A. Nagy, (Eds.), *Stable Isotopes in Ecological Research.* Springer-Verlag, New York.

Ehleringer, J.R & Monson, R.K. (1993). Evolutionary and ecological aspects of photosynthetic pathway variation. *Annual Review of Ecology and Systematics* 24: 411-439.

Ehleringer, J.R. & Osmond, C.B. (1989). Stable Isotopes. pp. 281-300. In: *Plant physiological ecology.* R.W. Pearcy, J. Ehleringer, H.A. Mooney & J.R. Rundel, (Eds.), Chapman and Hall, London.

Eller, B.M., Ruess, B.R. & Ferrari, S. (1993). Crassulacean acid metabolism (CAM) in the chlorenchyma and hydrenchyma of *Aloe* leaves and *Opuntia* cladodes: field determinations in an *Aloe*-habitat in southern Africa. *Botanica Helvetica* 103(2): 201-205.

Evans Ogden, L.J., Hobson, K.A. & Lank, D.B. (2004). Blood isotopic ($\delta^{13}C$ and $\delta^{15}N$) turnover and diet-tissue fractionation factors in captive dunlin (*Calidris alpina pacifica*). *Auk* 121(1): 170-177.

Ferguson-Lees, J. & Christie, D.A. (2001). *Raptors of the world.* London, Christopher Helm.

Flade, M, Diop, I., Haase, M., Le Nevé, A., Oppel, S., Tegetmeyer, C., Vogel, A. & Salewski, V. (2011). Distribution, ecology and threat status of the Aquatic Warblers *Acrocephalus paludicola* wintering in West Africa. *Journal of Ornithology* 152(S1): S129-S140.

Fleming, T.H., Tuttle, M.D. & Homer, M.A. (1996). Pollination biology and the relative importance of nocturnal and diurnal pollinators in three species of Sonoran Desert columnar cacti. *Southwestern Naturalist* 41: 257-269.

Friedman, I. & O'Neil, J.R. (1977). Compilation of stable isotope fractionation factors of geochemical interest. In: M. Fleischer, (Editor). *Data of geochemistry.* (6th Edition). Geological Survey Professional Paper 400-KK. US Department of the Interior. United States Government Printing Office, Washington.

Frost, S.K. & Frost, P.G.H. (1981). Sunbird pollination of *Strelitzia nicolai. Oecologia* 49: 379-384.

Fry, B. (1988). Food web structure on Georges Bank from stable C, N, and S isotopic compositions. *Limnology and Oceanography* 33: 1182-1190.

Fry, B., Brand, W., Mersch, F.J. Tholke, K. & Garritt, R. (1992). Automated analysis system for coupled $\delta^{13}C$ and $\delta^{15}N$ measurements. *Anal. Chem.* 64: 288-291.

Fry, B., Joern, A. & Parker, P.L. (1978). Grasshopper food web analysis: use of carbon isotope ratios to examine feeding relationships among terrestrial herbivores. *Ecology* 59(3): 498-506.

Fry, C.H. (2001). Family Meropidae (Bee-eaters). In: *Handbook of the Birds of the World (Volume 6): Mousebirds to Hornbills.* J. del Hoyo, A, Elliott & J. Sargatal, (Eds.), Lynx Edicions, Barcelona, Spain.

Gannes, L.Z., O'Brien, D.M. & Martínez del Rio, C. (1997). Stable isotopes in animal ecology: assumptions, caveats, and a call for more laboratory experiments. *Ecology* 78(4): 1271-1276.

Gartrell, B.D. (2000). The nutritional, morphologic, and physiologic bases of nectarivory in Australian birds. *Journal of Avian Medicine and Surgery* 14(2): 85–94.

Ginn, H.B. & Mellville, D.S. (1983). *Moult in birds. BTO Guide 19.* British Trust for Ornithology, Tring.

Global Raptor Information Network. (2008). Species account: Amur Falcon *Falco amurensis.* Downloaded from http://www.globalraptors.org on 19 December 2008.

Gonfiantini, R. (1978). Standards for stable isotope measurements in natural compounds. *Nature* 271: 534.

Gordo, O. (2007). Why are bird migration dates shifting? A review of weather and climate effects on avian migratory phenology. *Climate Research* 35:37-58.

Graves, G.R., Romanek, C.S. & Navarro, A.R. (2002). Stable isotope signature of philopatry and dispersal in a migratory songbird. *Proc. Natl. Acad. Sci.* 99(12): 8096–8100.

Greenberg, R. & Marra, P.P. (Eds.). (2005). *Birds of two worlds: the ecology and evolution of migration.* Johns Hopkins University Press, Baltimore.

Hahn, S., Bauer, S. & Liechti, F. (2009). The natural link between Europe and Africa - 2.1. billions birds on migration. *Oikos* 118:624-626.

Hartley, A.R. (2003). A milestone in migration studies – answering and raising questions: The recently published Migration Atlas provides a valuable source of ideas for future research on bird movements. *Bird Study* 50(1):94.

Hedenström, A. & Pettersson, J. (1987). Migration routes and wintering areas of Willow Warblers *Phylloscopus trochilus* (L.) ringed in Fennoscandia. *Ornis Fennica* 64: 137-143.

Hedenström, A., Lindström, A.A. & Pettersson, J. (1995). Interrupted moult of adult Willow Warblers *Phylloscopus trochilus* during autumn migration through Sweden. *Ornis Svecica* 5(2): 69-74.

Herremans, M. (1994). Fifteen years of migrant phenology records in Botswana: a summary and prospects. *Babbler* 28: 47-68.

Herremans, M. (1997a). Willow Warbler *Phylloscopus trochilus*. pp. 254-255. In: *The atlas of southern African birds - Vol. 2. Passerines.* J.A. Harrison, D.G. Allan, L.L. Underhill, M. Herremans, A.J. Tree, V. Parker & C.J. Brown, (Eds.), BirdLife South Africa, Johannesburg.

Herremans, M. (1997b). Spotted Flycatcher *Muscicapa striata*. pp. 332-333. In: *The atlas of southern African birds - Vol. 2. Passerines.* J.A. Harrison, D.G. Allan, L.L. Underhill, M. Herremans, A.J. Tree, V. Parker & C.J. Brown, (Eds.), BirdLife South Africa, Johannesburg.

Herrera, L.G., Hobson, K.A., Martínez, J.C. & Méndez, C.G. (2006). Tracing the origin of dietary protein in tropical dry forest birds. *Biotropica* 38: 735-742.

Herrera, L.G., Hobson, K.A., Rodríguez, M. & Hernandez, P. (2003). Trophic partitioning in tropical rain forest birds: insights from stable isotope analysis. *Oecologia* 136:439-444.

Hobson, K.A. (1987). Use of stable-carbon isotope analysis to estimate marine and terrestrial protein content in gull diets. *Canadian Journal of Zoology* 65: 1210-1213.

Hobson, K.A. (1999). Tracing origins and migration of wildlife using stable isotopes: a review. *Oecologia* 120(3): 314-326.

Hobson, K.A. (2005). Stable isotopes and the determination of avian migratory connectivity and seasonal interactions. *The Auk* 122(4): 1037-1048.

Hobson, K.A. (2011). Isotopic ornithology: a perspective. *Journal of Ornithology* 152: S49-S66.

Hobson, K.A., Alisaukas, R.T. & Clark, R.G. (1993). Stable-nitrogen isotope enrichment in avian tissue due to fasting and nutritional stress: implications for isotopic analyses of diet. *Condor* 95: 388-394.

Hobson, K.A., Aubry, Y. & Wassenaar, L.I. (2004a). Migratory connectivity in Bicknell's thrush: locating missing populations with hydrogen isotopes. *Condor* 106: 905-909.

Hobson, K.A. & Bairlein, F. (2003). Isotopic fractionation and turnover in captive garden warblers (*Sylvia borin*): implications for delineating dietary and migratory associations in wild passerines. *Canadian Journal of Zoology* 81: 1630-1635.

Hobson, K.A., Bowen, G.J., Wassenaar, L.I., Ferrand, Y. & Lormee, H. (2004b). Using stable hydrogen and oxygen isotope measurements of feathers to infer geographical origins of migrating European birds. *Oecologia* 141: 477–488.

Hobson, K.A. & Clark, R.G. (1992a). Assessing avian diets using stable isotopes I: turnover of ^{13}C in tissues. *Condor* 94: 181-188.

Hobson, K.A. & Clark, R.G. (1992b). Assessing avian diets using stable isotopes II: factors influencing diet-tissue fractionation. *Condor* 94: 189-197.

Hobson, K.A. & Clark, R.G. (1993). Turnover of ^{13}C in cellular and plasma fractions of blood: implications for non-destructive sampling in avian dietary studies. *Auk* 110:638-641.

Hobson, K.A., deMent, S.H., Van Wilgenburg, S.L. & Wassenaar, L.I. (2009). Origins of American Kestrels wintering at two southern U.S. sites: An investigation using stable-isotope (δD, δ^{18}O) methods. *J. Raptor Res.* 43: 325-337.

Hobson, K.A., McFarland, K.P., Wassenaar, L.I., Rimmer, C.C. & Goetz, J.E. (2001). Linking breeding and wintering grounds of Bicknell's Thrushes using stable isotope analyses of feathers. *The Auk* 118(1):16-23

Hobson, K.A., Piatt, J.F. & Pitoccelli, J. (1994). Using stable isotopes to determine seabird trophic relationships. *Journal of Animal Ecology* 63: 786-798.

Hobson, K.A. & Wassenaar, L.I. (1997). Linking breeding and wintering grounds of neotropical songbirds using stable hydrogen isotopic analysis of feathers. *Oecologia* 109: 142-148.

Hobson, K.A. & Wassenaar, L.I. (1999). Stable isotope ecology: an introduction. *Oecologia* 120: 312-313.

Hobson, K.A. & Wassenaar, L.I. (2008). *Tracking animal migration with stable isotopes*. Elsevier, Amsterdam, The Netherlands.

Hockey, P., Dean, R., Ryan, P. & Maree, S. (Eds.), (2005). *Roberts' birds of southern Africa.* (7th Edition). John Voelcker Bird Book Fund, Cape Town, pp 1-1296.

Hoefs, J. (2009). *Stable isotope geochemistry* (6th Edition). Springer-Verlag, Berlin. pp. 1-285.

Holmes, R.T. (2007). Understanding population change in migratory songbirds: long-term and experimental studies of Neotropical migrants in breeding and wintering areas. *Ibis* 149(S2): 2-13.

Holmgren, N. and Hedenström, A. 1995. The scheduling of molt in migratory birds *Evolutionary Ecology* 9(4): 354 – 368.

Hopcroft, C. (1984). A study of the Willow Warbler in South Africa. *Safring News* 13: 10-17.

Hunter, C. & Hunter, M. (1970). Recovery of ringed Spotted Flycatcher (R654). *WBC News* 70: 16.

Inger, R. & Bearhop, S. (2008). Applications of stable isotope analyses to avian ecology. *Ibis* 150: 447-461.

Irwin, M.P.S. (1981). *The birds of Zimbabwe*. Quest, Salisbury.

Jacot Guillarmod, A., Jubb, R.A. & Skead, C.J. (1979). Field studies of six southern African species of *Erythrina*. *Annals of the Missouri Botanical Garden* 66: 521-527.

Jenkins, A.R. (2005). Amur Falcon *Falco amurensis*. pp. 552-553. In: *Roberts' birds of southern Africa*. (7th Edition). P.A.R. Hockey, W.R.J. Dean & P.G. Ryan, (Eds.), The Trustees of the John Voelcker Bird Book Fund, Cape Town.

Jenni, L., Winkler, R & Degen, T. (1994). *Moult and ageing of European passerines*. Academic Press, London.

Jeppe, B. (1969). *Southern African aloes*. Purnell and Sons, Cape Town, pp. 1-144.

Johnson, D.N. (2005). Spotted Flycatcher *Muscicapa striata*. pp. 919-920. In: *Roberts' Birds of Southern Africa*. (7th Edition). P.A.R. Hockey, W.R.J. Dean & P.G. Ryan, (Eds.), The Trustees of the John Voelcker Bird Book Fund, Cape Town.

Kelly, J.F. (2000). Stable isotopes of carbon and nitrogen in the study of avian and mammalian trophic ecology. *Canadian Journal of Zoology* 78: 1-27.

Kelly, J.F., Atudorei, V., Sharp, Z.D. & Finch, D.M. (2002). Insights into Wilson's Warbler migration from analyses of hydrogen stable-isotope ratios. *Oecologia* 130: 216-221.

Kimura, M., Clegg, S.M., Lovette, I.J., Holder, K.R., Girman, D.J., Milá, B., Wade, P. & Smith, T.B. (2002). Phylogeographical approaches to assessing demographic connectivity between breeding and overwintering regions in a Nearctic–Neotropical warbler (*Wilsonia pusilla*). *Molecular Ecology* 11(9): 1605-1616.

Klaassen, M., Lindström, A., Meltofte, H. & Piersma, T. (2001). Arctic waders are not capital breeders. *Nature* 413: 794.

Kluge, M., Knapp, I., Kramer, D., Schwerdtner, I. & Ritter, H. (1979). Crassulacean acid metabolism (CAM) in leaves of *Aloe arborescens* Mill. *Planta* 145: 357-363.

Kok, O.B., Van Ee, C.A. & Nel E.G. (1991). Daylength determines departure date of the Spotted Flycatcher *Muscicapa striata* from its winter quarters. *Ardea* 79: 63-66.

Larson, K.W. & Hobson, K.A. (2009). Assignment to breeding and wintering grounds using stable isotopes: a comment on lessons learned by Rocque et al. *Journal of Ornithology* 150: 709-712.

Lee-Thorp, J.A., Sealy, J.C. & van der Merwe, N.J. (1989). Stable carbon isotope ratio differences between bone collagen and bone apatite, and their relationship to diet. *Journal of Archaeological Science* 16:585-599.

Lee-Thorp, J.A. & Thalma, A.S. (2000). Stable light isotopes and environments in the southern African Quaternary and late Pliocene. In: *The Cenozoic of southern Africa*. Partridge, T.C. & Maud, R.R. (Eds.), Oxford Monographs on Geology and Geophysics No. 40. Oxford University Press, Oxford, pp. 236-251.

Loske, K-H. & Lederer, W. (1988). Moult, weight and biometrical data for some Palaearctic passerine migrants in Zambia. *Ostrich* 59: 1-7

Lott, C.A. & Smith, J.P. (2006). A geographic-information-system approach to estimating the origin of migratory raptors in North America using hydrogen stable isotope ratios in feathers. *Auk* 123: 822-835.

Maclean, G.L. (1990). *Ornithology for Africa*. University of Natal Press, Pietermaritzburg, pp. 1-270.

Maclean, G.L. (1993). *Roberts' birds of southern Africa*. (6th Edition). John Voelcker Bird Book Fund, Cape Town, pp. 1-871.

Markman, S., Pinshow, B. & Wright, J. (1999). Orange-tufted sunbirds do not feed nectar to their chicks. *Auk* 116:257-259.

Markman, S., Pinshow, B., Wright, J. & Kotler, B.P. (2004). Food patch use by parent birds: to gather food for themselves or for their chicks? *Journal of Animal Ecology* 73(4): 747-755.

Marloth, R. (1915). *The flora of South Africa, with special synoptical tables of the genera of higher plants. Vol. IV. Monocotyledons*. Darter Brothers and Co., Cape Town, pp. 1-208.

Martínez del Rio, C., Wolf, N., Carleton, S.A. & Gannes, L.Z. (2009). Isotopic ecology ten years after a call for more laboratory experiments. *Biological Reviews* 84: 91–111.

McCutchan, J.H., Lewis, W.M., Kendall, C. & McGrath, C.C. (2003). Variation in trophic shift for stable isotope ratios of carbon, nitrogen and sulphur. *Oikos* 102: 378-390.

McKechnie, A.E. (2004). Stable isotopes: powerful new tools for animal ecologists. *South African Journal of Science* 100: 131-134.

McKechnie, A.E., Wolf, B.O. & Martínez del Rio, C. (2004). Deuterium stable isotope ratios as tracers of water resource use: an experimental test with rock doves. *Oecologia* 140: 191-200.

Meehan, T.D., Lott, C.A., Sharp, Z.D., Smith, R.B., Rosenfield, R.N., Stewart, A.C. & Murphy, R.K. (2001). Using hydrogen isotope geochemistry to estimate the natal latitudes of immature Cooper's Hawks migrating through the Florida Keys. *The Condor* 103(1): 11-20.

Mendelsohn, J.M. (1997). Eastern Redfooted Kestrel *Falco amurensis*. pp. 262-263. In: *The atlas of South African birds. Volume 1: Non-passerines*. J.A. Harrison, D.G. Allan, L.G. Underhill, M. Herremans, A.J. Tree, V. Parker & C.J. Brown, (Eds.), BirdLife South Africa, Johannesburg.

Meyburg, B.-U., Ellis, D. H., Meyburg, C., Mendelsohn, J. M. & Scheller, W. (2001). Satellite tracking of two Lesser Spotted Eagles, *Aquila pomarina*, migrating from Namibia. *Ostrich* 72: 35-40.

Meyburg, B.-U., Mendelsohn, J.M., Ellis, D.H., Smith, D.G., Meyburg, C. & Kemp, A.C. (1995). Year-round movements of a Wahlberg's Eagle *Aquila wahlbergi* tracked by satellite. *Ostrich* 66: 135-140.

Minagawa, M. & Wada, E. (1984). Stepwise enrichment of ^{15}N along food chains: further evidence and the relation between δ^{15}N and animal age. *Geochimica et Cosmochimica Acta* 48: 1135-1140.

Mizutani, H., Fukuda, M. & Kabaya, Y. 1992. ^{13}C and ^{15}N enrichment factors of feathers of 11 species of adult birds. *Ecology* 73: 1391-1395.

Napier, K.R., Purchase, C., McWhorter, T.J., Nicolson, S.W. & Fleming, P.A. (2008). The sweet life: diet sugar concentration influences paracellular glucose absorption. *Biology Letters* 4: 530-533.

Nichols, J.D. & Kaiser, A. (1999). Quantitative studies of bird movement: a methodological review. *Bird Study* 46(S1): S289-S298.

Oatley, T.B. (1964). The probing of aloe flowers by birds. *Lammergeyer* 3: 2-8.

Oatley, T.B. (1995). Selected recoveries reported to SAFRING: July 1994 - December 1994. *Safring News* 24: 27-38.

Oatley, T.B. & Skead, D.M. (1972). Nectar feeding by South African birds. *Lammergeyer* 15: 68-69.

Oatley, T.B. (2001). Garden visitors - Sunbirds. *Africa Birds and Birding* 6(3): 26-31.

O'Leary, M.H. (1981). Carbon isotope fractionation in plants. *Phytochemistry* 20: 553.

Oppel, S., Pain, D.J., Lindsell, J.A., Lachmann, L., Diop, I., Tegetmeyer, C., Donald, P.F., Anderson, G., Bowden, C.G.R., Tanneberger, F. & Flade, M. (2011). High variation reduces the value of feather stable isotope ratios in identifying new wintering areas for aquatic warblers *Acrocephalus paludicola* in West Africa. *Journal of Avian Biology* 42(4): 342-354.

Pain, D., Green, R.E., Gieβing, B., Kozulin, A., Poluda, A., Ottosson, U., Flade, M. & Hilton, G.M. (2004). Using stable isotopes to investigate migratory connectivity of the globally threatened aquatic warbler *Acrocephalus paludicola*. *Oecologia* 138:168-174

Park, R. & Epstein, S. (1960). Carbon isotope fractionation during photosynthesis. *Geochimica et Cosmochimica Acta* 21: 11.

Park, R. & Epstein, S. (1961). Metabolic fractionation of ^{13}C and ^{12}C in plants. *Plant Physiology* 36(2): 133-138.

Pearson, S.F., Levey, D.J., Greenberg, C.H. & Martínez del Rio, C. (2003). Effects of elemental composition on the incorporation of dietary nitrogen and carbon isotopic signatures in an omnivorous songbird. *Oecologia* 135: 516-523.

Perkins, S.E. & Speakman, J.R. (2001). Measuring natural abundance in ^{13}C in respired CO_2: variability and implications for non-invasive dietary analysis. *Functional Ecology* 15: 791-797.

Peterson, B.J. & Fry, B. (1987). Stable isotopes in ecosystem studies. *Annual Review of Ecology and Systematics* 18: 293-320.

Phillips, D.L. & Gregg, J.W. (2003). Source partitioning using stable isotopes: coping with too many sources. *Oecologia* 136: 261-269.

Pinnegar, J.K. & Polunin, N.V.C. (1999). Differential fractionation of $\delta^{13}C$ and $\delta^{15}N$ among fish tissues: implications for the study of trophic interactions. *Functional Ecology* 13: 225-231.

Podlesak, D.W. & McWilliams, S.R. (2006). Metabolic routing of dietary nutrients in birds: effects of diet quality and macronutrient composition revealed using stable isotopes. *Physiological and Biochemical Zoology* 79: 534-549.

Podlesak, D.W., McWilliams, S.R. & Hatch, K.A. (2005). Stable isotopes in breath, blood, feces and feathers can indicate intra-individual changes in the diet of migratory songbirds. *Oecologia* 142: 501-510.

Post, D.M. (2002). Using stable isotopes to estimate trophic position: models, methods, and assumptions. *Ecology* 83(3): 703-718.

Procházka, P., Hobson, K. A., Karcza, Z. & Kralj, J. (2008). Birds of a feather winter together: migratory connectivity in the Reed Warbler *Acrocephalus scirpaceus*. *Journal of Ornithology* 149: 141-150.

Procházka, P., Reif, J., Horák, D., Klvaňa, P., Lee, R.W. and Yohannes, E. (2010). Using stable isotopes to trace resource acquisition and trophic position in four Afrotropical birds with different diets. *Ostrich* 81: 273-275.

Ramos, R., González-Solís, J. & Ruiz, X. (2009). Linking isotopic and migratory patterns in a pelagic seabird. *Oecologia* 160: 97–105.

Reed, R.A. & Wells, S.J.C. (1957). Ringing. *WBC News* 23: 1-2.

Reynolds, G.W. (1969). *The aloes of South Africa*. AA Balkema, Cape Town, pp. 1-526.

Robbins, C.T., Felicetti, L.A. & Sponheimer, M. (2005). The effect of dietary protein quality on nitrogen isotope discrimination in mammals and birds. *Oecologia* 144: 534-540.

Roxburgh, L., Pinshow, B. 2002. Digestion of nectar and insects by Palestine Sunbirds. *Physiological and Biochemical Zoology* 75(6): 583–589.

Rubenstein, D.R. & Hobson, K.A. (2004). From birds to butterflies: animal movement patterns and stable isotopes. *Trends in Ecology and Evolution* 19(5): 256-263.

Ryan, P.G. (2005). Pink-backed Pelican *Pelecanus rufescens*. pp. 615-617. In: *Roberts' birds of southern Africa*. (7th Edition). P.A.R. Hockey, W.R.J. Dean & P.G. Ryan, (Eds.). The Trustees of the John Voelcker Bird Book Fund, Cape Town.

Sabat, P. & Martínez del Rio, C. (2006). Inter- and intraspecific variation in the use of marine food resources by three *Cinclodes* (Furnariidae, Aves) species: carbon isotopes and osmoregulatory physiology. *Zoology* 105(3): 247-256.

Safring unpubl. data. 2010. Animal Demography Unit. University of Cape Town. South Africa

Salewski, V., Altwegg, R., Erni, B., Falk, K.H., Bairlein, F & Leisler, B. (2004). Moult of three Palaearctic migrants in their West African winter quarters. *Journal of Ornithology* 145(2): 109-116.

Salewski, V., Bairlein, F & Leisler, B. (2002). Remige moult in Spotted Flycatcher (*Muscicapa striata*) on its West African wintering grounds. *Vogelwarte* 41: 301-303

Schäfer, S. (2003). *Studie an einer mongolischen Brutpopulation des Amurfalken* (Falco amurensis *Radde, 1863)*. Unpublished Dipl. Biol. thesis. Martin-Luther Universität Halle-Wittenberg. Halle/Saale, Germany.

Simmons, R.E., Barnard, P., Dean, W.R.J., Midgley, G.F., Thuiller, W. & Hughes, G. (2004). Climate change and birds: perspectives and prospects from southern Africa. *Ostrich* 75(4): 295-308.

Skead, C.J. (1967). *The sunbirds of southern Africa*. AA Balkema, Cape Town, pp. 1-351.

Smith, B.N. & Epstein, S. (1971). Two categories of $^{13}C/^{12}C$ ratios for higher plants. *Plant Physiology* 47: 380-384.

Snow, D.W. & Perrins, C.M. (1998). *The birds of the western Palaearctic. Concise Edition. Vol. 2: Passerines*. Oxford University Press, Oxford.

Spottiswoode, C.N. (2005). Blue swallow *Hirundo atrocaerulea*. pp. 752-754. In: *Roberts' birds of southern Africa*. (7th Edition). P.A.R. Hockey, W.R.J. Dean & P.G. Ryan, (Eds.), The Trustees of the John Voelcker Bird Book Fund, Cape Town.

Svensson, L. (1992). *Identification guide to the European passerines*. Svensson, Stockholm.

Symes, C.T. (2010). The sweet option: the importance of *Aloe marlothii* for opportunistic avian nectarivores. *Bulletin of the African Bird Club* 17(2): 178-187.

Symes, C.T., Downs, C.T. & McLean, S. (2001). Seasonal occurrence of the malachite sunbird, *Nectarinia famosa*, and Gurney's sugarbird, *Promerops gurneyi*, in KwaZulu-Natal, South Africa. *Ostrich* 72(1&2): 45-49.

Symes, C.T., McKechnie, A.E., Nicolson, S.W. & Woodborne, S.M. (2011). The nutritional significance of a winter-flowering succulent for opportunistic avian nectarivores. *Ibis* 153: 110-121.

Symes, C.T., Nicolson, S.W. & McKechnie, A.E. (2008). Response of avian nectarivores to the flowering of *Aloe marlothii*: a nectar oasis during dry South African winters. *Journal of Ornithology* 149: 13-22.

Symes, C.T. & Woodborne, S. (2009). Trophic level delineation and resource partitioning in a South African afromontane forest bird community using carbon and nitrogen stable isotopes. *African Journal of Ecology* 48: 984-993.

Symes, C.T. & Woodborne, S. (2010). Migratory connectivity and conservation of the Amur Falcon *Falco amurensis*: a stable isotope perspective. *Bird Conservation International* 20: 134-148.

Symes, C.T. & Woodborne, S. (2011). Variation in carbon and nitrogen stable isotope ratios of remiges in a moulting White-bellied Sunbird *Cinnyris talatala*. *Ostrich* 82(3): 163-166.

Szép, T., Møller, A. P., Vallner, J., Kovács, B. & Norman, D. 2003. Use of trace elements in feathers of Sand Martin *Riparia riparia* for identifying moulting areas. *Journal of Avian Biology* 34(3): 307-320.

Tarboton, W. & Allan, D. (1984) *The status and conservation of birds of prey in the Transvaal*. Transvaal Museum Monograph. Vol. 3. Transvaal Museum, Pretoria.

Tarboton, W.R., Kemp, M.I. & Kemp, A.C. (1987). *Birds of the Transvaal*. Transvaal Museum, Pretoria.

Tiainen, J. (1981). Timing of the onset of postnuptial moult in the Willow Warbler *Phylloscopus trochilus* in relation to breeding in southern Finland. *Ornis Fennica* 58(2): 56-63.

Tree, A.J. (1989). Recent reports. *Honeyguide* 35: 122-127.

Tree, A.J. (1990). Notes on sunbird movements and nectar sources in Zimbabwe. *Honeyguide* 36: 171-182.

Tree, A.J. (1995). Recent reports. *Honeyguide* 41: 242-255.

Tsahar, E, Martínez del Rio, C, Izhaki, I. & Arad, Z. (2005). Can birds be ammonotelic? Nitrogen balance and excretion in two frugivores. *Journal of Experimental Biology* 208: 1025-1034.

Tsahar, E., Wolf, N., Izhaki, I., Arad, Z. & Martínez del Rio, C. (2008). Dietary protein influences the rate of ^{15}N incorporation in blood cells and plasma of yellow-vented bulbuls (*Pycnonotus xanthopygos*). *Journal of Experimental Biology* 211: 459-465.

Underhill, L. (1994). Tailenders: Spotted Flycatcher from Sweden. *Birding South Africa* 46: 96.

Underhill, L.G. (1997). European Bee-eater *Merops apiaster*. pp. 660-662. In: *The atlas of southern African birds - Vol. 1. Non-passerines*. J.A. Harrison, D.G. Allan, L.L. Underhill, M. Herremans, A.J. Tree, V. Parker & C.J. Brown., (Eds.), BirdLife South Africa, Johannesburg.

Underhill, L.G. & Oatley, T.B. (1994). Spotted Flycatcher from Sweden. *Birding in Southern Africa* 46: 96.

Underhill, L.G., Prŷs-Jones, R.P., Dowsett, R.J. & Herroelen, P., Johnson, D.N., Lawn, M.R., Norman, S.C., Pearson, D.J. & Tree, A.J. (1992a). The biannual primary moult of Willow Warblers *Phylloscopus* trochilus in Europe and Africa. *Ibis* 134(3): 286-297.

Underhill, L.G., Prŷs-Jones, R.P., Harrison, J.A. & Martínez, P. (1992b). Seasonal patterns of occurence of Palaearctic migrants in southern Africa using atlas data. *Ibis* 134 (Suppl. 1): 99-108.

Urban, E.K., Fry, C.H. & Keith, S. (1997). *The birds of Africa. Vol. 5*. Academic Press, London.

Vanderklift, M.A. & Ponsard, S. (2003). Sources of variation in consumer-diet δ^{15}N enrichment: a meta-analysis. *Oecologia* 136: 169-182

Vanhinsbergh, D., Fuller, R.J. & Noble, D.G. (2003). *An analysis of changes in the populations of British woodland birds and a review of possible causes*. Research Report 245. BTO, Thetford.

Vogel, J.C., Fuls, A. & Ellis, R.P. (1978). The geographical distribution of Kranz grasses in South Africa. *South African Journal of Science* 74: 209-215.

Voigt, C.C., Baier, L., Speakman, J.R. & Siemers, B.M. (2008). Stable carbon isotopes in exhaled breath as tracers for dietary information in birds and mammals. *J. Exp. Biol.* 211: 2233–2238.

Wakelin, J., McKechnie, A.E. & Woodborne, S.M. (2011). Stable isotope analysis of migratory connections in a threatened intra-African migrant, the Blue Swallow (*Hirundo atrocaerulea*). *Journal of Ornithology* 152: 171-177.

Wassenaar, L.I. & Hobson, K.A. (1998). Natal origins of migratory monarch butterflies at wintering colonies in Mexico: new isotopic evidence. *Proc. Natl. Acad. Sci.* 95: 15436–15439.

Wassenaar, L.I. & Hobson, K.A. (2000). Improved method for determining the stable-hydrogen isotopic composition (δD) of complex organic materials of environmental interest. *Environ. Sci. Technol.* 34: 2354-2360

Wassenaar, L.I. & Hobson, K.A. (2001). A stable-isotope approach to delineate geographical catchment areas of avian migration monitoring stations in north America. *Environ. Sci. Technol.* 35: 1845-1850.

West, J.B., Bowen, G.J., Cerling, T.E. & Ehleringer, J.R. (2006). Stable isotopes as one of nature's ecological recorders. *Trends in Ecology and Evolution* 21(7): 408-414.

West, J.B., Bowen, G.J., Dawson, T.E. & Tu, K.P., (Eds.). (2010). *Isoscapes: Understanding movement, pattern, and process on earth through isotope mapping.* Springer Dordrecht, Heidelberg, Germany.

Williamson, K. (1972). Reversal of normal moult sequence in the Spotted Flycatcher. *British Birds* 65: 50-51.

Wolf, B.O. & Martínez del Rio, C. (2000). Use of saguaro fruit by white-winged doves: isotopic evidence of a tight ecological association. *Oecologia* 124: 536-543.

Wolf, B.O. & Martínez del Rio, C. (2003). How important are columnar cacti as sources of water and nutrients for desert consumers? A review. *Isotopes in Environmental and Health Studies* 39: 53-67.

Wolf, B.O., Martínez del Rio, C. & Babson, J. (2002). Stable isotopes reveal that saguaro fruit provides different resources to two desert dove species. *Ecology* 83: 1286-1293.

Wolf, B.O. & Hatch, K. (2011). Aloe nectar, birds and stable isotopes; opportunities for quantifying trophic interactions. *Ibis* 153: 1-3.

Yohannes, E., Biebach, H., Nikolaus, G. & Pearson, D.J. (2009). Passerine migration strategies and body mass variation along geographic sectors across East Africa, the Middle East and the Arabian Peninsula. *J Ornithol* 150: 369-381.

Yohannes, E., Hobson, K.A. & Pearson, D.J. (2007). Feather stable-isotope profiles reveal stopover habitat selection and site fidelity in nine migratory species moving through sub-Saharan Africa. *J Avian Biol* 38: 347-355.

Yohannes, E., Hobson, K.A., Pearson, D.J. & Wassenaar, L.I. (2005). Stable isotope analyses of feathers help identify autumn stopover sites of three long-distance migrants in northeastern Africa. *Journal of Avian Biology* 36: 235-241.

Yohannes, E., Lee, R.W. & Jochimsen, M.C. (2011). Stable isotope ratios in winter-grown feathers of Great Reed Warblers *Acrocephalus arundinaceus*, Clamorous Reed Warblers *A. stentoreus* and their hybrids in a sympatric breeding population in Kazakhstan. *Ibis* 153(3): 502–508.

13

The Impact of Shelterbelts on Mulch Decomposition and Colonization by Fauna in Adjacent Fields

M. Szanser
Polish Academy of Sciences, Centre for Ecological Research, Łomianki
Poland

1. Introduction

There is an increasing need of planting the strips of perennial habitats (shelterbelts, grass margins and others) into an agricultural landscape. Such strips play invaluable role in ecosystem functioning by reducing wind erosion, modifying climatic conditions, reducing eutrophication of ground waters and enhancing an organic matter content in the soil of adjacent arable fields etc. (Harper et al., 2005, Karg et al., 2003, Kostro-Chomać, 2003, Laurance et al., 2002, Prusinkiewicz et al., 1996). These changes are reflected also in the composition and abundance of the fauna (Bowen et al., 2007, Olechowicz, 2007, Szanser, 2003). The significance of relations between faunal and plant diversity, landscape structure and soil carbon dynamics is described in many studies (Dauber et al., 2003, Marshall et al., 2006).

The objective of this paper is to evaluate, in the field experiments, the effect of mid-field woody shelterbelts on litter decomposition rate and the numbers and biomass of litter and soil inhabiting macrofauna. Data concerning litter and soil fauna of the studied habitats are compared. The question was how far into the fields an effect of shelterbelts could reach.

2. Materials and methods

2.1 Study site and experimental design

The study was carried out in village of Turew, west Poland (16º45' to 16 º 50' E and 52 º 00' to 52 º 06' N). In the study region, prevail Glossoboric Hapludalf soils on loamy sand over light loam and Arenic Hapludalf soils on slightly loamy sand over light loam (Marcinek, 1996). They are slightly acid, poor in organic matter, with low water holding capacity.

The experiments were performed along the transect of the forest strip planted in 1993 (S), in the ecotone (E) and in the field 10 (F 10 - 15) and 50 (F 50) meters far from the tree line in 1999-2000 (I experiment) and 2003-2004 (II experiment). In the second experiment additional studies were performed in the control field (CF 50) located in deforested area. During the study time barley and wheat were cropped on field adjacent to strip and wheat and barley on field in deforested area(control field). Mulch was obtained from grass cocksfoot (*Dactylis glomerata*) cut in autumn in every year of the study time which had to simulate the input of decaying plants to soil. Grass litter was laid out on a uniform, poor in organic matter substrate (sand mixed with clay) (I experiment) or on soil (II experiment) in each study site. Grass was placed in modified litterbags (PVC rings with drilled holes and with top and bottom covered with a nylon gauze of a mesh size of 1 mm) per 10 g in each (Szanser, 2000, 2003).

2.2 Sample collecting and processing

Samples of litter (I and II experiment) and soil underlying it (II experiment) from the middle of a wood strip (S) and along the ecotone and field transect were taken on 26 September and 21 October 2004 after 11 months of litter exposure in I and II experiments, respectively (Table 1). Analyses from earlier studies revealed that macrofauna colonized litter intensively in later stages of its decomposition so only data for 11th months since litter exposure are compared (Szanser, 2003, unpubl.).

	I Experiment	II Experiment
Soil	Introduced sandy substrate	Local soil
Time of sampling since litter exposure	11 months	11 months
Sites along transect	Shelterbelt – S Ecotone at shelterbelt edge – Es [1] Ecotone at field edge – Ef [1] Field 10m since ecotone – F 10 [2] Field 50m since ecotone – F 50	Shelterbelt – S Ecotone - E Field 15m since ecotone – F 15 [2] Field 50m since ecotone – F 50
Field in deforested area	-	CF 50
Parameters analysed		
litter decomposition		
Faunal composition and biomass	+	+
Mass of invertebrate remnants	+	+

1 - values obtained from Es and Ef sites were recalculated as a mean for comparing to data from E in II experiment.
2 - values obtained from F 10 and F 15 sites compared between I and II experiment.

Table 1. Experimental design and parameters analysed. Treatments applied: S – forest strip, E – ecotone, F 10 – field 10 m from the strip, F 15 – field 15 m from the strip, F 50 – field 50 m from the strip, CF 50 – Control field 50 m not surrounded by the forest strip.

8 - 14 litter and soil samples were taken from every treatment on each sampling occasions. Soil samples for assessments of invertebrate organic matter input and macrofauna biomass were taken to the depth of 0-1 and 0-10 cm, respectively with a soil corer 100 cm^2 in area.

2.3 Plant and invertebrate materials

2.3.1 Litter mass loss

Litter mass was dried at 65ºC before the weight loss was determined using the gravimetric method.

2.3.2 Macro- and microarthropod collection

Temperature gradient Thompson, Kempson, Lloyd and Gheraldi's apparatus was used for extraction of macro- and microarthropod fauna (Górny & Grüm, 1993).

Macrofauna was divided into two trophic groups:

1. predatory macrofauna: spiders, chilopods, beetles (Carabidae and Staphylinidae).
2. other macrofauna mostly saprophagous mainly dipteran larvae, Symphyla, non-predatory beetles (Coleoptera).

Mesofauna: springtails (Collembola), mostly saprophagous, were extracted from the same litter samples.

2.3.3 Input of invertebrate remnants into litter and underlying substrate.

The input of the remnants of invertebrate origin (exuviae, cocoons, other remants) was assessed in the litter and soil samples underneath the litter by hand-sorting and using the stereoscope microscopy.

2.4 Statistical analysis

Statistical analyses of results (one-way $ANOVA$ and regression) were performed using Statistica 8.0. software (StatSoft, Inc. (2007). Differences in the biomass of animals and their remnants were analysed using multivariate $ANOVA$, the Tukey's test at P <0.05. Relationships between the biomass of fauna and the distance from the shelterbelt and between predators and prey were tested by the correlation significance with $ANOVA$.

3. Results

3.1 litter decomposition

The amounts of litter decomposed during the study periods varied. In average 48.2 and 63.1% of the exposed litter decomposed during 11 months in I and II experiments, respectively along the study transect. Litter decomposition rate along the transect forest-ecotone-centre of the field proceeded differently in both experiments. Remaining mass of the litter was similar in the ecotone (E) after 11 months in both experiments while it was significantly lower in the strip (S) and field (F 10-15 and F 50) at the end of II comparing to I experiment (Table 2). Interestingly grass litter decomposed differently in the fields comparing to strip and ecotone sites in every experiment. Significantly greater by 23.4% amounts of decomposing litter were left in the field comparing to strip and ecotone during the I experiment. On the other side significantly smaller by 19.8% mass of remaining litter was left in the field comparing to strip and ecotone during the II experiment. The mass of remaining litter in the centre of the control field located in deforested area was similar as in the field adjacent to forest strip in the II experiment (Table 2). Obtained data show that the litter decomposition does not depend on the location in the transect.

	S	E	F 10-15	F 50	F C
I experiment [1]	465.8 [a]	458.1 [a]	540.3 [b]	599.7 [b]	n.a. [2]
	(22.19)	(24.79)	(23.75)	(16.88)	
II experiment	387.61 [a]	428.78 [a]	339.02 [b]	310.19 [b]	308.76 [b]
	(23.36)	(14.79)	(32.14)	(23.45)	(18.52)
Ratio II/I	0.83	0.94	0.63	0.52	
	n = 20,	n = 20,	n = 18,	n = 19,	-
	$F = 4.648$	$F = 1.121$	$F = 21.225$	$F = 87.151$	
	$P <0.045$	$P <0.286$	$P <0.0003$	$P <0.0000000$	

1 - modified after Szanser 2003
2 - not analysed

Table 2. Mass of remaining litter (g dry weight m^{-2}) after 11 months of its exposure in I and II experiment and relation between both experiments. Sites as in Table 1. Different letters denote significant differences between experimental treatments at $P <0.05$ (Tukey's test) for $n = 8$-10. SE – standard error in parentheses.

3.2 Macrofauna biomass

Significant decrease of the macrofauna biomass in the litter along the transect from strip to the field was observed during the I experiment while this trend was not significant in the II experiment. The ratio of macrofaunal biomass in strip and ecotone to the field was 3.98 and 1.63 respectively in I and II experiments (Table 3). Lower was the biomass of litter dwelling macrofauna in the middle of field located in deforested area comparing to the field adjacent to forest strip being respectively 0.564 and 0.733 g dry weight m^{-2} but the difference was not significant (Table 4).

Macrofauna group	Mass ratio: S, E / F 10-15, F 50 and significance of differences
	I experiment [1]
Total biomass	3.98 $P < 0.00000$, $F = 53.49$
Nonpredatory biomass	6.19 $P < 0.04$, $F = 4.616$
Predatory biomass	1.41 $P < 0.001$, $F = 17.15$
	II experiment
Total biomass	1.63 $P < 0.327$, $F = 0.986$
Nonpredatory biomass	3.76 $P < 0.199$, $F = 1.709$
Predatory biomass	1.01 $P < 0.987$, $F = 0.00$

1 - modified after Szanser 2003

Table 3. Biomass ratio of macrofaunal groups (S, E/F 10-15, F 50) after 11 months of the I and II experiments ($n = 40$) assessed by *ANOVA*. Sites as in Table 1.

There were no differences in predatory biomass between strip - ecotone and field being similar for both group of sites: 0.538 and 0.542 g dry weight m^{-2} during the II experiment contrary to the findings in I experiment (Table 4, Szanser 2003).

There was significantly lower by 9.6 time (P <0.04) predatory biomass in the litter on the field in deforested area comparing to the centre of field adjacent to strip during the II experiment (Table 4).

	S	E	F 10	F 50	F C
I experiment [1]					
Total biomass	3.596 [a]	2.637 [a]	0.785 [b]	0.782 [b]	n.a.
	(0.50)	(0.30)	(0.13)	(0.14)	
Nonpredatory biomass	1.221 [a]	1.014 [a]	0.100 [b]	0.262 [b]	n.a.
	(0.34)	(0.31)	(0.03)	(0.07)	
Predatory biomass	2.375 [a]	1.623 [a]	0.686 [b]	0.520 [b]	n.a.
	(0.35)	((0.13)	(0.13)	(0.14)	
II experiment					
Total biomass	2.058 [a]	0.582 [b]	0.930 [b]	0.733 [b]	0.564 [b]
	(0.79)	(0.21)	(0.40)	(0.20)	(0.31)
Nonpredatory biomass	1.214 [a]	0.161 [b]	0.113 [b]	0.258 [b]	0.514 [b]
	(0.75)	(0.08)	(0.07)	(0.08)	(0.32)
Predatory biomass	0.824 [a]	0.259 [b]	0.601 [b]	0.475 [b]	0.05 [c]
	(0.20)	(0.12)	(0.34)	(0.18)	(0.03)
Ratio 2004/2000 total biomass	0.57	0.22	1.19	0.94	-
Nonpredatory biomass	0.99	0.16	1.13	0.98	-
Predatory biomass	0.35	0.16	0.88	0.91	-

1 – modified after Szanser 2003

Table 4. Macrofauna biomass (g dry weight m^{-2}) in litter after 11 months of I and II experiments. Sites as in Table 1. Different letters denote significant differences between experimental treatments at P <0.05 (Tukey's test) for n = 8-10. SE – standard error in parentheses.

The nonpredatory biomass was by 3.8 times lower on field comparing to strip and ecotone but the difference was not significant. Similarly there was no difference in nonpredatory biomass between fields in deforested area and adjacent to strip belt.

The negative correlation between the biomass of predators and non-predators dwelling the litter along the transect from strip to the centre of the field was observed in both experiments. It was only significant for carabids and their prey; collembolans and dipteran larvae during the I experiment (Szanser 2003) while between the total predatory and nonpredatory biomass (r = -0.843, P<0.01) for all treatments including field in deforested area during the II experiment.

Differences between treatments in faunal biomass were greater in the soil underlying exposed litter than in the litter itself. Significantly by 4.3 and 5.7 times higher was the fauna biomass in strip comparing to ecotone (and adjacent field (Table 5). There were no significant differences in faunal biomass either between ecotone and field and between F 10 and F 50 sites. Interestingly, significantly lower by 5.0 times was the fauna biomass in soil of the field located in deforested area comparing to the field adjacent to forest strip (Table 5).

	S	E	F 10-15	F 50	F C
II experiment	2.063 [a]	0.475 [b]	0.367 [b]	0.360 [b]	0.072 [c]
	(0.51)	(0.25)	(0.11)	(0.12)	(0.04)

Table 5. Macrofauna biomass (g dry weight m^{-2}) in soil layer 0-10cm under the litter after 11 months of II experiment. Sites as in Table 1. Different letters denote significant differences between experimental treatments at P <0.05 (Tukey's test) for n = 8-10. SE – standard error in parentheses.

The experiment revealed significant predator-prey relationships within the transect. It may be concluded that the presence of shelterbelts adjacent to field leads to the increase of macrofaunal biomass comparing to the field not surrounded by strip. The stimulating effect of mulching on faunal biomass was clearly seen mainly in the litter along the transect with strip comparing to the field without strip area. Soil under the litter in ecotone and fields was colonized to much lesser extent comparing to litter.

3.3 Input of invertebrate remnants

Input of dead invertebrate mass to the litter and soil under litter decreased significantly from the shelterbelt towards the field center in both experiments. The correlations between dead invertebrate mass and distance from the centre of the strip to field were found to be r = -0.425, P <0.0021 (Szanser 2003) and r = -0.907, P <0.05, respectively in I and II experiment. There were no differences in remaining mass of invertebrate remnants in forest strips and ecotones between two experiments. On the opposite there was considerably smaller amount of invertebrate remnants on fields during the II experiment comparing to I (Table 6). Input of invertebrate remnants into the litter and soil under litter was lower in deforested area comparing to the field adjacent to forest strip, but the differences between treatments were not significant (Table 6).

Input of animals remains into the soil can be quite high and reflects the fauna ability to penetrate the fields.

	S	E	F 10-15	F 50	F C
I experiment [1]	0.859 [a]	0.462 [a]	0.302 [a]	0.183 [a]	-
	(0.39)	(0.33)	(0.26)	(0.13)	
II experiment	1.04 [a]	0.893 [a]	0.234 [a]	0.049 [a]	0.037 [a]
	(0.55)	(0.42)	(0.2)	(0.02)	(0.01)
Ratio II/I	1.21	1.93	0.73	0.27	-

1 – modified after Szanser 2003

Table 6. Invertebrate matter mass (g dry weight m^{-2}) in litter after 11 months of I and II experiments. Sites as in Table 1. Differences between experimental treatments tested at P <0.05 (Tukey's test) for n = 11-14. SE – standard error in parentheses.

4. Discussion

The grass litter decomposition was different in the fields during two experiments while it proceeded similarly in the ecotone area though considerably greater in average litter mass loss was observed during II comparing to I experiment. Decomposition of forest litter placed along the studied transect proceeded faster on field comparing to forest and ecotone sites but the litter was exposed only for a few weeks (Bernacki unpubl.) comparing to the long-term grass decomposition in presented study. On the other side slower decomposition processes in field comparing to forest and ecotone transect were found for buried litter samples (Bernacki 1994). Differences in litter decomposition in fields could not be attributed to temperature and soil water content changes (Bernacki unpubl.). Mean precipitation did not differ between study periods being 607 and 616 mm for I and II experiment, respectively (after Bernacki, unpubl.). The differences between both experiments could arise from different substrata underlying the exposed litter as sandy-clay poor in organic matter substrate and natural soil were used in I and II experiment, respectively. Litter placed on poor uniform sandy-clay substrate decompose slower comparing to litter on natural soil (Szanser and Bogdanowicz 1997). It also seems that the differences in crop chemical composition being different during consecutive seasons might influence the process of mulch litter decomposition. Barley and wheat were cultivated during the I and II experiments, respectively. Chemical composition of crop biomass impacts its decomposition and carbon sequestration (Johnson et al., 2007, Wang et al., 2004). Use of fertilizers and herbicides might determine the way of litter disappearance processes in studied fields (Szanser 2003). It may be concluded from these statements that the presence of forest strip did not influence the litter decomposition in fields.

Mean mass of macroarthropods in exposed litter was higher than that found in soil and turf in the same site (Olechowicz 2004a,b, 2007). The differences between biomass of aboveground dwelling fauna along the studied transect and in the experimental mulch in my experiments were of 8 and 2 times respectively during the I experiment and 127 and 3.5 times during the II experiment. Lower predatory biomass data for control field than in site adjacent to forest belt are corroborated by lower average body mass, mean patrolling intensity and diversity index of spiders in the same studied field in deforested area comparing to fields adjacent to forest belts (Kajak 2007). It is known that introduction of semi-natural habitats into arable fields enhances the development of predatory arthropods' assemblages in these fields (Asteraki *et al.*, 2002, Marshall & Mooney 2002, Marshall *et al.*, 2006, Ryszkowski *et al.*, 1993, Schmidt & Tscharntke 2005).

It seems that the applied litter stimulated the number of animals. Dwelling the exposed litter seemed to reflect not only the numbers of animals in the environments but was the indicator of fauna ability to disperse from forest to the field. The lack of correlation between the mass of remaining litter and either fauna biomass and mass of invertebrate remnants was found in both experiments. This finding implies that the main reason for colonizing the exposed mulch by fauna was not searching for food resources but rather for shelters.

It must be pointed out that longtime field experiments show the effects of processes which confirm that colonization of area without strips was not so intensive as in the case of field with adjacent strip. Thus then introduced habitats of permanent vegetation are the reservoirs of fauna enabling it to colonize adjacent fields to far distances. The experimental studies indicated that establishment of strips margins in arable fields may enhance ecosystem services.

5. Conclusions

In general the results suggest that (1) the decomposition rate of grass litter proceeded similarly in the forest strip and its ecotone zone as in the fields, so the presence of forest strip did not influence mulch decomposition in the fields. It can be concluded basing on the comparison of the data of two experiments and on the comparison of the field located in deforested area and the field adjacent to the forest strip; (2) the biomass of macrofauna in the soil under the introduced litter was lower in deforested area than in the field adjacent to forest strip; (3) also the biomass of predators colonizing the introduced litter was clearly lower in the field in deforested area comparing to the field adjacent to the strip; (4) input of invertebrate remnants into the litter and soil under litter was lower in deforested area comparing to the field adjacent to forest strip but differences were not significant.

6. Acknowledgments

This research was supported by a grant from the Polish State Committee for Scientific Research, project 6 P04F 028 16.

7. References

Asteraki, E.J., Hart, B.J., Ings, T.C. and Manley W.J. (2004). Factors influencing the plant and invertebrate diversity of arable field margins. *Agriculture Ecosystems and Environments*, Vol. 102, No. 2, (April 2004), pp. 219–231, ISSN 0167-8809.

Bernacki, Z., 1994, Influence of main types of ecotones on decomposition of organic matter in agricultural landscape. (In: Functional Appraisal of Agricultural Landscape in Europe: Proceedings of III International MAB Seminar Błażejewko 1992, Eds., Ryszkowski, L., Bałazy, S.), ZBŚRiL PAN, Bonami, Poznań, (no information 1994), pp. 143-151, ISBN 83-85274-03-0.

Bogdanowicz, L. and Szanser M. (1997) The decomposition of *Dactylis glomerata* grass litter in meadow ecosystems differing in age and manner of use. *Ekologia Polska*, Vol. 45, No. 3-4, (September 2007), pp. 647-663, ISSN: 0070-9557.

Bowen, M. E., C. A., McAlpine, A.P.N., House and G.C., Smith, (2007). Regrowth forests on abandoned agricultural land: A review of their habitat values for recovering forest fauna. *Biological Conservation*, Vol. 140, No 3-4, (December 2007),pp.273-296, ISSN 0006-3207.

Dauber, J., Hirsch, M., Simmering, D., Waldhardt, R., Otte, A. & Wolters, V. (2003). Landscape structure as an indicator of biodiversity: matrix effects on species richness. *Agriculture Ecosystems and Environment*, Vol. 98, No. 1-3, (September 2003), pp.321-329, ISSN 0167-8809.

Górny, M. and Grüm, L. (1993). *Methods in Soil Zoology*. Elsevier/Tokyo and PWN-Polish Scientific Publ., Amsterdam, London, New Jork/Warszawa, (January 1993), pp. 459, ISBN-13: 978-0444988232.

Harper, K.A., Macdonald, S.E., Burton, P.J., Chen, J., Brosofske, K.D., Saunders, S.C., Euskirchen, 429 E.S., Roberts, D., Jaiteh, M.S. and Esseen, P.-A., (2005). Edge

Influence on Forest Structure and Composition in Fragmented Landscapes. *Conservation Biology*, Vol. 19,No. 3, (June 2005), pp. 768-782, ISSN 0888-8892.

Johnson, J. M. F., Barbour, N. W. and Weyers, S. L. (2007). Chemical Composition of Crop Biomass Impacts Its Decomposition. *Soil Science Society of America* Journal, Vol. 71, No. 1, (January 2007), pp.155-162, ISSN 1435-0661.

Karg, J., Kajak, A. and Ryszkowski, L. (2003). Impact of young shelterbelts on organic matter content and development of microbial and faunal communities of adjacent fields. *Polish Journal of Ecology*, Vol. 51, No. 3, (September 2003), pp. 283–290, ISSN 1505-2249.

Kajak, A. (2007). Effects of forested strips on spider assemblages in adjacent cereal fields: Dispersal activity of spiders. *Polish Journal of Ecology*, Vol. 55, No 4, (October 2007), pp. 691–704, ISSN 1505-2249.

Kostro-Chomać, A. (2003). Zmiany zawartości materii organicznej gleby pod wpływem pasów leśnych [Changes in soil organic matter content as a consequence of introduction of wooded shelterbelts]. Ph. Doctor`s Thesis, Agricultural University, Warszawa (in Polish).

Laurance, W.F., Lovejoy, T.E., Vasconcelos, H.L., Bruna, E.M., Didham, R.K. and Stouffer, P.C. (2002). Ecosystem decay of Amazonian forest fragments: a 22-year investigation. *Conservation Biology*, Vol. 16, No. 3, (June 2002), pp. 447 605-618, ISSN 0888-8892.

Marcinek, J. (1996) Soils of the Turew Agricultural Landscape (In: *Dynamics of Agricultural Landscape*, Ed. L., Ryszkowski, French, N.R., Kędziora, A.,), PWRiL, Poznań, (no information 1996), pp.19-26, ISBN 83-85274-03-0.

Marshall, E.J.P. and Moonen, A.C. (2002). Field margins in northern Europe: their functions and interactions with agriculture. *Agriculture Ecosystems and Environments*, Vol. 89, No. 1-2, (April 2002), pp. 5–21, ISSN 0167-8809.

Marshall, E.J.P., West, T.M. and Kleijn, D. (2006). Impacts of an agri-environment field margin prescription on the flora and fauna of arable farmland in different landscapes. *Agriculture Ecosystems and Environments.*, Vol. 113, No. 1-4, (April 2006), pp.36–44, ISSN 0167-8809.

Olechowicz, E. (2004a). Communities structure of soil-litter macrofauna in shelterbelt and adjacent crop field. *Polish Journal of Ecology*, Vol. 52, No. 2, (June 2004), pp.135–153, ISSN 1505-2249.

Olechowicz, E. (2004b). Soil-litter macrofauna in the mixed forest and midfield shelterbelts of different age (Turew area, West Poland). *Polish Journal of Ecology*, Vol. 52, No. 4, (October 2004), pp. 405–420, ISSN 1505-2249.

Olechowicz, E. (2007). Soil and litter macrofauna in shelterbelts and in adjacent croplands: changes in community structure after tree planting. *Polish Journal of Ecology*, Vol. 55, No. 4, (October 2007), pp.647–664, ISSN 1505-2249.

Prusinkiewicz, Z., Pokojska, U., Józefkowicz-Kotlarz, J. and Kwiatkowska, A. (1996). Studies on the functioning of biogeochemical barriers (In: *Dynamics of an agricultural landscape*. Ed. L. Ryszkowski, French, N.R., Kędziora, A.), PWRiL, Poznań, (no information 1996), pp. 110-119, ISBN 83-85274-03-0.

Ryszkowski, L., Karg, J. and Bernacki, Z. (2003). Biocenotic function of the mid-field woodlots in West Poland: study area and research assumption. *Polish Journal of Ecology*, Vol. 51, No 3, (September 2003), pp. 269–281, ISSN 1505-2249.

Schmidt, M.H. and Tscharntke, T. (2005). The role of perennial habitats for Central European farmland spiders. *Agriculture Ecosystems and Environments*, Vol. 105, No. 1-2, (January 2005), pp. 235–242, ISSN 0167-8809.

Szanser, M. (2000). Effect of macroarthropods patrolling soil surface on decomposition rate of grass litter (*Dactylis glomerata*) in a field experiment – Polish Journal of Ecology Vol. 48, No. 4, (October 2000), pp. 283-297, ISSN 1505-2249.

Szanser, M. (2003). The effect of shelterbelts on litter decomposition and fauna of adjacent fields: In situ experiment. *Polish Journal of Ecology*, Vol. 51, No. 3, (September 2003), pp. 309-321, ISSN 1505-2249.

STATISTICA (data analysis software system), version 8.0. www.statsoft.com).

Wang, W.J., J.A., Baldock, R.C., Dalal, and Moody P.W. (2004). Decomposition dynamics of plant materials in relation to nitrogen availability and biochemistry determined by NMR and wet-chemical analysis. *Soil Biology and Biochemistry*, Vol. 36, No 12, (December 2004), pp.2045–2058, ISSN 0038-0717.

14

New Technologies for Ecosystem Analysis Planning and Management

Pietro Picuno, Alfonso Tortora, Carmela Sica and Rocco Capobianco
University of Basilicata
Italy

1. Introduction

Planning the rural environment is one of the most intriguing examples of technical challenge where a multi-disciplinary approach plays a crucial role. The agricultural production, both food and non-food, the social role of rural settlements, the state and diffusion of the infrastructural networks, the rural architectonic heritage that in many countries constitutes a major positive value, should be appropriately considered and sinergically interlaced for a sound planning of agricultural biosystems.

Human activities impose a transformation of the extra-urban land that may lead to the modification of the frail equilibrium of whole ecosystems. Sound planning strategies should be therefore pursued, employing a multidisciplinary approach that should take into account geographical, environmental and landscape factors as variables interacting among themselves and with the social and economic aspects. In order to simultaneously analyse all these properties, tools able to manage, interpret and integrate several data are necessary.

Extra-urban land planning must pursuit, as a main goal, environmental sustainability, since sustainable development, in EU countries, has been perceived by social awareness and sensibility and is constantly been considered by new laws and regulations whose attempt is natural resources protection (Toccolini et al., 2006). In this scenario an accurate analysis of performing variations and a global monitoring of ecosystems is necessary in order to propose environmental protection politics. The farmer as a producer has traditionally been in focus when changes in agricultural landscapes are studied. Decisions on husbandry, rotational systems, machinery, fertilisation and pest management do indeed affect the landscape in crucial ways, and landscape dynamics cannot be understood if the farmer's decision making and the surrounding technology, socio-economics, and organisational structure are ignored (Primdahl, 1999). However, normally the farmer is not the only decision maker. Often he farms leased land and the owner may be an equally important actor concerning landscape changes.

Farmers are important agents in rural landscape management as they modify landscape elements to suit their needs (Kristensen et al., 2001). The industrialisation and intensification of agriculture over the last 50 years had a negative impact on landscape diversity and habitat values. During the last two decades, farmers have become increasingly engaged in landscape activities, to maintain or create habitats on their property: Kristensen (2003) analyzed the value of traditional housing and settlements in the countryside. For the sustainable development of rural settlements at least four characteristics should be protected: balance between nature and built-up area; historic traditional entities; local communities; the countryside as an own culture (Ruda, 1998).

New leading technologies could be adequately introduced for an improved analysis of the rural environment. Remote sensing techniques could be employed for the monitoring of agricultural land variation. Geographical Information System (G.I.S.) are excellent tools for landscape modelling and three-dimensional analysis. They allow an easy digitalisation of geographical information and coverage structure, as well as facilitate graphical representation. An evaluation of the aesthetic impact produced on the rural environment becomes therefore possible, paving the way for landscape simulations and possible minimizations of the landscape impact.

G.I.S. allows an easy digitalisation of geographical information and coverage structure, as well as facilitating graphical representation (Hernández et al., 2004/a). Vigiak, Sterk, Warren, Hagen (2003) presented a model able to integrate windbreak shelter effects into a Geographic Information System (G.I.S.). The G.I.S. procedure incorporates the 1999 version windbreak sub-model of the Wind Erosion Prediction System (WEPS) (Hagen, 1991). Windbreak shelter is modelled in terms of friction velocity reduction, which is a function of wind speed and direction, distance from the barrier, windbreak height, porosity, width, and orientation.

A specific landscape analysis conducted by Capobianco, Tortora, Picuno (2004) by a G.I.S. approach has shown how positive results of the applied agronomic practises, in terms of CO_2 fixation, have been able to contrast heavy emissions of greenhouse effect gases in the atmosphere by urban settlements. Through the implementation of a Digital Terrain Model (DTM), enriched with the drape of land cover pictures, the authors evaluated in a scenic way the morphological and vegetation variations of agro-forestry landscape. The digitalization of historical cartography, moreover, simulated an hypothetical and virtual historical jump backwards (Tortora et al., 2006), enabling the analysis of the natural and anthropic changes of rural land, able to affect the structure of whole ecosystems. A similar approach, using a Geographical Information System and Image Processing techniques, was used by Tortora & Picuno (2008) in order to analyze the aesthetic impact that the use of plastic coverings produces on the rural environment, so enabling landscape simulation and examining possible minimizations of the landscape impact.

A wide spread of crops covered with plastic, can damage the visual landscape, although they are detectable through Remote Sensing techniques Remote Sensing techniques. Some scientific efforts were recently conducted in order to allow a better monitoring and planning of these uses. Using a field spectrometer, Levin et al. (2007). studied the spectral properties of a sample of polyethylene sheets and various nets used in Israel through the detection of three major absorption features around 1218 nm, 1732nm and 2313 nm. Carvajal et al. (2006) presented a methodology able to detect greenhouses from 2.44 m pixel size QuickBird image, based on an Artificial Neural Network algorithm. Thanks to the information introduced through training sites, they "teach" to the mathematical model to classify the image considering its radiometric and wavelet texture properties. Classification accuracy was evaluated using multi-source data, comparing results including and non-including wavelet texture analysis. The Authors concluded that some texture analysis can not improve classification accuracy but if one choose correctly parameters and texture model, it can become better.

The effectiveness of image processing application to the classification of crop shelter sites and its accuracy was verified by Arcidiacono & Porto (2008). An intensively-cultivated area located in South-Eastern Sicily (Italy) was selected to perform the localization of greenhouses and other type of shelters, like vineyards coverings. The chosen study area was pick out from a digital orthophotograph file of the Sicilian territory. This image was georeferenced and then analyzed by an image processing software. A methodology based on supervised classification of the image was found as the most adequate to the classification of crop shelters. According to this methodology, suitable classes were selected on the basis of signatures related to specific sample areas. The classification was then refined by using neighbourhood and contiguity analysis algorithms. The results of the analysis allowed to recognize and localize the crop shelters and to quantify their planimetric area. The latter was also compared with the attributes of georeferenced feature classes based on visual recognition.

Finally, Capobianco & Picuno (2008) implemented remote sensing techniques for an analysis of the rural land use with special attention paid to greenhouse and other application of plastics in protected cultivation, inside a study area located near the coast border between the Italian Regions of Basilicata and Apulia, where plastics in agriculture is widely used. The analysis was realized using Thematic Mapper of multitemporal Landsat images through Supervised classification, image processing, vectorialization and G.I.S. tools. For the study were used Band 7 (2.08 - 2.35 μm), band 5 (1.55 - 1.75 μm), and band 3 (0.63 - 0.69 μm) ion with other cartographic information. The results that were obtained enable the possibility to create a routine in IDL and ENVI software for the auto-detection of the plastic covers.

Solid modelling techniques, moreover, could contribute to the analysis and planning of the rural environment. The implementation of a Digital Terrain Model (DTM), enriched with the drape of land cover pictures, enables the evaluation in a scenic way of the morphological and vegetation variations of agro-forestry landscape. The digitalization of historical cartography, finally, allowed (Picuno et al., 2011) the simulation of an hypothetical and virtual historical jump backwards, so facilitating the analysis of the natural and anthropic changes of rural land, able to affect the structure of the agricultural biosystems.

2. Analysis of plasticulture landscapes in Southern Italy through remote sensing and solid modeling techniques

2.1 Foreword

The use of plastic covers for protecting cultivation in wide rural areas sets remarkable technical problems connected with the effects that large extensions of agricultural land covered with continuous cladding material may determine on aesthetic pollution of the rural landscape and on negative environmental impacts on water cycle, air and agricultural soil of the agro-ecosystem. The distribution of plastic cover for greenhouse or other protected cultivation applications has been analyzed using thematic mapping of multi-temporal satellite images with unsupervised classification, image processing, and vectorialization.

According to recent data (Scarascia-Mugnozza et al., 2008), the annual consumption of plastic film in Italy for protected cultivation amounts to more than 150,000 tons (Tab. 1). Protection from hail, wind, snow, or strong rainfall in fruit-farming and ornamentals, together with the realization of a confined airspace with better microclimatic conditions, is the most common case. Plastic films are widely diffused for covering greenhouse, low and medium tunnel, and for soil mulching, while shading nets for greenhouses, or nets for a modification of the microenvironment, are employed.

ITALIAN REGION	Greenhouse	Middle/low	Covering	Mulch	TOTAL
Piedmont	977	528	0	1,481	2,986
Trentino	230	120	0	2	352
Lombardy	3,080	3,932	0	2,087	9,099
Friuli	345	96	0	209	650
Venetia	3,358	3,760	0	2,822	9,940
Liguria	805	288	0	283	1,376
Emilia-Romagna	2,875	4,152	0	7,375	14,402
Tuscany	2,185	960	0	1,746	4,891
Marche	1,334	600	0	1,073	3,007
Umbria	575	240	0	332	1,147
Latium	7,634	3,120	450	2,732	13,936
Abruzzo	780	600	1,050	1,538	3,968
Molise	85	54	20	427	586
Campania	8,520	3,040	90	4,355	16,005
Apulia	1,950	2,920	13,000	7,385	25,255
Basilicata	563	1,020	1,160	1,285	4,028
Calabria	713	360	200	1,946	3,219
Sicily	18,166	3,370	9,000	4,054	34,590
Sardinia	500	190	30	2,261	2,981
TOTAL (tons)	**54,675**	**29,350**	**25,000**	**43,393**	**152,418**

Table 1. Estimation of the agricultural plastic film (tons) used in the Italian Regions in Year 2002.

Film or net for the crop protection against meteorological agents, virus-vector insects and birds are used as standalone covers or in connection with typical structures for the growing of arboreal cultivation (e.g. vineyard or kiwi) as in Southern Italy, where the well-known technique of "tendone" growing is widely diffused. These large protected agricultural structures with continuous coverage can be studied (Picuno & Scarascia-Mugnozza, 1990) with reference to three different levels of analysis of their influence on the rural environment and the agricultural landscape:

- micro-scale: the influence on the microclimate of the air contained inside the envelop realised by the cover may modify environmental parameters, like temperature, relative humidity, carbon dioxide level, etc;
- meso-scale: the modification in the distribution of pollen, insects, birds, etc. may influence the global characteristics of the agro-ecosystem;
- macro-scale: the visual quality of rural landscape may be significantly altered by a heavy diffusion of artificial coverings.

The use of plastic covers sets significant problems connected with the effect that large extensions of agricultural land covered with continuous cladding material may determine on the visual perception of the rural landscape (Scarascia-Mugnozza et al., 1999), with significant landscape variations and negative consequences on the rural environment. In the present study both an application of remote sensing techniques for the detection of wide protected cultivation and a corresponding analysis enabling the realisation of a virtual model of a part of rural land through the suitable manipulation of the photographic image of the study area were conducted. The aesthetic impact generated by the plastic coverings was therefore evaluated, based on these results. On the same virtual model, moreover, it is possible to include further elements that could be successively introduced in the rural land, such as new tensile structures for arboreal cultivation covered with plastic sheets. The assessment of design parameters, like geometry, material and colour, able to minimize their aesthetic and environmental impact becomes therefore possible.

2.2 Materials and methods

The landscape analysis has been conducted in an area of Southern Italy, where structures for fruit farming covered with plastic net or film are widely employed.

2.2.1 Study area

The study area is located near the coastal border between Basilicata and Apulia Regions (Southern Italy, close to the Jonian Sea). The total surface of the study area covers about 3080 km^2, including the Jonian Coast from Metaponto to Taranto (40°39′41″ N; 16°58′15″ E; center area), the city of Matera at west, the town of Casamassima at North, and the town of Alberobello at East (Fig. 1). This area is characterized by a strong agricultural vocation, due to the wide diffusion of vineyards grown using the traditional "*tendone*" technique, a grape cultivation system similar to the pergola, with a supporting structure that may be covered with a plastic sheet. Other arboreal cultivation (i.e., kiwi, olive and fruit) and horticultural crops are widely diffused in this area too. Due to the need to limit numerical calculation, a smaller zone (about 100 hectares) inside this Study Area was selected in order to perform there the three-dimensional virtual simulations.

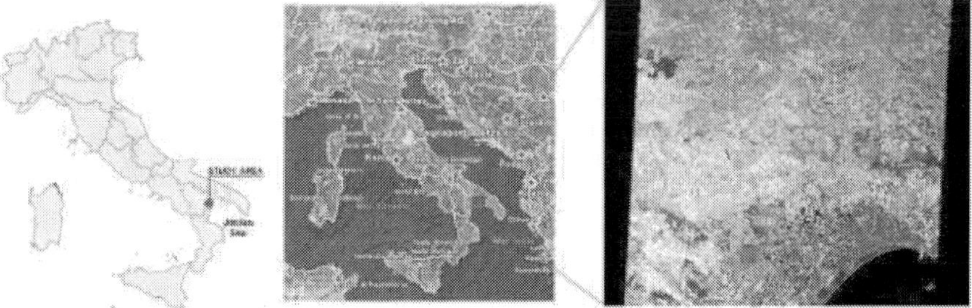

Fig. 1. The Study Area with the indication of the restricted area where the virtual solid modelling was performed.

2.2.2 Remote sensing of protected cultivation

The detection of the land use, with specific reference to greenhouse and other protected structures, has been obtained by Sabins (1996) using Thematic Mapping of multi-temporal Landsat images with Unsupervised classification, image processing, vectorialization and G.I.S. analysis. For the validation of the Landsat classification, a new approach using some SAR (Synthetic Aperture Radar) image (ERS 1 and ERS 2 satellite) was studied and experienced. In this study the multitemporal Landsat images used were: Year 1990; Year 1992; Year 1994; Year 1998; Year 2000.

The Landsat images were geographically corrected through an Envi's Registration procedure ("Image to Image", 20 Ground Control Point for each data). The Landsat data have been used for a correct vision of the band 7 (2.08 - 2.35 µm), band 5 (1.55 - 1.75 µm), and band 3 (0.63 – 0.69 µm) associated with other cartographic information and radiometric and thermal properties of greenhouse covering materials (Papadakis et al., 2000). In Fig.2a the three RGB visible bands referred to one of the images that were used for the experimental analysis are reported. The components in the infrared band reflective are showed, in Fig. 2b, by using false colour with reference to the same image as in Fig.2a. The difference between the two categories of coverage (greenhouse in blue and urbanized in purple) is here more evident. This aspect demonstrates the utility to have multispectral images with infrared components. The best combination to highlight this contrast on the display is the following: the Red colour represent the Band 7 (Spectral Resolution 2.08 - 2.35 µm), Green colour the Band 5 (Spectral Resolution 1.55 - 1.75 µm) and in Blue colour the Band 3 (Spectral Resolution 0.63 - 0.69 µm, visible red light). The Band 3, takes into account absorbed radiation of the chlorophyll that it can be helpful to highlight covers with net may be wrongly considered as a plastic film if observed in real colours (Jensen, 2005). Before carrying out the analysis and classification, it was necessary to monitor the field matching remote sensed information. In particular the following types of areas were investigated: wasteland; built environment; arable land; vegetation; plasticulture. Considering the agricultural activities inside the study area, in the category "plasticulture" both greenhouse and vineyard covered for advancing or postponing crop yield were included.

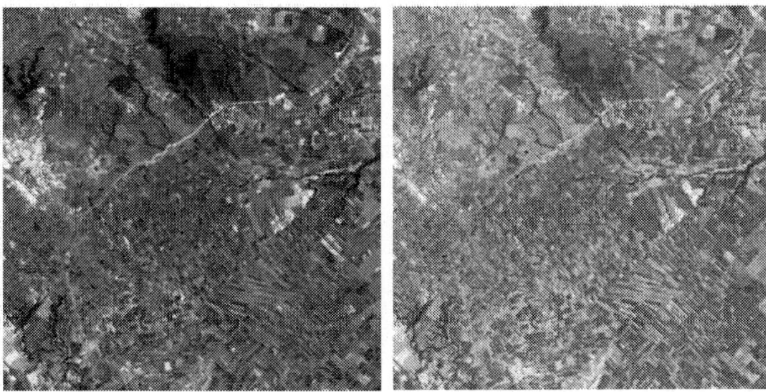

Fig. 2a. Landsat Image: True colours. Fig. 2b. Landsat Image: False colours (RGB: band 7,5,3).

The method for the acquisition of "spectral signatures" of different land use (Fig. 3) started at a stage when achieved knowledge on training areas is spent to know the spectral response of the different types of coverage. "Spectral signatures" are formed by the various statistical parameters (mean, covariance, etc.) related to the values of spectral components corresponding to the types of coverage of interest.

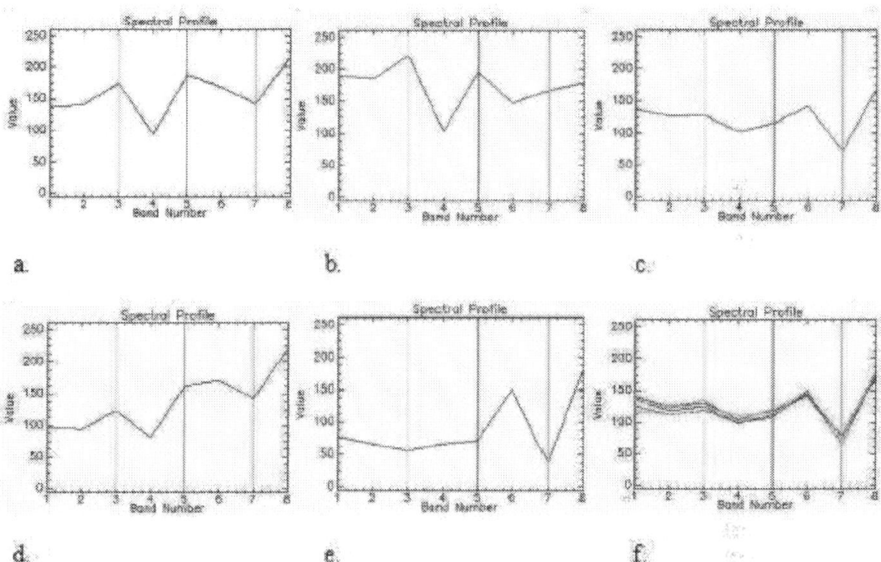

Fig. 3. Spectral signature of: a. wasteland; b built environment, c. greenhouse; d. arable land; e. vegetation; f. plasticulture. (Y-axis in a conventional RGB value).

In the present case Supervised classification techniques were used in order to classify, as a first attempt, the different land uses in ENVI Software. In Supervised classification, spectral signatures are developed starting from specific locations in the image. These specific locations are known as ROI (Region Of Interest), and they are defined by the user. The mathematical procedure assigns a label to each pixel in the image, according to their similarity to the class statistical signature. There are many different decision rules which can be applied in a Supervised classification methodology (Tso & Mather, 2009). In the present study the Parallelepiped method has been used. Parallelepiped classification uses a simple decision rule in order to classify multispectral data.

The decision boundaries form an adimensional parallelepiped classification in the image data space. The dimensions of the parallelepiped classification are defined based upon a standard deviation threshold from the mean of each selected class. If a pixel value lies above the low threshold and below the high threshold for all n bands being classified, it is assigned to that class. If the pixel value falls in multiple classes, ENVI assigns the pixel to the last class matched. The Maximum Standard deviation from the Mean used in the present work was 1,3 (expressed in the conventional RGB scale).

For the validation of the Landsat classification, in the present work a new approach using some SAR (Synthetic Aperture Radar) image (ERS 1 / ERS 2 satellite) was also studied and applyed with negative results. The data inferred by the classifications were then compared with SAR ERS PRI Type, ESA I-PAF images. The ERS-1, 2 satellites are devoted to global measurements of sea wind and waves, ocean and ice monitoring, coastal studies and land sensing using active and passive microwave remote sensing systems.

ERS-1 was launched in July 1991 and ERS-2 in April 1995. ERS-1 uses a synthetic aperture radar (SAR), an instrument able to acquire images of ocean, ice and land regardless of cloud and sunlight conditions. Unlike optical images, radar images are formed by coherent interaction of the transmitted microwave with the targets. Hence, it suffers from the effects of speckle noise which arises from coherent summation of the signals scattered from ground scatterers distributed randomly within each pixel. A radar image appears more noisy than an optical image. The speckle noise is sometimes suppressed by applying a speckle removal filter on the digital image before its display and further analysis. Single radar image is usually displayed as a grey scale image, such as the one shown in Figure 4.

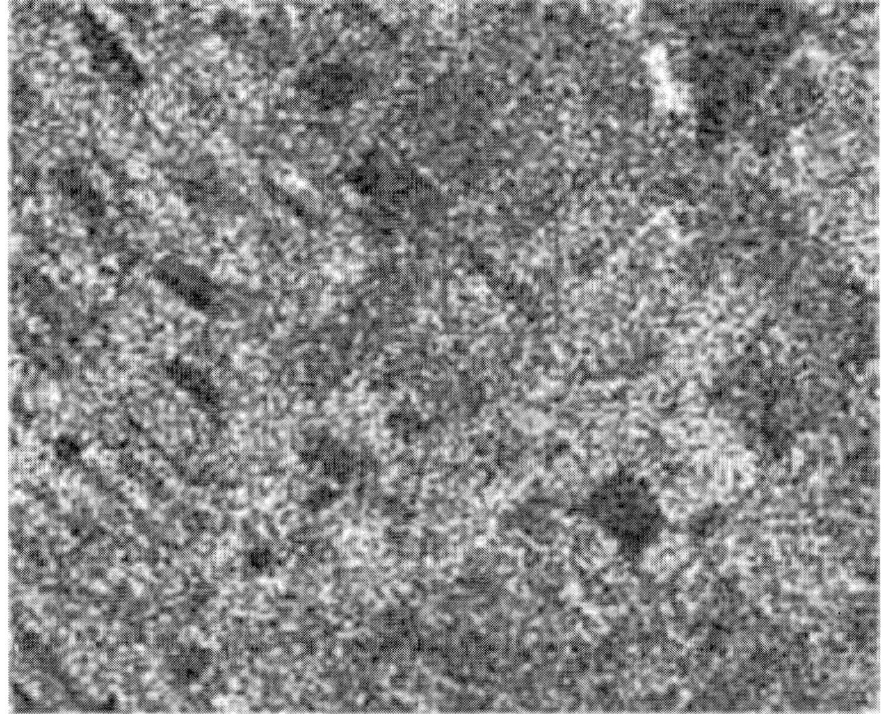

Fig. 4. SAR Image of the study area.

The intensity of each pixel represents the proportion of microwave backscattered from that area on the ground, which depends on a variety of factors: types size, shape and orientation of the scatterers in the target area; moisture content of the target area; frequency and polarisation of the radar pulses; incident angles of the radar beam.

Interpreting a radar image is not a straightforward task. It very often requires some familiarity with the ground conditions of the imaged areas. As a useful rule of thumb, the higher the backscattered intensity, the rougher is the surface being imaged. Flat surfaces such as paved roads, runways or calm water normally appear as dark areas in a radar image since most of the incident radar pulses are specularly reflected away. A more specific analysis was conducted on plastic coverings, but this analysis showed that these areas don't have a really significant interaction with the electromagnetic waves generated by the satellite ERS (c Band, Frequency 5.3 GHz, Wavelenght 6 cm). The plastic material is permeable to the electromagnetic waves so causing, as the only response, a reduction of real humidity detectable by the SAR. This phenomenon does not, therefore, support the use of radar for this type of investigation. After classification, the images obtained in GeoTiff, were imported and processed using a G.I.S. software (GeoMedia - Intergraph Corporation). The image has undergone a process of transformation and subsequent GRID elaboration in vector format (shp). With the Geoprocessing's "Dissolve" procedure it was possible to aggregate features based on specified attributes, then representing only shell plastic cover over the study area.

The result of the analysis of radar images is that you can are not applicable for this type of analysis.

2.2.3 Landscape impact simulations

The evaluation of the landscape impact produced by the tendone structures located in the area of study has been possible by implementing, through a G.I.S. procedure, a three-dimensional land modelling with the overlap of some photographic images where protected crops were clearly detectable. The Digital Terrain Model (DTM) was produced using the numerical cartography and extrapolating the graphic elements (contour and spot heights) that characterize the land altimetry. The individuation on the study area of the crops covered with plastic film or net was performed, followed by their location on the aero-photographic support (Scarascia-Mugnozza et al., 1999).This land three-dimensional virtual model is representing the real situation, proposed through the virtual photographic image of the study area (Fig. 5).

New plastic-covered structures were then inserted in the virtual model through solid extrusion, according to the crops that are currently grown in the study area and that would be potentially covered for an increase of the crop production. All data DTM, aerial images, tensile structures and extruded cover were analyzed through specific G.I.S. extensions, in order to visualize the virtual model of the study area with the agricultural structures. The cover was modelled using a specific "off-set" procedure, considering the height of the tensile structure equal to 3.0 m. The structure was modelled in a clear-grey (grey 20%) colour, considering the natural colour for this material (the posts are usually concrete-made). Regarding the covering material, in order to reproduce the hatch of the aerial image, a specific colour scale was implemented.

Fig. 5. Three-dimensional visualization of the restricted study area.

Among different conventional colour scales (i.e., RGB for PC monitor; CMY and CMYK for printer; RAL scale for paints; PANTONE, CIE and CIELAB for different use), considering the variability of the material colour, a colorimeter (Portable Colorimeter Minolta CR300) using the CIELAB scale for measuring the colour was finally selected. The scale colour used by the CIELAB colorimeter (Garcìa et al., 2003) was then converted in RGB value using the OpenRGB (Logicol 2008 version 2.10.91215), available from http://www.easyrgb.com/index.html.

The hatch of the structure was generated using the tendone real geometric characteristics, both for colour and for texture. The following cladding materials were considered for virtual simulation:

- UTILITY PRO 2.5″. Anti-hail HDPE net, colour black or white, produced by TENAX (Fig. 6a).
- OMBRAVERDE 50″. Shading net, warp in green colour, weft in black colour, produced by ARRIGONI (Fig. 6b).

The shading factor plays a central role in the virtual model; this has been implemented as the factor of transparency by calculating the inverse of the shading factor, obtained by the technical leaflet of the material (e.g., in case of UTILITY PRO 2.5, that has a shading factor of 6-12%, the factor of transparency was 91%).

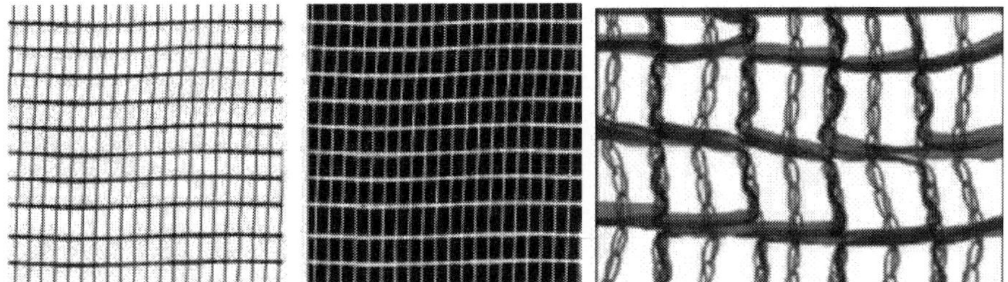

Fig. 6a. UTILITY PRO 2.5 black and white net. 6b. OMBRAVERDE 50 net.

2.3 Results and discussion

Whit satellite remote sensing analysis it was possible to obtain different multi-temporal thematic maps of surfaces covered with plastic film inside the study area. Comparing different chronological information, it is possible to deduce a sharp increase in the diffusion of protected crops in the study area, from about 65 Ha in Year 1990 to more than 680 Ha in Year 2000. In Figure 7 this trend is graphically shown, together with the formulation of a suitable interpolation equation. Most of the changes took place along the Jonian.

The impacts on agriculture of this wide use of plastics were evaluated using the method of Sorenson (CIE Cause/Impact/Effect), by filling the matrix of Leopold in simplified form through the application of coaxial matrices (Dal Sasso et al., 2007).

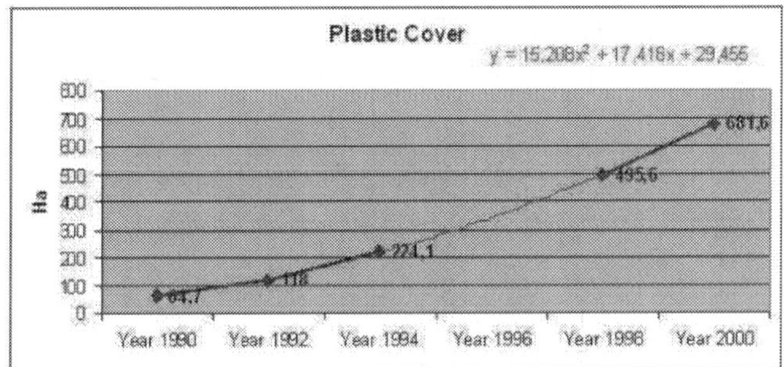

Fig. 7. Development trend of plastic-covered areas.

This method puts in relation and in logical sequence respectively the planning actions, the causal factors, the environmental components and mitigations, enabling a quick identification of the main impacts of the plastic covers and the evaluation of their effects on the agro-environmental system. In this way, a threshold range was identified as the value whose produced impacts are unacceptable, that was obtained through the calculation of the number of boxes resulting form the application of the matrix (Tab. 2, Tab.3, Tab. 4).

Assuming the actual status quo (i.e. a surface covered with plastic equal to ~ 3.3% of the total study area) as the unit value, the impact values corresponding to an increasing percentage of plastic cover were estimated: through an interpolation curve of data obtained from a coaxial matrices, was calculated as the threshold value, equal 10%, as a maximum percentage value of the plastic cover above which the impacts on the agricultural landscape should be considered unacceptable.

PROJECT ACTIONS	Employment Land	Water Drawing	Removal of Natural Resources	Liquid Discharges	Solid Waste	Visual Obstruction	Atmospheric Emissions	Induced Traffic	Interference with the Flow of Water	Compaction and Soil Proofing	TOTAL
Supporting Structure						•	•	•		•	4
Coverage's Type			•		•	•		•	•		5
Irrigation or Fertigation		•	•	•					•		4
Shielding and Shadowing			•			•					2
Phytosanitary Treatments		•	•	•			•				4
Fertilization			•		•						2
Working Land			•					•	•		3
Transplanting and Harvesting					•			•			2
Maintenance			•		•	•		•			4
TOTAL	0	2	7	2	4	4	2	5	3	1	30
CAUSAL FACTORS											TOTAL

Table 2. Coaxial matrices: Project Actions/Causal Factors

Employment Land	Water Drawing	Removal of Natural Resources	Liquid Discharges	Solid Waste	Visual Obstruction	Atmospheric Emissions	Induced Traffic	Interference with the Flow of Water	Compaction and Soil Proofing	TOTAL	ENVIRONMENTAL COMPONENTS
				•		•	•			3	Atmosphere
	•		•					•		3	Acquatic Environment (Surface and Groundwater)
•	•	•	•	•				•	•	7	Soil and Subsoil
•			•	•			•			4	Vegetation and Flora
•			•	•			•			4	Fauna
•			•	•		•		•		5	Ecosystems
			•	•		•				3	Public Health
							•			1	Noise and Vibrations
										0	Ionizing and not Ionizing Radiations
•				•	•	•				4	Landscape
•					•	•				3	Cultural and Historical Heritage
		•				•	•			3	Socio-Economic Context
6	2	2	6	7	2	6	5	3	1	40	TOTAL

Table 3. Coaxial matrices: Causal Factors/Environmental Components

ENVIRONMENTAL COMPONENTS	Compromising Ecosystems	Pollution of Surface Water	Groundwater Pollution	Reduction of Water Potential	Landscape-Environmental Degradation	Damage of Vegetation and Fauna	Atmospheric Pollution	Thermal Pollution	Waste and Scrap Outputs	Removal of Natural Resources	Health Risks	Socio-Economic Impact	Traffic Alteration	Loss of Open Spaces	TOTAL
Atmosphere	•						•						•		3
Acquatic Environment (Surface and Groundwater)	•	•	•	•						•				•	6
Soil and Subsoil	•				•				•	•					4
Vegetation and Flora	•	•			•	•				•			•		6
Fauna	•	•			•	•	•		•	•				•	8
Ecosystems	•	•			•		•	•	•			•	•	•	9
Public Health	•	•	•				•				•	•	•		7
Noise and Vibrations													•		1
Ionizing and not Ionizing Radiations															0
Landscape	•	•			•		•			•				•	6
Cultural and Historical Heritage					•									•	2
Socio-Economic Context		•			•							•	•		4
TOTAL	8	7	2	1	7	2	5	1	3	5	1	3	6	5	56

IMPACTS

Table 4. Coaxial matrices: Environmental Components/Impacts

The corresponding solid modelling simulation are shown in figures 8-9-10-11. Figure 8 shows the solid modelling of the actual situation (3.3% of the total analyzed land surface), whereas the other pictures represent new protected structures, inserted on the basis of the hypothesis that in future some new crops, actually cultivated in this area, would be covered with a plastic material (anti-hail, shading, insect net or film, etc.) with covering percentages equal to 7% and 10% of the total surface (Picuno, 2005).

For each one of these virtual situations, the use of different covering plastic nets was also considered and the corresponding impacts on the visual quality of the agricultural landscape were evaluated.

Moving inside the model it has been therefore possible to evaluate the effects determined on the aesthetic properties produced by the structures covered with plastic material on the surrounding landscape, with a notable influence on the visual perception for an observer of the landscape (Fig. 12).

The use of coloured nets registered recently an increased spreading, thanks to their special radiometric characteristics able to specifically filter some components of the solar radiation. The consequent alteration of the spectrum of bright colours using very intense colorations (as the fiery red, electric blue, gold yellow, etc.) determines special positive effects on crop growth such as dwarfism or, conversely, gigantism effect. The use of such nets would further change the visual perception of the agricultural landscape (Fig. 13).

Fig. 8. Solid modelling of the study area in the actual real situation - i.e. about 3.3% of the surface covered with plastic net clear- gray (gray 20%) colour.

Fig. 9. Virtual solid modelling of the study area for different percentages of surface (3.3%, 7% and 10%) covered with plastic net OMBRAVERDE 50.

Fig. 10. Virtual solid modelling of the study area for different percentages of surface (3.3%, 7% and 10%) covered with plastic net UTILITY PRO 2.5 white.

Fig. 11. Virtual solid modelling of the study area for different percentages of surface (3.3%, 7% and 10%) covered with plastic net UTILITY PRO 2.5 black.

Fig. 12. Detail of the three-dimensional virtual view of the study area covered with 10% of plastic material.

Fig. 13. Thee-dimensional simulation with three different ChromatiNet.coloured-nets.

3. Historical cartography and G.I.S. for the analysis of carbon balance in rural environment: A study case in Southern Italy

3.1 Foreword

An environmental analysis conducted by a G.I.S. approach shows how positive results of the applied agronomic practises, in terms of CO_2 fixation, could be able to contrast heavy emissions of greenhouse effect gases in the atmosphere by urban settlements. Using a geographic information system applied to historical maps in order to assess the environmental impact of land use transformation, with a special emphasis on the atmospheric carbon dioxide balance (Tortora et al., 2002). The analysis was focused on the transformation of a rural area in Southern Italy along the last 138 years, due to the change of land use through the introduction of corn and fruit orchards increasingly substituting olive trees and forested surfaces.

3.2 Materials and methods

The study area has been chosen, among other ecosystemically homogeneous regions, due to the availability of historical cartography starting from Year 1859, in order to assess land use change among a temporally long enough period.

This area, covering about 1500 ha, is located in Southern Italy, mainly in Bernalda e Pisticci municipalities of the Basilicata Region, nearby the Ionian Sea coastline (Fig.14). The thermoregulation effect of the sea, the aspect, the soil structure and the socio-economic context determined and still are determining the vegetation species pattern of this area.

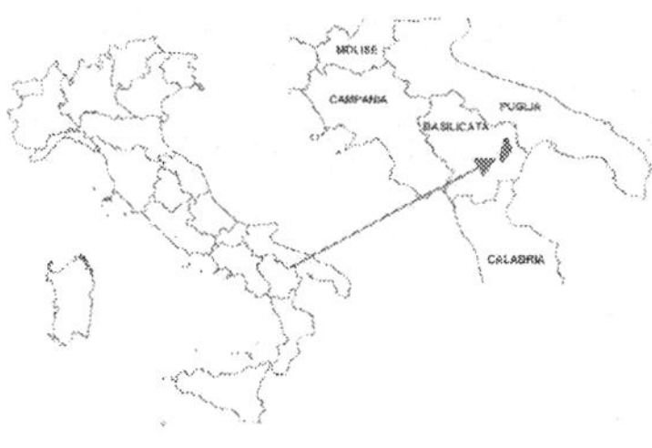

Fig. 14. Study area

The morphology is mainly constituted by low hills, suitable for corn and olive production. The spread of mechanization technology for crop cultivation and harvesting together with, especially in the late '800, the development of industrial wheat transformation techniques serving the pasta and bakery products industry (Dal Sasso & Picuno, 1996), caused remarkable changes in the whole land asset.

3.2.1 Cartography

Land use change in the study area was examined over four different time periods: Years 1859, 1873, 1957 and 1997. The geographical information dated Year 1873 and 1957 was collected, in a 1:50.000 and 1:25.000 scale, using the historical maps of the Italian Geographic Military Institute (IGMI). Digital orthophotos - dated 1997 - were used together with a

1:5.000 scale technical map of Bernalda area obtained from a recent aerial photogrammetric survey.

The older time level (Year 1859) has been analyzed using an antique map created on Marquis Peres de Navarrete's demand (Fig. 15). This map represents most of Bernalda municipality land, especially the northern zone above the built-up area, within the borders of adjacent communal areas, describing the rural landscape of that time. It constitutes a complete and consistent cartographic support, endowed with planimetric coordinates and enriched with thematic information about the land use at that time.

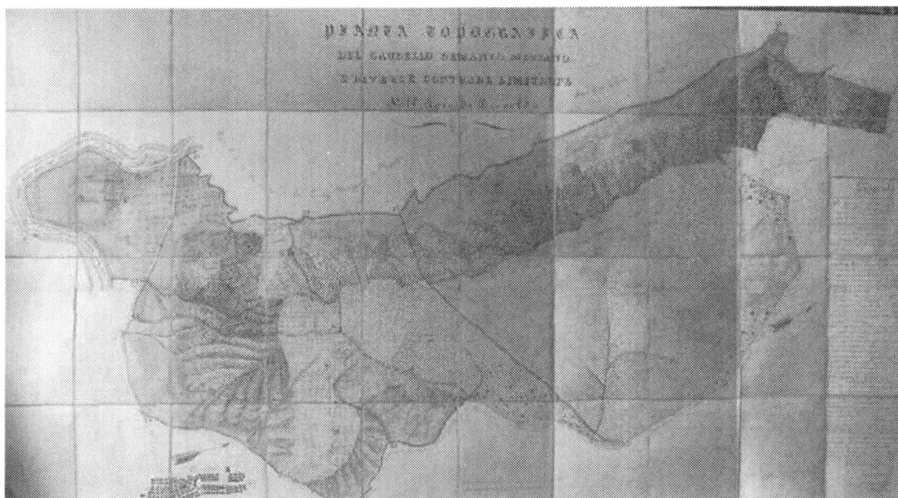

Fig. 15. Historical Cartography (Year 1859) of the Study Area

Land use maps, chronologically based on different periods, have been obtained by the interpretation of the different cartographic supports recovered; every layer has been then classified in 9 classes: woodland, shrubland, arable land, orchards, olive groves, vineyard, meadows, urban, wasteland.

1997 land use classification was based on the interpretation of recent photographs; field data were collected, in order to obtain more accurate results and to localize the coordinates of control points, with the use of a GPS Mod. Garmin IIIplus.

Data retrieved have been implemented in a land information database and processed in a multi-temporal G.I.S. (GeoMedia - Intergaph) where each Year has been considered as an homogenous chronological layer. In this way, all the information about land orography and use were organized on four different layers (corresponding to Years 1859, 1873, 1957 and 1997), constitued by data characterized by the same time level. Then, through spatial overlay, the processed input layers sequence resulted in output vegetation temporal dynamism data.

Finally, two temporally different maps, one dated 1879 and the other being the contemporaneous one, have been also 3-D digitalized and processed, in order to analyse the different elevation attributes and vegetation cover condition between the two considered periods: morphological changes are strictly connected to the evolution of vegetation covers and, consequently, to different soil protection.

3.2.2 Image processing

In order to input the historical cartography, the maps were geo-referenced through a sequence of rectification and referencing procedures; especially for the iconographic map dated 1859, control points on the map at known locations were located and projected on the modern overlaid maps (Capobianco et al., 2004). The geographic framework has been achieved using georeferencing and pixel resampling tools through an affine transformation six-parameter dependent (Della Maggiore et al., 2002).

Using spatial analysis functions the map of Year 1859 was appropriately correlated to altimetry (DEM) of Year 1873; therefore, the land use here reported was associated with the visualization of land use as in the present orto-photos, so obtaining the historic reconstruction of Year 1859 landscape. This enabled an evaluation of the aesthetic changes of the study area in terms of both morphologic and vegetation variation of the agro-forestry landscape. Figure 16 shows three-dimensional reconstruction of 1859 landscape compared with the actual situation.

Fig. 16. Comparison of three-dimensionally reconstructed landscape (1859 – 1997)

3.3 Results and discussion

3.3.1 Land use

Figure 17 represents land use maps obtained for the four time periods produced on the basis of the thematic information contained in the historical maps that were retrieved. Based on polygonal topology, these maps represent for each data the state of landscape; the comparison obtained through a crossed over interpretation of the output maps has enabled the analysis of land changes from 1859 to present days, covering a time period of 138 years, giving information on historic persistence of soil use typologies along with their time-driven modifications. Dominant soil use typologies of the site have been grouped in order to better compare output data through a more evident highlighting of variations in time. Visualizing data in graphs (Graph 1) we can observe an increase of crop growing and a reduction of forested surfaces, that have now almost disappeared leaving their place to urban areas and fruit orchards.

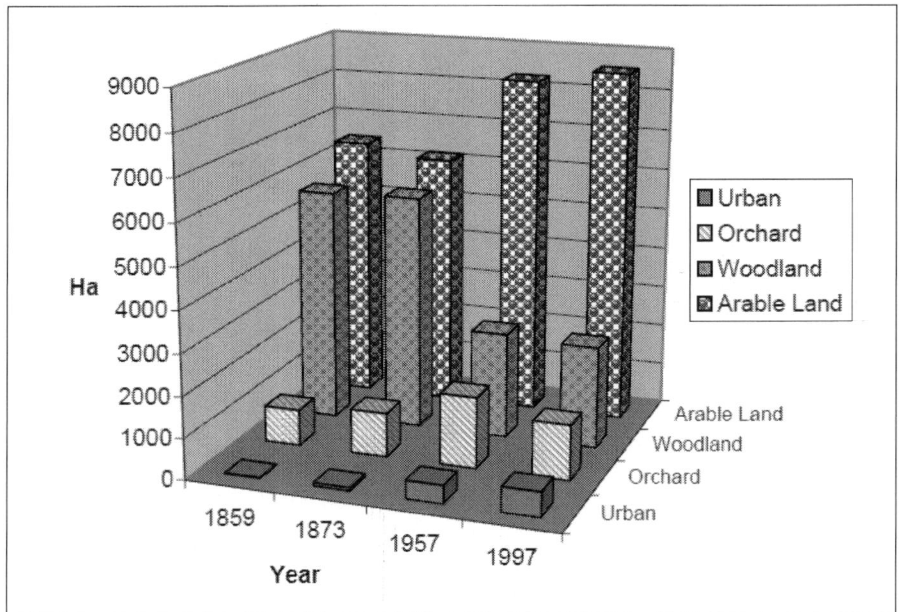

Graph 1. Evolution of principal land use over the four time periods

Historic dynamism of vegetation and morphology has evolved together with the effects of technology on cartographic production quality; this aspect is testified by the characteristics of the historical planimetric cartography of Year 1859, almost an iconography with very low metric precision, arriving to modern cartography, in paper or digital format, that has greater metric precision together with more accurate information.

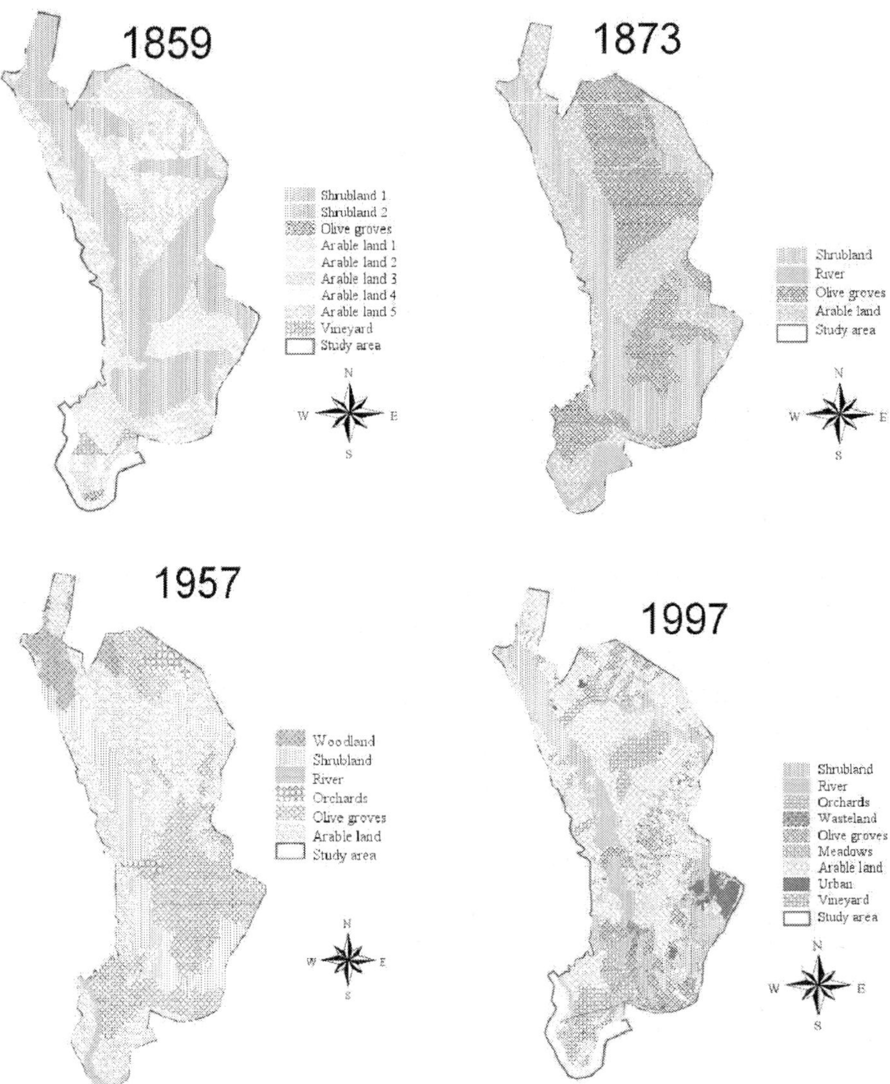

Fig. 17. Thematical land use maps for the four different time periods

3.3.2 Carbon dioxide balance

With the aim to quantify the effect of land use changes on the environment, with special emphasis on air quality, we estimated the CO_2 time variation connected with the use of the crops (Woodland, Shrubland, Arable land, Orchards, Olive groves and Urban) in the study area reported in the different chronological informative levels (Years 1859, 1873, 1957 and 1997).

CO_2 sequestration rates were calculated through adopting the user-friendly CO2FIX V.2 model (Masera et al., 2003), tool for the dynamic estimation of the carbon sequestration potential of forest management, agroforesty and afforestation projects. CO2FIX V.2 is a multi-cohort ecosystem-level model based on carbon accounting of forest stands, including forest biomass, soils and products. Carbon stored in living biomass is estimated with a forest cohort model that allows for competition, natural mortality, logging, and mortality due to logging damage. Soil carbon is modeled using five stock pools, three for litter and two for humus. The dynamics of carbon stored in wood products is simulated with a set of pools for short, medium and long lived products, and includes processing efficiency, re-use of by-products, recycling, and disposal forms. The CO2FIX V.2 model estimates total carbon balance of alternative management regimes in both even and uneven aged forests, and thus has a wide applicability for both temperate and tropical conditions. (Masera et al., 2003). The CO2FIX model was developed as part of the "Carbon sequestration in afforestation and sustainable forest management" (CASFOR) project, which was funded by the European Union INCO-DC program. (Mohren et al., 1999).

CO2FIX V 2.0 is a carbon book-keeping model that simulates stocks and fluxes of carbon in (the trees of) a forest ecosystem, the soil, and (in case of a managed forest), the wood products. It simulates these stocks and fluxes at the hectare scale with time steps of one year. For an extensive description of carbon dynamics in forest ecosystems, and the role of forests in the global carbon cycle see Kauppi et al. (2001). Some of the results of CO2FIX have been used in the IPCC 1995 climate change assessment.

In order to initialise the model, different analysis parameters were used. The assumptions that were made were consistent with the software input characteristics (Mohren et al., 1999) and the local area characteristics (Capobianco et al., 2004). For forestry area the following characteristics were used: tree species, area, age, dominant height, standing volume, growth class and the coordinate of the stand.

Woodland in the study area is represented by highly degraded coppice forest with prevailing Quercus ilex. Rotation length is 30 years with maximum biomass in the stand equal to 130 Mg/ha. The allocation factor for foliage, branches and root production were copied from existing CO2FIX runs for comparable species. The turnover (annual rate of mortality of the biomass component) was evaluated in 0.3 for foliage, 0.06 for branches and 0.05 for roots. The soil organic matter compartment consists of dead wood, litter layers and stable humus in the soil. On the basis of this analysis, a total carbon stock ranging from 17 to 70 Mg/ha and an average atmospheric carbon sequestration approximately equal to 4.40 MgC/ha/yr were estimated.

In the study area, the orchard areas are generally orange groves with rare presence of apricot trees. For the purpose of CO_2 calculation, the orchard area was compared to coppice forest with a rotation of 20 years and periodical removal of organic matter through agronomic practices like pruning, comparable to a turnover (annual rate of mortality of the biomass component) of 0.3 for foliage, 0.07 for branches and 0.04 for roots. In an orchard, carbon balance depends on the intrinsic structural and morphological characteristics of each species and it is also influenced by population density, rearing system, and especially on the canopy and aboveground and underground woody organisms. Moreover, in the case of young plantation, canopy has to provide for a relatively small amount of branches and roots and, consequently, primary production is net and the surplus of organic matter increases every year up to maturity when dry matter increases over time and subsequently tends to zero (Xiloyannis et al., 2005). Based on such a principle, it is possible to estimate the average yearly sequestration of atmospheric carbon as being equal to 7.25 MgC/ha/yr for the orchards, to 2.75 MgC/ha/yr for shrubland and to 3.6 MgC/ha/yr for arable land. On the other hand, urban areas represent a source of CO_2 emission from both municipal and industrial combustion; a yearly amount of 15.0 MgC/ha/yr of CO_2 release into the atmosphere was therefore estimated on the basis of a report on the environmental state of Basilicata Region (AA.VV., 2000). All the above-mentioned values of average atmospheric carbon sequestration were adopted for each one of the four time periods (Years 1859, 1873, 1957 and 1997). The data resulting from the implementation of the G.I.S. gave the values reported in Table 5 expressed in terms of areas occupied by the different vegetation typologies and, applying their respective CO_2 sequestration rates, in terms of absolute values of annual sequestration of CO_2. The balance of CO_2 does not include the effects of the agricoltural machinery, supplies and transportation on CO_2: in woodland these factors are almost absent, while in case of orchard and arable land they depend strongly by crop techniques, and in some cases are negligible.

Year	Woodland		Orchard		Arable land		Urban		Total	
	Area [Ha]	Annual balance of CO_2 [MgC/ha/yr]	Area [Ha]	Annual balance of CO_2 [MgC/ha/yr]	Area [Ha]	Annual balance of CO_2 [MgC/ha/yr]	Area [Ha]	Annual balance of CO_2 [MgC/ha/yr]	Area [Ha]	Annual balance of CO_2 [MgC/ha/yr]
1859	5638	24807	890	6452	6440	23184	17	-255	12985	54188
1873	5691	25040	1080	7830	6153	22150	79	-1185	13003	53835
1957	2555	11242	1731	12549	8270	29772	438	-6570	12994	46993
1997	2474	10885	1339	9707	8582	30895	598	-8970	12993	42517

Table 5. Annual balance of CO_2 in the study area.

Examining Table 5, in the investigated scheme it is clear that the greatest changes in land use occurred after the establishment of large orchard grown areas and mainly consisting of orange tree plantations. The percentage rise in arable land was equally considerable with increases as high as 30-40 %, to the detriment of woodland and shrubland. Olive groove reached its peak in the late 19th century until after the First World War, since it was one of the early livelihood sources of farm families at that time. As a result of the different performance in terms of CO_2 fixation and relative to the investigated study area, all these land changes caused progressive decrease in carbon dioxide sequestered by biotic agents embedded in the soil. We can argue that the sequestration of land carbon in Year 1859 was higher than in more recent periods, and that during time the land carbon balance worsened: the cultivation conversion occured during time caused a constant loss of CO_2 fixation value (Mohren et al., 1999), while heavy emission of greenhouse effect gas in the atmosphere by urban settlements were at the same time increasingly growing.

This pattern could be considered a typical situation also for many other areas located in Southern Italy or even elsewhere, and this approach seems that could be considered as a useful tool for the planning and management of rural landscape and environment: the study case showed that a sound planning in agricultural activities could significantly contrast the release in the atmosphere of CO_2 deriving from the diffusion of anthropic activities.

4. The optimization of the management of agricultural plastic waste in Italy using a geographical information system

4.1 Foreword

The extensive and expanding use of plastic material in the Italian agriculture for several diversified application (e.g., tunnel and greenhouse covering film, mulching film, silage bag, irrigation pipe, agrochemical containers, etc.), results in increased accumulation of plastic waste in rural areas (Scarascia-Mugnozza, 1995). It constitutes an environmental and economic problem (Sica, 2000), because agricultural plastic waste, if abandoned along rivers and/or in rural areas, may cause severe damages for the landscape environment, agricultural soil, air, shallow and deep water (Picuno & Scarascia-Mugnozza, 1994). If burnt in open and uncontrolled sites it may damage the human health because harmful substances could be released during the combustion: due to inefficiencies of open combustion, emissions from open burning are much greater per mass of material burned than emissions from controlled incineration (e.g., 20 times as for dioxin; 40 times as for particulate matter - Travis & Nixon, 1991). Plastic incineration produces anyway large CO_2 emissions (about 3 Kg of CO_2 per Kg of Polyethylene). Unfortunately, the abandonment and burning are practices still frequently in use in Italy, although they are against the law, while only a part of agricultural plastic waste is collected and recovered in a controlled way by the National Consortium "PolieCo", that has the task to collect, transport and direct them toward the final disposal, that is the mechanical recycling. In the year 2010, the Consortium recycled over 350,000 tons of post-consume PE (over the 35 % of the PE articles placed on the market) (PolieCo, 2010); they include APW, but it is very difficult to quantify them. The mismanagement of the agricultural plastic packaging waste (APPW) creates more acute problems because: they are not always properly cleaned before being disposed and the APPW management scheme has been established yet in many Countries. In order to analyze the current situation, to estimate APPW streams (quantity, temporal and spatial distribution, etc.) and existing technologies (specific disposal solutions applied) a Program called "Design of a common agrochemical plastic packaging waste management scheme to protect natural resources in synergy with agricultural plastic waste valorisation (acronym AGROCHEPACK), started in the 2010, under the frame of a Transnational Cooperation Programme Mediterranean (MED). The Program is carried out by Picuno & Sica , of the DITEC Dept., and other Project Partners of Italy, Greece, Cyprus, France and Spain. Fortunately, farmers are becoming more aware of two important problems: direct damage to the environment and reduction of non-renewable resources. Farmers must follow sound procedure for the collection and disposal of the agricultural plastic waste because they are "secondary raw material".

Since Geographical information systems are currently employed in order to optimize the flux of materials and goods, a Geographical Information System (G.I.S.) ad hoc designed may reveal as a tool suitable for the management both of the rural land and the agricultural plastic waste flux (Scarascia-Mugnozza et al., 2006). In fact, many studies demonstrate the applicability of G.I.S. in the agricultural sector. A G.I.S. optimal routing model was proposed (Ghose et al., 2006) to determine the minimum cost/distance efficient collection paths for transporting solid wastes to the landfill. The model can be used as a decision support tool by municipal authorities for efficient management of the daily operations for transporting solid wastes, load balancing within vehicles, managing fuel consumption and generating work schedules for the workers and vehicles.

The disposal in landfill is the waste destination method with the largest demand for land, while land is a resource whose availability has been decreasing. Shortage of land for landfills is a problem frequently cited in the literature as a physical constraint. Leao et al. (2001) presented a method to quantify the relationship between the demand and supply of suitable land for waste disposal over time using a G.I.S. and modelling techniques. Based on projections of population growth, urban sprawl and waste generation the method can allow policy and decision-makers to measure the dimension of the problem of shortage of land into the future. The procedure can provide information to guide the design of programs to reduce and recover waste, leading to a better use of the land resource.

Basnet et al. (2002) developed a G.I.S.-based manure application plan for the specific application of animal waste to agricultural fields. Sites suitable for animal waste application were identified using a G.I.S. based weighted linear combination (WLC) model. The degree of land suitability for animal waste application was determined using a range of social, economic, environmental and agricultural factors.

In the agricultural sector, management strategies were planned through a G.I.S. approach since its attitude in synthesising complex land relations (Toccolini, 1998). The location in Almeria, a semi-arid Spanish region, of areas with best attitude for intensive horticultural production in greenhouses was analyzed in relation to risks connected with aquifer salinization caused by an indiscriminate use of underground water (Ayala et al., 1999). The analysis of agricultural-forestry land evolution phenomena and the related environmental impacts (Langaas, 1995) and the definition of several sustainable development indicators for the monitoring and planning process and the definition of land use and attitude (Manera et al., 2001) were investigated by means of G.I.S. procedure too.

The present section describes the implementation of a Geographical Information System, at regional scale, in order to contribute to the analysis of agricultural plastic waste production, flux, collection and disposal in Italy. The results enabled the analysis and planning of agricultural plastic waste fluxes, together with the possibility to investigate different development scenarios and to consider new planning strategies for the management of agricultural plastic waste.

4.2 Materials and methods

The amount of the post-consume agricultural plastic film, particularly mulching (LDPE) and greenhouse (LDPE and/or EVA) film, and the correspondent amount of the produced waste in one year in each Italian Region was estimated by analysis of statistical data. This result has been obtained considering the useful-life, expressed in years, of the different typologies without considering: a) the loss of material during the exercise and removal phases and b) the increase in weight, above all for mulching films, due to contamination (soil and sand, organic material, humidity, etc.) during both their application and/or their collection and storing in the farm areas. Four Italian Regions were specifically analyzed; in particular, Campania and Sicily Regions have been chosen among those characterized by higher densities of protected structure (greenhouse and tunnel) while Apulia has been chosen both for its mulched areas (inside greenhouse or in open field) and wide diffusion of covering vineyards; the fourth Region (Emilia Romagna), chosen because characterized by medium-high amounts of greenhouse, meddle/low tunnel and mulching films. Actually, the study is proceeding, so the Regional data refer to an only zone, included in the province of Modena. In Apulia, Campania and Sicily Regions the total amounts of plastic consumption is respectively equal to 16.6 %, 10.5 % and 22.7 % of the total National consumption of film for protected cultivation. There, the survey has been carried out at provincial level through an identification of the main agricultural zones (Figg. 18, 19).

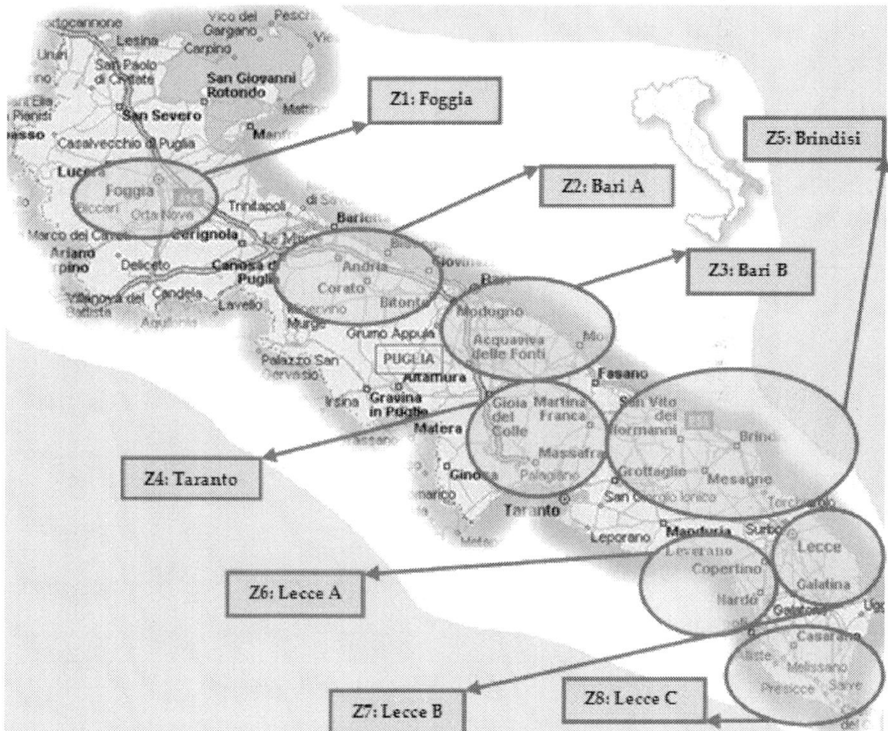

Fig. 18. Apulia Region and analyzed municipal farming areas.

The data that were collected regarded the plastic covered areas (hectares) according to the type of application (greenhouse, tunnel, mulching, etc.), the cultivation (vegetable, flower, fruit), the type of plastic (LDPE, EVA, PVC, etc.), the consumptions (tons) of these materials, the amount of agricultural plastic waste (produced and collected), the collection system (condition, transportation means, cost, dirtiness) These data were implemented in a relevant database and processed in a G.I.S. (GeoMedia - Intergaph) assuming as a cartographic base a geographic map at European scale. The main agricultural zones of Apulia, Campania and Sicily Regions were geographically individuated in order to connect them to the first collection areas and recycling centres on the national territory.

Fig. 19. Campania Region and analyzed municipal farming areas.

4.3 Results and discussion

The results enabled the analysis and planning of agricultural plastic waste fluxes, together with the possibility to investigate different development scenarios and to consider new planning strategies for the management of agricultural plastic waste.

In the Geographical Information System at regional scale, implemented in order to contribute to the analysis of agricultural plastic waste production, flux, collection and disposal in Italy, it is possible, by pressing on an agricultural area, to examine the data introduced in the database relating to the area in consideration (Fig. 20). Three different layers were created: the first one is represented by the geographical basis, the second layer shows the main agricultural areas using a specific type of plastic materials (Fig. 21), the third layer reports indication about the number of the collection areas and the main plastic waste recycling firms, associated to the Consortium "PolieCo".

The implemented G.I.S. enables the followings possibilities:

- to localize the main agricultural areas characterized by intensive production with plastic material;
- to know both the quantities of surface (hectares) and the consumption (tons) of plastics;
- to analyse the specific type of plastic material and to know the generated type of agricultural plastic waste and the quantities in each zone;
- to study the stream of agricultural plastic waste from the farms to collection areas in order to transport the agricultural plastic waste towards those more close and/or more easily to reach;
- to propose the enlargement or the creation of a new collection areas in barycentric zone as regards to the areas that produce high amount of agricultural plastic waste, through a previous economic analysis;
- to direct the stream of agricultural plastic waste from the collection areas towards pre-established recycling firms in order to optimize the flow of material, avoiding the block of little recyclers due to lack of material, or the stop of the working process of big firms due to scarcity of plastic waste.

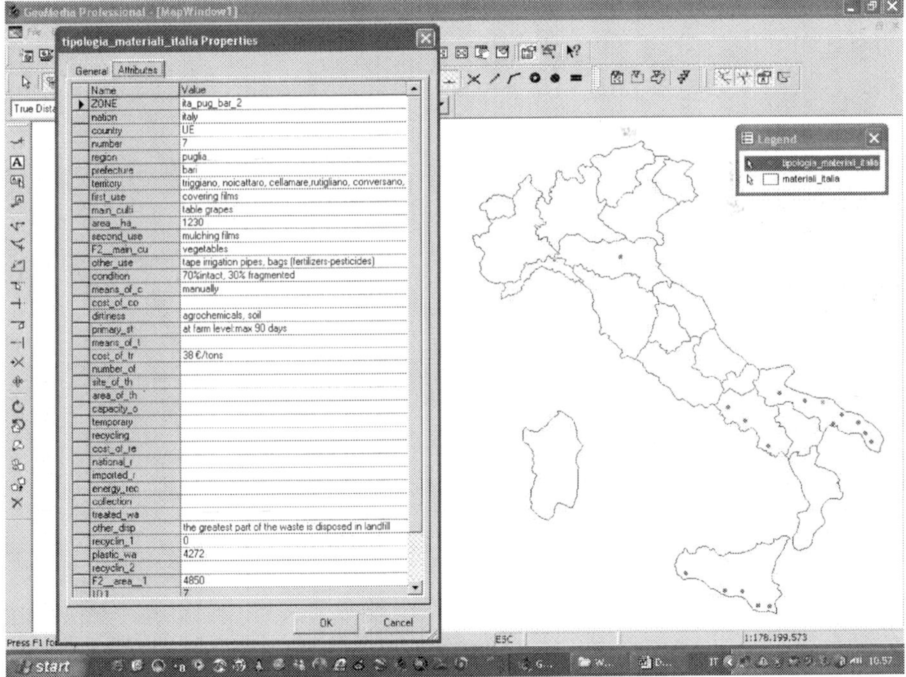

Fig. 20. Agricultural data included in the database of the G.I.S.

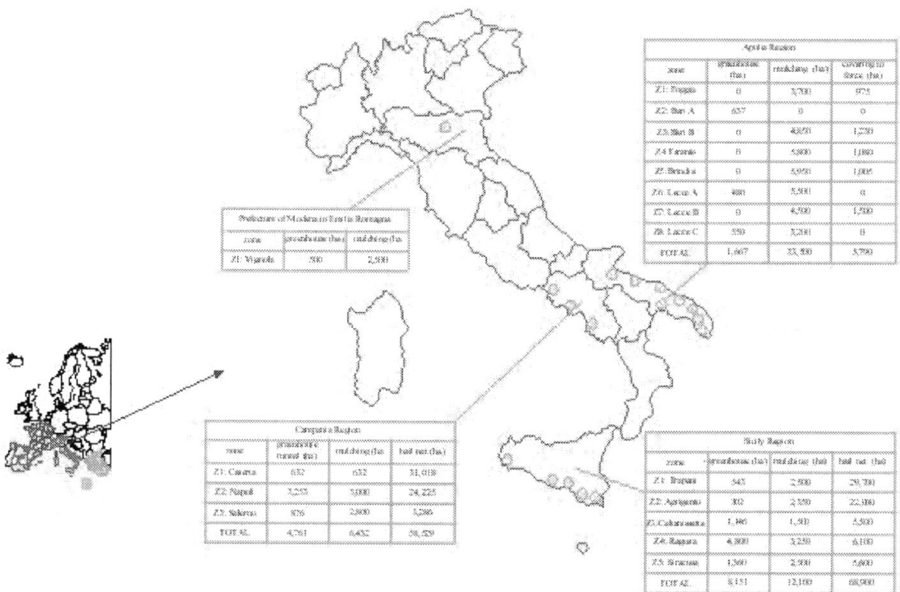

Fig. 21. Agricultural data included in the database of the G.I.S.

Different development scenarios may be therefore examined. Introducing the results in an optimal Decision Support System it would be moreover possible to analyse and plan agricultural plastic waste fluxes. Besides, different importance degree ("weighing" operation of informative levels) can be attributed both to each farm producing large amount of plastic waste and to the different parameters that characterise the viability net. For the attribution of weights to different informative levels the influence of each single factor should be preliminarily identified (Manera et al., 2001).

5. The utilization of a geographical information system (G.I.S.) for the valorisation of typical products from marginal areas

5.1 Foreword

The economy of marginal areas is frequently compromised by the inadequacy of the transport system, the lack of co-operation between farms, and by the insufficient distribution of their typical products, whose valorisation may be a factor of growth for lands that, due to orographical and geographical handicaps, are often delayed in their economic development. In those areas the problem that more frequently arises is difficulty in planning land development due to the lack of or poor knowledge and classification of all possible information, together with the inadequate capability to get new information and possibility to simultaneously analyse many data.

From this point of view, the use of a Geographical Information System (G.I.S.) appears to be a very useful tool because it allows matching information of geographical level (terrain height, gradient, slope orientation, soil utilisation, structures and infra-structures etc.) with pasture characteristics (pasture aromatic herbs, grass percent coverage, nutritional values, etc.).

G.I.S. and image processing method was employed for an application in land use planning with reference to an internal area of Basilicata Region (Southern Italy), well known for its typical food products (sheep and goat cheese), with the aim to individuate new areas, that may be devoted to pasture, with the best characteristic and highest potential performance able to contribute for an increase of quantity and a standardisation of quality in production of Pecorino cheese.

The Geographical Information System that was implemented, through a crossing among its numerous informative levels, enabled us to obtain thematic maps with specific uses with the aim to locate areas destined to pasture. Through image processing, a different degree of importance both at any value of the single theme and to the different obtained themes has been attributed by weights, with particular care for pasture herbs and environmental load capacity of the different areas. The re-sampling of these informative level led to a final new thematic map named "Pasturage capability map" where areas with higher productive capabilities and with the best botanical characteristic are highlighted.

Sheep and goat raising plays a major role in animal breeding in Italy, not only in terms of economic weight of production but also of the related social aspects. Ewe's and goat milk products greatly differ in their characteristics - often original indeed – and their diversity is closely linked to the peculiarities of the growing areas and the production techniques in many cases related to old and consolidated traditions. In Basilicata region (Figure 1), in particular, most of cropping and livestock farms are located in mountain areas. On one hand, this further aggravates the problem of marketing, on the other hand, emphasises the different characteristics of production.

The G.I.S. (Geographic Information System) is a support tool increasingly applied in agricultural and forest land, for its analysis (Gomarasca, 1995; Manera et al., 2000; Zucca et al., 1998), planning (Brunschwig et al., 2000; Coulter et al., 2000) and management (Wade et al., 1998; Weber et al., 2001), and for the development of forecast models to support decision-making programming processes (Ayala et al., 1999).

Through the use of G.I.S. and image processing, have been identifying the main characteristics of the surfaces to be used for pasture by sheep and goat-raising farms producing Pecorino cheese in a study area (Fig. 22) situated north-west of the Province of Potenza. It covers 31 municipalities included in the specifications for the production of "Pecorino cheese of Filiano" where agriculture has been since long one of the major subsistence factors for the resident population.

Fig. 22. Basilicata Region and study area

5.1.1 Materials and methods

A G.I.S. and image processing method was employed for an application in land use planning with reference to an internal area of Basilicata Region (Southern Italy), well known for its typical food products (sheep and goat cheese), with the aim to individuate new areas, that may be devoted to pasture, with the best characteristic and highest potential performance able to contribute for an increase of quantity and a standardisation of quality in production of Pecorino cheese.

1. This result was obtained by organising the operations into the following steps:
    *   Identification and filing of basic informative layers;
    *   administrative boundaries;
    *   villages and towns;
    *   road network;
    *   geo-lithological data;
    *   hydrological data;
    *   altimetric data;
    *   vegetational data;
    *   temperature-rainfall data;
    *   data on sheep-goat raising farms.
2. Homogenisation and integration of basic informative layers

LAYER	QUALITY
Elevation belt map (DEM – Digital Elevation Model) Slope map Exposure map	MORPHOLOGICAL
Permeability map Temperature-rainfall map Phyto-climatic belt map	CLIMATIC
Land use Map of the areas presently used for pasture	VEGETATIONAL
Farm distribution map	ANIMAL HUSBANDRY

3.   Spatial overlay

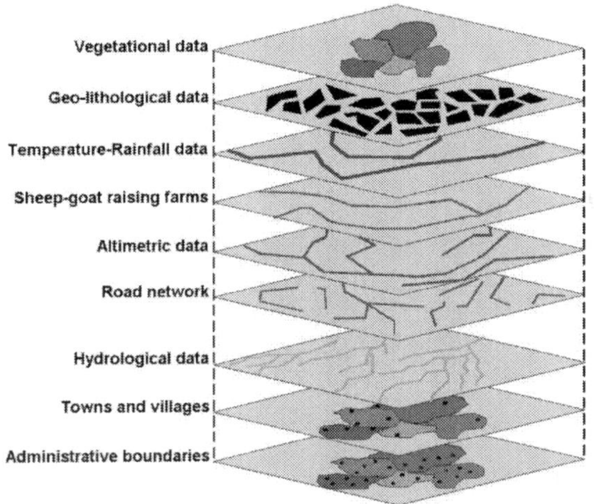

Filing being completed, the data were first sampled, attributing a different importance degree ("weighing" operation of informative levels) both to every class of each single theme and to the different thematic levels obtained, in order to characterise the area of higher yield capability. For the attribution of weights, the influence of each single factor was preliminarily identified and weights were thus attributed to the different informative levels, quantifying them according to the estimated fodder yield. The homogenisation and integration of data thus being completed, the layers were adequately processed to obtain a "capability " map, where the derived layers were gathered into distinct classes of "quality " subsequently used into new re-sampling operations where themes were processed through multiplicative algorithms (Manera et al., 2000). The result of simultaneous processing (Fig. 23) of informative levels is a summarising map called Grazing capability map.

5.1.2 Results and discussion

The reliability of the new informative level was first checked by overlaying the capability map with the location of the sheep-goat raising farms. Overlay highlighted that the highest number of small raising farms falls within areas belonging to lower capability classes, whereas farms with a greater number of heads are located in areas identified as having a greater parametric grazing capability value (Fig. 24).

Then, since the characteristics of more or less valuable pasture can be defined through agronomic and nutritional parameters, some of the said parameters were surveyed on a sample of farms falling within the study area in order to fully check the reliability of the proposed model.

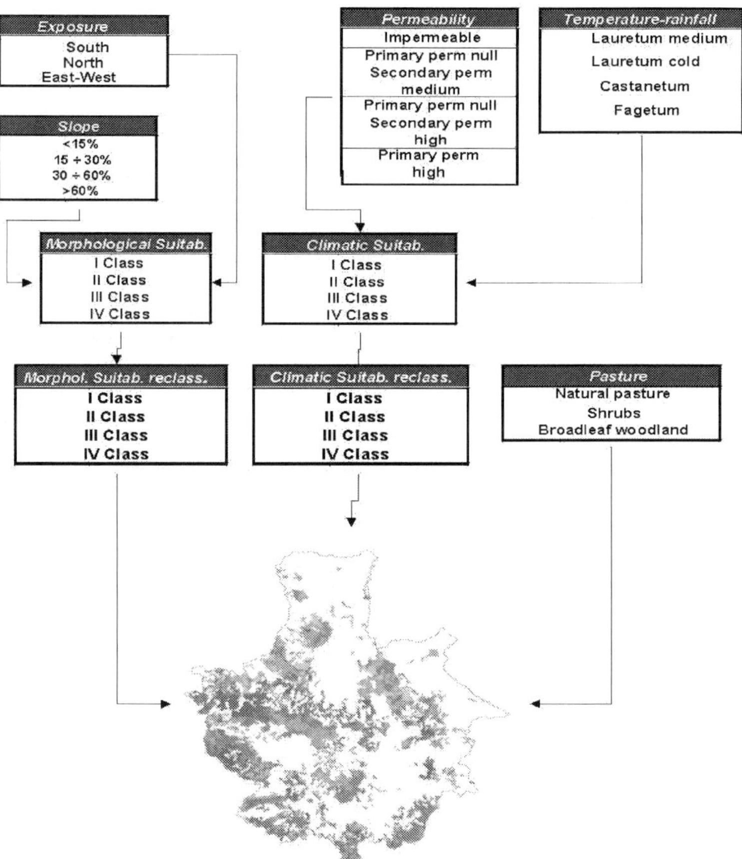

Fig. 23. Summarising scheme and grazing capability map

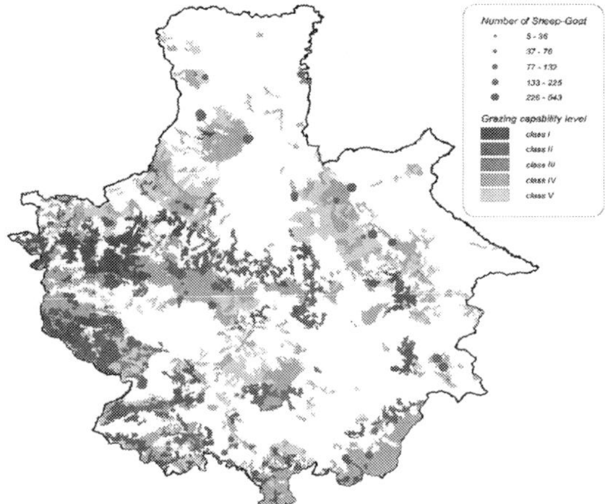

Fig. 24. Grazing capability and position of sheep-goat raising farms

These data were collected in collaboration with the Research Unit for the Extensive Animal Husbandry CRA-ZOE (Potenza), and concerned both some sheep-goat raising farms and the corresponding pasture they use. The agronomic data on pasture were grouped into five classes, in order to make them comparable with previous data treatments and overlay them with the grazing capability map.

Spatial overlay (Fig. 25), showed that the values of the FMU parameter (equal to the energy contained in milk produced from an intake of 1 Kg of standard barley) increases for higher

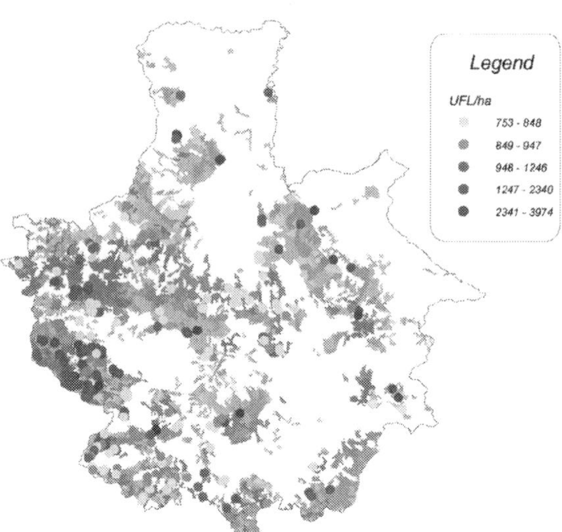

Fig. 25. Spatial overlay

capability classes, thus confirming the better grazing suitability of those areas identified as having higher capability.

Such processing was made by considering the distribution of qualitative parameters individually (dm, rp, rf) and it showed that, as for the FMU, also for dry matter the maximum values coincide with the classes of higher capability, whereas this was not the case for the two other parameters (rp, rf). Such a difference is probably due to the variability of the species present in the grazing turf and, in particular, in the associations of legumes and grasses.

Finally, also considering that the qualitative/quantitative characteristics of pastures were surveyed in a special period of the vegetative season and that the results of processing could be thereby disturbed, the parameter of the food supplement supplied to herds to make up any nutritional deficit independently of the vegetative season were analysed too (Graph 2).

Through the data analysis, the relationship between the types of food supplements and grazing capability classes was pointed out (Tab. 6).

	V CLASS	IV CLASS	III CLASS	II CLASS	I CLASS	TOT
Oat	6	6	0	5	0	17
Oat, barley, maize, broad beans	7	3	1	1	1	13
Oat, barley	4	1	1	3	3	12
Barley, maize, broad bean	0	4	0	3	0	7
Oat, barley, broad bean	0	2	1	0	0	3
Maize, barley	1	2	0	0	0	3
No food supplement	1	2	0	0	0	3
Oat, maize	0	0	0	2	0	2
Oat, maize, barley	1	0	0	1	0	2
Barley	2	0	0	0	0	2
TOT	22	20	3	15	4	**64**

Table 6. Farm distribution per grazing capability class on the basis of food supplements supplied

The wider diffusion of oat as food supplement, is because such fodder is more widely available in smaller farms with subsequently lesser heads, whereas in farms with a larger number of heads, combinations of oat and barley, maize and broad beans are more largely used.

Finally, based on the results obtained so far and considering that, on one hand, large agricultural surfaces are presently used for cereal growing and benefit from the economic compensation granted by the European Union that is supposed to cease in the near future and, on the other hand, pasture areas may increase due to the re-launching of Pecorino cheese, the "propensity to grazing" was assessed in the cereal-grown areas.

This further operation consisted in overlaying the grazing capability map (Fig. 23) with the data relative to climatic and morphological suitability of areas classified as predominantly cereal growing in the land use map.

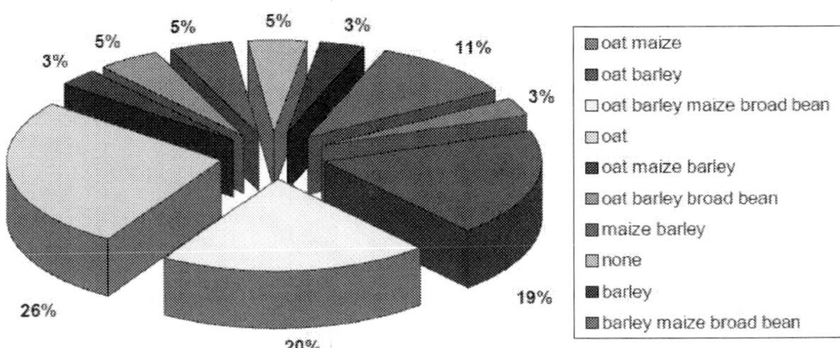

Graph 2. % distribution of farms per type of food supplement supplied

The result of such further processing, reported in Figure 26 as "grazing propensity map", allowed highlighting additional capability classes thereby confirming again that most of the study area shows a marked grazing suitability.

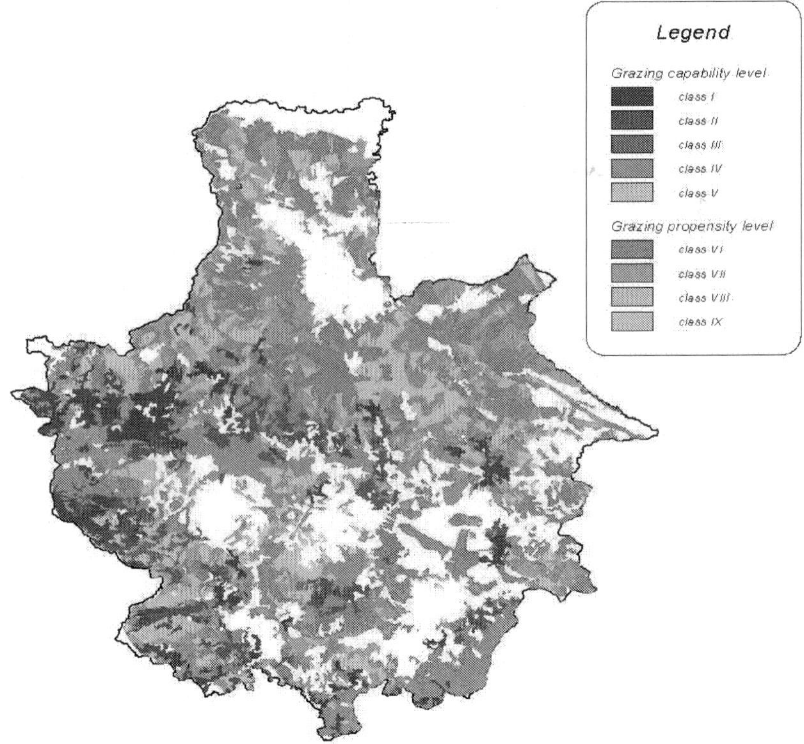

Fig. 26. Grazing propensity map

6. Conclusions

Agriculture constitutes in some areas of Europe the traditional and in most cases still the principal source for supporting the local economies; it plays, therefore, a central role, that should be adequately planned and managed, in order to avoid a confused and uncontrolled development of the rural land. The development of recent technologies for the automatic detection and positioning of greenhouse distribution enables an accurate analysis, that may be helpful for the definition of sound policies for the planning of the rural land and environment and a sound management of the agricultural landscape.

The results that were obtained in this paper enable the possibility to create a routine in IDL and ENVI software for an auto-detection of greenhouse and other protected structures. This tool can be used for monitoring the variation of rural land use and support better environmental and landscape planning policies. Moreover, through G.I.S., Image Processing techniques and landscape analysis procedures implementing coaxial matrices, areas that may be more environmentally endangered, in terms of the impact that large permeable coverings produce on natural cycles (water, soil, air, etc.) of the agro-ecosystem, could therefore be critically analyzed.

New structures for protected cultivation covered with wide surface of plastic material could alter in a significant way the agricultural landscape, that constitute one of the most important heritage of these fragile territories (Hernandez et al., 2004/b). If the phenomenon detected in the study area of the present research should continue with the same trend, an increase of the areas covered of more than 10% in the next years could be predictable. The present research showed how the impact of plastic-covered agricultural structures may be evaluated and potentially mitigated, so contributing to the preservation of the formal aesthetic characteristics of the rural landscape.

Finally, the concept of a "threshold" limit for the quantity of plastic-covered agricultural structures could also been considered in extra-urban planning: this limit would represent the maximum value that, from an aesthetic point of view, also incorporating shape and colour of the materials employed for crop protection, could be tolerated in an intensive agricultural context.

A Geographical Information System procedure for the analysis and planning of the collection and disposal of agricultural plastic waste, implemented at Italian level, may be enlarged introducing new fields and/or data of other Nations of the European Union, too.

Thanks to its attitude for synthesizing of complex land relations, the proposed G.I.S. may be improved including an analysis of the road network, in order to optimize the localization of the recycling centers, the optimal distances from them to the principal areas subject to intensive use of agricultural plastic films and the useful time and the average speed to cover these distances. Also the data regarding the distribution along different seasons of the year could be introduced in the G.I.S., in order to best fit the material flow towards the recycling centers.

The elaboration performed in this research work allowed producing a synthesis informative tool that reports the suitability of pasture for fodder production. (conclusion of the 2.4)

Extra-urban land planning must pursuit, as a main goal, environmental sustainability. A sustainable rural development, at least in European countries, has been perceived by social awareness and sensibility and is constantly been considered by new laws and regulations whose attempt is the natural resources protection.

In this scenario an accurate analysis of performing variations and a global monitoring of ecosystems seem necessary in order to propose environment protection politics, crucial element for a sound planning of extra-urban land and for a sustainable growth of the civilized World.

This analysis has shown how the results of the applied agronomic practises, in terms of CO_2 fixation, would be able to contrast heavy emissions of greenhouse effect gases in the atmosphere by urban settlements, demonstrating how a correct rural-site management could efficiently balance environmental pollution determined by the human development. The use of this approach for other environmental factors, such as water, soil etc., would lead to a more comprehensive understanding of landscape development dynamics through its principal environmental components, contributing to the proposal of production oriented politics that achieve compensation of natural balance alterations, and a real application of the concept of sustainable development. The results showed that the cultivation conversion caused a loss of CO_2 fixation value, that was accompanied by heavy emission of greenhouse effect gas in the atmosphere by urban settlements too. A sound rural land management should efficiently balance environmental pollution determined by the economic development; the methodology employed in the present case study could properly be transported into other areas, and the resulting analysis extended to different rural context.

7. Acknowledgment

The work has been realized under the "AGROCHEPACK Project" (Program Reference 2G-MED09-015) funded by the Programme MED, Priority Axis 2: Protection of environment and promotion of a sustainable territorial development. Objective 2-1: Protection and enhancement of natural resources and heritage.

8. References

Arcidiacono, C. & Porto, S.M.C. (2008). Image Processing for the classification of crop shelters. *Acta Hortic*, ISSN: 0567-7572, Vol.801, pp. 309-316.

Ayala, R.; Becerra, A.; Iribarne, L.F.; Bosch, A. & Díaz, J.R. (1999). G.I.S. System as a Decision Support Tool for Agricultural Planning in AridZones of Spain. *Agricultural Engineering International: the CIGR Ejournal*, Vol. I, 1999.

Basilicata Region (2000). *Basilicata environmental Year 1999*.

Basnet, B.B.; Apan, A.A. & Raine, S.R. (2002). Geographic information system based manure application plan. *Journal of Environmental Management*, Vol.64, pp. 99-113.

Brunschwig G. (2000) – Terrois d'élevage laitier du Massif central: Identification & caractérisation (in French) – *ENITA de Clermont Ferrand*, Collection estudes n.6, Lempdes (France).

Capobianco, R.L. & Picuno, P. (2008). Remote sensing and thematic mapping of protected cultivation in Southern Italy agricultural landscapes. *Proceedings of the AgEng2008 Conference: Agricultural Engineering & Industry* Exhibition, Crete (Greece), 23-25 June.

Capobianco, R.L.; Tortora, A. & Picuno, P. (2004). Analisi del territorio, dell'ambiente e del paesaggio rurale mediante tecniche di modellizzazione spaziale ed image processing di cartografie storiche (Analysis of rural land, environment and landscape through spatial modelling techniques and hystorical cartography image processing). *Rivista di Ingegneria Agraria*, Vol.35, No.3, pp. 71-77. [in Italian]

Carvajal, F.; Crisanto, E.; Aguilar, F.J.; Aguera, F. & Aguilar M.A. (2006). Greenhouses detection using an artificial neural network with a very high resolution satellite image. *Proceedings of the ISPRS Technical Commission: II Symposium,*.

Coulter, L.; Stow, D.; Hope, A.; O'Leary, J.; Turner, D.; Longmire, P.; Peterson, S. & Kaiser, J. (2000). Comparison of High Spatial Resolution Imagery for Efficient Generation of GIS Vegetation Layers. *Journal of the American Society for Photogrammetry and Remote Sensing (PE&RS)*, Vol.66, No.11, pp. 1329–1336.

Dal Sasso, P. & Picuno, P. (1996). Il recupero funzionale dei fabbricati agro-industriali di valore storico nel contesto territoriale (Functional recovery of agro-food buildings of histrorical value in the land cointext). *Rivista di Ingegneria Agraria*, Vol.2, pp. 111-121. [in Italian]

Dal Sasso, P.; Scarascia-Mugnozza, G. & Marinelli, G. (2007). Dalle matrici ambientali le proposte per la mitigazione dell'impatto delle serre. (From environmental matrices the proposals for the mitigation of greenhouse impact). *Colture protette*, Vol.2, pp. 63-73. [in Italian]

Della Maggiore; R., Fresco, R.; Mura, E. & Perotto, E. (2002). Historical cartography Georeferencing. In: *Proceedings of the 2002 ASITA Italian Congress*, Vol. I, pp. 671-677.

Ghose, M.K.; Dikshit, A.K. & Sharma, S.K. (2006). A G.I.S. based transportation model for solid waste disposal – A case study on Asansol municipality. *Waste Management*, Vol.26, pp. 1287–1293.

Gomarasca, M.A. (1995). Environmental evaluation with GIS and remote sensing techniques. *Genio Rurale*, Vol.58, No.5, pp. 65–71. [in Italian]

Hagen, L. J. (1991). A wind erosion prediction system to meet user needs. *Jour. Soil and Water Conserv*, Vol.46, No.2, pp. 106-111.

Hernández, J.; García, L. & Ayuga, F. (2004/a). Integration methodologies for Visual Impact Assessment of Rural Buildings by Geographic Information Systems. *Biosystems Engineering*, Vol.88, No.2, pp. 255–263.

Hernández, J.; García, L. & Ayuga, F. (2004/b). Assessment of the visual impact made on the landscape by new buildings: a methodology for site selection. *Landscape Urban Plan*, Vol.68, pp. 15–28.

Kauppi, P. & Sedjo, R. (2001). Technological and economic potential of options to enhance, maintain, and manage biological carbon reservoirs and geo-engineering. In: Metz, B., et al. (Eds.), *Climate Change 2001: Mitigation*. IPCC, Cambridge University Press, New York, Chapter 4, pp. 303–353.

Kristensen, S.P. (2003). Multivariate analysis of landscape changes and farm characteristics in a study area in central Jutland - Denmark. *Ecol. Model.*, Vol.168, pp. 303-318.

Kristensen, S.P.; Thenail, C. & Kristensen, L. (2001). Farmers' involvement in landscape activities: An analysis of the relationship between farm location, farm characteristics and landscape changes in two study areas in Jutland - Denmark. *J Environ Manage*, Vol.61, pp. 301-318.

Langaas, S. (1995). The spatial dimension of indicators of sustainable development: the role of geographic information system (G.I.S.) and cartography. In: B. Moldan and S. Billharz, (Eds.), *Sustainability indicators: Report of the Project on Indicators of Suistainable Development*. SCOPE. Wiley, New York, pp. 33-39.

Leao, S.; Bishop, I. & Evans, D. (2001). Assessing the demand of solid waste disposal in urban region by urban dynamics modelling in a G.I.S. environment. *Resources, Conservation and Recycling*, Vol.33, pp. 289-313.

Levin, N.; Lugassi, R.; Ramon, U.; Braun, O. & Ben-Dor, E. (2007). Remote sensing as a tool for monitoring plasticulture in agricultural landscapes. *Int J Remote Sens.*, Vol.28, pp. 183-202.

Manera, C.; Picuno, P. & Tortora, A. (2001). The utilisation of a Geographical Information System (G.I.S.) for the valorisation of typical products from marginal areas. *Proceedings of Agribuilding 2001*, Campinas - SP, (Brazil), 3-6 September, pp. 223-233.

Manera, C.; Picuno, P. & Tortora, A. (2000) – Use of a GIS for the exploitation of typical cheese products from marginal areas - *Proceedings of the Italian congress on: Exploitation of local and land resources in the framework of rural development policy*, 14–17 June, Matera (Italy). [in Italian]

Masera, O.R.; Garza- Caligaris, J.F; Kanninen, M.T.; Karjalainen, T.; Liski, J.; Nabuurs, G.J.; Pussinen, A.; de Jong, B.H.J. & Mohren, G.M.J. (2003). Modeling carbon sequestration in afforestation, agroforestry and forest management projects: the CO2FIX V.2 approach. *Ecological Modelling*, Vol.164, No.2-3, pp. 177-199.

Mohren, G.M.J.; Garza Caligaris, J.F.; Masera, O.; Kanninen, M.; Karjalainen, T.; Pussinen, A. & Nabuurs, G.J. (1999). CO2FIX for Windows: a dynamic model of the CO2-fixation in forests, Version 1.2. *IBN Research Report* 99/3, pp. 33.

Papadakis, G.; Briassoulis, D.; Scarascia-Mugnozza, G.; Vox, G.; Feuilloley, P. & Stoffers, J.A. (2000). Radiometric and thermal properties of, and testing methods for, greenhouse covering materials. *J Agr Eng Res.*, Vol.77, pp. 7-38.

Picuno, P. (2005). Impatti estetici e ambientali di coperture agricole realizzate con materiali permeabili. (Aesthetic and environmental impacts of agricultural coverings realized with permeable materials). *Proceedings of the AIIA2005 National Congress: L'ingegneria agraria per lo sviluppo sostenibile dell'area mediterranea.* [in Italian]

Picuno, P. & Scarascia-Mugnozza, G. (1990). L'impatto sul territorio rurale delle coperture plastiche di tendoni per vigneto. (The effect of plastic covers for vine cultivations on rural environment)). *Proceedings of the 2nd Seminar of the 2nd Section of A.I.G.R. (Italian Association of Agricultural Engineering).* [in Italian]

Picuno, P. & Scarascia-Mugnozza, G. 1994. The management of agricultural plastic film wastes in Italy. *Proceedings of the International Agricultural Engineering Conference*, Bangkok (Thailand), 6-9 December, pp. 797-808.

Picuno P., Tortora A., Capobianco R.L. (2011). Analysis of plasticulture landscapes in Southern Italy through remote sensing and solid modelling techniques. *Landscape and Urban Planning*, Vol.100 , No1-2, pp. 45-56..

Primdahl, J. (1999). Agricultural landscape as places of production and for living in owner's versus producer's decision making and the implications for planning. *Landscape Urban Plan*, Vol.46, pp. 143-150.

Ruda, G. (1998). Rural buildings and environment. *Landscape Urban Plan*, Vol.41, pp. 93–97.

Sabins, F.F. (1996). *Remote Sensing, Principles and interpretation*. W.H. Freeman and Company. New York, U.S.A.

Scarascia-Mugnozza, G. (1995). Sustainable greenhouse production in Mediterranean climate: a case study in Italy. *MEDIT*, Vol.6, No.4, pp. 48-53.

Scarascia-Mugnozza, G.; Picuno, P. & Sica, C. (2006). Innovative solution for the management of the agricultural plastic waste. *Proceedings of World Congress 2006 "Agricultural Engineering for a Better World"*. Bonn (Germany), September 3-7.

Scarascia-Mugnozza, G.; Sica C. & Picuno P. (2008). The Optimization of the Management of Agricultural Plastic Waste in Italy Using a Geographical Information System. *Acta Hortic*, ISSN: 0567-7572, Vol.801, pp. 219-226.

Scarascia-Mugnozza, G.; Vox, G. & Picuno, P. (1999). Metodologie di analisi territoriale in relazione all'uso di plastiche per le coltivazioni protette. (Land Analysis methodologies relating with the use of plastic for protected cultivation). Proceedings of the AIIA National Congress: on: The protected crops. Agronomic aspects, spatial and technical-constructive), Ragusa (Italy), 24-26 June, pp. 313-326. [in Italian]

Sica, C. (2000). Environmental problems connected to the use of plastic materials in protected cultivation (in Italian). Doctoral Thesis, Technical-Economical Department, University of Basilicata, Potenza (Italy).

Toccolini, A. (1998). Analysis and planning of agricultural-forestry systems through G.I.S. (in Italian). *Raisa, Ricerche avanzate per innovazioni nel sistema agricolo*. Franco Angeli Editore, Milano (Italy).

Toccolini, A.; Fumagalli, N. & Senes, G. (2006). Greenways planning in Italy: the Lambro River Valley Greenways System. *Landscape Urban Plan*, Vol.76, pp. 98–111.

Tortora, A.; Capobianco, R.L. & Picuno, P. (2006). Historical Cartography and GIS for the Analysis of Carbon Balance in Rural Environment: a Study Case in Southern Italy. *Agricultural Engineering International: the CIGR Ejournal*, Vol. VIII, Manuscript BC 06 003.

Tortora, A.; Picuno, P. & Capobianco, R.L. (2002). Historical cartography for analysis of the rural land evolution. In: *Proceedings of the 2002 ASITA Italian Congress*, Perugia, Italy. Vol II, pp. 1961-1966.

Tortora, A. & Picuno, P. (2008). Analysis of the effect on rural landscape of wide coverings for crop growing. *Acta Hortic*, ISSN: 0567-7572, Vol.801, pp. 325-332.

Travis, C.C. & Nixon, A.G. (1991). *Human exposure to dioxin. Environmental Science Technology*. pp. 17-30.

Tso, B. & Mather, P.M. (2009). *Classification Methods for remotely sensed data*. 2nd edition. CRC press.

Vigiak, O.; Sterk, G.; Warren, A. & Hagen, L. (2003). Spatial modeling of wind speed around windbreaks, *J Agr Eng Res*, Vol.52, pp. 273-288.

Wade, T.G.; Schultz, B.W.; Wickham, J.D & Bradford, D.F. (1998) – Modeling the potential spatial distribution of beef cattle grazing using a Geographic Information System – *Journal of Arid Environments*, Vol.38, No.2, pp. 325–334.

Weber, R.M. & Dunno-Glenn, A. (2001) – Riparian Vegetation Mapping and Image Processing Techniques, Hopi Indian Reservation, Arizona – *Journal of the American Society for Photogrammetry and Remote Sensing (PE&RS)*, Vol.67, No.2, pp. 167–170.

Xiloyannis, C.; Sofo, A.; Nuzzo, V.; Palese, A.M.; Celano, G.; Zukowskyj, P. & Dichio, B. (2005). Net CO2 storage in Mediterranean olive and peach orchards. *Scientia Horticulturae*, Vol.107, No1, pp. 17-24.

www.easyrgb.com/index.html

Zucca, C.; Loda, B.; D'Angelo, M.; Madrau, S.; Percich, L. & Previstali, F. (1998) – The evaluation of land suitability for grazing improvement: the GIS contribution (in italian) – *Genio Rurale*, Vol.61, No3, pp. 47–56.

15

Extreme Climatic Events as Drivers of Ecosystem Change

Robert C. Godfree
CSIRO Plant Industry, Canberra, ACT
Australia

1. Introduction

It is difficult to stand in the cliff dwellings that lie in the shallow caves that line the canyon walls of Mesa Verde National Park in south-eastern Colorado and not be amazed. These impressive structures were built in the late 12th and early 13th centuries by pre-Columbian native American Anasazi people, who, at that time, occupied much of the Four Corners region of south-western USA. Some cliff dwellings were of exceptional size, containing over 150 rooms and lodging 100 or more people, and were supported by the farming of maize on the surrounding semi-arid mesa tops (Benson *et al.* 2007a) which today are covered in forests of piñon pine and juniper. Perhaps more interesting, however, is that this phase of occupation represented just one stage in the development and decline of Anasazi culture in the southwest. Anasazi populations waxed and waned repeatedly over time, and most of the even bigger, multi-storey stone houses (great houses) located across the central San Juan Basin, for example, were abandoned in the mid 12th century (Benson *et al.* 2007a,b). Construction activity in the wetter, more favourable northern San Juan Basin then increased (Lekson & Cameron 1995; Benson *et al.* 2007a), but by the late 13th century the Mesa Verde cliff dwellings, along with other Anasazi population centres in the Four Corners region, had also been abandoned for areas closer to the Northern Rio Grande region of New Mexico (Ahlstrom *et al.* 1995).

What caused these large scale human migrations? The most likely scenario is that Anasazi populations, which over time had become increasingly sedentary and dependent on maize for provision of dietary needs (Benson *et al.* 2007b), were primarily responding to movement of the climatic niche in which maize could be grown (Petersen 1994). Reconstructed climate data suggest that between AD 900 and AD 1300 the south-western United States was affected by elevated aridity and protracted drought that exceeded in severity anything that has been observed in the centuries since (Cook *et al.* 2004; Stahle *et al.* 2007), and particularly severe multi-decadal drought, apparently linked to changes in the Pacific Decadal Oscillation (MacDonald & Case 2005), occurred during AD 1135-1170 and AD 1276-1297. Around this time the central and northern San Juan Basin Anasazi cultures, respectively, declined (Benson *et al.* 2007a), and it is likely that agricultural collapse (Benson *et al.* 2007a), coupled with breakdown in societal structure (Benson *et al.* 2007b), ultimately led to the depopulation of the entire region.

This and other historical examples in which drought has played a pivotal role in socioeconomic and cultural decline (e.g., Weiss & Bradley 2001; Acuna-Soto *et al.* 2002, 2005; Endfield *et al.* 2004; Hodell *et al.* 2005) underscore the capacity for extreme climatic events to threaten the very fabric of society. Modern agricultural systems are not immune from similar pressures; recently the severe 1997-2009 "Millennium Drought" in south-eastern Australia (Whitaker 2006; Bond *et al.* 2008) caused massive agricultural decline (Pook *et al.* 2009), extensive job losses (Mpelasoka *et al.* 2008), and real declines in household consumption, wages, and gross regional product (Horridge *et al.* 2005). What makes these cases of added importance is that the frequency of extreme events is expected to increase under anthropogenic climate change (Tebaldi *et al.* 2006; Planton *et al.* 2008), and changes in precipitation and temperature extremes are already being observed around the world (e.g., Collins *et al.* 2000; Easterling *et al.* 2000; Goswami *et al.* 2006). Significant impacts on human societies and the natural world are expected if such trends continue (Easterling *et al.* 2000).

Recently, there has been a significant increase in research focusing on the impact of extreme climatic events, and, more broadly of climate change, on natural and agro-ecosystems (e.g., Easterling *et al.* 2000; Meehl *et al.* 2000; Walther *et al.* 2002; Tubiello *et al.* 2007). Extreme events can have severe and often disproportional effects (Gutschick & BissiriRad 2003) on a wide range of animal and plant groups (e.g., Dudley *et al.* 2001; Morecroft *et al.* 2002; Martinho *et al.* 2007), with population-level changes to extinction rates, range movement, behaviour and reproduction observed in a range of different ecosystems (reviewed in Easterling *et al.* 2000; Parmesan *et al.* 2000). However, such examples raise further questions, to which we at present only have a rudimentary and fragmented understanding. For example, which ecosystems are most sensitive to extreme climatic events, and to what type of events? How extreme, and over what timeframes, do climatic conditions have to be to cause significant mortality among plant species? Over what timeframes can community-level compositional change occur? Most importantly, can we predict the nature and magnitude of change in plant communities that are affected by different kinds of extreme events? These questions are more than just academic: the ecosystem services that plant communities provide underpin both human societies and biodiversity alike.

The objectives of this chapter are twofold. First, by drawing on a range of case studies, I assess the conditions under which extreme climatic events are likely to rapidly alter the structure and composition of natural plant communities. I focus specifically on the impacts of extreme drought and heatwaves, since both are expected to increase in severity in coming decades as climate change alters the probability distribution of temperature- and precipitation-related climatic variables (Meehl *et al.* 2000; Hennessy *et al.* 2008; Planton *et al.* 2008); the impacts of other extreme events are discussed elsewhere (see Easterling *et al.* 2000; Parmesan *et al.* 2000; Holmgrem *et al.* 2006). I then test some of these ideas by reporting the results of a field study which compares the demographic responses of three semi-arid Australian grassland species to drought, with a specific focus on using a better understanding of the specific roles of habitat heterogeneity, species characteristics, and drought severity to predict the likely impact of extreme events on plant communities under climate change.

2. Characterisation of extreme events

It is widely understood that ecological stresses often reflect statistical extremes rather than climatic means or variances (Gaines & Denny 1993), and the responses of individuals and populations to stressful abiotic conditions is, in many cases, non-linear and sensitive to discrete thresholds (Easterling *et al.* 2000; Beniston & Stephenson 2004). As such, accurate quantification of the likelihood and intensity of extreme climatic events is essential. Unfortunately, this task is made difficult by the fact that no single definition of what constitutes an extreme event actually exists, and the impacts of unusual climatic situations on ecosystems have not traditionally been studied in a systematic manner (Gutschick & BassiriRad 2003; Smith 2011). As noted by Beniston & Stephenson (2004), extremes can be defined in terms of rarity, intensity or impact, although individually these definitions may fail to capture the critical features of climatic variability that impact on plant and animal populations. Smith (2011) suggests that an extreme climatic event should be defined as a rare or unusual climatic period that alters ecosystem structure and/or function well outside what is considered normal variability, and many of the studies discussed below inherently use this concept, despite the fact that the rarity of the climatic conditions being studied is often not formally described.

From a quantitative standpoint, a range of approaches are used to characterise extreme events. The Intergovernmental Panel on Climate Change (IPCC 2011) defines a climate extreme as "the occurrence of a value of a weather or climate variable above (or below) a threshold value near the upper (or lower) ends of the range of observed values of the variable" (p. 2). Indices derived using this approach usually quantify the duration or frequency of events which exceed a specific temperature or precipitation threshold, for example total number of frost days, growing season length, heat wave duration, and number of consecutive dry days (e.g., Tebaldi *et al.* 2006). Another commonly-used method is to define extreme events as those occurring within a certain percentile range (often the 5th, 10th, 90th or 95th percentiles) of a climatological distribution within a given timeframe (see Bell *et al.* 2004). Hennessey *et al.* (2008) define exceptional droughts as those of one year in duration and occurring, on average, once every 20 years (i.e., a 5% probability of occurring within a given year; Katz *et al.* 2005).

The standard statistical approach for quantifying climatic variation is to fit a probabilistic model to a given climatic data set and then to evaluate the likelihood (and severity) of specific climatic events based on the associated probability density function. In recent years a cohesive statistical theory of extreme events has emerged (Coles 2001). These techniques, and their application to ecological problems, have been discussed in detail elsewhere; readers are directed to Katz & Brown (1992), Gaines & Denny (1993), Katz *et al.* (2005), and Resnick (2007) for details. The basic statistical approach can be visualised in Fig. 1; for further discussion see Meehl *et al.* (2000) and IPCC (2011). Consider a normally-distributed climatic variable, such as temperature (Fig. 1a). For this probability density function (PDF), the top 5% of values, which may be classified as extreme, lie 1.65 standard deviation (σ) units above the mean (PDF I; Fig. 1a). A shift in climate alters the frequency and severity of extreme events, with the magnitude of change depending on the location and shape of the new distribution (Fig. 1b). In PDF II (Fig. 1b), the distribution undergoes only a mean shift with variability (σ; the standard deviation) remaining constant, this results in an increase in the frequency of climatic events of a given magnitude (from 5% in PDF I to 26% in PDF II). Similarly, the incidence of extreme events at the other end of the distribution declines. A change in the shape of the distribution (increase in σ in PDF III, Fig. 1b) results in a further increase in the frequency of extreme events (PDF III, Fig. 1b), indeed more than an equivalent change in the mean (Katz & Brown 1992; Meehl *et al.* 2000).

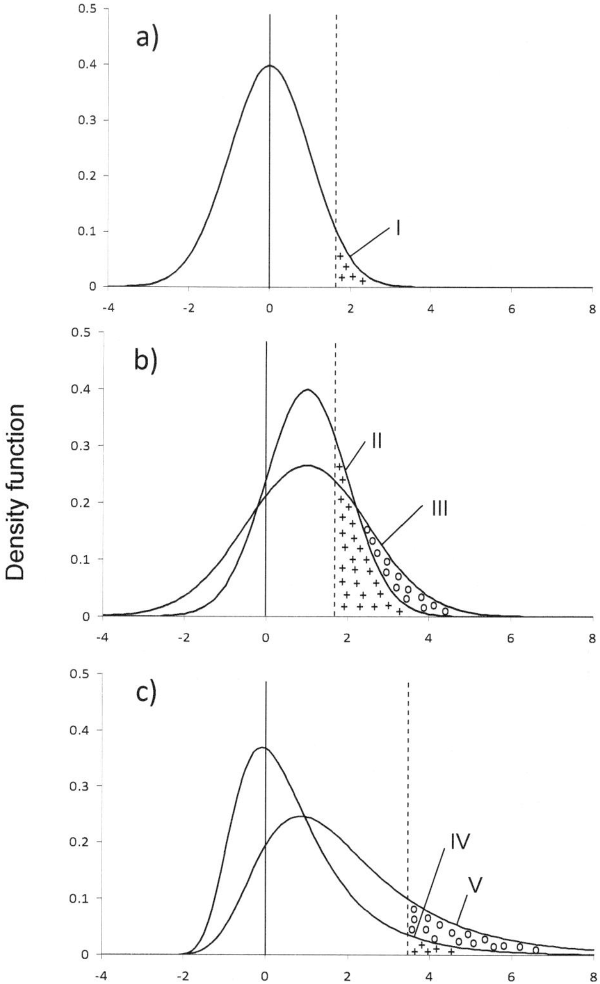

Fig. 1. Frequency of extreme events for a given hypothetical climatic variable showing variation in probability related to change in shape and location of the probability density function (PDF). a) normal distribution (PDF I; $\mu = 0$, $\sigma = 1$) with the 5% most extreme events occurring 1.65 or more standard deviation units from the mean (vertical dashed line). b) Increase in frequency of extreme events following increases in the climatic mean (PDF II; $\mu = 1$, $\sigma = 1$) and standard deviation (PDF III; $\mu = 1$, $\sigma = 1.5$). c) Generalised extreme value (GEV) distribution with location parameter (μ) = 0, scale parameter (σ) = 1, and shape parameter (ξ) = 0.1 (PDF IV) showing the larger tail that characterises the GEV PDF, with the 5% most extreme events lying above the vertical dashed line. An increase in the parameter σ to 1.5 (PDF V) increases the spread of the distribution and the probability of extreme events that lie in the upper tail. The shape of the GEV function is determined by ξ (see Katz *et al.* 2005 for details).

While some climatic variables can be approximately normally distributed, many others have heavy tails with significant skew towards large values (Gaines & Denny 1993; Gutschick & BassiriRad 2003). For example, precipitation-related variables tend to be non-normal, and are often modelled using gamma (Watterson & Dix 2003) or lognormal (Cho *et al.* 2004) distributions. Model selection is critical, since the objective is to adequately characterise the distributional tails. One family of distributions widely used for modelling extreme values, especially maxima, is the generalised extreme value (GEV) probability density function, which produces light-, heavy- and bounded-tailed distributions (Katz *et al.* 2005). A typical heavy-tailed (Fréchet) GEV distribution is shown in Fig. 1c (PDF IV). Shifts in location (μ), scale (σ) and shape (ξ) parameters that define the GEV distribution all impact on the frequency and severity of extreme events (Fig. 1c, PDF V).

3. Extreme climatic events as ecosystem drivers: The evidence

3.1 Demographic change and plant mortality

Probably the most important mechanism by which extreme climatic events (ECEs) can drive rapid ecosystem change is by causing extensive mortality in dominant or keystone plant species. The loss of such species can result in rapid, cascading compositional change in plant communities, in turn affecting the population viability of faunal and floral community associates (Gitlin *et. al.* 2006) and the ecosystem services that they provide (Walker *et al.* 1999; Kremen 2005). Unfortunately, ECEs are difficult to study within a pre-planned experimental and statistical framework (Buckland *et al.* 1997), and so little or no information exists on the responsiveness of most plant communities to ECEs of varying severity. However, a range of studies provide insight into the dramatic and persistent impacts of prolonged abiotic stress on plant community composition and structure.

The classic studies of Albertson and Weaver (1944, 1945) on prairie and woodland ecosystems in North America were among the first, and most comprehensive, to document the impacts of protracted, multi-year drought on plant communities. During the 1930-1940 "dustbowl" years, much of the central US experienced record- or near-record low rainfall, high temperatures, high evaporation, and declining soil moisture. In the most extreme years (e.g., 1934, 1936, 1939), rainfall was up to 40% below normal, summer maximum temperatures were 3-6°C above normal, evaporation exceeded that of non-drought years by up to 33%, and the water table fell by one metre or more (Albertson & Weaver 1945). This drought was one of the three most extreme to affect North America since 1900 (Cook *et al.* 2004), although not as severe as previous megadrought periods that occurred in the 12th to 16th centuries (Woodhouse & Overpeck 1998; Stahle *et al.* 2007).

The most striking effect of this drought was the high level of mortality that occurred in virtually all plant species across the landscape. Mortality of dominant and subordinate tree species (e.g., *Ulmus americana, Populus sargentii, Celtis occidentalis, Salix* spp.) exceeded 50% across a broad range of topoedaphic habitats with distinctly different hydrological regimes, most notably in dry ravines and along intermittent creeks (Albertson & Weaver 1945). In prairie communities perennial grasses such as *Andropogon scoparius, Koeleria cristata* and *Poa pratensis* suffered up to 80-90% mortality, while subordinate grasses and forbs were almost eliminated (Albertson & Weaver 1944). The catastrophic loss of groundcover resulted in intense wind erosion, with dust accumulating to depths of 2 feet (60 cm) in sheltered locations (Albertson & Weaver 1945). This dust exacerbated the severity of water stress experienced by plants during the drought by preventing rain infiltration into the soil, effectively blocking moisture from reaching the rhizosphere (Albertson & Weaver 1945). Insect attack also increased the impact of drought on tree populations (Albertson & Weaver 1945; Mattson & Haack 1987).

Other studies report similar effects during prolonged periods of exceptionally severe drought. Record dry conditions experienced during 2006-2007 at a semi-arid grassland site in south-eastern Australia resulted in 90% or higher mortality of the dominant grass species *Austrostipa aristiglumis* (Godfree *et al.* 2011). Here, the most acute period of rainfall deficiency occurred during the middle of a decade-long drought. Similarly, Edwards & Krockenberger (2006) reported 64% mortality among seedlings of rainforest species during the 2002 ENSO event in north-eastern Australia, during which rainfall at their study site was only 36% of average. On a much larger spatial scale, mortality of piñon pine (*Pinus edulis*) in western North America varied between 40% and >90% in response to drought during 2000-2003 (Breshears *et al.* 2005). The severity of this drought, which was among the three driest in the past century, was exacerbated by high temperatures (Breshears *et al.* 2005).

Similar levels of drought-induced mortality have been observed in Canadian aspen forests (Hogg *et al.* 2008), beech forests (Peterken & Mountford 1996), and in forests globally (for review see Allen *et al.* 2010). Indeed, forest mortality in response to drought and heatwaves is so common that Allen *et al.* (2010) concluded that no forest biome is invulnerable to climate change, even in systems that are not thought to be water-limited. However, while droughts do not have to be of unprecedented or record severity to cause significant mortality in plant communities, it is also clear that not all droughts result in high levels of plant mortality (e.g., Condit *et al.* 1995; Fensham and Holman 1999; references in Allen *et al.* 2010); and not all generate detrimental, lasting effects on plant populations (e.g., Morecroft *et al.* 2002; Yurkonis & Meiners 2006).

Briefly, other extreme climatic events can also have strong, direct impacts on ecosystems by inducing injury or mortality in plants. High temperatures, especially those exceeding 55°C, when Rubisco activity, electron transport and overall photochemical performance becomes impaired (Kappen 1981; Musil *et al.* 2009) are known to be detrimental to plants. Simulated heatwaves in which temperatures approach or exceed these temperatures have been shown to result in canopy decline and mortality of succulent plant species (Musil *et al.* 2005); in this study mortality in heated treatments ranged from 33-74% compared with 7-38% in controls that experienced 5.5°C cooler daytime temperatures. Such extremes are probably most likely to occur in arid and semi-arid systems where lower-canopy species lack cover, and may in part explain the loss of low-stature mesophyllic grasses and forbs following death of overstorey plants in drought-affected grasslands and woodlands (Albertson and Weaver 1944, 1945). Studies conducted in tundra environments have also resulted in physiological impairment and mortality in cold-adapted species (Marchand *et al.* 2006); numerous examples exist in other systems (e.g., Van Peer *et al.* 2001; Groom *et al.* 2004, Larcher *et al.* 2010).

3.2 Community composition and structure

During drought, a range of physiological, demographic and environmental factors interact to determine the impact of extreme water deficiencies on specific plant individuals and species. McDowell *et al.* (2008) recently reviewed these mechanisms; briefly, the key drivers of plant mortality under moisture stress are thought to be carbon starvation, the activity of biotic agents, and hydraulic failure, with the relevance of each depending on the intensity and duration of stress. Plants adopt a range of mechanisms to tolerate or resist drought, with intraspecific and interspecific variation found in water use efficiency (Farquhar *et al.* 1989), dormancy (Oram 1983), dehydrin expression (Volaire *et al.* 2001), extraction of water at low soil water potential (Volaire and Lelièvre 2001), senescence of aerial tissue (Volaire *et al.* 1998; Bolger *et al.* 2005), root structure (Van Splunder *et al.* 1996) and resource allocation (Aronson *et al.* 1993) to name a few. Indeed, most species use different physiological and anatomical mechanisms to protect tissue during periods of moisture stress (Scott 2000).

Plant survival also depends strongly on spatial heterogeneity in the landscape. Variation in soil moisture occurs at a range of spatial scales (Buckland *et al.* 1997; Gitlin *et al.* 2006; Dobrowski 2011; Godfree *et al.* 2011), and because plant mortality is non-linearly related to soil moisture content, microscale variation in water availability can critically influence plant survival under extreme drought (Godfree *et al.* 2011). At larger spatial scales, hydrological variation and the frequency of soil drought is a key driver of species assortment and community composition (Oberbauer & Billings 1981; Buckland *et al.* 1997; Yurkonis & Meiners 2006), a process which reflects variation in the ability of species to recover from drought (e.g., Tilman & El Haddi 1992; Stampfli & Zieter 2004) as much, or more, than tolerance of drought itself (Gutschick & BassiriRad 2003).

Given these sources of variation, it is not surprising that plant species tend to show highly differential mortality when placed under extreme drought in natural settings. During drought in the 1930's, Albertson and Weaver (1944, 1945) observed that survival among tree species in ravine environments ranged from 64% (*Celtis occidentalis*) to only 30% (*Ulmus americana*), and drought hardy species, such as *Juniperus virginiana*, had much lower mortality (1-37%) than all other co-occurring species, even persisting as monospecific stands on the most xeric sites (Albertson & Weaver 1945). In prairie communities, all species were affected but some, including *Andropogon furcatus*, suffered much lower mortality than others (e.g., *Andropogon scoparius*, *Stipa spartea*) which experienced over 80% mortality. Survivorship was linked strongly to depth of the rooting system, since during the drought soil water deficiencies gradually moved lower in the soil profile (Albertson & Weaver 1944). Interestingly, over the entire drought period the composition of the prairie community was in constant flux (Albertson & Weaver 1944). Interspecific variation in mortality has also been observed in numerous other studies (Tilman & El Haddi 1992; Condit *et al.* 1995; Gitlin *et al.* 2006), and in response to other climatic stressors (e.g., Stiles 1930; Barua *et al.* 2003; Henry & Molau 2003; Marchand *et al.* 2006; many others).

Under extreme, prolonged drought, rapid changes in vegetation composition can be persistent or effectively permanent. Perhaps the best documented example occurred during extreme drought in the 1950's, when the ecotone separating *Pinus ponderosa* forest from *Pinus edulis-Juniperus monosperma* woodland moved by 2 km or more within only five years. This change has persisted for at least 40 years (Allen & Breshears 1998). One of the most significant persistent changes observed by Albertson & Weaver (1944) was the expansion of wheat grass (*Agropyron smithii*), which was favoured by moist springs and dry summers, at the expense of other species such as *Andropogon furcatus*. Elsewhere, Gitlin *et al.* (2006)

showed that continuation of extreme drought conditions in the southwestern United States experienced during 2002 would likely result in significant change to the composition and structure of entire plant communities and ecosystems. Similarly, Tilman & El Haddi (1992), in a controlled experiment, showed that post-drought recovery of grassland vegetation was not accompanied by a significant recovery in species richness, and suggested that post-drought recruitment limitation may determine the richness of prairie plant communities. Here, drought with a return interval of approximately 50 years was sufficient to significantly alter local species richness (Tilman & El Haddi 1992). Drought can also alter competitive relationships among species, resulting in stratification of plant communities along hydrological gradients (Buckland *et al.* 1997) or promotion of invasion by fast-growing annuals (White *et al.* 2001).

4. Case study: Change in composition of an Australian semi-arid grassland during and following extreme drought

The case studies above provide overwhelming evidence that extreme climatic events, and especially drought, can drive rapid changes in plant community composition by causing differential rates of mortality and recovery among plant species. A central remaining challenge to ecologists, however, is to understand when, and under what threshold conditions, abiotic stress will lead to changes of large magnitude (McDowell *et al.* 2008), knowledge that will be essential if we are to accurately predict the impact of climate change on natural vegetation globally. In this section I aim to improve our understanding of such processes by drawing conclusions from a study in which I investigate the impact of a multi-year period of exceptional drought on a semi-arid Australian grassland ecosystem.

4.1 Background: The "Millenium Drought" in Australia

Between 1997 and 2009 much of south-eastern Australia was affected by an extremely severe and protracted drought known as the "Millennium Drought" (Whitaker 2006; Bond *et al.* 2008). During this period, annual rainfall was well below average, especially during autumn (March-May), while maximum and minimum temperatures were at or near record highs (Murphy & Timbal 2008). Over the past half century mean maximum and minimum atmospheric temperatures have increased over most of Australia (Nicholls *et al.* 2004; Nicholls 2006) and extreme temperature events have become more frequent (Collins *et al.* 2000). Recent droughts also may have become more severe due to increased evaporation associated with warmer temperatures (Nicholls 2004; Cai & Cowan 2008). Severe drought is a regular phenomenon in Australia (Ummenhofer *et al.* 2009), and there is a strong suggestion that rising atmospheric greenhouse gas concentrations have played a role in the development of rainfall deficiencies and elevated temperatures across southern parts of Australia in recent decades (Timbal *et al.* 2006; Murphy & Timbal 2008).

The most extreme droughts in Australia impact on regional agriculture, hydrology, ecosystem function, and the population dynamics of flora and fauna. The 1997-2009 drought was no exception: extreme low rainfall and high temperatures (Cai & Cowan 2008) during the drought resulted in the near total loss of surface water and a persistent decline of groundwater (Leblanc *et al.* 2009), which in turn reduced agricultural productivity significantly (Horridge *et al.* 2005; Pook *et al.* 2009). Large impacts were observed on aquatic flora and fauna (Bond *et al.* 2008; references therein), with extensive tracts of the key riparian tree species *Eucalyptus camaldulensis* dying in the lower Murray-Darling Basin (Cunningham *et al.* 2009). While Australian ecosystems are commonly perceived to be resilient to drought (e.g., Bond *et al.* 2008), their ability to recover from events of this severity and duration, or the increasingly extreme events of the future (Mpelasoka *et al.* 2008), remains open to question (Godfree *et al.* 2011).

4.2 Study objectives

The objective of this study was to improve understanding of the role of extreme climatic events as drivers of rapid ecosystem change by comparing the landscape-level responses of different plant species to drought across multiple grassland habitats, and investigate whether the observed phytosociological changes could be explained by simple predictors including statistical quantification of drought severity, topography, and knowledge of the broader range distributions and habitat affinities of the study species. I specifically addressed the following hypotheses:

1. prolonged drought of unprecedented severity will result in high mortality across a range of grassland plant species;
2. mortality will be highest in the most xeric sites and mesic low-lying habitats will act as refugia during drought;
3. post-drought recruitment and recovery will be the primary drivers of post-drought community composition and structure; and
4. species with ranges that extend further into drier regions will have higher survival and recruitment than species with more mesic distributions.

Evidence supporting these hypotheses might indicate that at least some basic principles could be generally applied to the study of extreme events to improve prediction of their impacts on vegetation systems.

4.3 The study system

The study was conducted in a high quality remnant semi-arid grassland located approximately 30 km to the east of West Wyalong in central NSW (Fig. 2). The choice of a semi-arid biome reflects the general view that these ecosystems are highly susceptible to shifts in climate (Allen & Breshears 1998; Holmgren *et al.* 2006). The topography of the 34 hectare site is mainly flat with extensive treeless plains dissected by a series of small creeks that incise up to 2-3 m below the surrounding terrain. While low, this topographic heterogeneity does generate a range of habitat types characterised by different floral assemblages (illustrated in Fig. 2). Prior to 2006, grasslands dominated by the tussock grass *Austrostipa aristiglumis* (F.Muell.) S.W.L.Jacobs & J.Everett (plains grass) and *Panicum prolutum* F.Muell [= *Walwhalleya proluta* (F.Muell.) Wills & J.J.Bruhl] (rigid panic) occurred on flat and mesic low-lying terraces and gullies, while xeric, sloping terrain was dominated by the small perennial shrub *Leiocarpa panaetioides* (DC.) Paul G. Wilson (woolly buttons). Natural grasslands and grassy woodlands dominated by *A. aristiglumis* and other grassland species once occurred widely across inland NSW (Benson *et al.* 1997), but unfortunately most have been degraded by overgrazing, cultivation, and weed invasion and the remaining areas have now been listed as critically endangered under the *Environment Protection and Biodiversity Act 1999* (Threatened Species Scientific Committee 2008). *A. aristiglumis, P. prolutum* and *L. panaetioides* all grow on the western slopes and plains regions of NSW, Queensland and Victoria, but *P. prolutum* and *L. panaetioides* extend further into the drier, semi-arid zone than *A. aristiglumis*, and *L. panaetioides* occurs in arid habitats in far western NSW (Fig. 3).

Fig. 2. Location and characteristics of the semi-arid grassland study site in south-central NSW, Australia. The location of the site is shown in the inset as a red star. The main habitat types along with the study species *Leiocarpa panaetioides*, *Austrostipa aristiglumis* and *Panicum* spp. (mainly *P. prolutum* with some *P. decompositum*) are labelled on the photograph taken looking north-east near the centre of the study site. The terrace habitat lies around 2 m in elevation below the *Austrostipa* flat.

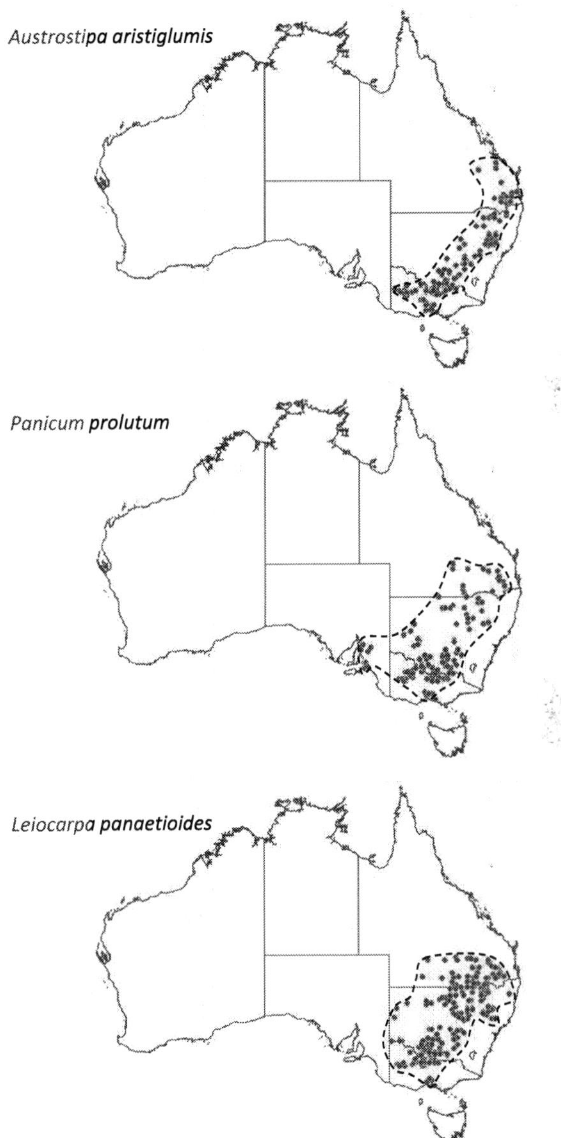

Fig. 3. Approximate distribution of the three study species in Australia. Dots represent locations where herbarium records exist for each species (some incorrect records or those representing adventive occurrences have been removed). Data derived from Australia's Virtual Herbarium, Council of Heads of Australasian Herbaria Inc. (http://chah.gov.au/avh/public_query.jsp)

4.4 Climatic conditions during 2006-2007

Like most of south-eastern Australia, the study site was affected by chronic drought between 2001 and 2009, with particularly severe conditions occurring during 2006 and 2007. At Wyalong Post Office (S 33.93°, E 147.24°), the nearest high quality meteorological station to the study site, only 181 mm of rain fell in 2006 – easily the driest year since at least 1900 and 62% below the 1900 to 2009 average of 474 mm (Fig. 4a). Rainfall was also low during 2007 (356

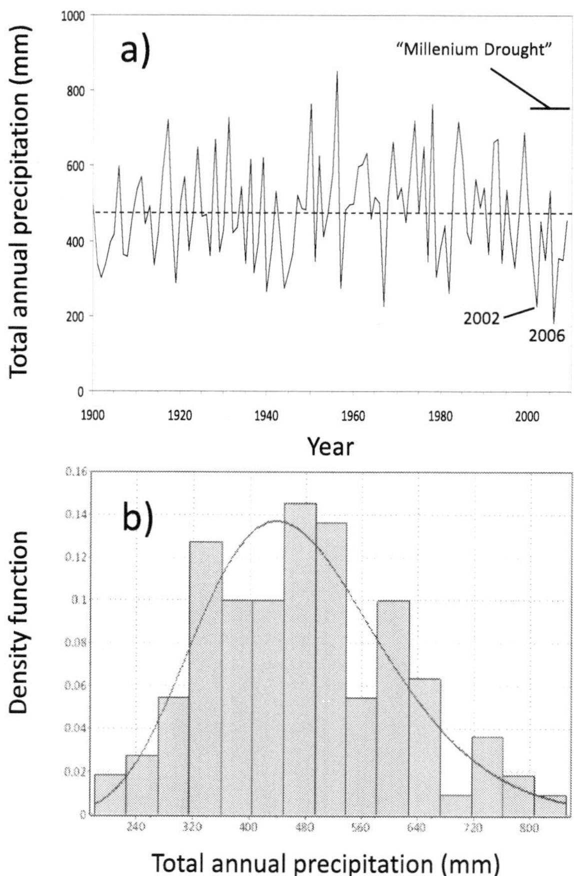

Fig. 4. a) Total annual rainfall at Wyalong Post Office, NSW, Australia, 1900-2009. The extreme drought years of 2006 and 2002 are indicated, along with the timeframe of the "Millenium Drought", a protracted period of rainfall deficiencies experienced across much of south-eastern Australia. b) Total annual rainfall fit with gamma distribution (α = 12.50, β = 38.08). Annual rainfall data were obtained from the Australian Bureau of Meteorology's Patched Point Dataset (available at http://www.longpaddock.qld.gov.au/silo/). Approximately 87% of data were station data with the remaining being interpolated daily observations (mainly pre-1950).

mm) and 2008 (350 mm), and collectively over the period 2001-2009, annual rainfall averaged only 362 mm, 24% below the long term average (Fig. 4a). Temperatures during this period were also at or near record levels (see below), which exacerbated drought severity (see Nicholls 2004).

4.5 Field surveys

The exceptionally dry conditions of 2006 and 2007 presented an ideal opportunity to quantify the responses of *A. aristiglumis*, *P. prolutum* and *L. panaetioides* to acute water deficiencies in different habitat types across the study site. For each species, I quantified rates of population mortality and recruitment across six different topoedaphic habitat types (for detailed description of habitat types see Godfree *et al.* 2011; Fig. 2) based on surveys conducted in three representative transects which spanned the study site. I estimated mortality rates based on counts of dead and live adult plants, and recruitment rates based on counts of juvenile plants that had established in 2007. Further details (for *A. aristiglumis*), along with soil water data documenting the severity of the drought, are provided in Godfree *et al.* (2011).

Survey data were used to estimate pre-drought population densities of each species in each habitat type (based on adult plant density), while the post-drought population density was determined based on total counts of surviving adult and juvenile plants. Species survival rates were determined based on estimated pre- and post-drought adult plant densities, and adult plant replacement rates were calculated as the number of recruits per number of dead plants recorded at the time of the survey. The areal contribution of each habitat type to the total site area was determined based on the total intercepted length (m) of each habitat across all three transects.

4.6 Demographic change in response to drought

Prior to the major mortality event that occurred in late 2006, *A. aristiglumis* dominated the more mesic and flat habitats at the study site, while *P. prolutum* was most abundant in the *Panicum* flats habitat (Fig. 5a). In drier habitats (south- and north-facing slopes) both grasses had much lower densities, being largely replaced by the more xerophytic shrub *L. panaetioides* (Fig. 5a). By late 2007, however, the population density of *A. aristiglumis* and *P. prolutum* had changed considerably (Fig. 5b), reflecting drought-induced mortality followed by a major recruitment event in autumn and winter 2007.

During the most extreme phase of the drought mortality of *A. aristiglumis* and *P. prolutum* occurred in all habitats, with populations in the more xeric, sloping habitats suffering losses of 90% or more (Fig. 6a). *P. prolutum* suffered >65% mortality in all habitats and was eliminated from north-facing slopes (the most xeric habitat), while *A. aristiglumis* survival exceeded 30% only in the most mesic terrace environments (Fig. 6a). In contrast, *L. panaetioides* survival was at least 70% in all but one habitat (north-facing slopes) and no plants died in the terrace and gully habitats (although density was low to start with). Mean site survival, averaged across habitat types, was 21% for *A. aristiglumis*, 17% for *P. prolutum*, but 79% for *L. panaetioides*.

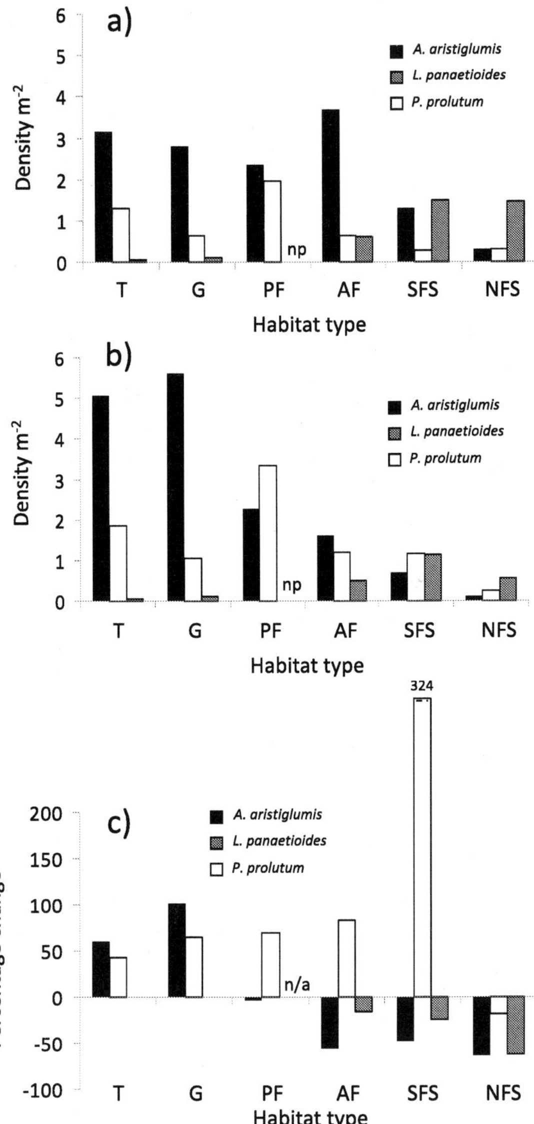

Fig. 5. Demographic changes observed across the study site July 2006-December 2007. a) Estimated pre-drought population density (July 2006). b) Post-drought population density (December 2007). c) Percentage change in density July 2006 to December 2007. T = terrace, G = gully, PF = *Panicum* flat, AF = *Austrostipa* flat, SFS = south-facing slope, NFS = north-facing slope. Habitat types are arranged from most mesic (terraces) to most xeric (north-facing slopes). np = species not present in the habitat type; n/a = not applicable.

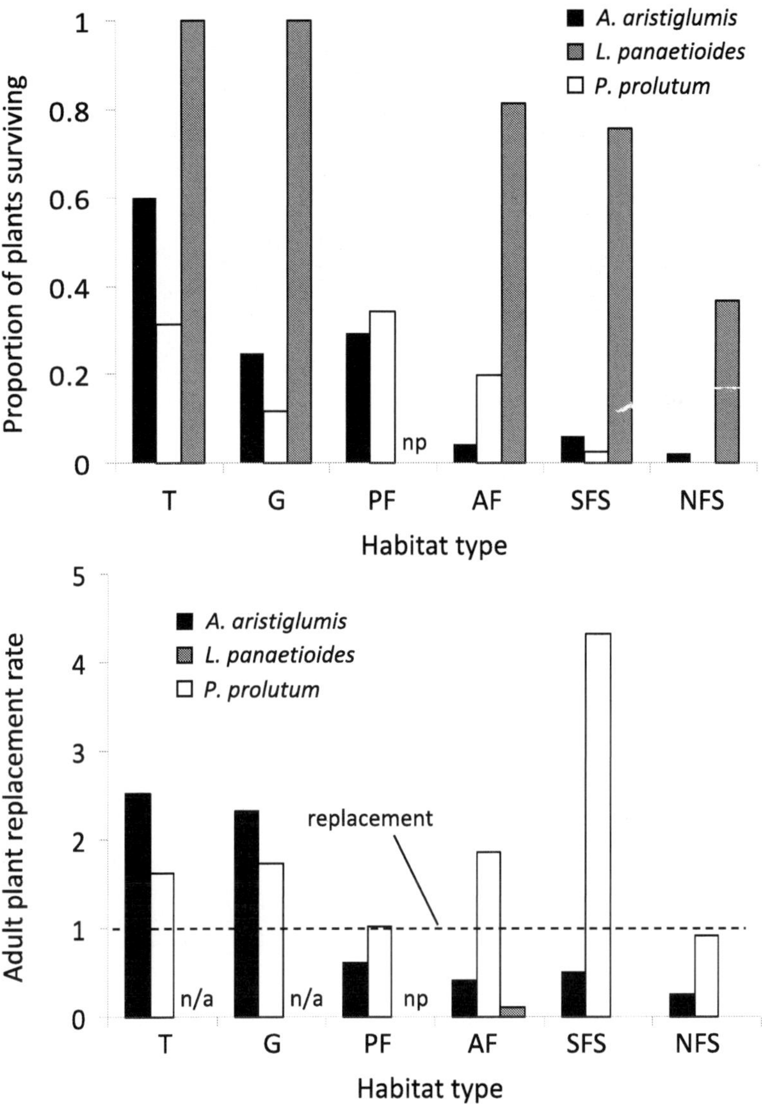

Fig. 6. Survival and recovery of plant populations observed across the study site July 2006-December 2007. a) Survival rate of adult plants present in July 2006. b) Rate of replacement of dead adult plants by recruits. The dotted line indicates the replacement rate (1) where numbers of new recruits exactly equals that of dead adult plants. T = terrace, G = gully, PF = *Panicum* flat, AF = *Austrostipa* flat, SFS = south-facing slope, NFS = north-facing slope. Habitat types are arranged from most mesic (terraces) to most xeric (north-facing slopes). np = species not present in the habitat type; n/a = not applicable.

Spatially, each of the habitats accounted for a different proportion of the area of the study site, and the impact of the drought on the total population size of each species reflected both mortality rate and the spatial extent of each habitat in which it occurred. For *A. aristiglumis*, only 16% of all plants at the site survived the drought, since high mortality (96%) was observed in the extensive *Austrostipa* flats habitat. *P. prolutum* survival was highest (34%) in habitats where its original density was also highest (2.0 m^{-2}), and so overall survival (27%) exceeded that of *A. aristiglumis*. The relatively high total survivorship of *L. panaetioides* (73%) also strongly reflected generally high survivorship across habitat types.

Following rain in autumn 2007, significant recruitment of *A. aristiglumis* and *P. prolutum*, but not *L. panaetioides*, occurred across the study site. Recruits exceeded the number of drought-killed plants in gullies and terraces (*A. aristiglumis*) and gullies, terraces, *Austrostipa* flats and south-facing slopes (*P. prolutum*). Recruitment was minimal in *L. panaetioides*, and in no habitats did recruits fully replace plants that succumbed to the drought (i.e., adult plant replacement rate < 1; Fig. 6b). These differences in mortality and recruitment over the 2006-2007 study period resulted in a significant spatial redistribution of plant species across the study site, and a landscape-level change in community composition. By late 2007, *A. aristiglumis* was predominantly restricted to terrace and gully habitats, where post-drought populations were actually larger than pre-drought populations (Fig. 5b), and was virtually absent from xeric habitats (Fig. 5b). *P. prolutum*, in contrast, increased in abundance in most habitats, and became co-dominant with *A. aristiglumis* in *Austrostipa* flat and south-facing slopes habitats, and increased its dominance in the *Panicum* flats habitat (Fig. 5b). *L. panaetioides* declined in all habitats, but maintained dominance on the most xeric north-facing slope environments (Fig. 5c). In reality, however, the decline of all species in this habitat (Fig. 5c) left it essentially bare (Fig. 2), a condition which has been largely maintained for at least 3 years since (R. Godfree, personal observation).

4.7 Implications for predicting the impacts of extreme events

The results of this work, and those published previously (Godfree *et al.* 2011), support the hypothesis that extreme climatic events can significantly reconfigure landscape-scale vegetation mosaics within relatively short timeframes via direct mortality of established plants. While other studies have reported very high plant mortality during extreme drought (e.g., Albertson and Weaver 1944, 1945; Allen & Breshears 1998; Breshears *et al.* 2005; Gitlin *et al.* 2006; Edwards & Krockenberger 2006), the >90% mortality rates observed in this study in two community dominant species do appear to be unusually high. Perhaps this ultimately reflects the magnitude of rainfall deficiencies observed at the study site – 2006 was the driest year in at least a century, and, surprisingly, 20% drier than the next driest year (2002). Heavy mortality can have a range of important demographic and genetic consequences for the long-term fitness of plant populations and species, and if the frequency of events such as the one described here increase under projected climatic change (Meehl *et al.* 2000; Hennessy *et al.* 2008; Planton *et al.* 2008), the consequences for the conservation of native vegetation are likely to be significant.

As a result of the 2006-2007 drought, populations of two of the three study species (*A. aristiglumis* and *L. panaetioides*) shifted lower in the landscape, with the dominance of *A. aristiglumis* declining in all but the most mesic terraces and gully environments. Consistent with hypothesis 2, all three species suffered the greatest mortality in the more xeric environments, and at the height of the drought, live plants were restricted almost entirely to mesic refugial habitats (with the exception of *L. panaetioides*, which although completely defoliated, did survive in modest numbers in drier areas). Interestingly, however, there was also some evidence that pre-drought habitat suitability was not a good predictor of drought survival or post-drought recruitment. For example, *Austrostipa* and *Leiocarpa* both declined most in the habitats in which their pre-drought populations were most dense (Figs. 5a, 6a), and recruitment of *P. prolutum* was poorer in the *Panicum* flats habitat than in most other habitats. In the case of *A. aristiglumis* and *P. prolutum* it is possible that terrace habitats were actually more suitable for these species (the plants present, although of lower density, were much larger than in drier habitats). This did not seem to apply to *L. panaetioides* - plants were large and abundant on the most xeric sites. Perhaps *L. panaetioides* is competitively excluded from mesic habitats by *A. aristiglumis* and *P. prolutum* and drought alters this competitive hierarchy, a pattern that has been observed elsewhere (White *et al.* 2001). Regardless, mesic refugia, albeit small in spatial scale (Godfree *et al.* 2011) clearly play a key role in ensuring survival of a range of species in a given plant community during extreme drought, including those that are adapted to drier conditions.

The pattern of change among species observed at the study site did not support the hypothesis that post-drought recruitment and recovery are the primary drivers of post-drought community composition for all species, the demographic responses of which were highly idiosyncratic. The final distribution of *P. prolutum* mainly reflected strong post-drought recruitment across multiple habitats, but *A. aristiglumis* was most abundant in terrace habitats due to significant post drought recruitment and high mortality. *Leiocarpa* density depended almost solely on high drought survivorship, and the final composition of the vegetation found in the more xeric habitats primarily reflected the drought hardy nature of this species. The presence of such complex patterns is perhaps not surprising given the diversity of strategies displayed by plants for ensuring survival through drought and other abiotic stresses (e.g., Barrett 1998; Mal & Lovett-Doust 2005; McDowell *et al.* 2008) but it does indicate that post-drought community composition jointly reflects the processes of mortality and recovery in heterogeneous environments.

Finally, from a practical point of view, the species-level responses observed here do partly support the hypothesis that population behaviour in response to drought can be predicted by their broad climatic envelopes (see McDowell *et al.* 2008; references therein). Drought survival of *A. aristiglumis* was much lower than that of the more arid-adapted (see Fig. 3) *L. panaetioides* in all habitats, with differences in survivorship between the two species being greatest in the more xeric, sloping habitat types (Fig. 6a). Differences between *A. aristiglumis* and *P. prolutum* were not as clear, since survival of *A. aristiglumis* was actually higher than that of *P. prolutum* in three habitats (terraces, gullies and north-facing slopes; Fig. 6a), which does not appear to be consistent with the fact that *P. prolutum* is capable of surviving in much drier regions than *A. aristiglumis* (Fig. 3). On the other hand, *P. prolutum* populations did perform better overall than *A. aristiglumis*, mainly as a result of a higher rate of post-drought recruitment from the seedbank (Fig. 6b). Perhaps the presence of a large persistent seedbank, rather than high drought survivorship, explains why *P. prolutum* grows in areas that are considerably drier than *A. aristiglumis* can tolerate (e.g., far western NSW and Queensland; Fig. 3). Such traits are known to be linked to population fitness and reproductive assurance in arid environments (Auld *et al.* 1995; Facelli *et al.* 2005).

4.8 Drought severity and species responses

This study has shown that extreme drought drives changes to ecosystem structure and composition by impacting on mortality and recruitment of plant populations, and that these processes may be broadly predictable given an understanding of drought severity and community composition. But this observation begs the questions: how extreme do droughts need to be to result in changes of this magnitude, and how much might drought severity increase under anthropogenic global warming?

Let us return to the study system at hand. As mentioned, drought conditions at the study site during 2006, when only 181 mm of rain fell, can reasonably be described as being of unprecedented severity with respect to the 1900-2009 instrumental record (Fig. 4a). Based on historical annual rainfall data for Wyalong, NSW, fit with a gamma probability distribution (Fig. 4b), an annual rainfall of 181 mm has a predicted return interval of 453 years, and although care needs to be taken when such low probabilities are involved, it is obviously an exceptionally rare occurrence indeed. As noted previously, the mortality rates observed in this study are consistent with the impacts of exceptional drought observed in other systems. But an event like 2006 is highly unusual, and somewhat less extreme events are much more likely to occur, even under climate change. Unfortunately, we have little evidence beyond anecdotal sources whether less severe droughts have had similar effects on this grassland vegetation.

In 2002, when 225 mm of rain fell at Wyalong (the second driest year on record, return interval = 84 years) I observed mortality of around 50% of *A. aristiglumis* at a nearby grassland site, but mortality was lower in wetter, low-lying sites. This might, however, reflect the fact that 2006 occurred after many years of drought, in contrast to 2002. During 2002-2009 significant tree death occurred in central NSW, including around the study area, but similar events have been observed previously in NSW, for example during the 1896-1902 drought (McKeon *et al.* 2004). We simply do not understand the exact conditions that resulted in the observed changes at the study site, beyond the fact that many weeks of dry weather occurred during spring and summer at the end of a very dry year in the middle of a decade-long drought. Perhaps we can at best speculate that substantial mortality of natural, minimally disturbed grassland in the study region is most likely to occur when extremely dry years (roughly 250 mm, around 50% of average; return interval 40 years) occur during an extended period of below-average rainfall. There might have been as many as 3 to 5 instances of such conditions over the past 110 years (Fig. 4a). However, it may require extremely rare events, like 2006, to generate the high levels of mortality observed in this study.

A further complication is that the actual level of water stress experienced by a plant population is a function of evapotranspirational demand relative to water availability. The extremely dry conditions experienced in central NSW during the Millenium Drought were exacerbated by high temperatures (Cai & Cowan 2008), which suggests that consideration of rainfall deficiencies alone would underestimate the severity of the drought. Data from Wagga Wagga AMO (S 35.16°, E 147.46°; Fig. 7), the nearest station with suitable observations for estimating potential evapotranspiration (ETo), show the severity of 2006 in terms of low rainfall (Fig. 7a), high ETo (Fig. 7b), and very low atmospheric water balance (AWB; calculated here as annual rainfall – ETo). Indeed, the AWB was considerably lower in 2006 than any year since at least 1970 (Fig. 7c). Given that 2006 was drier at Wyalong than at Wagga Wagga compared with other years, these data support the contention that moisture stress experienced at the study site during 2006 was the most extreme in many decades, if not the last century.

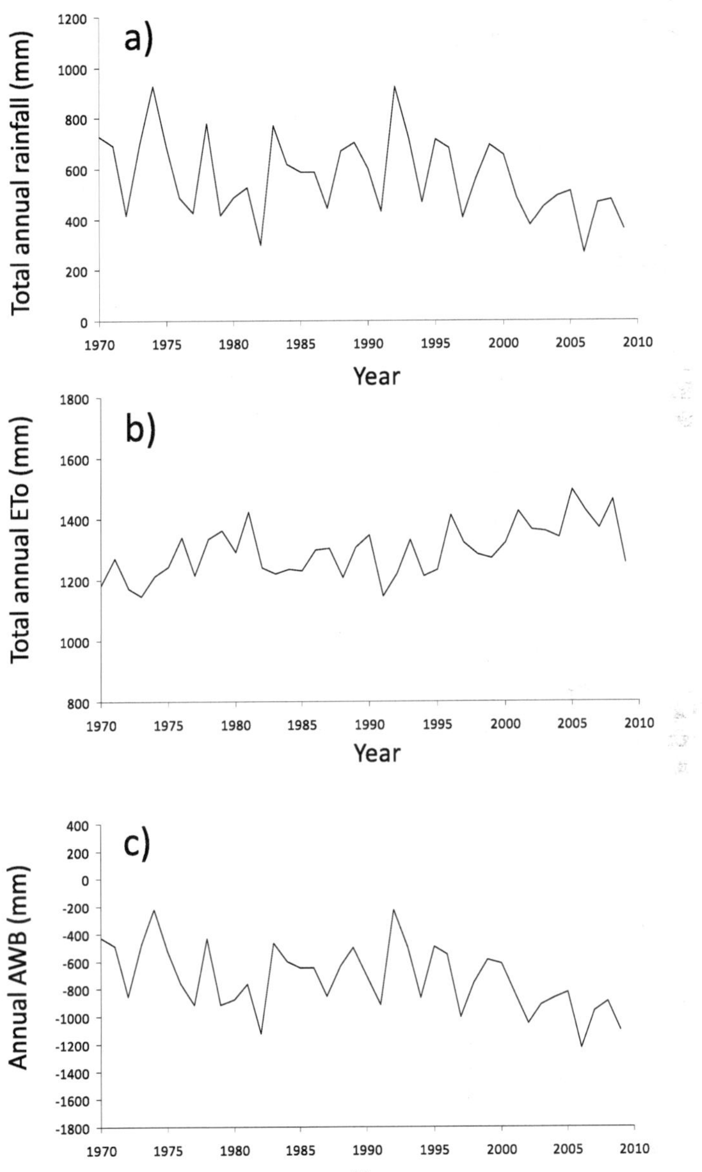

Fig. 7. Climatic conditions experienced at Wagga Wagga, NSW. a) Total annual rainfall (R) 1970-2009. b) Estimated total annual potential evapotranspiration (ETo). c) Annual atmospheric water balance (AWB), determined as AWB = R - ETo. Based on these data, AWB during 2006 was the lowest since at least 1970.

4.9 Predictions under climate change

We may conclude this study by making some very rough guesses as to the possible impacts of drought on grassland vegetation at the study site under climate change. First, we can estimate changes in drought severity at the study site based on the projections of global circulation models (here I use 50th percentile, medium emissions projections; see www.climatechangeinaustralia.gov.au). Current estimates for changes in precipitation in the study region by the year 2070, relative to the 1980-1990 baseline period, are for declines of 0%, 3.5%, 15%, and 15% for summer, autumn, winter and spring respectively, with changes in ETo of 6%, 10%, 14%, and 3%. Historically, mean annual AWB at Wagga Wagga (1970-2009) is -720 mm (Fig. 7c). If we use the extremely simple approach of modifying the observed 1970-2009 data according to these projections, we obtain a 2070 estimate for annual AWB of -860 mm, a 19% increase. Under the current climate regime (1970-2009) a year like 2006 (AWB = -1229 mm) has a return interval of 62 years, but by 2070, the return interval becomes 19 years. This suggests that, if a similar condition holds at Wyalong, the frequency of years in which significant mortality might occur in native plant communities could increase by around three-fold, possibly resulting in persistent shifts in vegetation composition similar to those observed in this study. However, given the low mortality of the three study species in terrace habitats even under the conditions experienced during 2006, their long-term persistence at the study site, albeit in a possibly restricted manner, seems virtually certain.

A final line of evidence supports this prediction. Nyngan, NSW, which lies approximately 450 km north of Wyalong, is an approximate 2070 climate analogue for the study site. All three study species occur at, or near Nyngan (although *A. aristiglumis* is restricted to very mesic riverine habitats), and *P. prolutum* and *L. panaetioides* occur much further west in drier areas. As mentioned, however, it is exceedingly difficult to make accurate predictions of this kind, and to account for mitigating factors such as atmospheric CO_2 enrichment (Koch *et al.* 2004), population-level evolution for drought tolerance, competition (White *et al.* 2001) or many other potentially important factors (see Godfree *et al.* 2011) that are known to affect the response of plant species to drought.

5. Conclusions

Understanding the role of extreme climatic events as drivers of contemporary and future vegetation change is one of the greatest challenges that ecologists face today. Extreme events are difficult to study, and the responses of plants, species and communities to abiotic stress are contingent on a broad array of physiological, demographic and landscape-scale process that are often nonlinear in nature. In this paper I have provided evidence that extreme climatic events, and especially drought, have the capacity to rapidly alter the structure and composition of plant communities, with the magnitude of change roughly reflecting the statistical severity of the conditions. The majority of droughts that cause the highest rates of mortality among extant plant populations seem to be exceptionally rare events, occurring only on multi-decadal or century timescales, although quantifying the exact relationship between drought severity and plant mortality clearly needs further research.

The results of surveys conducted in a semi-arid grassland site in Australia suggest that changes in community structure and composition following drought reflect the processes of both mortality and post-drought recruitment and recovery, and that the demographic responses of species to extreme water stress are highly idiosyncratic. Nonetheless, drought performance of individual species does appear to be at least partly predictable based on the nature of their climatic envelopes. The data also suggest that, if drought is severe enough, topoedaphic refugia are crucial for the survival of a broad suite of species, not just those that favour mesic habitats. Finally, quantification of the statistical distribution of rainfall and atmospheric water balance in the semi-arid study region suggests that relatively modest changes in rainfall and evaporation could lead to large changes in the frequency and severity of extreme drought in coming centuries.

Much remains to be understood about the mechanisms and conditions under which extreme climatic events act as drivers of vegetation mortality, and the specific characteristics of populations, species and communities that predispose them to rapid abiotically-driven change. The overall objective of ecology is to develop theory that usefully predicts phenomena in nature, and working towards development of a theory that improves our understanding of the relationship between extreme climatic events and vegetation change is certain to be a fruitful area of ongoing ecological research.

6. Acknowledgments

I would like to thank Brendan Lepschi for assistance with data collection and acquisition of herbarium records.

7. References

Acuna-Soto R, Stahle DW, Cleaveland MK and Therrell MD (2002) Megadrought and megadeath in 16th Century Mexico. *Revista Biomédica* 13: 289-292.

Acuna-Soto R., Stahle D. W., Therrell M. D., Gomez Chavez S. and Cleaveland M. K. (2005) Drought, epidemic disease, and the fall of the classic period cultures in Mesoamerica (AD 750-950). Hemorrhagic fevers as a cause of massive population loss. *Medical Hypotheses* 65: 405-409.

Ahlstrom RVN, Van West CR, and Dean JS (1995) Environmental and chronological factors in the Mesa Verde-Northern Rio Grande Migration. *Journal of Anthropological Archaeology* 14: 125-142.

Albertson FW and Weaver JE (1944) Nature and degree of recovery of grassland from the great drought of 1933 to 1940. *Ecological Monographs* 14: 393-479.

Albertson FW and Weaver JE (1945) Injury and death or recovery of trees in prairie climate. *Ecological Monographs* 15: 393-433.

Allen CD, Macalady AK, Chenchouni H, Bachelet D, McDowell N, Vennetier M, Kitzberger T, Rigling A, Breshears DD, Hogg EH, Gonzalez P, Fensham R, Zhang Z, Castro J, Demidova N, Lim J-H, Allard G, Running SW, Semerci A and Cobb N (2010) A global overview of drought and heat-induced tree mortality reveals emerging climate change risks for forests. *Forest Ecology and Management* 259: 660-684.

Allen CD and Breshears DD (1998) Drought-induced shift of a forest-woodland ecotone: Rapid landscape response to climatic variation. *Proceedings of the National Academy of Sciences of the USA* 95: 14839-14842.

Aronson J., Kigel J. and Shmida A. (1993) Reproductive allocation strategies in desert and Mediterranean populations of annual plants grown with and without water stress. *Oecologia* 93: 336-342.

Auld TD (1995) Soil seedbank patterns of four trees and shrubs from arid Australia. *Journal of Arid Environments* 29: 33-45.

Barrett SCH (1998) The evolution of mating strategies in flowering plants. *Trends in Plant Science* 3: 335-341.

Barua D, Downs CA and Heckathorn SA (2003) Variation in chloroplast small heat-shock protein function is a major determinant of variation in thermotolerance of photosynthetic electron transport among ecotypes of *Chenopodium album*. *Functional Plant Biology* 30: 1071-1079.

Bell JL, Sloan LC and Snyder MA (2004) Regional changes in extreme climatic events: a future climate scenario. *Journal of Climate* 17: 81-87.

Beniston M and Stephenson DB (2004) Extreme climatic events and their evolution under changing climatic conditions. *Global and Planetary Change* 44: 1-9.

Benson JS, Ashby EM and Porteners MF (1997) The native grasslands of the Riverine Plain, New South Wales. *Cunninghamia* 5: 1-48.

Benson L, Petersen K and Stein J (2007a) Anasazi (pre-Columbian native-American) migrations during the middle-12th and late-13th centuries – were they drought induced? *Climatic Change* 83: 187-213.

Benson LV, Berry MS, Jolie EA, Spangler JD, Stahle DW and Hattori EM (2007b) Possible impacts of early-11th-, middle 12th-, and late-13th-century droughts on western Native Americans and the Mississippian Cahokians. *Quaternary Science Reviews* 26: 336-350.

Bolger TP, Rivelli AR and Garden DL (2005) Drought resistance of native and introduced perennial grasses of south-eastern Australia. *Australian Journal of Agricultural Research* 56: 1261-1267.

Bond NR, Lake PS and Arthington AH (2008) The impacts of drought on freshwater ecosystems: and Australian perspective. *Hydrobiologia* 600: 3-16.

Breshears DD, Cobb NS, Rich PM, Price KP, Allen CD, Balice RG, Romme WH, Kastens JH, Floyd ML, Belnap J, Anderson JJ, Myers OB and Meyer CW (2005) Regional vegetation die-off in response to global-change-type drought. *Proceedings of the National Academy of Sciences of the USA* 102: 15144-15148.

Buckland SM, Grime JP, Hodgson JG and Thompson K (1997) A comparison of plant responses to the extreme drought of 1995 in northern England. *Journal of Ecology* 85: 875-882.

Cai W and Cowan T (2008) Evidence of impacts from rising temperatures on inflows to the Murray-Darling Basin. *Geophysical Research Letters* 35, L07701, doi:10.1029/2008GL033390.

Cho H-K, Bowman KP and North GR (2004) A comparison of gamma and lognormal distributions for characterising satellite rain rates from the Tropical Rainfall Measuring Mission. *Journal of Applied Meteorology* 43: 1586-1597.

Coles S (2001) An introduction to statistical modelling of extreme values. Springer, London, UK. 208p.

Collins DA, Della-Marta PM, Plummer N and Trewin BC (2000) Trends in annual frequencies of extreme temperature events in Australia. *Australia Meteorological Magazine* 49: 277-292.

Condit R, Hubbell SP and Foster RB (1995) Mortality rates of 205 neotropical tree and shrub species and the impact of a severe drought. *Ecological Monographs* 65: 419-439.

Cook ER, Woodhouse CA, Eakin CM, Meko DM and Stahle DW (2004) Long-term aridity changes in the Western United States. *Science* 306: 1015-1018.

Cunningham SC, Mac Nally R, Read J, Baker PJ, White M, Thompson JR and Griffioen P (2009) A robust technique for mapping vegetation condition across a major river system. *Ecosystems* 12: 207-219.

Dobrowski SZ (2011) A climatic basis for microrefugia: the influence of terrain on climate. *Global Change Biology* 17: 1022-1035.

Dudley JP, Criag GC, Gibson DStC, Haynes G and Klimowicz J (2001) Drought mortality of bush elephants in Hwange National Park, Zimbabwe. *African Journal of Ecology* 39: 187-194.

Easterling DR, Meehl GA, Parmesan C, Changnon SA, Karl TR and Mearns LO (2000) Climate extremes: observations, modeling, and impacts. *Science* 289: 2068-2074.

Edwards W and Krockenberger A (2006) Seedling mortality due to drought and fire associated with the 2002 El Niño event in a tropical rain forest in north-east Queensland, Australia. *Biotropica* 38: 16-26.

Endfield GH, Fernández Tejedo I and O'Hara Sl (2004) Drought and disputes, deluge and dearth: climatic variability and human response in colonial Oaxaca, Mexico. *Journal of Historical Geography* 30: 249-276.

Facelli JM, Chesson P and Barnes N (2005) Differences in seed biology of annual plants in arid lands: a key ingredient of the storage effect. *Ecology* 86: 2998-3006.

Farquhar GD, Ehleringer JR and Hubick KT (1989) Carbon isotope discrimination and photosynthesis. *Annual Reviews of Ecology and Systematics* 40: 503-537.

Fensham RJ and Holman JE (1999) Temporal and spatial patterns in drought-related tree dieback in Australian savanna. *Journal of Applied Ecology* 36: 1035-1050.

Gaines SD and Denny MW (1993) The largest, smallest, highest, lowest, longest, and shortest: extremes in ecology. *Ecology* 74: 1677-1692.

Gitlin AR, Sthultz CM, Bowker MA, Stumpf S, Paxton KL, Kennedy K, Muñoz A, Bailey JK and Whitham TG (2006) Mortality gradients within and among dominant plant populations as barometers of ecosystem change during extreme drought. *Conservation Biology* 20: 1477-1486.

Godfree R, Lepschi B, Reside A, Bolger T, Robertson B, Marshall D and Carnegie M (2011) Multiscale topoedaphic heterogeneity increases resilience and resistance of a dominant grassland species to extreme drought and climate change. *Global Change Biology* 17: 943-958.

Goswami BN, Venugopal V, Sengupta D, Madhusoodanan MS and Xavier PK (2006) Increasing trend of extreme rain events over India in a warming environment. *Science* 314: 1442-1445.

Groom PK, Lamont BB, Leighton S, Leighton P and Burrows C (2004) Heat damage in schlerophylls is influenced by their leaf properties and plant environment. *Ecoscience* 11: 94-101.

Gutschick VP and BassiriRad H (2003) Extreme events as shaping physiology, ecology, and evolution of plants: toward a unified definition and evaluation of their consequences. *New Phytologist* 160: 21-42.

Hennessy K, Fawcett R, Kirono D, Mpelasoka F, Jones D, Bathols J, Whetton P, Stafford Smith M, Howden M, Mitchell C and Plummer N (2008) An assessment of the impact of climate change on the nature and frequency of exceptional climatic events. Commonwealth of Australia; available online at www.daff.gov.au.

Henry GHR and Molau U (2003) Tundra plants and climate change: the International Tundra Experiment (ITEX). *Global Change Biology* 3: 1-9.

Hodell DA, Brenner M and Curtis JH (2005) Terminal Classic drought in the northern Maya lowlands inferred from multiple sediment cores in Lake Chichancanab (Mexico). *Quaternary Science Reviews* 24: 1413-1427.

Hogg EH, Brandt JP and Michaelian M (2008) Impacts of a regional drought on the productivity, dieback and biomass of western Canadian aspen forests. *Canadian Journal of Forest Research* 38: 1373-1384.

Holmgren M, Stapp P, Dickman CR, Gracia C, Graham S, Gutiérrez JR, Hice C, Jaksic F, Kelt DA, Letnic M, Lima M, López BC, Meserve PL, Milstead WB, Polis GA, Previtali MA, Richter M, Sabaté S and Squeo FA (2006) Extreme climatic events shape arid and semiarid ecosystems. *Frontiers in Ecology and the Environment* 4: 87-95.

Horridge M, Madden J and Wittwer G (2005) The impact of the 2002-2003 drought on Australia. *Journal of Policy Modeling* 27: 285-308.

IPCC (2011) Summary for policymakers. In: Intergovernmental Panel on Climate Change Special Report on Managing the Risks of Extreme Events and Disasters to Advance Climate Change Adaptation [Field CB, Barros V, Stocker TF, Qin D, Dokken D, Ebi KL, Mastrandea MD, Mach KJ, Plattner G-K, Allen S, Tignor M and Midgley (eds.)]. Cambridge University Press, Cambridge, United Kingdom and New York, NY, USA.

Kappen L (1981) Ecological significance of resistance to high temperature. Pages 439-474 in: OL Lange, PS Nobel, CB Osmond & H Ziegler (eds.) Physiological Plant Ecology I: Response to the Physical Environment. Encyclopedia of Plant Physiology, New Series, Vol. 12A. Springer-Verlag, New York.

Katz RW and Brown BG (1992) Extreme events in a changing climate: variability is more important than averages. *Climatic Change* 21: 298-302.

Katz RW, Brush GS and Parlange MB (2005) Statistics of extremes: modeling ecological disturbances. *Ecology* 86: 1124-1134.

Koch PL, Diffenbaugh NS and Hoppe KA (2004) The effects of late Quaternary climate and pCO2 change on C4 plant abundance in the south-central United States. *Palaeogeography, Palaeoclimatology, Palaeoecology* 207: 331-357.

Kremen C (2005) Managing ecosystem services: what do we need to known about their ecology? *Ecology Letters* 8: 468-479.

Larcher W, Kainmüller C and Wagner J (2010) Survival of high mountain plants under extreme temperatures. *Flora* 205: 3-18.

Leblanc MJ, Tregoning P, Ramillien G, Tweed SO and Fakes A (2009) Basin-scale, integrated observations o the early 21st century multiyear drought in southeast Australia. *Water Resources Research* 45, W04408, doi: 10.1029/2008WR007333, 2009.

Lekson SH and Cameron CM (1995) The abandonment of Chaco Canyon, the Mesa Verde Migrations, and the reorganisation of the Pueblo world. *Journal of Anthropological Archaeology* 14: 184-202.

Lind PR, Robson BJ and Mitchell BD (2006) The influence of reduced flow during a drought on patterns of variation in macroinvertebrate assemblages across a spatial hierarchy in two lowland rivers. *Freshwater Biology* 51: 2282-2295.

MacDonald GM and Case RA (2005) Variations in the Pacific Decadal Oscillation over the past millennium. *Geophysical Research Letters* 32: L08703, doi:10.1029/2005GL022478.

Mal TK and Lovett-Doust J (2005) Phenotypic plasticity in vegetative and reproductive traits in an invasive weed, *Lythrum salicaria* (Lythraceae), in response to soil moisture. *American Journal of Botany* 92: 819-825.

Marchand FL, Verlinden M, Kockelbergh F, Graae BJ, Beyens L and Nijs I (2006) Disentangling effects of an experimentally imposed extreme temperature event and naturally associated desiccation on Arctic tundra. *Functional Ecology* 20: 917-928.

Martinho F, Leitão R, Viegas I, Dolbeth M, Neto JM, Cabral HN and Pardal MA (2007) The influence of an extreme drought event in the fish community of a southern Europe temperate estuary. *Estuarine, Coastal and Shelf Science* 75: 537-546.

Mattson WJ and Haack RA (1987) The role of drought in outbreaks of plant-eating insects. *Bioscience* 37: 110-118.

McDowell N, Pockman WT, Allen CD, Breshears DD, Cobb N, Kolb T, Plaut J, Sperry J, West A, Williams DG and Yepez EA (2008) Mechanisms of plant survival during drought: why do some plants survive while others succumb to drought? *New Phytologist* 178: 719-739.

McKeon GM, Cunningham CM, Hall WB, Henry BK, Owens JS, Stone GS and Wilcox DG (2004). Chapter 2: Degradation and recovery episodes in Australia's rangelands: an anthology. Pages 87-172 in Pasture degradation and recovery in Australia's rangelands: learning from history; G McKeon, W Hall, B Henry, G Stone and I Watson (eds.). Queensland Department of Natural Resources, Mines and Energy, Queensland, Australia. 256p.

Meehl GA, Karl T, Easterling DR, Changnon S, Pielke Jr R, Changnon D, Evans J, Groisman PY, Knutson TR, Kunkel KE, Mearns LO, Parmesan C, Pulwarthy R, Root T, Sylves RT, Whetton P and Zwiers F (2000) An introduction to trends in extreme weather and climate events: observations, socioeconomic impacts, terrestrial ecological impacts, and model projections. *Bulletin of the American Meteorological Society* 81: 413-416.

Morecroft MD, Bealey CE, Howell O, Rennie S and Woiwood IP (2002) Effects of drought on contrasting insect and plant species in the UK in the mid-1990s. *Global Ecology and Biogeography* 11: 7-22.

Mpelasoka F, Hennessy K, Jones R and Bates B (2008) Comparison of suitable drought indices for climate change impacts assessment over Australia towards resource management. *International Journal of Climatology* 28: 283-1292.

Murphy BF & Timbal B (2008) A review of recent climate variability and climate change in southeastern Australia. *International Journal of Climatology* 28: 859-879.

Musil CF, Schmeidel U and Midgley GF (2005) Lethal effects of experimental warming approximating a future climate scenario on southern African quartz-field succulents: a pilot study. *New Phytologist* 165: 539-547.

Musil CF, Van Heerden PDR, Cilliers CD and Schmiedel U (2009) Mild experimental climate warming induces metabolic impairment and massive mortalities in southern African quartz field succulents. *Environmental and Experimental Botany* 66: 79-87.

Nicholls N (2004) The changing nature of Australian droughts. *Climatic Change* 63: 323-326.

Nicholls N (2006) Detecting and attributing Australian climate change: a review. *Australian Meterological Magazine* 55: 199-211.

Nicholls N, Della-Marta P and Collins D (2004) 20th century changes in temperature and rainfall in New South Wales. *Australian Meteorological Magazine* 53: 263-268.

Oberbauer SF and Billings WD (1981) Drought tolerance and water use by plants along an alpine topographic gradient. *Oecologia* 50: 325-331.

Oram RN (1983) Ecotypic differentiation for dormancy levels in oversummering buds of *Phalaris aquatica* L. *Botanical Gazette* 144: 544-551.

Parmesan C, Root TL and Willig MR (2000) Impacts of extreme weather and climate on terrestrial biota. *Bulletin of the American Meteorological Society* 81: 443-450.

Peterken GF and Mountford EP (1996) Effects of drought on beech in Lady Park Wood, an unmanaged mixed deciduous woodland. *Forestry* 69: 125-136.

Petersen KL (1994) A warm and wet little climatic optimum and a cold and dry little ice age in the southern Rocky Mountains, U.S.A. *Climatic Change* 26: 243-269.

Planton S, Déqué M, Chauvin F and Terray L (2008) Expected impacts of climate change on extreme events. *Comptes Rendus Geoscience* 340: 564-574.

Pook M, Lisson S, Risbey J, Ummenhofer CC, McIntosh P and Rebbeck M (2009) The autumn break for cropping in southeast Australia: trends, synoptic influences and impacts on wheat yield. *International Journal of Climatology*: 29: 2012-2026.

Resnick SI (2007) Heavy-tail phenomena: probabilistic and statistical modelling. Springer, New York, NY. 404p.

Scott P (2000) Resurrection plants and the secrets of eternal leaf. *Annals of Botany* 85: 159-166.

Smith MD (2011) An ecological perspective on extreme climatic events: a synthetic definition and framework to guide future research. *Journal of Ecology* 99: 656-663.

Stahle DW, Fye FK, Cook ER and Griffin RD (2007) Tree-ring reconstructed megadroughts over North America since A.D. 1300. *Climatic Change* 83: 133-149.

Stampfli A and Zieter M (2004) Plant regeneration directs changes in grassland composition after extreme drought: a 13-year study in southern Switzerland. *Journal of Ecology* 92: 568-576.

Stiles W (1930) On the cause of cold death in plants. *Protoplasma* 9: 459-468.

Tebaldi C, Hayhoe K, Arblaster JM and Meehl GA (2006) Going to extremes. An intercomparison of model-simulated historical and future changes in extreme events. *Climatic Change* 79: 185-211.

Threatened Species Scientific Committee (2008) Commonwealth Listing Advice on natural grasslands on basalt and fine-textured alluvial plains in northern New South Wales and southern Queesland. Available online at http://www.environment.gov.au

Tilman D & El Haddi A (1992) Drought and biodiversity in grasslands. *Oecologia* 89: 257-264.

Timbal B, Arblaster JM and Power S (2006) Attribution of the late-twentieth-century rainfall decline in southwest Australia. *Journal of Climate* 19: 2046-2062.

Tubiello FN, Soussana J-F and Howden SM (2007) Crop and pasture response to climate change. *Proceedings of the National Academy of Sciences of the USA* 104: 19686-19690.

Ummenhofer CC, England MH, McIntosh PC, Meyers GA, Pook MJ, Risbey JS, Gupta AS and Taschetto AS (2009) What causes southeast Australia's worst droughts? *Geophysical Research Letters* 36: doi. 10.1029/2008GL036801

Van Peer L, Nijs I, Bogaert J, Verelst I and Reheul D (2001) Survival, gap formation, and recovery dynamics in grassland ecosystems exposed to heat extremes: the role of species richness. *Ecosystems* 4: 797-806.

Van Splunder I, Voesenek LACJ, Coops H, Devries XJA and Blom CWPM (1996) Morphological responses of seedlings of four species of *Salicaceae* to drought. *Canadian Journal of Botany* 74: 1988-1995.

Volaire F and Lelièvre F (2001) Drought survival in *Dactylis glomerata* and *Festuca arundinaceae* under similar rooting conditions. *Plant and Soil* 229: 225-234.

Volaire F, Conéjero G and Lelièvre F (2001) Drought survival and dehydration tolerance in *Dactylis glomerata* and *Poa bulbosa*. *Australian Journal of Plant Physiology* 28: 743-754.

Volaire F, Thomas H and Lelièvre F (1998) Survival and recovery of perennial forage grasses under prolonged Mediterranean drought. I. Growth, death, water relations and solute content in herbage and stubble. *New Phytologist* 140: 439-449.

Walker B, Kinzig A and Langridge J (1999) Plant attribute diversity, resilience, and ecosystem function: the nature and significance of dominant and minor species. *Ecosystems* 2: 95-113.

Walther G-R, Post E, Convey P, Menzel A, Parmesan C, Beebee TJC, Fromentin J-M, Hoegh-Guldberg O and Bairlein F (2002) Ecological responses to climate change. *Nature* 416: 389-395.

Watterson IG and Dix MR (2003) Simulated changes due to global warming in daily precipitation means and extremes and their interpretation using the gamma distribution. *Journal of Geophysical Research* 108: doi:10.1029/2002JD002928.

Weiss H and Bradley RS (2001) What drives societal collapse? *Science* 291: 609-610.

Whitaker R (2006) Australia's natural disasters. New Holland Publishers (Australia), French's Forest, NSW. 239p.

White TA, Campbell BD, Kemp PD and Hunt CL (2001) Impacts of extreme climatic events on competition during grassland invasions. *Global Change Biology* 7: 1-13.

Woodhouse CA and Overpeck JT (1998) 2000 years of drought variability in the central United States. *Bulletin of the American Meteorological Society* 79: 2693-2714.

Yurkonis K. A. and Meiners S. J. (2006) Drought impacts and recovery are driven by local variation in species turnover. *Plant Ecology* 184: 325-336.

16

Primary Producers of the Barents Sea

Pavel Makarevich, Elena Druzhkova and Viktor Larionov
Murmansk Marine Biological Institute
Russia

1. Introduction

1.1 Morphologic features of the relief, hydrologic regime and types of water masses

Studies of patterns of production processes in marine arctic ecosystems have a long history. Over the last decades this branch of hydrobiology has been demonstrating a significant increase in research activity. The main driven factor here is rational use of biological resources increasing year by year. In this respect, the Barents Sea has always been among bodies of water experiencing intensive exploitation of their living resources: fish, invertebrates and marine mammals. Issues of both sustainable exploitation of fish and other living resources and their conservation and restoration need more profound and thorough ecological research in areas subject to heavy exploitation by man. Studies of production processes in the ocean have gained particular importance in the last years of the XX century in the light of a significant increase in man-caused pressure on marine ecosystems, including those in the Arctic and Northern Basin of the World Ocean.

The Barents Sea is fairly considered the most productive body of water among Arctic shelf seas. Duration of the primary production cycle determined for different areas of the Barents Sea enables the annual primary production of this water body to be estimated as equal to 33 tonnes $C/km^2/yr$ (tonnes of carbon per square kilometer per year), or 562 tonnes (wet weight) per square km per year. Such high productivity is possible on the one hand due to a full range of pelagic and bottom floral communities, phytocoenoses (macrophytes, microphytobenthos, phytoplankton, and cryoflora), and on the other hand due to specific conditions of flow and transformation of biogenic elements (Makarevich & Druzhkova, 2010).

Oceanographically, the Barents Sea is a unique natural body of water with complicated geomorphology and hydrology. The pelagic zone of the Barents Sea presents an aggregation of water masses, each with its specific type of a pelagic ecosystem. The Barents Sea lies at the junction of the Arctic and Atlantic Oceans, therefore it contains two major types of waters, Arctic and Atlantic water masses. Their interconnections form the general system of anticyclonic macrocirculation when arctic waters dominate in the north and waters of the Atlantic Ocean origin prevail in the south. The zone of contact between these two water masses called the Polar Front presents a natural structural border which divides not only two types of waters and mixing regimes but also two basic types of pelagic communities, arctic and subarctic, and two classes of their annual production cycles.

The latter, however, by no means are single formations. Geographical position of the Barents Sea determines its complicated hydrodynamic structure and the system of constant currents which form circulations of a smaller scale (although some of them can occupy large areas, for instance the basin of a so-called Pechora Sea, the south-eastern Barents Sea). In coastal areas, such factors as river and glacial run-off, tidal events, high isolation of inlets and bays, etc. can produce specific abiotic conditions in smaller water areas and therefore relatively independent assemblages of primary producers. However, each of them possesses features characteristic of one of the two aforementioned types, i.e. arctic or subarctic.

Parallel to this, all pelagic ecosystems of the Barents Sea shelf area can bathymetrically be divided into two groups: 1) ecosystems of the open shelf, and 2) coastal ecosystems, which in their turn also greatly differ in the course of annual cycles of autotrophic organisms and processes of the primary production of organic matter. Besides, complicated seabed relief of the Barents Sea includes such forms as deepwater troughs (in the northern and north-eastern Barents Sea). However, they do not present separate biotopes for primary producers as the major representatives of the latter, phytoplanktonic organisms, mainly inhabit the upper 50-meter layer of the water column with single individuals living lower than 100 meters. Other primary producers, microphytobenthos and macrophytes, do not live at great depths at all.

All the traits of the Barents Sea geomorphology and water dynamics described above do not have direct impacts on assemblages of autotrophic organisms but are determinative for the factors that directly affect their structure and functioning, first of all the thermal regime. The degree of heat uptake and recoil differs among water areas. Thus, the non-freezing part of the Barents Sea does not freeze exactly due to compensation of the heat lost from the surface by the heat advected with the Atlantic Ocean waters (Kudlo, 1970). However, the temporal structure of consecutive changes of hydrological parameters, mainly the parameters of water temperature, is common for all areas of the Barents Sea. This temporal structure presents a series of periods changing each other through the year, which are called hydrological seasons. The hydrologic spring starts with the beginning of the thermocline and the formation of positive heat exchange between the sea and atmosphere and usually lasts about one and a half month (April – May). Summer is characterized by distinct stratification (June – July) and lasts 2.5 – 3.5 months. Autumn (September) as opposite to spring starts with the formation of the negative heat budget and simultaneous destruction of temperature stratification and lasts 1.5 – 3 months. The hydrologic winter is characterized by complete homogeneity of waters and vertical distribution of major hydrological and hydrochemical parameters. This is the longest season in the Barents Sea which lasts 5 to 6.5 months.

The annual production cycle forms according to the same principles as the hydrologic one, i.e. it consists of stages characterized by particular biologic parameters. In pelagic ecosystems, the basis for such division is the structure of the succession cycle of the phytoplankton cell abundance, the major primary producer. The range and consecutiveness of succession stages in different pelagic ecosystems are mostly common, although some stages may drop out. However, the terms of the beginning and duration of each stage differ greatly across ecosystems (table 1).

Stage of the hydrologic cycle	Phytoplankton succession stages				
	Shelf ecosystems of the open sea		Coastal ecosystems		
	Subarctic	Arctic	Subarctic	Arctic	Estuarine
Winter	Dormant stage	Dormant stage; cryoflora bloom	Dormant stage	Dormant stage; cryoflora bloom	Dormant stage; cryoflora bloom
Winter-spring transition period	Spring succession cycle	Early spring succession cycle	Early spring succession cycle	Spring succession cycle	Early spring succession cycle
Spring	Balanced stage	Late spring succession cycle	Late spring succession cycle	Balanced stage	Late spring succession cycle*
Spring-summer transition period		Balanced stage	Early summer succession cycle*		Balanced stage
Summer			Balanced stage		
Summer-autumn transition period	Autumnal succession cycle*	Phytocoenosis degradation stage	Autumnal succession cycle	Phytocoenosis degradation stage	Phytocoenosis degradation stage
Autumn	Phytocoenosis degradation stage	Phytocoenosis degradation stage; cryoflora bloom	Phytocoenosis degradation stage	Phytocoenosis degradation stage; cryoflora bloom	Phytocoenosis degradation stage; cryoflora bloom
Autumn-winter transition period			Early winter nannoplankton maximum		

* – facultative stage of the cycle

Table 1. Annual cycle of the phytoplankton cell abundance in different ecosystems of the Barents Sea.

The thermal regime is also one of the major factors determining a specific structure of the annual cycle of pelagic algal cenoses in coastal ecosystems and high total productivity of the latter: increased water temperature throughout the year determines higher speed of biochemical reactions, and production and decomposition of organic matter. Still other factors also play a significant role. Little depths allow other organisms beside phytoplankton, such as macrophytes and microphytobenthos, to vegetate actively. Smaller volumes of seawater, tidal events and wind mixing intensify exchange processes. Freshwater runoff supplies primary producers with biogenic elements.

Thus, the Barents Sea evidently occupies a special place among arctic seas due to a range of specific natural conditions. The unique geographic position, large extent of the water area both in longitudinal and latitudinal directions, and abundance of shallow coastal zones result in extraordinary high diversity of abiotic environmental factors. This high diversity determines the composition of biota, including assemblages of primary producers, and patterns of all biologic processes, mainly the structure of the annual production cycle.

1.2 Zoning of the Barents Sea water area according to the oceanologic structure of its pelagic zone

All the aforementioned features of the distribution of abiotic parameters enable the Barents Sea water area to be divided into four distinct zones (excluding estuaries), each with its peculiar type of the pelagic ecosystem (fig. 1). These types are characterized first of all by the seasonal course of phytoplankton cell abundance, the main primary producer, and correspondence of its succession stages to phases of the annual hydrological cycle (table 1).

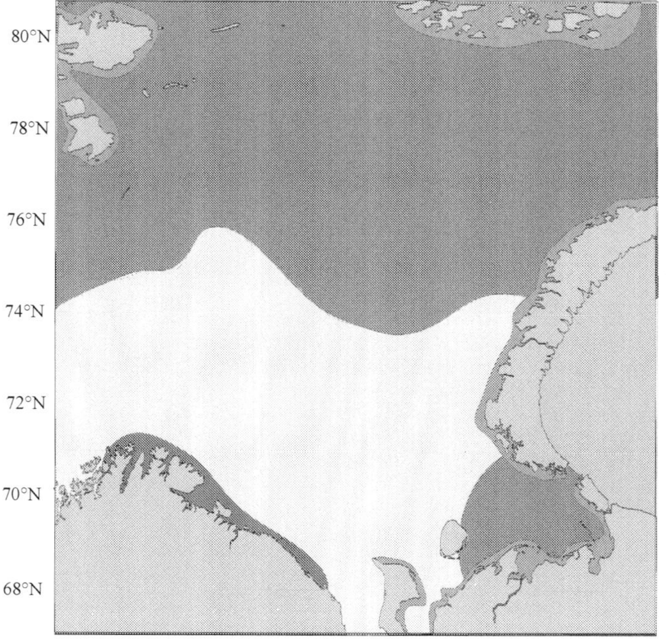

Fig. 1. Pelagic ecosystems of the Barents Sea. Legend: ☐ – subarctic shelf, ■ – arctic shelf, ▨ – subarctic coastal ecosystems, ☐ – arctic coastal ecosystems

Shelf pelagic ecosystems of the open sea are spatially the most common type in the Barents Sea. The biotope area of these ecosystems is limited by shelf edge fronts on the side of the high seas and by systems of shallow fronts of different types on the side of coasts. The formation of the seasonal pycnocline in the warm period of the year is the most universal feature of the structural organization of the water column in the shelf zone (Bouden, 1988). At the same time subarctic and arctic areas have principle differences in the way of forming the seasonal stratification. The thermal stratification is characteristic of the subarctic shelf. The water column of the arctic shelf is stratified mainly due to freshening of the upper layer as a result of the sea ice cover melt.

In fact, the whole open area of the Barents Sea subarctic shelf is occupied by waters of the Atlantic Ocean origin, which determines the homogeneity of the taxonomic and temporal structure of microphytoplankton assemblages. As mentioned above, the temporal structure of the hydrologic cycle in this area (as in other areas free of seasonal ice cover) is determined by regimes of warming of the sea surface.

2. Pelagic ecosystems of the outer shelf zone

Shelf pelagic ecosystems of the open sea are spatially the most common type in the Barents Sea. The biotope area of these ecosystems is limited by shelf edge fronts on the side of the high seas and by systems of shallow fronts of different types on the side of coasts. The formation of the seasonal pycnocline in the warm period of the year is the most universal feature of the structural organization of the water column in the shelf zone (Bouden, 1988). At the same time subarctic and arctic areas have principle differences in the way of forming the seasonal stratification. The thermal stratification is characteristic of the subarctic shelf. The water column of the arctic shelf is stratified mainly due to desalination of the upper layer as a result of ice cover melt.

2.1 Subarctic shelf

In fact, the whole open area of the Barents Sea subarctic shelf is occupied by waters of the Atlantic Ocean origin, which determines the homogeneity of the taxonomic and temporal structure of microphytoplankton assemblages. As mentioned above, the temporal structure of the hydrologic cycle in this area (as in other areas free of seasonal ice cover) is determined by regimes of warming of the sea surface.

The annual hydrological cycle of the open subarctic shelf presents the most basic type typical of the whole moderate zone (table 1).

The biologic spring begins in March when quantitative characteristics of the phytocenosis stably increase, mainly at the expense of intensive growth of the diatomic complex. In May, diatoms form the first vernal maximum of microphytoplankton bloom when not only quantitative parameters but also species diversity of planktonic community reach their maximum annual values (more than 1 mil cells per liter in numbers and more than 1 mg/l in biomass) (fig. 2A).

This period lasts 2 – 3 weeks. Almost all the phytoplankton of the subarctic shelf area is presented in this period by arcto-boreal neritic forms. The composition of the phytoplankton here consists mainly of dominant species of colonial diatoms of the genus *Thalassiosira* (*T. angulata, T. antarctica, T. cf. gravida, T. hyalina, T. nordenskioeldii*), genus *Nitzschia* (*N. arctica, N. cylindrus, N. grunowii*), genus *Navicula* (*N. granii, N. pelagica, N. vanhoeffenii*) and genus *Chaetoceros* (*C. cinctus, C. curvisetus, C. debilis, C. diadema, C. fragilis, C. furcellatus, C. holsaticus, C. socialis* and other species), and the flagellate species *Phaeocystis pouchetii*.

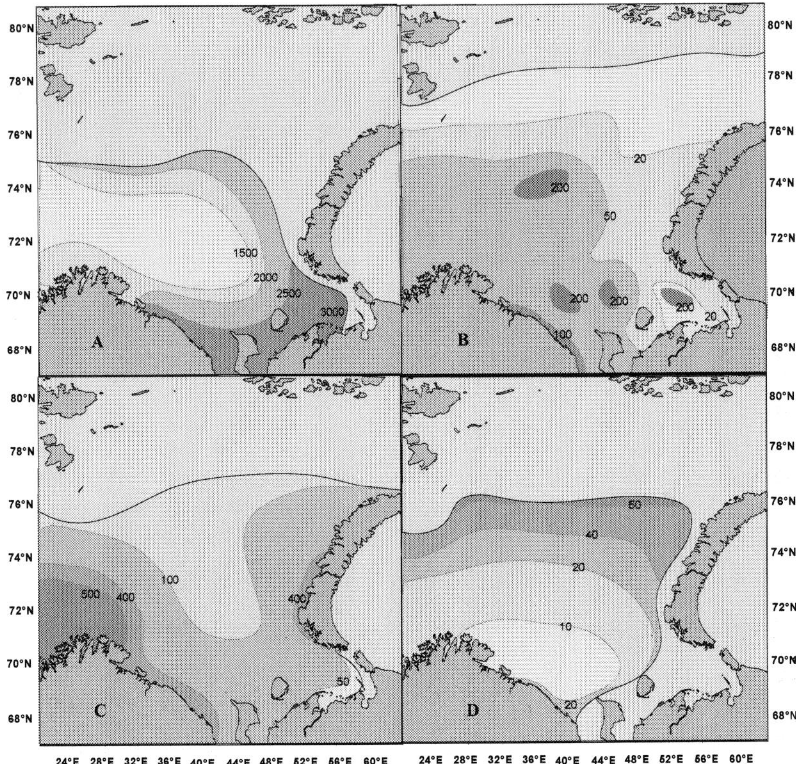

Fig. 2. Distribution of microphytoplankton biomass (µg/l) in the Barents Sea. A – spring, B – summer, C – autumn, D – winter

At the end of May/beginning of June the primary production activity of pelagic algae rapidly decreases and vertical redistribution of phytoplankton biomass occurs. The dominant position in upper layers is occupied by flagellate species (mainly Cryptophyceae, Prymnesiophyceae, Chrysophyceae and Dinophyta, sometimes Prasinophyceae and Raphidophyceae). Diatomic algae sink to the pycnocline forming the phytoplankton subsurface maximum at the end of the vernal period.

Rapid depletion of contents of biogenic elements in the upper layer and the trophic activity of zooplankton lead to the formation of the summer balanced stage of phytoplankton annual cycle when abundance and biomass of planktonic algae remain stable (fig. 2B). Populations of autotrophic flagellates of different taxa (Dinophyta, Chrysophyceae and Prymnesiophyceae) start playing a dominant role within microalgae assemblages. This structural condition of the Barents Sea outer shelf pelagic ecosystem remains until the early autumn period (until August). In September, a distinct autumnal maximum of the phytoplankton biomass presented by large dinoflagellates (mainly *Ceratium* spp.) forms in the south-western Barents Sea (fig. 2C).

Then, numbers and biomass of the microphytoplankton community gradually decrease reaching their winter values by the beginning of November. Spatial distribution of quantitative parameters of planktonic algae is quite homogeneous, which is determined by substantial taxonomic homogeneity of the pelagic phytocenosis on the whole water area. Transition to the winter stage in the shelf pelagic zone occurs during the autumn-winter transition period (presumably the second decade of November – first decade of December). The hibernal stage of dormancy lasts from November until February. Maximum desintegrity of the phytoplankton community is a distinctive feature of this period. This stage is characterized by low values of both quantitative parameters (fig. 2D) and species diversity of the pelagic algal cenosis. Its background forms at this stage are represented mostly by *Protoperidinium* species.

Thus, total duration of the active vegetation period of planktonic microalgae of the subarctic pelagic zone lasts approximately 8 months. The course of the seasonal dynamics of phytocenoses as a whole is characterized by a rapid increase in numbers and biomass of organisms, vernal maximum and a consequent decrease. Then these values increase again as the summer species complex starts forming, first of all due to increased inflow of warm Atlantic waters entering the Barents Sea from the west and rich in biogenic matter. After that levels of microalgae constantly decrease and the dormant stage begins.

2.2 Arctic shelf

The Barents Sea arctic shelf can spatially be subdivided into two main zones: (1) deep-water outer shelf zone (northern Barents Sea beyond the Polar Front) and (2) shallow nearshore shelf zone (Kanin–Kolguev shallow area and Pechora Sea) (fig. 1).

Deep-water outer shelf zone. The hydrological year in the ice-covered arctic pelagic zone begins with the formation of ice edge zones (table 1, fig. 2D). These zones are formed by a complex of microalgae, in which early spring neritic diatoms and some colonial flagellates, such as *Phaeocystis pouchetii* and *Dinobryon balticum*, occupy the dominant position (Hansen et al., 1990).

As the ice cover melts it desalinates the upper layer of water causing stratification of the water column. In such a stratified state the water column of the arctic pelagic zone remains during the open sea period until the beginning of active vertical autumn-winter mixing. As a result, in the summer period (fig. 2C) after the end of the spring bloom, microalgae biomass is redistributed vertically, with its subsurface maximum in the pycnocline zone formed mainly at the expense of *Phaeocystis pouchetii* and *Thalassiosira* spp.

As farther north to higher latitudes, the autumnal maximum gradually drops out from the annual cycle structure of the algal cenosis and the seasonal curve of the microphytoplankton abundance acquires single-peak character.

Shallow nearshore shelf zone. The main feature of the south-eastern Barents Sea (Pechora Sea) is its shallowness. Due to this the whole water column in the cold period of the year presents the hibernal mixed layer surface to bottom. During this period, pennate diatoms dominate within the phytoplanktonic community. *Coscinodiscus cf. stellaris, Amphiprora kjellmanii, Cylindrotheca closterium, Gyrosigma fasciola, Navicula/Plagiotropis* spp., *Nitzschia grunowii* and *Pleurosigma stuxbergii* are the most common floristic elements. It should be noted that at this stage the biomass of this complex is formed mainly by single non-colonial algae. The only exclusion is *N. grunowii* that plays a rather insignificant role in quantitative respect being the constant component of the flora. The phytoplankton succession cycle in the Pechora Sea begins in March with activation of populations of early-spring neritic diatoms typical for the ice-edge bloom (*N. grunowii, Achnanthes taeniata, Thalassiosira* spp., *Chaetoceros* spp.). Abundance levels of pelagic phytocenosis reach their maximum annual values by April (fig. 2A). In the warm period of the year, a distinct mosaic structure of water masses and heterochrony of the seasonal development of microphytoplankton hamper the analysis of spatial and temporal structure of the pelagic community. Numbers of autotrophs may reach in summer 200 000 – 300 000 cells per liter with a biomass of 150 – 200 µg/l (fig. 2B). The dominant position is occupied by Bacillariophyta algae, species of the genera *Nitzschia* and *Skeletonema*, and by large dinoflagellates. During the autumnal degradation of the algal cenosis, numbers of microalgae vary within 500 – 3 000 cells per liter with a biomass of 2 – 50 µg/l (fig. 2C). The main contribution into the structure of the community in this period belongs to centric diatomic algae *Paralia sulcata*.

The runoff of the Pechora River is undoubtedly the most important factor affecting phytoplankton abundance. In the summer period (June – July), levels of phosphates and dissolved silicon in Pechora Bay are an order of magnitude higher than those in the open part of the Pechora Sea. Apparently, constant inflow of biogenic matter into the bay with good mixing of water masses due to a relative shallowness of the basin ensures a high level of primary production of pelagic microalgae. The generated organic matter is only half utilized within Pechora Bay. The rest of it is carried out into the Pechora Sea where it is assimilated within the water area of local circulation that occupies the central part of the sea. As a result, the open part of the Pechora Sea also demonstrates high values of phytoplankton abundance.

3. Coastal ecosystems

3.1 Microphytoplankton annual succession cycle

3.1.1 Subarctic coastal ecosystems

As a whole the annual cycle of microphytoplankton of the coastal pelagic zone is much more complicated compared to that of the outer shelf zone. Thus, in subarctic coastal ecosystems two maximums of microphytoplankton cell abundance occur at the end of winter and in spring, i.e. the early-spring one and the late-spring one. Besides, the summer phase of balanced numerical abundance, which comes after the formation of the seasonal stratification, also begins with the early-spring maximum and ends with the autumnal one. As farther to the north-east, early-summer and autumn cycles become occasional and form only in certain years (Druzhkov et al., 1997b) (table 1).

The spring activity of phytoplankton (second decade of March) begins when early-spring diatomic forms, *Thalassiosira hyalina, T. cf. gravida, Navicula pelagica, N. septentrionalis, Nitzschia grunowii* and *Amphora hyperborea*, appear in coastal pelagic waters. Numbers of cells in this period are little and vary within several dozens to several hundreds of cells per liter (Larionov, 1997). The first spring maximum of phytoplankton cell abundance, the most universal ecological event in the coastal zone, occurs in the middle of April and is formed by early-spring neritic arcto-boreal diatomic genera *Thalassiosira, Chaetoceros, Navicula* and *Nitzschia*. Phytoplankton numerical abundance reaches its maximum which remains within several days. During the early-spring bloom, numbers of phytoplankton vary within several hundreds of thousands to two million cells in a liter with a biomass of 1 to 3 mg/l. The core of the community is concentrated in this period in the upper ten-meter layer. This first maximum of phytoplankton cell abundance is formed by species *Thalassiosira cf. gravida, T. nordenskioeldii, Chaetoceros socialis, C. furcellatus* and *Navicula vanhoeffenii*.

The second spring maximum (end of May – beginning of June) is associated with the continental runoff and varies from year to year in terms of the beginning, numbers, and taxonomic composition depending on the terms of the continental runoff maximum. In most cases, it repeats the first spring maximum, probably with reduced number of dominants. However in years when volumes of the continental runoff are the lowest, the species *Phaeocystis pouchetii* dominates in the bloom with highest registered levels of cell numbers and biomass of 8 mil cells/l and 1.7 mg/l, respectively (Druzhkov, Makarevich, 1999). Spatially, the wave of the spring bloom starts from the Cape of Svyatoy Nos and spreads to the west.

In the summer period (end of June – end of August) the role of dinophyte microalgae increases in the phytoplanktonic community. At the same time cosmopolite forms take the place of arcto-boreal ones, while panthalassic and oceanic algae take the place of neritic species. It should be noted that the summer phase of the balanced cell abundance is the most variable. All possible ecological scenarios can be combined into two main types: 1) summer abundance minimum of the community consisting of small pennate diatoms and unarmored (naked) dinophlagellates and 2) formation of one succession cycle at the end of July with a subsequent decrease in activity until the end of the summer period. In this case the early-summer maximum is as a rule monospecific as it forms at the expense of a mass bloom of a single planktonic species *Skeletonema costatum*, which accounts for more than 80 % of phytoplankton biomass in this period.

Autumnal succession cycle (from the middle of September till the beginning of October) is casual and usually associated with the appearance of spring diatomic forms in the pelagic zone (Druzhkov et al., 1997b; Kuznetsov et al., 1996). Diatomic algae of the genus *Chaetoceros* and dinophyte algae of genera *Ceratium, Dinophysis* and *Protoperidinium* dominate in the pelagic zone in this period. Numbers of cells are less than 2 000 per liter with a biomass of less than 5 µg/l.

During the whole winter period (middle of November – middle of March) the phytoplankton community rests in dormancy. In the pelagic zone, it mostly consists of large oceanic dinophyte algae of the cosmopolite and arcto-boreal origin. The concentrations vary within several cells to dozens of cells per liter. Species *Ceratium longipes, C. tripos), Dinophysis norvegica* and *Protoperidinium depressum* form the basis of the dominant complex.

3.1.2 Arctic coastal ecosystems

The main feature of arctic nearshore ecosystems of the open type is the seasonal occurrence of fast shore ice. As in pelagic ecosystems of the outer shelf zone, the whole seasonal dynamics of the development of coastal phytoplanktonic ecosystems (especially in case of polar archipelagos where the role of the continental runoff is insignificant) is fully determined by the seasonal dynamics of the ice cover.

Apart from subarctic coastal ecosystems, in coastal waters of the so-called Pechora Sea (southeastern Barents Sea), the early-summer and autumnal succession cycles completely drop out from the structure of the annual cycle of the floral cenosis. The hydrologic year in the Pechora Sea begins at the end of February when populations of neritic diatoms become active. The early-spring cycle, as in the outer shelf zone, transforms into the ice-edge bloom, though in this case it occurs near the edge of the fast shore ice but not the pack ice. In March the coastal algal community of the Pechora Sea enters the initial stages of the spring bloom and in some areas is characterized by intensive growth (with a two- and threefold increase in biomass, up to 500 µg/l) of populations of early-spring colonial forms of centric diatoms. By April, levels of the phytoplankton cell numbers and biomass reach their maximum annual values of more than 500 000 cells per liter and more than 2 mg per liter, respectively. Diatomic algae belonging to genera *Thalassiosira*, *Chaetoceros*, *Navicula* and *Pleurosigma* as before remain dominant species. The late-spring bloom in this body of water forms somewhat later (in June) when the continental runoff reaches its maximum volumes (Grönlund et al., 1996, 1997; Druzhkov et al., 1997a; Kuznetsov et al., 1997).

In coastal zones of arctic archipelagos, the phytoplankton active vegetation period is even shorter (less than three months) and the annual succession cycle of phytoplankton cell abundance actually lacks the early-spring, late-spring and autumnal stages. Single studies in coastal waters off the Franz Josef Land Archipelago enable an assumption to be made that the ice-edge bloom does not occur here at all. Intensive ice melting during the April – June period, simultaneous inflow of fresh melt water from glaciers and snowfields, and wind-induced and wave-induced mixing of water must have negative effects on populations of pelagic microalgae. The spring stage of the algal community (June – August) gradually passes into the summer one (August – September) after which the degradation stage begins. Taxonomic composition of the community, according to field observations made in August, was quite unvaried and consisted mostly of diatomic algae of the genus *Chaetoceros* and dinoflagellate algae of the genus *Protoperidinium*. However, at the same time the numerical abundance maximum was made by the species *Dinobryon balticum* while the biomass was mostly made by large-sized forms *C. decipiens* and *P. ovatum*. In whole, the values of these parameters were quite low and varied within 3 500-89 700 cells per liter and 28.0-40.2 µg/l, respectively. Representatives of microphytobenthos, namely pennate diatoms of genera *Navicula*, *Pleurosigma* and some other, have many times been found in the pelagic zone together with typically planktonic organisms (especially in shallow areas and areas exposed to impact of the ice melt). Production of the pelagic floristic cenosis, according to samples taken in the upper layer of the water column at different sites of the Franz Josef Land coastal waters, made up on average 20.2 µgC/m³/day varying within 16.1-50.4 µgC/m³ during the day (Kuznetsov & Shoshina, 2003). At the same time, when this parameter was calculated for 1 m² with the use of the equation created for this very region (Bul'yon, 1985), an average value of 113.4 µgC/m³/day was obtained which is consistent with the data for the same period in the subarctic coastal zone (along the Eastern Murman Coast, Kola Peninsula).

3.1.3 Estuarine ecosystems

In typical cases, the main dynamic characteristics of microphytoplankton in boreal and arctic estuarine ecosystems with the constant stratification are close to those of outer shelf ecosystems with the exclusion that the stratification here is driven by the continental runoff but not by thermal factors (Svendsen, 1986). It should be noted that due to intensive eutrophication of continental waters, the inflow of biogenic elements into estuaries loses its cyclic manner and therefore the whole warm period of the year is characterized by a single large spring-summer-autumn maximum with an unexpressed structure of peaks.

Kola Inlet is a typical example of an estuarine ecosystem in the Barents Sea. The activity of primary producers in Kola Inlet can remain at a rather high level even during the whole period of the polar night. Although the abundance of phytoplankton cells at this time demonstrates a clear decrease in December and the beginning of January, still the concentration of phytoplankton reaches a value of 10^3-10^4 cells per liter, i.e. 1-2 orders higher than in nearshore marine ecosystems (Druzhkov et al., 1997b). Which is the most important here is that the phytoplankton community at this time mostly consists of autotrophic organisms: diatomic algae, and chlorophyll-bearing flagellates. Apparently, the reason of such a phenomenon is that biologic processes in the warm period of the year are more intensive in estuarine zones constantly exposed to freshwater runoff from large rivers, the Tuloma River in the case of Kola Inlet. This enables algae to accumulate a considerable production potential.

The spring development of phytoplankton in Kola Inlet begins in March and is characterized by the complete dominance of diatomic algae. The bloom peak occurs in the second half of April with cell numbers of up to 1 mil cells per liter and a biomass of 1.5 mg per liter and is represented by three diatomic species: *Thalassiosira* cf. *gravida*, *T. nordenskioeldii* and *Chaetoceros socialis*. During May, the abundance and biomass decrease down to 100 000-200 000 cells per liter and 300-400 µg/l, respectively. At the same time the structure of the algal coenosis changes as new species appear, mainly dinophyte algae, and the community demonstrates the greatest species diversity. This transition period ends by July when the pelagic phytocoenosis enters the summer phase of its succession cycle with cell numbers of 3 000-4 000 cells per liter and a biomass of 30-40 µg/l (Makarevich, 2007). Shares of diatomic algae (mainly species of the genus *Chaetoceros*) and dinoflagellate algae are much alike in the community. Shares of golden, green, and euglena algae are much lesser. The latter however are capable of forming sites of a mass bloom with an increase in cell abundance 3 to 4 orders of magnitude but only in separate local nearshore areas of Kola Inlet (Trofimova, 2009). In September, the autumnal stage of the algal coenosis succession cycle begins when species with mixotrophic and heterotrophic feeding types start dominating in the pelagic zone. At this time, a distinct dominance of oceanic forms, mostly dinophyte species, is observed in the phytoplankton community. In whole, the change of the spring phase by the summer one and of the summer phase by the autumnal one happens by means of the inflow of a microalgae complex of the Atlantic Ocean origin into Kola Inlet with marine coastal waters. The winter succession stage is characterized by a decrease in activity of phytocenosis: quantitative parameters demonstrate the lowest values of the whole annual cycle (Makarevich, 2007).

A similar but somewhat different picture is observed in Pechora Bay covered with ice a considerable part of the year. Apparently, due to little depths and the active hydrodynamic regime in this body of water, which homogenizes the density structure of the pelagic biotope surface to bottom, the phytoplankton bloom does not occur in the form of shaped "classic" ice-edge spots as ice melts, but develops in the whole water column in areas becoming free of ice as it breaks up and is carried out to the open part of the sea. A distinct spring maximum does not form in Pechora Bay. After the spring bloom the estuarine ecosystem passes into the balanced cell abundance stage (table 1). The beginning of the intensive numerical growth of microalgae in the mouth of the Bay falls onto the end of April and the first decade of May. Centric diatoms of genera *Thalassiosira* and *Chaetoceros*, the pennate algae *Nitzschia grunowii*, and the golden algae *Phaeocystis pouchetii* completely dominate in the community at this time. The maximum registered phytoplankton numbers in this body of water is 1.4 mil cells per liter with a biomass of 2.7 mg/l (Makarevich, 2007). It should be noted that a rapid increase of these values in Pechora Bay may happen not only due to the intensive cell fission, but also owing to the incoming of the microalgae hibernating in bottom sediments, which is typical for shallow estuarine zones (McQuid, Hobson, 1995).

The summer season in Pechora Bay begins in June when diatomic algae *Aulacoseira granulata, Asterionella formosa, Skeletonema costatum, Chaetoceros wighamii* and *C. constrictus* still dominate in the pelagic algal coenosis. In July the composition of the community slightly changes as diatomic algae *C. compressus, Paralia sulcata* and *Rhizosolenia setigera* appear in the dominant complex. Dinoflagellates of the genus *Protoperidinium* and the golden algae *Dinobryon balticum* start playing a significant role. In whole, concentration and biomass levels of phytoplankton during summer vary within a rather wide range, 120 000 to 850 000 cells per liter and 200 to 1 100 µg per liter, respectively (Makarevich, 2007). At the end of August the pelagic algal coenosis of Pechora Bay enters the autumnal stage. During this stage, quantitative parameters of numerical abundance and biomass decrease to several tens of thousands of cells per liter and tens of micrograms per liter, respectively. The species diversity is characterized by the complete dominance of a single species of microalgae and these dominant species are different at different sites of the water body. Among such dominant species are *Paralia sulcata, Asterionella formosa, A. kariana* and *Nitzschia seriata*. At the same time over the whole water area, particularly in the mouth of the bay, the largest part in the quantitative structure of the algal coenosis belongs to dinophyte algae.

Like in the outer shelf zone, as farther north, the autumnal cycle gradually drops out from the succession structure. In arctic stratified estuaries covered with ice large part of the year (which spring phytoplankton and cryoflora are unfortunately very poorly studied), there is probably only one succession cycle, the spring one, which follows the cryoflora bloom.

In the coastal pelagic zone of outlet glaciers of the Novaya Zemlya Archipelago there is the most simple type of the seasonal succession of the arctic coastal phytoplankton with the only maximum in the structure of the annual cycle, the spring one (ice-edge bloom) (Druzhkov et al., 1997b). Thus, there are only three stages in the vegetation season of microalgal communities, i.e. fast ice cryoflora bloom, the spring maximum, and the balanced cell abundance stage. The similar situation is obviously happens in the coastal waters of the

Svalbard Archipelago (Węslawski et al., 1988, 1991). In high arctic latitudes, an extreme variant of the annual cycle of pelagic algae is possible, with the development of the ice-associated flora only.

It should be noted once again that the sub-ice bloom described above is registered in coastal areas directly close to estuarine zones and is not observed in open parts of water bodies. However, analyzing data of planktonic studies at the central part of the Barents Sea basin and in coastal waters off the Novaya Zemlya Archipelago we can suppose the existence of such a phenomenon there. Already in May, in that area near the ice edge, there are phytocoenoses at different stages of the vernal bloom. In June almost the whole eastern part of the Barents Sea contains the summer complex of pelagic microalgae. This suggests the rapid development of the succession cycle, which can obviously be explained by a short vegetation period of phytoplankton in the area free of ice during only a short period of the year. It should be mentioned that this factor causes different specific manifestations at different areas of the arctic basin. In high latitudes, single stages simply drop out from the succession cycle. In the eastern Barents Sea, microalgae belonging to different seasonal complexes are registered simultaneously in the pelagic zone, i.e. one stage of the succession clashes with another. In the region described, especially near the shores of the Novaya Zemlya Archipelago, where environmental conditions enable autotrophic organisms to function over a longer period of time, a phenomenon called *sinking of bloom* is observed when communities of phytoplankton at earlier stages of the seasonal succession occur in deeper layers of the water column (Larionov, 1995).

However, both qualitative and quantitative compositions of phytocoenoses in this region distinctly differ from those in the aforementioned coastal and near-estuarine zones. Although the latter also do not remain long under the condition of open water, communities of microalgae in these zones are characterized by much greater species diversity and high quantitative parameters. At the same time, pelagic algal assemblages of the eastern Barents Sea rarely include more than ten species, and the biomass (less than 1 mg/l) is formed mostly by a single species, the diatomic alga *Thalassiosira antarctica*. The summer complex is also characterized by low quantitative values and is often formed by only several representatives of dinophyte algae: large-sized species of the genus *Protoperidinium*, and sometimes by Chrysophyceae and Cryptophyceae organisms. In the southern and western Barents Sea, such a state more likely corresponds to the autumnal stage of the succession of the algal coenosis.

3.2 Nanophytoplankton annual succession cycle

The conceptual scheme of the annual succession cycle of nanophytoplanktonic algae (unicellular organisms 2-20 μm in size belonging to different taxonomic groups) forms like that of the microphytoplankton, i.e. in correspondence with the stages of the seasonal dynamics of the hydrophysical structure and with the stages of the microphytoplankton succession cycle (Druzhkov et al., 1997b) and the stages of the production cycle of northern seas (Fyodorov, 1987; Fyodorov et al., 1988; Smirnov et al., 1989). The graph of the annual course of biomass of both size groups of algae in the Barents Sea coastal zone (fig. 3) suggests distinct correlation between them and at the same time definite autonomy of the nanoplanktonic component of the pelagic algal cenosis.

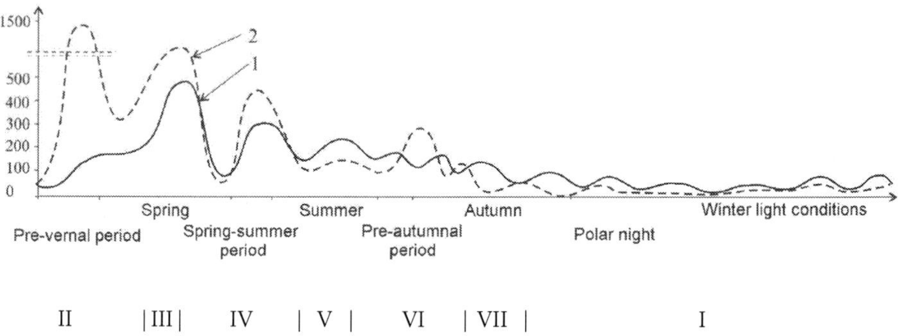

Fig. 3. Major dynamic trends (polynomial smoothing) of biomass (µg/l) change of nanophytoplankton (1) and microphytoplankton (2) in the pelagic zone of the Central Murman Coast (Kola Peninsula) in different seasons (averaged data over 6 years of observations). Roman numerals stand for the stages of the annual succession cycle (see in the text)

The annual cycle of the nanoplanktonic community includes the following periods.

I The winter minimum stage. The winter season in the Subarctic (the longest stage of the annual cycle) can be divided into two stages: the period of the polar night (December – January) and the following period of the winter light regime (February – March). In general, the annual minimum of the phytocoenosis falls on the winter season which is characterized by the absolute dominance of nanophytoplankton in the coastal pelagic zone. During the polar night, the biomass of organisms of this group rapidly decreases and their size structure grows smaller. Rapid cooling of the water column and increased turbulent activity act as the main ecosystem regulators determining the dynamics of this community during the period of the polar night. In the second half of winter, processes of accumulation of winter supplies of biogenic matter no longer occur in the pelagic zone. The role of the main regulator passes to light, which causes the development of the primary production processes in the coastal ecosystem at the expense of nanoalgae populations dominating at this time.

II Spring activation stage. In the pre-vernal period, the early spring succession cycle of microphytoplankton starts developing in the coastal pelagic zone under conditions of the light optimum, relative stabilization of the coastal water column, maximum levels of biogenic elements, and maximum screening from negative influence of the ultraviolet. Evident activation of nanophytoplankton still does not enable it to occupy the leading position. Perhaps, this fact can first of all be explained not by competitive interactions with microphytoplankton, but by restrictive influence of distinct turbulence, low temperatures, and microzooplankton predation pressure.

III   Spring maximum stage. During this period which is characterized by rapid warming of coastal waters and maximum levels of the continental runoff, the second succession cycle of the pelagic algal coenosis forms in the coastal pelagic zone. In contrast to the previous stage, at this one maximums of biomass of microphytoplankton and nanophytoplankton develops simultaneously (synchronically). Besides, nanoalgae make a substantial contribution into primary production processes.

IV   Early summer maximum stage. The biomass maximum in the spring-summer transition period, like in the previous one, also forms synchronically in both assemblages. The primary production activity of the phytocoenosis is very high at this stage. Intensive photosynthesis is accompanied by rapid transformation of biogenic elements into dissolved and shaped organic matter. Apparently, this is the very period of the annual cycle that ends the primary synthesis phase. Then the pelagic ecosystem enters the mixed synthesis stage which is characterized by the balance of primary production and heterotrophic processes and a complicated hierarchical system of redistribution of organic matter within heterotrophic components of planktonic biota.

V   Balanced cell abundance stage. Development of the nanophytoplanktonic community at this stage is possible according to two scenarios. The first one is a rather long period of balanced cell abundance (thus in 1989 this phase lasted through the whole summer and summer-autumn periods). During this period, quantitative parameters of the phytocoenosis, which is characterized by the absolute dominance of nanoalgae, change in an oscillatory manner. These oscillations, on the one hand, can be explained by competitive interactions between microphytoplankton and nanophytoplankton assemblages. On the other hand, they can be caused by intensive predation pressure from microzooplankton and mesozooplankton assemblages, which demonstrate the highest numerical abundance at that time. The period of balanced abundance is characterized by a distinct decline in photosynthetic activity of the coastal phytocoenosis, decreased reproductive rates of pelagic algae and/or substantially increased predation pressure from phytoplankton-eating organisms, and by intensive accumulation of organic substances in the fraction of suspended matter. Ultraviolet radiation being at its maximum during this period can have a substantial effect on the structure of algal coenosis. Thereupon, nanoplanktonic flagellates, more resistant to negative impact of this factor, account for the major part of the community. Processes of recycling prevail in the hydrochemical regime of the pelagic zone at this stage. A great role in these processes by all appearances belongs to nanoplanktonic algae actively consuming dissolved organic forms of biogenic elements. Afterwards, in the early autumnal period, the inflow of shelf waters rich in biogenic matter into the pelagic zone turns the nanophytoplanktonic community out of the balanced state and stimulates a new burst of its primary production activity, i.e. the formation of the autumnal maximum. At the same time, the nanofraction of the phytocoenosis keeps functioning in an oscillatory regime with a general trend to gradual attenuation of activity. The other scenario of the summer development of the nanophytoplanktonic community is the formation of a distinct maximum. In coastal waters of the Central Murman Coast (Kola Peninsula) such a scenario occurs in years with decreased hydrodynamic activity of coastal waters (like for instance in 1987). At this period, nanoalgae demonstrate both high photosynthetic and heterotrophic activities taking part in processes of the secondary synthesis of organic matter and recycling of biogenic elements.

VI   Activity attenuation stage. In the autumnal period, both nanophytoplankton and microphytoplankton assemblages undergo a gradual decrease in abundance progressing in an oscillatory manner. Parallel to this runs the process of disintegration of the phytocoenosis floristic structure developing on the background of rapid cooling of coastal waters and increased hydrodynamic activity. Mass decay of pelagic algae stimulates processes of re-mineralization of dead organic matter in the coastal zone and accumulation of biogenic elements in mineral fractions. This is the very period when the main winter structural trend in the development of the nanophytoplanktonic community takes shape, i.e. nanophytoplanktonic algae gradually grow smaller in size. The main factors regulating the community structure at this period are likely the light regime and hydrodynamic (turbulent) activity of the pelagic biotope.

VII Pre-hibernal maximum stage. At the end of November and the beginning of December, in the Barents Sea coastal zone, there is a distinct increase in activity of phytoplanktonic community, which in this autumn-winter transition period mostly consists of nanophytoplankton. Although in subarctic latitudes, when the polar night is drawing near, extremely low levels of solar radiation in the pelagic zone hamper an evident manifestation of this phase. It's hard to tell by now what structural transformations in the coastal biotope cause this late outburst of activity of the nanophytoplanktonic community. Determining this stage and describing its ecologic characteristics is the task for future studies.

3.3 The role of other groups of primary producers in the coastal zone

3.3.1 Cryoflora

As it was mentioned before, a substantial contribution into primary production processes in the Barents Sea area covered with pack ice belongs to the cryoflora. An average annual production of sea-ice algal community per day makes up to 30 mgC/m²/day, within which phytoplankton accounts for 18 mgC/m²/day and 12 mg C/m²/day is the share of the cryoflora (Gosselin et al., 1997). Thus, if the ice cover area and the duration of daylight hours are taken into account, the primary production of the cryoflora can be estimated at 3 mil tons of carbon, i.e. about 5 % of the whole primary production of the Barents Sea pelagic zone.

However, results of studies in arctic coastal zones carried out over the last twenty years enable the conclusion to be made that the major part of this volume is produced by algae of the fast ice biotope. In August, the primary production of the cryoflora off the Franz Josef Land Archipelago made up 42.7 mgC/m²/day (Kuznetsov et al., 1994). Moreover, samples were made already at the heterotrophic stage of the ice-associated algal community succession when the respiration exceeded the photosynthesis level more than two orders of magnitude. A supposition can be made that during the spring exponential growth of sea-ice microalgae abundance, this value may be an order higher. Optimal light regime for the vegetation of the cryoflora remains in these latitudes within four months (from May till August). During this time the Barents Sea ice breaks up and the area of the ice cover decreases reaching the minimal values. Thus, the production of Barents Sea ice algae may make up to 6-8 gC/m²/year (Kuznetsov & Shoshina, 2003).

3.3.2 Macrophyte algae

The main difference between processes of the primary organic matter synthesis in coastal ecosystems of the Barents Sea and those in outer deep-water shelf areas is that the major producers of organic matter in coastal areas are not pelagic algal communities but macrophytes with annual production of ca. 630 000 tons in wet weight which is equal to ca. 66 000 tons of organic carbon per year. Calculated for a unit of area of Barents Sea coastal regions, the phytobenthos production makes up 1.2-2.3 kgC/m²/year, i.e. it can exceed the production of phytoplankton an order of magnitude (Romankevich & Vetrov, 2001).

The littoral zone of the Eastern Murman Coast (Kola Peninsula) is the habitat for macrophytes belonging to three major taxonomic groups: brown, red, and green algae. Each group numbers a dozen of species; still the most common are *Laminaria saccharina*, *Palmaria palmata* and *Ulvaria obscura*. During the polar night (from the beginning of December till the third decade of January) vital functions of algae is characterized only by respiration processes. In a month after the end of the polar night, the speed of the photosynthesis starts exceeding the consumption of oxygen. The vegetation period lasts eight months a year. Reduction of photosynthesis to the respiration level falls on the end of September. After that destruction processes prevail in the bottom phytocoenosis.

According to experimental studies made in bays of the Eastern Murman Coast (Kola Peninsula) over the last years, the production of brown algae averaged to 0.98 mgC/g/day (wet weight) with a range of 0.32 to 1.74 mgC/g/day (wet weight). The same average value for the green algae was slightly higher, 1.18 mgC/g/day (wet weight), and for the red algae still higher, 1.65 mgC/g/day (wet weight). Ranges of these values were also higher, 0.31-2.24 and 0.53-3.18 mgC/g/day (wet weight), respectively (Kuznetsov & Shoshina, 2003). However, it should be taken into account that brown algae absolutely dominate the macrophyte community in this area. The species *Laminaria saccharina* can account for 85-99 % of the total biomass (Propp, 1971). Having an average biomass of 8 kg/m² laminaria algae can make a pure production equal to 1 500 gC/m²/year (Kuznetsov & Shoshina, 2003).

As for arctic coastal ecosystems, available data of single studies off the Franz Josef Land Archipelago enable the production potential of these areas to be assessed very approximately. Taxonomic composition of bottom assemblages here is several times poorer than in the southern Barents Sea and mostly consists of different species. Among species typical for bays of the Eastern Murman Coast (Kola Inlet) only brown algae *Laminaria saccharina*, *Alaria esculenta*, and *Pilayella litoralis* are found here. The production of brown algae compared to coastal waters of the southern Barents Sea is 26 % lower and the production of green algae is 42 % lower. Respiration levels differ even more greatly, 58 % and 83 % lower for brown and green algae, respectively (Kuznetsov & Shoshina, 2003). That gives reason to suppose that the main factor determining the differences observed is low water temperature that to a greater extent affects the intensity of respiration of macrophytes than the process of photosynthesis.

3.3.3 Microphytobenthos

Microphytobenthic organisms also play an important role in coastal ecosystems though they somewhat yield to phytoplankton in productive capacity. The total annual biomass of the

Barents Sea microphytobenthos makes up 0.35 mil tons with the annual production of 5.3 mil tons of carbon (Vetrov & Romankevich, 2004).

In the littoral and upper sublittoral zones of the Eastern Murman Coast (Kola Peninsula), the flora of soft grounds is represented solely by diatomic algae. The dominant position in the community belongs to pennate diatoms which remain in bottom biocoenoses throughout the year (Bondarchuk & Kuznetsov, 1988). An assumption can be made that in the winter period when the photosynthesis is hampered by the absence of light, pennate diatoms turn to mixotrophic and heterotrophic types of feeding. The taxonomic list of organisms of this community includes, according to data of different studies, 65-70 species and forms of bottom diatoms (Korotkevich, 1960; Kuznetsov & Shoshina, 2003). In April and May, during the vernal bloom of phytoplankton, and in autumn at the end of the vegetation period of pelagic microalgae, samples of benthos contain typical planktonic species.

The beginning of active development of microphytobenthos in coastal waters of the southern Barents Sea occurs in April while the end falls on September or October. Studies carried out in different areas of the southern Barents Sea coastal waters showed a rather wide range of values of the bottom flora production, 80 to 500 $mgC/m^2/day$. Levels of the primary production were measured in areas with different depths. The maximum values were registered in shallow water (with depths less than 5 meters) and averaged 16.4 mgC/m^2 over the period from April till September with a range of 0.1 to 50.6 $mgC/m^2/hour$. According to calculations, taken in whole over the given period of time, this value exceeds the production of phytoplankton in the water column almost three orders of magnitude. At a depth of 10 m this value averaged 15.0 $mgC/m^2/hour$ (which is 1.2 times lower than the production level of pelagic algae) and at a depth of 17 m it made up 9.9 $mgC/m^2/hour$ on average (40 % of the phytoplankton production in the water column) (Kuznetsov & Shoshina, 2003).

Studies of microphytobenthos off the Franz Josef Land Archipelago are also of great interest. These studies showed that during the period of open water (from June till September) the community of bottom microalgae remains in the active photosynthesis stage and the level of photosynthesis to a considerable degree depends on depth. In shallow water at depths less than 5 m, where small grounded hummocks, icebergs, and fast shore ice have destructive impacts on biotopes, levels of the primary production are very low. The maximum levels are observed at depths of 7-20 m, at greater depths levels of photosynthesis sharply decrease (Kuznetsov & Shoshina, 2003). The values of the primary production of microphytobenthos here make up on average 40 to 65 $mgC/m^2/day$, with values of 170 to 235 mgC/m^2 registered in a single small area where fast shore ice is subject to destruction accompanied by the inflow of allochthonous organic matter into bottom sediments.

To sum it up, the following conclusions can be made. An analysis of calculation data well enables the contribution of coastal ecosystems of the Barents Sea into the total production balance of this water body to be estimated. The area of coastal waters is less than 3 % of the whole area of the Barents Sea. Still it accounts for 9-33 % (14 % on average) of the total organic matter in this body of water. Even more significant are differences between the productivity of coastal waters and waters of the deep-water outer shelf zone calculated for a unit of area. Calculations show that the productivity of coastal waters is 3 to 12 times higher and makes up 253 tons of carbon per square km a year versus 21.1-77 tons in the deep-water outer shelf zone. The productivity of shallow waters (with depths from 0 to 10 meters) makes up 1 222 tons of carbon per square km a year, which is 16 to 58 times higher of the corresponding values for the deep-water outer shelf zone.

4. Planktonic assemblages associated with the sea ice

4.1 Ice-edge bloom of pelagic microalgae

In previous chapters we described the so-called *ice-edge phytoplankton bloom* as the first stage of the annual cycle of Barents Sea pelagic algal assemblages or the main stage of the primary production formation in the pelagic zone of arctic coastal ecosystems (see above). In truth, this phenomenon has already been known for a hundred of years and was first recorded in reports of polar expeditions as long as the beginning of the XX century. Still it continues provoking interest among specialists. By all appearances, first researchers were very much amazed just by the fact of rapid growth of microalgae abundance in cold high latitudes right near the ice edge while in southward warmer open waters it was not observed. Ever since hydrobiologists have been discussing the connection between the impetus of the phytoplankton bloom and processes of the sea ice melt.

The most complete review of hypotheses on this issue proposed in the first quarter of the XX century was made by P. Shirshov (1937). Shirshov described all possible mechanisms of direct influence of sea ice melt on microalgae abundance growth including freshening of the upper water layer, inflow of nutrients from the ice surface with melt water, and even increased levels of carbonates and tri-hydrolic molecules reported in publications of that time. Using many examples Shirshov shows the impossibility to give a comprehensive explanation of how these factors, taken either separately or in combination, can stimulate the beginning of the spring bloom. P. Shirshov (1937) believes that the only factor able to act as a trigger is light or rather an increase in insolation as the sea ice melts. This point of view for a long time occupied the dominant position in hydrobiology although it was in a simplified manner treated as if it asserted that the state of the ice cover has no effect on the behavior of production processes in the pelagic zone. This idea has been many times criticized. It was pointed out that in the Barents Sea, which never completely freezes, an increase in insolation cannot serve as a comprehensive explanation of the abundance growth of phytoplankton just near the ice edge. In high arctic latitudes this increase in insolation is insignificant while in open waters of the southern Barents Sea, where the level of the solar radiation rapidly increases as the polar night comes to an end, such a rapid bloom as near the ice edge is not observed. This is why the appearance of a new concept was quite expected. The authors of this new concept took notice of one more phenomenon driven by the sea ice melt, namely the density jump caused by the freshening of sea water and separating the upper homogeneous water layer. Norwegian oceanologist H.U. Sverdrup (Sverdrup, 1953) introduced the notion of a so-called critical depth. According to Sverdrup's Critical Depth Hypothesis, when the lower boundary of the mixed homogenous layer locates below the critical depth, the bloom is impossible because the organic matter decomposing in the water column below this compensation level exceeds the organic matter produced by the photosynthesis in the layer above the compensation depth.

This concept was then developed and many times confirmed by other researchers when they observed the phytoplankton bloom in different areas of the Barents Sea (Rey et al.,

1987; Skjoldal et al., 1987; and other). Nowadays most biologists adhere to this concept though it is not able to explain, in particular, the fact of the bloom outburst both in coastal waters with little depths, where the whole water column locates above the critical depth, and in outer areas never being freshened. Thus, the question remains open and needs further discussion. The ice-edge phytoplankton bloom, an important stage in the succession cycle of Barents Sea primary producers, still needs thorough research.

4.2 Phytoplanktonic assemblages in water areas entirely covered with the sea ice

Among different aspects of influence of the sea ice on the development of primary producers in the Barents Sea pelagic zone, one more thing needs thorough attention. For a long period of studying arctic pelagic ecosystems, the winter season has always been considered a dormant stage when all vital processes in phytoplankton populations come to a stop. Only at the end of the 1970s, a rapid burst of studies on pelagic unicellular organisms with the heterotrophic type of feeding revealed inconsistency of orthodox opinions explaining processes of the organic matter re-mineralization solely by the activity of bacterioplankton. In truth, a complicated system of the heterotrophic metabolism functions in the pelagic zone in winter. This system includes different groups of mixotrophic and heterotrophic organisms belonging to several divisions and categories of microalgae and heterotrophic flagellates, which form the basis of the winter algal coenosis.

Results of studies carried out over the last decades in winter and at the beginning of spring in coastal arctic areas from nuclear powered ice-breakers are of particular interest in this respect. These studies show that over a large area, from the Kanin-Kolguev Shoal to the southern coast of the Pechora Sea (southeastern Barents Sea), the activation of production processes starts during the polar night already under the complete sea ice cover. Although the most part of studies was carried out in the Pechora Polynia, the biologic activity of the community cannot be explained solely by the occurrence of an open water area of the pelagic zone. Observations made at the same period of time in the Eastern Novaya Zemlya Polynia located approximately at the same latitude and thus having similar light conditions did not prove even the slightest signs of the spring awakening of phytoplankton populations (Makarevich & Druzhkova, 2010).

An analysis of the taxonomic structure and quantitative parameters of the community revealed that in the first half of February the core of the community consists of flagellates including unarmored (naked) forms. The composition of this group was almost homogeneous over the whole water area while the diatomic complex demonstrated diversity though yielding to the dynophyte complex in biomass. At the end of February, the seasonal state of pelagic floral assemblages in the Pechora Sea (southeastern Barents Sea) can be characterized as a stage of the primary activation of diatomic populations. Pennate forms dominate the community at this stage. Moreover, mostly single non-colonial algae account for the most part of the biomass. The most typical representatives of the flora at this stage are *Coscinodiscus cf. stellaris, Amphiprora kjellmanii, Cylindrotheca closterium, Gyrosigma fasciola, Navicula/Plagiotropis* spp. and *Pleurosigma stuxbergii*. In March, the community comes into the initial stage of the spring bloom and at some sites demonstrates a rapid growth of abundance of early-spring diatomic populations. As typical for arctic pelagic ecosystems at the beginning of the bloom, the increase of biomass happens mostly at the expense of species of the genus *Thalassiosira*: namely *T. antarctica, T. cf. gravida, T. hispida* and *T. hyalina*.

Thus, the sequence of initial stages of the phytoplanktonic community succession system in the Pechora Sea coastal zone corrected according to the latest data, first of all data on the active development process of microalgae under the sea ice, looks as follows (Makarevich & Druzhkova, 2010). I) winter stage (the first and the second decades of February): heterotrophic processing of organic matter, II) early spring stage: 1) activation of the early spring diatomic complex (from the third decade of February till the first decade of March): beginning of photosynthesis processes, 2) rapid growth of biomass of early spring diatomic algae (the second and the third decades of March): exponential growth of abundance.

5. Spatial and temporal dynamics of the primary production in the Barents Sea

In general, in the Barents Sea, zones of increased productivity correspond to seats of the phytoplankton bloom, i.e. to near-shore areas, ice edge zones, and streams of permanent currents of the Atlantic Ocean origin. The analysis of the spatial distribution of chlorophyll-*a* confirms this regularity. Thus in April 1985, in the southeastern Barents Sea, maximum concentrations of chlorophyll-*a* were registered in the coastal zone (21.8 mg/m^2) and the ice edge zone (19.6 mg/m^2) while in the central part of the area lower levels of chlorophyll-*a* were observed with an even distribution of this parameter. In the western Barents Sea the same year in April, chlorophyll-*a* levels varied from 6.2 mg/m^2 to 34.6 mg/m^2 with local maximums confined to streams of permanent currents of the Atlantic Ocean origin, namely the Central Branch of the Nordkapp Current (17 mg/m^2), the Murman Current (34.6 mg/m^2), and the Coastal Murman Current (28.1 mg/m^2) (Savinov & Bobrov, 1990).

Maximum levels of the primary production in the Barents Sea were registered over seabed elevations (Murman, Finnmarken, and Nordkinn Shoals), in the coastal zone, and in areas where waters of the Atlantic Ocean origin mix with Barents Sea waters (Bobrov, 1985).

As it was mentioned before, the beginning and duration of the phytoplankton vegetation cycle in different latitudinal zones of the Barents Sea in temporal aspect depend first of all on ice conditions, i.e. the duration of the active vegetation season shortens from 8-9 months in the southern Barents Sea to 2-3 months in its northern part. Simultaneous occurrence of phytoplankton assemblages, being at different succession stages, in the pelagic zone determines the formation of seasonal heterogeneities of the primary production distribution. At the same time, areas with increased productivity confined to frontal zones do not have seasonal character and are connected to upward water movement along the frontal surface.

Despite sufficient knowledge on the Barents Sea compared to other arctic seas, there are no reliable estimates of the annual primary production of the Barents Sea phytoplankton. The situation is hampered by a mixed hydrological structure and a diversity of seasonal changes of the productivity of various Barents Sea areas. As a result, estimates of the Barents Sea total production made by various researchers greatly differ. In a number of publications (Romankevich et al., 1982; Danyushevskaya et al., 1990) estimates of the primary production are much the same, 77 to 80 mil tons of organic carbon per year, while other publications suggest higher values of the Barents Sea total production, 100 to 150 mil tons of C$_{org}$ per year (Matishov & Drobysheva, 1994).

Based on the generalization of published data and the use of the existing notions of the typical seasonal course of the primary production, P. Makarevich (2007) makes the reconstruction of this parameter for the Barents Sea ecosystem (fig. 4). Over the period of 1964 to 2002 this parameter varied in a range of 23 $gC_{org}/m^2/year$ (1970) to 69 $gC_{org}/m^2/year$ (2002) with an average value of 44 $gC_{org}/m^2/year$, which well correlates with the estimate made in the last review (Romankevich & Vetrov, 2001): 44.5 gC_{org}/m^2.

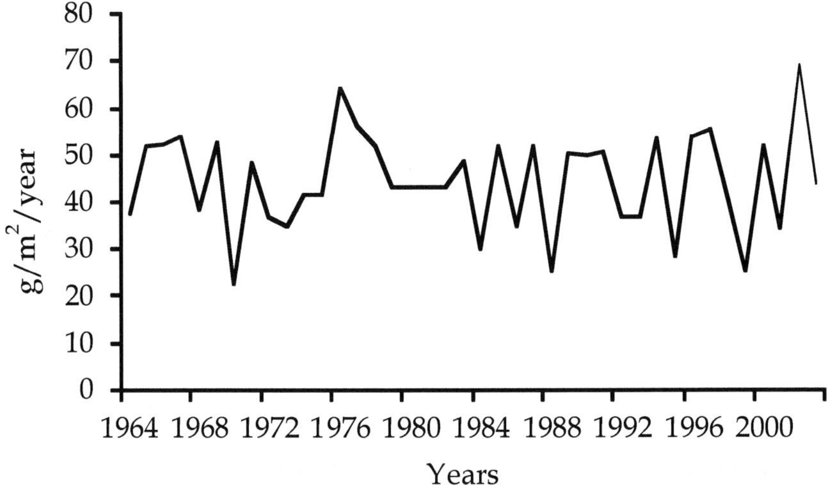

Fig. 4. Dynamics of the annual primary production in the Barents Sea over the period of 1964-2000 (reconstruction)

It is evident that both values and the character of the primary production distribution in the Barents Sea greatly differ across seasons. At the same time, there is distinct interrelation between spatial and temporal variability of this parameter, i.e. its seasonal dynamics appears similar in areas with a definite type of water masses. Thus, in waters of the Atlantic Ocean origin, in the outermost southwestern Barents Sea, the maximum of the specific primary production falls on the end of May and the beginning of June (2.5-3 mil tons per month); in July this value decreases (0.5 mil tons/month) while in August the second peak of productivity is observed (1.3 mil tons per month). The variations established are obviously determined first of all by changes in the zooplankton community, as the predation of zooplankton on phytoplankton in warm waters of the Atlantic Ocean origin is the main regulator of abundance in pelagic algal assemblages.

In the artic zone occupying the whole northern part of the Barents Sea, the only primary production maximum registered (1.2 mil tons per month) is determined by the aforementioned ice-edge bloom, which starts in June during the intensive ice melt and lasts through the whole warm period in a narrow stripe of water along the retreating edge of the ice cover. Besides, an area with increased productivity is registered off the southern coasts of the Franz Josef Land Archipelago where prevailing summer winds push the sea ice off the coast and release a space of open water occupied by an actively vegetating phytoplankton complex. A similar picture is observed in the southeastern Barents sea (often called the Pechora Sea), the zone of the maximum continental runoff. The period of high productivity here also relates to the sea ice melt and falls on June – September (0.8-1.3 mil tons per month), however the bloom is not confined only to the ice edge and occupies a larger water area.

The central part of the Barents Sea, from the Murman Coast to the Novaya Zemlya Archipelago, contains waters of mixed origin and therefore is characterized by a more complicated hydrological regime, which affects the course of seasonal change of productivity. Maximum values of the specific primary production are observed in April and make up 6.5 mil tons per month. Then in May a short recession period begins with a level of the specific primary production of 1.5 mil tons per month. This period is replaced in summer months by the period of a stable increase of the specific primary production level up to 5 mil tons per month. The main factor affecting the dynamics of the primary production in this water area is the concentration of biogenic elements in the water column. Usually registered in September, the autumn maximum (ca. 4 mil tons per month) forms in the southern coastal area of the water body and is determined by the activation of the spring phytoplankton when water is enriched with biogenic elements mainly due to the wind mixing and its temperature decreases.

According to Vinogradov et al. (2000), the total primary production of the Barents Sea calculated with the use of remote sensing data and average values for the defined ranges of the chlorophyll contents in surface waters of arctic seas (for the period from April till September) makes up 38.4 mil tons C_{org} per year. When the values for the months characterized by a continuous cloudiness are corrected, the annual production makes up 55 mil tons C_{org} per year. The level of the average annual primary production in the Barents Sea calculated in that way appears rather little, 0.12 gC_{org}/day (compared to 0.45 gC_{org}/day for the World Ocean). According to existing classifications, the Barents Sea corresponds to mesotrophic water bodies by the level of biological productivity: 0.1-0.5 gC_{org}/day (Romankevich & Vetrov, 2001).

6. References

Bobrov, Yu.A. (1985). Issledovanie pervichnoy produktsii v Barentsevom more, In: *Zhizn' i usloviya yeyo sustchestvovaniya v pelagiali Barentseva morya*, G.G. Matishov (Ed.), pp. 110-125, Izd. KF AN SSSR, Apatity, Russia (In Russian)

Bondarchuk, L.L. & Kuznetsov, L.L. (1988). Sezonnaya dinamika diatomovoy flory verkhney sublitorali Barentseva morya, In: *Novosti sistematiki nizshikh rasteniy*, Vol. 25, pp. 27-31, ISSN 0568-5435 (In Russian)

Bouden, K. (1988). Fizicheskaya okeanografiya pribrezhnykh vod, Mir, ISBN 5-03-000700-8, Moscow, Russia (In Russian)

Bul'yon, V.V. (1985). Aktivnost' mikroflory v pribrezhnykh vodakh Zemli Frantsa-Iosifa, In: *Biologicheskie osnovy promyslovogo osvoyeniya otkrytykh rayonov okeana*, pp. 101-108, Nauka, Moscow, Russia (In Russian)

Danyushevskaya, A.I., Petrova, V.I., Yashin, D.S. et al. (Eds). (1990). *Organicheskoye vestchestvo donnykh otlozheniy polyarnykh zon Mirovogo okeana*, Nedra, Leningrad, Russia (In Russian)

Druzhkov, N.V. & Makarevich, P.R. (1999). Comparison of the Phytoplankton Assemblages of the South-Eastern Barents Sea and South-Western Kara Sea: Phytogeographical Status of the Regions. *Botanica Marina*, Vol.42, pp. 103–115, ISSN 0006-8055

Druzhkov, N.V., Grönlund, L. & Kuznetsov, L.L. (1997a). The phytoplankton of the Pechora Sea, the Pechora Bay and the Cheshskaya Bay, In: *Pechora Sea Ecological Studies in 1992-1995 (Finnish-Russian Offshore Technology Working Group Report B13)*, pp. 41-51, FIMR, Helsinki, Finland

Druzhkov, N.V., Kuznetsov, L.L., Baytaz, O.N. & Druzhkova, E.I. (1997b). Sezonnyye tciklicheskiye protsessy v severoyevropeyskikh pribrezhnykh pelagicheskikh ekosistemakh (na primere Tsentral'nogo Murmana, Barentsevo more), In: *Plankton morey Zapadnoy Arktiki*, G.G. Matishov (Ed.), pp. 145-178, Izd. KNTs RAN, Apatity, Russia (In Russian)

Fyodorov, V.D. (1987). Aktual'noye i neaktual'noye v biologii. *Biologicheskie nauki*, No. 8, pp. 6-26, ISSN 0470-4606 (In Russian)

Fyodorov, V.D., Smirnov, N.A. & Fyodorov, V.V. (1988). Nekotorye zakonomernosti produtsirovaniya organicheskogo vestchestva fitoplanktnom. *Doklady AN SSSR*, Vol. 299, No. 2, pp. 506-508 (In Russian)

Gosselin, M., Legendre, L., Demers, S. & Ingram, R.G. (1997). Responses of sea-ice microalgae to climatic and fortnightly tidal inputs (Manitounuk Sound, Hudson Bay). *Canad. Journ. Fish. Aquat. Sci.*, Vol. 42, pp. 999-1006, ISSN 0706-652X

Grönlund, L., Kuznetsov, L.L. & Druzhkov, N.V. (1996). Hydrography, water chemistry, plankton production and plankton species composition in the Pechora Bay in 1995, In: *Pechora Bay Ecological Studies in 1995 (Finnish-Russian Offshore Technology Working Group Report B8)*, pp. 3-33, FIMR, Helsinki, Finland

Grönlund, L., Kuznetsov, L.L. & Druzhkov, N.V. (1997). Hydrology of the Pechora Sea, the Pechora Bay and the Cheshskaya Bay In: *Pechora Sea Ecological Studies in 1992-1995 (Finnish-Russian Offshore Technology Working Group Report B13)*, pp. 15-27, FIMR, Helsinki, Finland

Hansen, B., Berggreen, U.C., Tande, K.S. & Eilertsen, H.C. Post-bloom grazing by *Calanus glacialis, C. finmarchicus* and *C. hyperboreus* in the region of Polar Front, Barents Sea // Mar. biol. 1990. Vol. 104. P. 5-14.

Hesse, K.-J., Liu, Z.L. & Schaumann, K. (1989). Phytoplankton and fronts in the German Bight. *Scient. Mar.*, Vol. 53, No. 2-3, pp. 187-196, ISSN 0214-8358

Korotkevich, O.S. (1960). Diatomovaya flora litorali Barentseva moray, *Trudy MMBI*, No. 1(5), pp. 68-338 (In Russian)

Kudlo, B.P. (1970). Mnogoletnie izmeneniya temperatury vody u pribrezh'ya Murmana. In: Materialy mnogoletnikh issledovaniy severnogo basseyna (Trudy PINRO, No. 16, Pt I), K.A.Lyamin (Ed.), pp. 68-338, Murmansk, Russia (In Russian)

Kuznetsov, L.L. & Shoshina E.V. (2003). Fitotsenozy Barentseva moray (fiziologicheskie i strukturnye kharakteristiki), Izd. KNTs RAN, Apatity, Russia (In Russian)

Kuznetsov, L.L., Druzhkov, N.V. & Baytaz, O.N. (1996). Pribrezhnye ekosistemy pelagiali Barentseva morya. In: *Ekosistemy pelagiali morey Zapadnoy Arktiki*, G.G. Matishov (Ed.), pp. 73-84, Izd. KNTs RAN, Apatity, Russia (In Russian)

Kuznetsov, L.L., Grönlund, L. & Druzhkov, N.V. (1997). The annual production cycle in the Pechora Sea, the Pechora Bay and the Cheshskaya Bay, In: *Pechora Sea Ecological*

Studies in 1992-1995 (Finnish-Russian Offshore Technology Working Group Report B13), pp. 53-68, FIMR, Helsinki, Finland

Kuznetsov, L.L., Makarevich, P.R., & Makarov, M.V. (1994). Strukturno-produktsionnye pokazateli morskikh fitotsenozov. In: *Sreda obitaniya i ekosistemy Zemli Frantsa-Iosifa (arkhipelag i shel'f),* G.G. Matishov (Ed.), pp. 89-94, Izd. KNTs RAN, Apatity, Russia (In Russian)

Larionov, V.V. (1995). Fitoplankton pribrezhnoy zony Barentseva morya. In: *Sreda obitaniya i ekosistemy Novoy Zemli (arkhipelag i shel'f),* G.G. Matishov (Ed.), pp. 52-59, Izd. KNTs RAN, Apatity, Russia (In Russian)

Larionov, V.V. (1997). Obstchiye zakonomernosti prostranstvenno-vremennoy izmenchivosti fitoplanktona Barentseva morya. In: *Plankton morey Zapadnoy Arktiki,* G.G. Matishov (Ed.), pp. 65-127, Izd. KNTs RAN, Apatity, Russia (In Russian)

Makarevich, P.R. & Druzhkova, E.I. (2010). Sezonnye tsiklicheskie protsessy v pribrezhnykh planktonnykh al'gotsenozakh severnykh morey, Izd. YuNTs RAN, ISBN 978-5-902982-75-3, Rostov-na-Donu, Russia (In Russian)

Makarevich, P.R. (2007). Planktonnye al'gotsenozy estuarnykh ekosistem: Barentsevo, Karskoye i Azovskoye morya, Nauka, ISBN 978-5-02-035570-5, Moscow, Russia (In Russian)

Matishov, G.G. & Drobysheva, S.S. (1994). Obstchie zakonomernosti struktury i razvitiya morskikh ekosistem yevropeyskoy Arktiki, In: Evolyutsiya ekosistem i biogeografiya morey yevropeyskoy Arktiki, D.S. Pavlov (Ed.), pp. 9-30, Nauka, ISBN 5-02-026744, Sankt-Peterburg, Russia (In Russian)

McQuoid, V.R., Hobson, L.A. (1995). Importance of resting stages in diatom seasonal succession. *J. Phycol.,* Vol. 31, pp. 44-50, ISSN 1529-8817

Propp, M.V. (1971). Ekologia pribrezhnykh donnykh soobstchestv Barentseva morya, Nauka, Leningrad, Russia (In Russian)

Rey, F., Skjoldal, H.R. & Slagstad, D. (1987). Primary production in relating to climatic changes in the Barents Sea, In: *The effect of oceanographic conditions on distribution and population dynamics of commercial fish stocks in the Barents Sea,* H. Loeng (Ed.),pp. 29-46, IMR, Bergen, Norway

Romankevich, E,A., Danyshevskaya A.I., Belyayeva, A.N. & Rusanov, V.P. (Eds). (1982). *Biogeokhimiya organicheskogo vestchestva arkticheskikh morey,* Nauka, Moscow, Russia (In Russian)

Romankevich, E.A. & Vetrov, A.A. (2001). Tsikl ugleroda v arkticheskikh moryakh Rossii, Nauka, ISBN 5-02-026744, Moscow, Russia (In Russian)

Savinov, V.M. & Bobrov, Yu.A. (1990). Prostranstvennye neodnorodnosti raspredeleniya pervichnoy productsii Barentseva morya. In: *Ekologiya i biologicheskaya produktivnost' Barentseva morya,* G.G. Matishov (Ed.), pp. 9-17, Nauka, ISBN 5-02-004734-1, Moscow, Russia (In Russian)

Shirshov, P.P. (1937). Sezonnye yavleniya v zhizni fitoplanktona polyarnykh morey v svyazi s ledovym rezhimom. *Trudy Arktich. In-ta,* Vol. 82, pp. 47-111, (In Russian)

Skjoldal, H.R., Hassel, A., Rey, F. & Loeng, H. (1987). Spring phytoplankton development and zooplankton reproduction in the central Barents Sea in the period 1979-1984, In: *The effect of oceanographic conditions on distribution and population dynamics of commercial fish stocks in the Barents Sea,* H. Loeng (Ed.), pp. 59-89, IMR, Bergen, Norway

Smirnov, N.A., Fyodorov, V.V. & Fyodorov, V.D. (1989). Funktsional'noye ecologicheskoye opisanie sezonnogo razvitiya fitoplanktona Belogo morya. *Zhurnal obstchey biologii*, Vol. 50, No. 3, pp. 353-364, ISSN 0044-4596 (In Russian)

Svendsen, H. (1986). Mixing and exchange processes in estuaries, fjords and shelf waters, In: *The role of freshwater outflow in coastal marine ecosystems*, S. Skreslet (Ed.), pp. 13-45, Springer-Verlag, ISBN 3-540-16089-2, Berlin - Heidelberg, Germany

Sverdrup, H.U. (1953). On conditions for the vernal blooming of phytoplankton. *Journ. du Cons.*, Vol. 18, No. 3, pp. 287-295, ISSN 0020-6466

Trofimova, V.V. Fotosinteticheskie pigmenty fitoplanktona: Sutochnaya dinamika i prostranstvennoye raspredelenie, In: Kol'skiy zaliv: osvoyenie i ratsional'noe prirodopol'zovanie, G.G. Matishov (Ed.), pp. 86-108, Nauka, ISBN 978-5-02-035871-3, Moscow, Russia (In Russian)

Vetrov, A.A. & Romankevich, E.A. (2004). Carbon Cycle in the Russian Arctic Seas, Springer, ISBN 3-54-021477-1, Berlin, Germany

Vinogradov, M.E., Vedernikov, V.I., Romankevich, E.A. & Vetrov, A.A. (2000). Komponenty tsikla ugleroda v arkticheskikh moryakh Rossii: Pervichnaya produktsiya i potok C_{org} iz foticheskogo sloya. *Okeanologiya*, Vol. 40, No. 2, pp. 221-233, ISSN 0030-1574 (In Russian)

Węslawski, J.M., Kwaśniewski, S. & Wiktor, J. (1991). Winter in a Svalbard fjord ecosystem. *Arctic*, Vol. 44, pp. 115-123, ISSN 0004-0843

Węslawski, J.M., Zajaczkowski, M., Kwaśniewski, S., Jezierski, J. & Moskal, W. (1988). Seasonality in the Arctic fjord ecosystem: Hornsund, Spitsbergen. *Polar Res.*, Vol. 6, pp. 185-189, ISSN 0800-0395

17

Protecting Ecosystems from Underground Invasions – Seed Bank Dynamics in a Semi-Arid Shrub-Steppe

David R. Clements[1] and Lynne B. Atwood[2]
[1]*Trinity Western University, Langley, British Columbia*
[2]*Genoa Environmental, Cobble Hill, BC*
Canada

1. Introduction

Plant invasions threaten natural and managed ecosystems throughout the world (Hobbs & Humphries, 1995). Invasive plants reduce species diversity through competition with native plant species, leading to local reductions in populations of native species. According to E.O. Wilson: "on a global basis…the two great destroyers of biodiversity are, first habitat destruction and, second, invasion by exotic species" (Simberloff et al., 1997). In some cases invasive plants act more as "passengers" than "drivers" of ecological change in degraded ecosystems (MacDougall & Turkington, 2005). In either capacity invasive plants are a significant biotic element to consider in evaluating the integrity of a given type of ecosystem. The spread of non-native plants throughout the world has a homogenizing impact on regional floras, particularly given the tendency of certain invasive species in forming monocultures or near monocultures. Yet the seriousness of plant invasions is sometimes called into question (Larson, 2007) and because invasion biology is a relatively young field (Davis, 2009), further research is required to better assess the impacts of invasive species on ecosystem function.

Shrub-steppe ecosystems occur in North America in the rainshadow of the western mountain ranges. These habitats are characterized by relatively low preciptation and extremes of temperature with vegetation structure typified by scattered shrubs of various species such as *Artemesia* spp. (sagebrush) or *Purshia tridentata* (antelope-brush). Similar ecosystems are found throughout the world in high altitute temperate continental areas such as southwestern Russia, other parts of Asia, and South America. The South Okanagan Valley in British Columbia is representative of shrub-steppe ecosystems featuring *P. tridentata* prominently and is recognized federally as a biodiversity hot spot (Mosquin et al., 1995). Isolated remnant grasslands are home to Provincially and Federally listed endangered wildlife and form a key component of the South Okanagan Conservation Strategy (Bryan, 1996). A major contributor to the loss of biodiversity in these intermountain areas in British Columbia and throughout the Pacific Northwest of North America is the presence of extensive infestations of invasive species (Clements & Scott, 2011). Key invasive plants that degrade arid grasslands in this region include *Centaurea diffusa* (diffuse knapweed) and

Bromus tectorum (cheatgrass), which form large, spreading patches (Mack, 1981; Roché & Roché, 1999; Clements et al., 2007; Clements & Scott, 2011).

Restoration of such areas is a serious challenge; a long-history of grazing, invasion of non-native plants, other types of soil disturbance and habitat fragmentation in shrub-steppe habitats tends to simplify the ecosystem and reduce its integrity. Although Europeans arrived in the western U.S. decades earlier, it was during the 30 year period from the end of the U.S. civil war in 1865, that the steppe regions in western North America from British Columbia south to Nevada underwent a dramatic transformation from small isolated habitation by Europeans to the development of more permanent farmland and ranches (Meinig, 1968; Elliott, 1973). This resulted in the first serious weed invasion but many of the serious modern-day invaders such as *B. tectorum* did not arrive until around the turn of the 20th century, but quickly made up for lost time covering large areas of these interior grasslands by 1914 (Mack, 1981; Mack, 1986). The large scale movement and grazing of cattle, coupled with the introduction of horticultural crops such as apples and other fruit trees in the 20th century has continued to provide disturbed conditions condusive to the spread of non-native plants (Krannitz, 2008; Clements & Scott, 2011). The native grassland in this area is dominated by shrubs such as *P. tridentata* and perennial bunchgrasses such as *Pseudoroegneria spicata* (bluebunch wheatgrass) and others such as *Hesperostipa comata* (needle-and-thread grass) (Atwood & Scudder, 2003; Erickson, 2003). The soils may often be shallow and sandy, but in the absence of heavy grazing or other pronounced disturbances, a microbiotic crust, dominated by mosses and lichens, is maintained and provides valuable ecosystem services such as soil moisture retention (Loope & Gifford, 1972; Atwood & Krannitz, 2000).

A unique aspect of population dynamics in seed bearing plants is recruitment of populations over multiple time scales via seed banks. The invasive success of many non-native species is attributed to a persistent soil seed bank (Baker, 1974; Holm et al., 1977; Roché & Roché, 1999). Seed banks of invasive species can be both indicative of past disturbances and predictive of future weed population dynamics, and thus have great significance to restoration ecology, although there is considerable scope for further work in this area (Bakker et al., 1996). Many non-native invasive plants, particularly annuals, produce persistent seed banks and thus are very difficult to remove from an ecosystem once established, particularly if disturbances such as soil tillage (Clements et al., 1996), fire (Mandle et al., 2011) or grazing (Clements et al., 2007) occur regularly. Seed bank sizes and dynamics vary according to ecosystem type (Leck et al., 1989), and are dependent on climatic factors, disturbance regimes, edaphic conditions and plant community structure. Thus the seed bank dynamics must be understood in these particular contexts. Two invasive alien plant species of particular interest in this system as previously mentioned are *C. diffusa* and *B. tectorum*. Both of these exhibit seed banks in response to disturbance, with the seed bank of *C. diffusa* tending to be more persistent (Clements et al., 2007).

This chapter describes the results of a study examining various restoration treatments for the antelope brush ecosystem in the southern Okanagan and makes recommendations for the restoration of such systems. The objectives of the study, conducted from 1998-2003 were as follows:

1. measure temperatures under polyethylene sheets used in experimental solarization trials in the southern Okanagan,
2. evaluate survival of diffuse knapweed and seeds of other native and non-native species *in situ* within solarization treatments,
3. investigate effects of high temperatures on *Centaurea diffusa* (diffuse knapweed) and *Sporobolus cryptandrus* (sand dropseed) seed viability, using temperatures equivalent to those that would be experienced under solarization, and
4. evaluate an array of alternative restoration treatments involving control of non-native species or seeding of native species.

2. Study area

The study was conducted in a semi-desert shrub steppe ecosystem dominated by *Purshia tridentata* (antelope bitterbrush) in the southern, more arid section of the Okanagan Valley in British Columbia, Canada. This ecosystem occurs within the northernmost extension of the intermountain plateau which extends southward to Nevada. The southern Okanagan valley receives just over 300 mm of precipitation annually. Precipitation is bi-modal: early summer (June) and mid-winter (December – January) (Chilton 1988). The native bunchgrass community persists but is highly invaded by non-native grasses and forbs, particularly where heavy grazing has occurred (Atwood and Scudder, 2003; Clements et al., 2007; Krannitz, 2008). The soil is typically overlaid with a microbiotic crust comprised of a mixture of lichen, moss, liverworts, algae, fungi, and bacteria. Atwood and Krannitz (2000) found that five days after a rainfall event in this region, crusted soils retained an average of 31% of the initial soil water, while bare soils retained just 9.5%.

The research was conducted at the Osoyoos Desert Centre, a site located several km northwest of Osoyoos, British Columbia (Fig. 1). The Osoyoos Desert Society acquired 50-ha of shrub steppe with a *P. tridentata* system heavily invaded by non-native plants but still containing substantial native plant diversity. The *P. tridentata* shrub steppe is susceptible to livestock grazing, with heavy grazing resulting in reduction of native species and increased cover of invasive nonindigenous species (Krannitz, 2008). Historically, cattle grazed the Osoyoos Desert Centre site every year between March and June. Prior to the establishment of the Centre, about 40 cattle were removed from the site.

Fig. 1. Antelope brush (*Purshia tridenta*) landscape at the Osoyoos Desert Centre study site, near Osoyoos, British Columbia, Canada; photo courtesy of the Osoyoos Desert Centre

3. Research methods

In 1998, when the Osoyoos Desert Centre was established, plots were set up on the site in five replicates to examine the impact of various restoration treatments.

3.1 Restoration treatments

Six different restoration treatments were put in place utilizing 10 x 10 m plots at the Osoyoos Desert Centre site (Table 1).

Treatment	Year initiated
Solarization	1998
Native bunchgrass hayseeding	1998
Removal of livestock grazing (controls)	1998
Manual + chemical control of *Centaurea diffusa*	1999
Broadcast seeding of natural grasses	2000
Addition of native vesicular arbuscular mycorrhizae and native seed	2000

Table 1. Restoration treatments at the Osoyoos Desert Centre site, 1998-2003, showing treatment type and year initiated

3.1.1 Solarization

Solarization plots were randomly chosen from a subset of plots with a high percent cover of diffuse knapweed (two plots per replicate for a total of 10 plots). Solarization plots averaged 27± 12 SD% cover of diffuse knapweed. Polyethylene sheets were placed over the entire 10 × 10 m plot in 1998, and left in place for a minimum of two growing seasons. The plastic was removed from 5 of the plots in April 2000 (plastic removed from 1 plot per replicate). Vegetation data were collected from the plots before the plastic was put down and following its removal (June 2000 and 2002).

3.1.2 Native bunchgrass hayseeding

The hayseeding experiment was initiated on two plots per replication (10 plots total) in September 1998. As seed matured, seed heads and plant stalks were cut from four native bunchgrasses; *Aristida longiseta* (red three-awn), *Hesperostipa comata* (needle and thread grass), *Sporobolus cryptandrus* (sand dropseed), and *Pseudoroegneria spicata* (bluebunch wheatgrass). The plant material was collected from natural shrub-steppe communities within the South Okanagan Basin Ecosection. Approximately 200 litres of plant material (50 litres from each species) was distributed evenly over each 100 m² plot.

3.1.3 Removal of livestock grazing (controls)

In 1998, two randomly chosen 100-m² plots in each replication were established as control plots (10 plots total). Species identity and percent cover data were collected annually, in June, from 1998 to 2002. The control plots were monitored to document changes in the plant community without livestock grazing or restoration activity.

3.1.4 Manual and chemical control of *Centaurea diffusa*

The manual control of *C. diffusa* experiment was to determine the most effective time to hand weed *C. diffusa*, and whether weed density was related to the timing of the manual control. The first hand pulling was scheduled for early May, after which monthly treatments were scheduled if weed density was greater than 25% of the original *C. diffusa* cover.

The experiment for the chemical control of *C. diffusa* was implemented in 2000. *C. diffusa* plants in two 100-m^2 plots per replication (10 plots) were spot sprayed with an over-the-counter broadleaf herbicide, Killex, at the recommended label rate of 1.85 kg active ingredient per hectare in May 2000. Killex, a combination of 2,4-D, mecoprop, and dicamba was used in the chemical control experiment to determine if adequate control of *C. diffusa* could be obtained using a less expensive broad-leaf herbicide with less residual than the commonly used Tordon 22K (picloram).

3.1.5 Addition of native vesicular arbuscular mycorrhizae and native seed

Five 100-m^2 plots were divided into four 25-m^2 subplots and two treatments (Nurse plant inoculant and Soil-Root inoculant) and two control plots (no inoculant) were randomly established in each plot. The experimental plots were tilled, inoculated or not then seeded with the native grass seed mix at 28 kg per ha. The experiments were installed in the fall of 2000 and percent cover data for the seeded native grasses were collected in June 2001 and 2002.

3.1.6 Broadcast seeding of natural grasses

Broadcast seeding experiments were initiated in October 2000. The 100-m^2 plots were double split, producing four 25-m^2 subplots. One-half of the plots were tilled to mimic soil disturbance that would be associated with development projects. Shrubs remained, but existing herbaceous vegetation was cut and removed from the plot before tilling and the soil was packed after tilling. Standing herbaceous vegetation was also cut and removed from the no-till plots. Non-native species remaining in the plots were spot treated with the herbicide glyphosate applied at the full label rate.

The seed mix consisted of four perennial native bunchgrasses (*Aristida longiseta, Hesperostipa comata, Sporobolus cryptandrus,* and *Pseudoroegneria spicata*) and one annual agronomic grass *Lolium multiflorum* (annual ryegrass). All native grass seed used in the mix was collected from the area immediately adjacent to the research site. The native grasses were combined evenly in the mix (25% live seed per species) and seed rates were 28 kg per ha (1027 seeds per m^2) and 41 kg per ha (1504 seeds per m^2). Application rates were adjusted to account for the germination rate of the collected seed. Each seed rate was broadcast on one-half of the 100-m^2 plot and the soil was rolled after seeding.

3.2 Above-ground plant population monitoring

Plant species identity and percent cover data for all vascular plant species were collected annually for each of the 10 x 10 m plots, in June, from 1998 to 2002. Soil texture data collected from the plots in 1998 identified differences in soil texture between the replications. Replicates 1-4 contained significantly more silt and significantly less sand than Replicate 5. As a result, for many of the experiments, data from Replicates 1 to 4 were analysed separately from data collected from Replicate 5.

3.3 Seed bank studies

A hand coring device was used to collect 2.3 × 10 cm soil cores. In May of 1999, three cores were randomly taken from each 10 × 10 m plot sampled and bulked, except in the case of the solarization plots wherein the sampling procedure was repeated three times to obtain a larger sample of *C. diffusa* seeds (collecting a total of nine 2.3 × 10 cm cores from each plot).

Seeds were extracted from the soil using the soil flotation method (Malone, 1967). The soil was dispersed using an aqueous solution of sodium hexametaphosphate (50 g L^{-1}) and sodium bicarbonate (25 g L^{-1}). Magnesium sulfate (75 g L^{-1}) was added to the aqueous solution to extract the seeds by flotation. Each soil sample was mixed slowly with 400 ml of the chemical solution, and agitated for two minutes. The organic matter was decanted through a 1.25 mm sieve, with a finer 0.1 mm sieve below. Any seeds appearing on the 1.25 mm were collected. The solution was then re-mixed with the inorganic material to allow any remaining seeds to achieve flotation. This solution was then re-decanted twice more through the 0.1 mm sieve. The organic material trapped by the 0.1 mm sieve was placed to dry in a Petri dish for a minimum of one week before the seeds were counted.

Germination tests were done on seeds of the following species as they were extracted from the samples: *Sporobolus cryptandrus*, *Centaurea diffusa*, *Stipa comata* and *Bromus tectorum*. Seeds were placed into Petri dishes with a Whatcom filter paper and moistened with deionized water. The dishes were then placed in a growth chamber set at 14 hours light at 25°C and 10 hours dark at 15°C. After two weeks under these conditions, the germinated seeds were counted using a dissection microscope.

3.4 Experimental exposure of seeds to high temperatures

Temperatures within the solarization plots ranged as high as 78.0 C (Table 2). The maximum ambient temperature recorded was 39.9 C. The mean daily high temperature for the period when temperatures were recorded from July 27-Oct. 9 was 26.9 C.

Time period	Solarization temperatures			Ambient air temperatures	
	Highest Recorded	Mean daily max	# days above 40 C	Highest recorded	Mean daily max
August	75.1	64.3 (11.1)	28.4 (1.6)	39.9	30.8 (5.1)
September	62.5	52.9 (9.7)	21.2 (4.8)	30.88	24.3 (4.7)
July 27-October 9	78.0	57.3 (13.0)	56.2 (8.0)	39.9	26.9 (6.4)

Table 2. Temperatures (± standard deviation) at the soil surface under solarization compared to ambient air temperatures in 1999 at Osoyoos, British Columbia, Canada

The temperatures under solarization were marginally higher than temperatures of 60-70 C reported in Mississippi (Egley, 1983), 42-52 C recorded in Israel (Horowitz et al., 1983), and a 57 C maximum recorded in Syria (Linke, 1994). As indicated by the ambient temperatures, summer temperatures near 40 C are not uncommon in the southern Okanagan valley, and thus solarization treatments attain high temperatures. The peak temperatures usually occurred in early afternoon, but as the example readings from the first week of August indicate (Fig. 2), the duration of periods above 40 C frequently persisted at least 5 hours. On this basis, an experimental period of 5-h exposure was chosen for heat-shock experiments, along with a 1-h exposure for comparison.

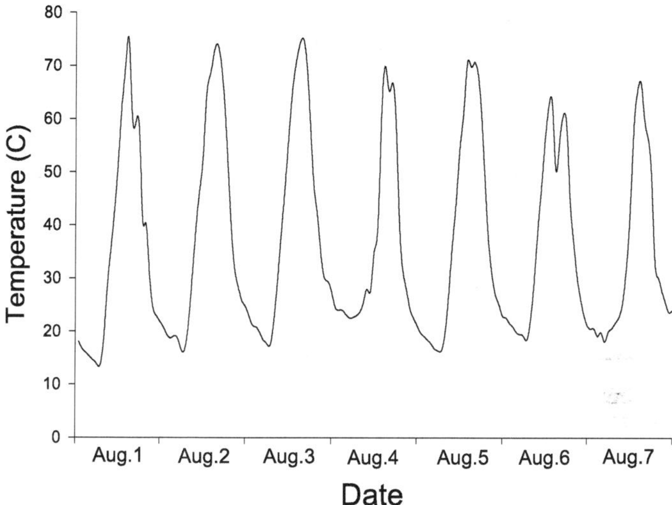

Fig. 2. A seven day sequence of hourly recorded temperatures within a solarization plot in the southern Okanagan valley of British Columbia, Canada (plot #507 which had the highest recorded temperatures)

Seeds of both the invasive alien species, *C. diffusa*, and the native species *S. cryptandrus* were tested for resilience to high temperatures that would be experienced under solarization treatments. Seeds for *C. diffusa* heat shock tests (Mangrich & Saltveit, 2000) were collected from the field in the fall and stored dry at room temperature prior to exposure to high temperatures in the laboratory. Seeds of *S. cryptandrus* for testing were randomly selected from the seed bank samples collected at the Osoyoos Desert Centre study site.

Seeds in Petri dishes were exposed to a given temperature for either 1 h or 5 h, with Petri dishes lined with dry or damp filter paper to signify dry or wet treatments, respectively. Temperatures tested for *C. diffusa* were 40, 50, 60, 70, 80, 90, and 100 C; temperatures tested for *S. cryptandrous* were 70 and 110 C. The 5 h period was to simulate the approximate duration of exposure to peak temperatures experienced daily under solarization in the field. After the high temperature exposure, seeds were removed from the oven and tested for germination (Fig. 3).

Seeds were tested for germination in a growth cabinet maintained at a day/night regime of 25 C/15 C, and a 14-h photoperiod, consistent with the germination requirements of diffuse knapweed (Nolan & Upadhyaya, 1988). Seeds were placed on moist filter paper. Protrusion of the radical by about 2 mm was the criterion for germination. After two weeks, ungerminated seeds were tested for viability using the tetrazoium method (Lakon, 1949; Van Waes & Debergh, 1986). A 1% 2, 3, 5-triphenyltetrazolium chloride (TTC) solution was made by dissolving 5 g of TTC in 500 ml of sterile distilled water. The pH was adjusted to 7 with 1 M NaOH. Seeds were dissected to expose the embryos and soaked in water prior to adding the tetrazolium solution. The embryo was evaluated for color change within 8 h.

White embryos that turned pink were recorded as viable. Seeds with embryos that failed to change color, or that were soft and showing signs of decay were evaluated as nonviable. In the case of seeds of *C. diffusa* exposed to 100 C, the embryos took on a distinctly different appearance, and thus the results for the tetrazolium test were not reported.

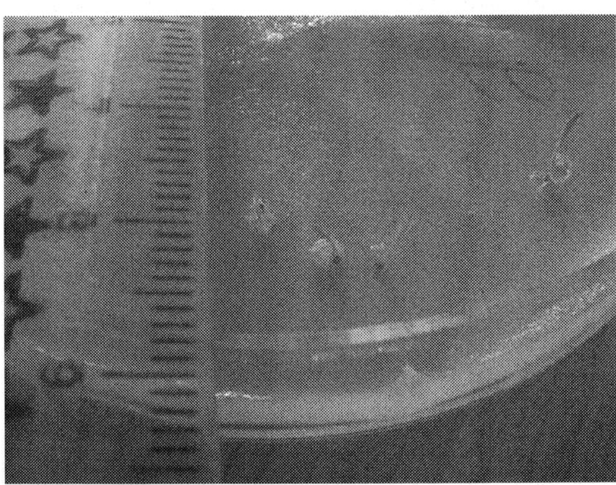

Fig. 3. Germinating seeds of *Sporobolus cryptandrus*; photo by Hannah Buschhaus

4. Restoration treatment results

More than 70 vascular plants were recorded in the above-ground communities in the Osoyoos Desert Centre plots during the 5-year study. Only 28 species were identified in the seed bank (Table 3), although several species in the seed bank were unidentified. This asymmetry between above-ground and seed bank communities is common to most ecosystems, including the antelope brush ecosystem (Clements et al., 2007).

The four dominant species in terms of density m^{-2} identified in the seed bank were *Sporobolus cryptandrus* (66-67% over the two sampling periods), *Centaurea diffusa* (10-14%), *Polygonum douglassii* (6-9%) and *Verbena bracteosa* (6%). This was similar to the pattern seen in another study which encompassed 10 sites in the southern Okanagan, but did not include the Osoyoos Desert Centre Site (Clements et al. 2007) with the exceptions that *B. tectorum* was more abundant at these other sites and *V. bracteosa* did not figure prominently. *Bromus tectorum* comprised 2% of the seed bank on average in the Osoyoos Desert Centre study; in the study spanning 10 sites *B. tectorum* seed comprised 21% of the seeds found in the seed bank (Clements et al. 2007).

Of these four dominant seed bank species, only *C. diffusa* was not native, and as long-term strategies for restoration of plant communities and associated ecosystems are developed, seed banking native species such as *S. cryptandrus*, *P. douglassii*, and *V. bracteosa* can be valuable facets of such a strategy, and represent significant potential species for seeding (Clements et al., 2007).

Species	Density m^{-2} in 1999	Density m^{-2} in 2002
Achillea millefolium	9	17
Arabis hoelboelii	9	0
Arenaria serpyllifolia	62	26
Astragalus purshii	9	0
Bromus tectorum	832	1223
Centaurea diffusa	6945	7147
Collinsia parviflora	0	245
Delphinium bicolor	18	35
Gypsophila paniculatum	1053	1276
Lewisia rediviva	0	9
Linaria dalmatica	88	262
Microsteris gracilis	327	122
Myosotis arvensis	0	26
Myosotis stricta	318	0
Phacelia linearis	9	0
Plantago patagonica	522	926
Polemonium micranthum	318	507
Polygonum douglasii	2875	6361
Potentilla recta	0	44
Pseudoroegneria spicata	53	253
Purshia tridentata	44	166
Rumex acetosella	257	87
Setaria virdis	0	17
Sporobolus cryptandrus	32176	47803
Verbascum thapsus	9	52
Verbena bracteosa	2796	4115
Vicia americana	0	17
Zygadenus venenosus	9	149

Table 3. Mean densities per m^2 for the 28 vascular plant seeds identified in the seed bank in the 10 x 10 m plots across all treatments at the Osoyoos Desert Centre site in 1999 and 2002.

4.1 Restoration treatment results by treatment

The results of the restoration treatments revealed some major changes in the plant community over the five year period, both in terms of the above-ground cover (Atwood & Scudder, 2003) and in terms of the seed banks among the six restoration treatments.

4.1.1 Solarization

In terms of above-ground vegetation, five weed species were recorded in the solarization plots in 1998 prior to the treatment. In June 2000, two months after the plastic was removed three of the species as well as two new weeds were found in the plots. The average cover of *C. diffusa*, and *Verbascum thapsus* (mullein) was greatly reduced from the 1998 level, but it was evident from the above-ground vegetation growth and seed banks (Fig. 4) that solarization had not killed the seeds. *Bromus tectorum* (cheatgrass) was the third species evident in the plots in June 2000, however seed from it and two new weed species that were found, *Meliotus alba* (sweet white clover) and *Sisymbrium loeselii* (Loesel's tumble= mustard), likely moved into the plots between April and June. *Agropyron cristatum* (crested wheatgrass) and *Tragopogon dubius* (yellow salsify) were recorded in the solarization plots in 1998 but were not evident in 2000.

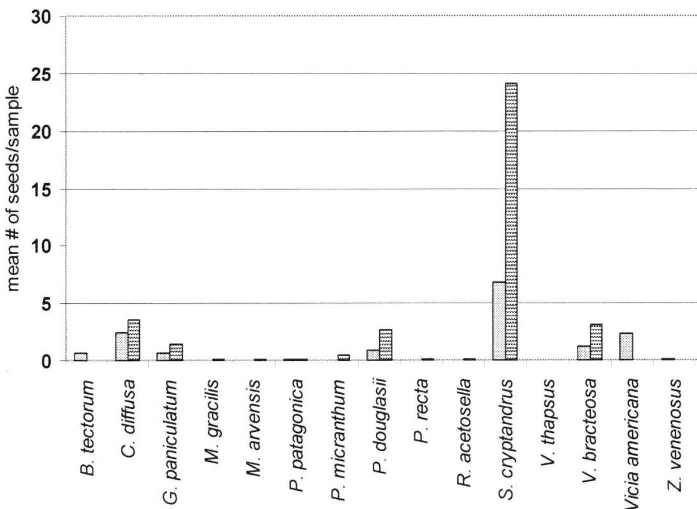

Fig. 4. Mean seed bank composition of solarization plots at the Osoyoos Desert Centre, comparing 1999 seed banks (solid bars) to 2002 seed banks (hatched bars); for full species names see Table 3

Seed bank analysis also indicated that seeds of many plant species were present both after one year of the treatment (1999) and after the solarization treatment in 2002 (Fig. 4). The same four species that dominated the seed bank throughout all plots were dominant in the solarization plots: *S. cryptandrus, C. diffusa, P. douglassii* and *V. bracteosa*. Interestingly all four species increased between 1999 and 2002 despite the solarization treatment, with *S. cryptandrus* exhibiting an increase from a mean of 7 seeds per sample to 24 seeds per sample. As was the case in the above-ground vegetation the invasive alien species *C. diffusa* and *B. tectorum* were present in the plots after the plastic was installed in 1999, although only *C. diffusa* seeds were found in 2002 samples.

Germination tests done on seeds of *B. tectorum, C. diffusa,* and *S. cryptandrus* revealed that some of the seeds were clearly viable despite the extremely high temperatures experienced under the plastic (Fig. 5). In fact, the pattern of germination among the three species, with *S. cryptandrus* exhibiting >40% germination, and *C. diffusa* and *B. tectorum* exhibiting much lower germination percentages was consistent with the pattern seen in all restoration treatments.

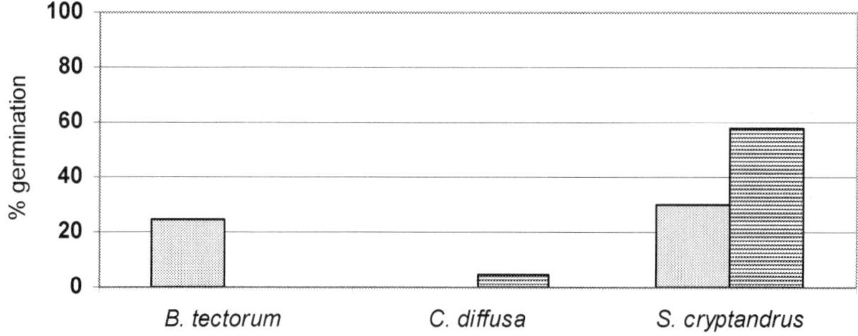

Fig. 5. Percent germination of seeds of three species (*Bromus tectorum, Centaurea diffusa* and *Sporobolus cryptandrus*) from the soil under solarization treatments at the Osoyoos Desert Centre site in 1999 (solid bars) and 2002 (hatched bars)

4.1.2 Native bunchgrass hayseeding

The hayseed material added new species to the hayseed plots. *Pseudoroegneria spicata* was not recorded in the hayseed plots in 1998 but in 2002 *P. spicata* accounted for about 1% of the cover. *Hesperostipa comata* was also newly recorded in Replicate 5 hayseed plots in 2002 where *H. comata* cover averaged of 9.0% ± 5.5% but its cover did not increase in Replicates 1-4. The hayseed material did not significantly increase the percent cover of *S. cryptandrus,* and the seed bank comparison between 1999 and 2002 showed an 80% decline in mean number of seeds of *S. cryptandrus* per sample.

4.1.3 Removal of livestock grazing (controls)

The removal of livestock had a marked effect on vegetation components across the site. Although there was no change in the average cover of shrubs during the five years, herb and native grass cover increased significantly (P < 0.05) and non-native (weed) cover decreased significantly (P < 0.05).

The largest reduction in weed cover occurred between 1998 and 1999, the year following the removal of cattle. Weed cover dropped 71% in Replicates 1 to 4 and 77% in Replicate 5 over the four years. *Centaurea diffusa* was the dominant weed on site in 1998 but in 2002, *Agropyron cristatum* (crested wheatgrass), which had been seeded by the former lessee, was the dominant non-native species. In terms of seed banks, small increases were seen in the native species *S. cryptandrus* but also in non-native species such as *C. diffusa* (Fig. 6).

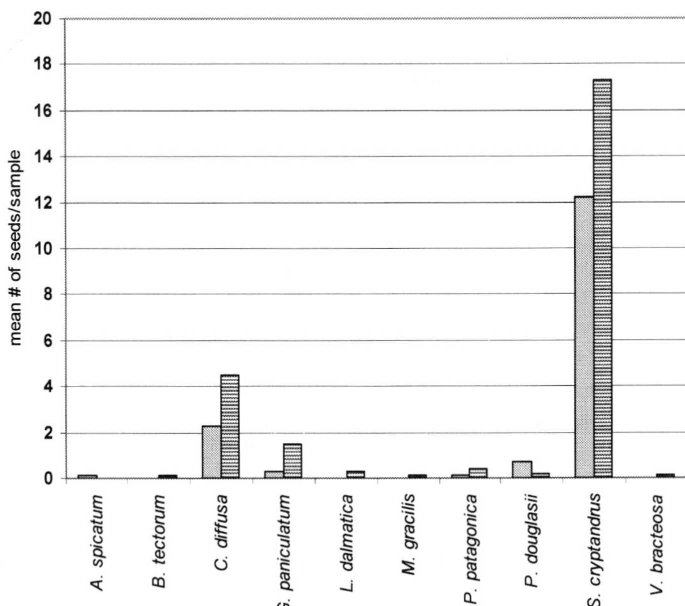

Fig. 6. Mean seed bank composition of control plots at the Osoyoos Desert Centre, comparing 1999 seed banks (solid bars) to 2002 seed banks (hatched bars); for full species names see Table 3; note *A. spicatum* = *P. spicata*

4.1.4 Manual and chemical control of *Centaurea diffusa*

The May 1999 hand weeding (manual treatment) of *C. diffusa* drastically reduced the average cover in the plots between 1999 and 2002 and there was a significant difference in the average cover between the manual treatment plots and control plots (no-treatment) in 1999 and 2000 as a result of the one hand weeding in May 1999 (P<0.05). However, *C. diffusa* was decreasing across the site and by 2001 there was no difference in the cover of *C. diffusa* between the plots that received treatment and those that did not. There was an average of 2.5 *C. diffusa* seeds per sample in the manual removal plots in 1999; the *C. diffusa* seed bank persisted through to 2002 at the same level.

Chemical control with Killex reduced *C. diffusa* cover from an average of 30.33% ± 2.74% in 2000 to 0.33% ± 0.21% in 2001. However, in 2001 there was no difference in the average cover of *C. diffusa* between plots treated with Killex and control plots (P > 0.05). Over the five years of the methods research project, *C. diffusa* decreased rapidly across the site. Over the long-term, the control plots (i.e., merely removing livestock) showed the same level of decline as the plots where *C. diffusa* was treated. Gayton (2011) showed that the period from 1998-2002 was the beginning of a long-term decline in populations of *C. diffusa* on the site due to predation by several insects introduced to the region to provide biological control, resulting in virtually zero percent cover on the site by 2009.

4.1.5 Addition of native vesicular arbuscular mycorrhizae and native seed

Although initially very little difference was seen due to the addition of native vesicular mycorrhizae, by 2001, there was a significantly higher average cover of *P. spicata* in the plots inoculated with VAM and this relationship was still evident in 2002. Cover of the native grasses did not change significantly between 2001 and 2002 except for *H. comata*, which more than doubled in this one year. The seed bank of *S. cryptandrous* in these plots increased three-fold between 1999 and 2002, but the 2002 level was just 12 seeds per sample.

4.1.6 Addition of native seed

Significantly greater cover of native grasses occurred in no-till versus tilled seedbed preparation (P<0.05). After one growing season, cover of seeded grasses in tilled plots averaged 9.13% ± 1.59 compared to 26.57% ± 2.99 in no-till plots.

After one growing season there were significant differences in the percent cover of the native grass seedlings, but it differed by species. There was little response in % cover by *P. spicata* or *S. cryptandrus* but *A. longiseta* and *H. comata* responded differently in the different replications; *S. cryptandrus* did tend to exhibit moderate increases from 1999 to 2002 in the seed bank in all seeding treatments. The average cover of *A. longiseta* was significantly higher in the sandy soils of Replicate 5 (P <0.05) and the average cover of *H. comata* was significantly higher in Replicates 1 to 4, which contained soils with a higher silt content (P <0.05).

4.2 Overall restoration treatment results

On the Osoyoos Desert Centre research site, locally collected natural grasses established successfully as a result of both hayseeding and broadcast seeding. Broadcast seeding was more effective than hayseeding on undisturbed soils and seeding rate (1027 seeds per m² versus 1504 seeds per m²) did not affect establishment. Further work is required to determine if the level of plant establishment is a reflection of the carrying capacity of the local soils, given their low moisture and nutrient availability (Wicklow, 1994) or the result of self-induced seed dormancy, which has limited germination in harsh environmental conditions (Halvorson & Lang, 1989; Allen et al., 1994; O'Keefe, 1996).

The hayseed appeared to repress one of the most common native grasses on the site, *S. cryptandrus*. The average cover of *S. cryptandrus* fell slightly in Replicates 1-4 over the four years as compared to an 18% increase in cover in the control plots. The cover of *S. cryptandrus* did not decrease in Replicate 5 where light availability was likely higher, even with the hayseed cover. Sabo et al. (1979) reported germination of *S. cryptandrus* increased with light availability. In contrast, the hayseed cover enhanced *P. spicata* and *H. comata* establishment. *Pseudoroegneria spicata*, absent from the research plots before seeding, only established in areas that received the hayseed mulch or vasicular arbuscular mycorrhizae (VAM) inoculant. All of the seeded grasses are mycorrhizal (Trappe, 1981) and VAM is particularly critical for the establishment of warm season grasses (Clapperton & Ryan, 2001), which would include *A. longiseta* and *S. cryptandrus*. To date, VAM colonization levels that will improve grass establishment are unknown.

Species establishment was also influenced by soil type. *Aristida longiseta* had higher establishment in sandier soils, while *H. comata* did best in siltier soils. Both species are promoted as drought tolerant species and yet the limited establishment of *A. longiseta* and restricted conditions favouring establishment of *H. comata* suggests that they were affected by dry conditions experienced in the South Okanagan during the interval of the study. Weaver (1968) found that *A. longiseta* decreased during extended droughts.

Solarization was not an effective weed control method for the primary weeds on the Osoyoos Desert site. *Centaurea diffusa* and *V. thapsus* germinated readily following the removal of the plastic, indicating the 75 C recorded under the plastic during treatment was not sufficient to kill the seeds (see also section 5). In addition, solarization resembles broadcast herbicide treatment, exposing large expanses of bare soil after treatment. Revegetating solarized areas with native species will also require a consistent and long-term weed control program.

Manual and chemical control of *C. diffusa* did reduce the weed component, however results were confounded by biological control agents, which were also onsite. In one-year weed cover in the plots monitored for the effect of the removal of livestock declined by about 75% and over five years there was a significant increase in native grass and herb cover. The rapid decline of *C. diffusa* was puzzling since the species is known to have an extensive and long-lived seed bank. Reduced soil disturbance is a factor, because *C. diffusa* did germinate in the tilled plots. The presence of few viable knapweed seeds on the site may be indicative of the successful result of *Sphenoptera jugoslavica* (a beetle utilized as a biological control agent), which occurs throughout the area. Similarly, throughout the region major declines in *C. diffusa* due to success of biological agents have been observed (Myers et al., 2009; Gayton, 2011). By 2002, *A. cristatum* was the dominant non-native species on the site.

5. Solarization and high temperature exposure

As well as the indications from both above-ground cover and seed bank sampling of the solarization plots, experimental exposure of seeds of *C. diffusa* and *S. crypandrus* to temperatures even higher than solarization temperatures served to confirm that seeds of both species are highly resilient to high temperatures.

5.1 Solarization and high temperature impacts on seeds of *Centaurea diffusa*

Fewer diffuse knapweed seeds per sample occurred in the non-solarization samples than in the solarization samples (P<0.05, student's t-test). There were 0.98 ± 1.1 SD and 2.2 ± 2.6 SD diffuse knapweed seeds per sample, in non-solarization and solarization samples, respectively. This amounted to a total of 71 seeds in the solarization plots and 39 seeds in the non-solarization plots. Three seeds from non-solarization plots and one seed from the solarization plots germinated. No other non-solarization plot seeds were found viable by the tetrazolium test, while 2 additional seeds from solarization plots were evaluated as viable. There was no difference (P<0.05) between mean viability of seeds per sample, which was 0.10 ± 0.31 SD and 0.06 ± 0.24 SD seeds per sample for solarization and non-solarization plots, respectively. The relatively low viability of diffuse knapweed seeds of 4% within solarization plots is similar to a value of 3% recorded in another study of seed banks in the southern Okanagan (Clements et al., 2007). Germination and viability tests indicate it is possible for diffuse knapweed seeds to remain viable despite exposure within solarization treatments to temperatures frequently ranging over 40 C with peaks greater than 70 C.

Germination did not differ with the length of the heat shock treatment, although there were differences in the percent of germination between wet and dry seeds (Table 4). There were no differences among seeds exposed to 40 C, whether wet or dry with all treatments exhibiting germination of 60% or higher. Non-heat shocked control germination percentages ranged from 40-70%. Germination was significantly reduced to less than 15% for wet seeds exposed to 50 C and no germination was observed for wet seeds exposed to temperatures of 60 C and higher. By contrast, dry seeds germinated following exposure of temperatures up to 90 C, although the germination percentage was significantly reduced. The germination percentage for dry seeds progressively declined with increasing temperatures. Germination of 21 ± 29 SD % and 18 ± 25 SD % was recorded after 1-h and 5-h exposures of dry seeds to 90 C, respectively. No germination of dry seeds was observed following exposure to 100 C, and the seed embryos exhibited a liquefied appearance.

Temperature (C)	5 h wet	1 h wet	5 h dry	1 h dry
40	60 (24) Aa	80 (13) Aa	83 (16) Aa	76 (23) Aab
50	12 (16) Bb	13 (28) Bb	52 (31) Aab	71 (34) Aabc
60	0 (0) Bb	0 (0) Bb	69 (28) Aab	90 (12) Aa
70	0 (0) Bb	0 (0) Bb	56 (38) Aab	56 (15) Abc
80	0 (0) Bb	0 (0) Bb	42 (23) Abc	44 (38) Acd
90	0 (0) Ab	0 (0) Ab	18 (25) Acd	21 (29) Ade
100	0 (0) Ab	0 (0) Ab	0 (0) Ad	0 (0) Ae

Table 4. Mean % germination (± standard deviation) for *Centaurea diffusa* seeds heat shocked for either 1 h or 5 h at various temperatures; means within a row followed by the same upper case letter are not significantly different at the 5% level; means within a column followed by the same lower case letter are not significantly different at the 5% level (Fisher's Protected LSD test)

Although no seeds incubated in a moist environment germinated above 50 C, the tetrazolium test indicated that some seeds were viable after exposure to higher temperatures, although percent viability was significantly lower for these treatments than for the dry incubated seeds (Table 5). Even after exposure to 90 C, seed viability of 2 ± 5 SD % and 12 ± 13 SD % was recorded in the 5-h wet and 1-h wet heat shock treatments, respectively. Viability was reduced at 60 C or higher, with less than 30 % of wet seeds exposed to temperatures of 60 C still viable. Within the dry heat shock treatments, viability was generally >90%, up to and including 80 C. Viability of dry incubated seeds was significantly reduced for 90 C heat shock treatments compared to lower temperatures, but still remained substantial at 38 ± 43 SD % and 54 ± 29 SD % for 1-h and 5-h exposures of dry seeds, respectively. Though the tetrazolium test does not predict seed viability with complete reliability, the indication that even more seeds were likely capable of germinating than actually germinated further supports the likelihood of large numbers of diffuse knapweed seeds surviving solarization treatments. Whether or not sub-lethal high temperatures actually induced seed dormancy, as was the case for *Sida spinosa, Amaranthus retroflexus, Abutilon theophrasti, Anoda cristata,* and *Ipomoea lacunosa* (Egley, 1990) cannot be inferred from the data recorded in this study, nor is it known whether high temperatures induce dormancy in diffuse knapweed seeds. As seen in other species, it is possible that temperatures of 50-60 C may break dormancy in some seeds, possibly stimulating germination and subsequently mortality of emerging seedlings (Rubin & Benjamin, 1984).

Temperature (C)	5 h wet	1 h wet	5 h dry	1 h dry
40	85 (20) Aa	96 (9) Aa	98 (5) Aa	96 (9) Aa
50	42 (48) Bb	66 (38) ABb	96 (9) Aa	94 (9) Aa
60	2 (5) Bc	6 (9) Bc	98 (5) Aa	96 (9) Aa
70	28 (24) Bbc	22 (33) Bc	98 (5) Aa	90 (17) Aa
80	22 (29) Bbc	10 (12) Bc	86 (22) Aa	96 (9) Aa
90	2 (5) Cc	12 (13) BCc	54 (29) Ab	38 (43) ABb

Table 5. Mean % viability (± standard deviation) for *Centaurea diffusa* seeds heat shocked for either 1 h or 5 h at various temperatures; percent viability includes seeds germinating under optimal conditions and seeds that were viable according to the tetrazolium test; means within a row followed by the same upper case letter are not significantly different at the 5% level; means within a column followed by the same lower case letter are not significantly different at the 5% level (Fisher's Protected LSD test)

The 1-h and 5-h *in vitro* tests in the current study only examined the effect of a single exposure to high temperatures. It is not known whether long-term temperature fluctuations over the season would increase mortality of diffuse knapweed seeds due to exposure to high temperatures. It is also unclear what effect condensation on the soil surface beneath the plastic has on seed germination and viability. Studies where seeds of other species were incubated at high temperatures over longer periods of time (i.e., one week or more) also recorded high survival rates of seeds (Horowitz et al., 1983; Egley, 1990). Although Horowitz et al. (1983) found that 2-4 wks of solarization provided control of many annual weed species, there were still some seeds that retained viability after 8 wks.

5.2 Solarization and high temperature impacts on seeds of *Sporobolus cryptandrus*

As shown in Fig. 5, seeds of *S. cryptandrus* maintained relatively high germination percentages even when exposed to the relatively high temperatures experienced in solarization treatments, with *S. cryptandrus* seeds from solarization treatments in 1999 exhibiting 30% germination, and *S. cryptandrus* seeds in 2002 exhibiting 58% germination. In the control treatments, *S. cryptandrus* seeds exhibited 10% and 33% germination in 1999 and

2002 respectively; in the hayseeding experiments *S. cryptandrous* seeds exhibited 34 and 66% germination, respectively.

As in the case of *C. diffusa*, seeds of *S. cryptandrus* maintained a high level of viability, when heat shocked at temperatures experienced under solarization (i.e., 70 C) and at even higher temperatures (110 C). Of the 82 seeds of *S. cryptandrus* we heat shocked at 70 C, 32% were viable; of the 45 seeds of *S. cryptandrus* we heat shocked at 110 C, 16% were viable. Unlike the *C. diffusa* seeds that were heat shocked which were collected the previous fall from seed heads on the plants, the source of seeds for the *S. cryptandrus* heat shock trails was seeds extracted from the soil at the Osoyoos Desert site. Thus, the moderate level of viability at 70 C and yet substantial viability at 110 C was observed despite other factors that already would have lead to decline in seed viability in the soil environment prior to heat shocking.

6. Implications for restoration of ecosystems

6.1 Solarization as a means of managing invasive species

The temperatures experienced under the polyethylene under field conditions were extremely high, even by comparison to other locations, such as Mississippi, where 70 C is unlikely to be observed on a frequent basis (Egley, 1983). The survival of viable *C. diffusa* seeds under these conditions, and after heat-shock treatments of even higher temperatures under laboratory conditions renders the elimination of the *C. diffusa* seed bank by solarization unlikely. It should also be considered that some *C. diffusa* seeds may be located deep enough within the soil profile to avoid the extreme temperatures at the soil surface (Rubin and Benjamin 1984; Standifer et al., 1984). Standifer et al. (1984) found seed mortality due to solarization decreased with depth in the soil, and was primarily effective in the top 5 cm. As observed elsewhere, solarization would reduce the seed bank population to some degree while exhibiting its primary impact on above-ground plants (Horowitz et al., 1983; Egley, 1990). Furthermore, solarization may have limited impact on perennial or biennial species, whose growth may actually be stimulated by solarization (Stapleton & DeVay, 1986; Sauerborn et al., 1989; Linke, 1994). Given that *C. diffusa* is a biennial or short-lived perennial, it would be interesting to investigate the effect of solarization on its perennating tissues.

The long-term ramifications of the small reduction in seed viability of *C. diffusa*, following a short-term exposure to excessive temperatures, on plant community dynamics are unclear. Without re-seeding native plant species, *C. diffusa* could re-establish from the seed bank and nearby seed sources, particularly in the wake of the disturbance caused by the removal of above-ground vegetation by solarization. With sufficient re-seeding of desired plant species, the impact of emerging *C. diffusa* seedlings may be minimal, particularly in light of the relatively low percentage of *C. diffusa* seeds that remain viable in the seed bank (Clements et al., 2007). The scale of solarization treatments is also a critical issue. The efficiency of solarization in minimizing weed interference has been demonstrated in horticultural crops (Braun et al., 1988; Horowitz et al., 1983; Jacobsohn et al., 1980; Linke, 1994). However, restoring large areas of natural habitat is more difficult, particularly in the face of disturbance-adapted weeds like *C. diffusa* that are abundant over large areas and possess persistent seed banks resilient to high temperatures. Solarization is an efficient weed control technique in horticultural crops (Jacobsohn et al. 1980; Horowitz et al. 1983; Braun et al. 1988; Linke 1994) and may be an inexpensive and labour-saving tool for reducing weeds in natural habitats protected from chemical treatments or disturbed soils where weeds dominate. This study suggests solarization will not control diffuse knapweed in the South Okanagan, however many issues require further study.

6.2 Recommendations for managing shrub-steppe ecosystems incorporating knowledge of seed bank dynamics

The importance of removing or at least reducing disturbance by livestock was clear from this study as even apart from other restoration measures, livestock removal was effective in increasing the native species component in these shrub-steppe plant communities. Likewise in a study of 10 sites with varying grazing regimes in the same region of the southern Okanagan valley, Krannitz (2008) found that more bare soil was associated with livestock grazing, with particularly large impacts seen in sandy versus rocky sites, and in areas not protected by antelope brush shrubs. The presence of bare soil meant that there was a corresponding loss in microbiotic crust cover, and this meant reduced overall health of the native plant community primarily due to lack of moisture retention without crust (Atwood & Krannitz, 2000; Krannitz, 2008).

Another mechanism by which differences in microbiotic crust may impact plant communities is through variations in seed bank dynamics depending on degree of crust cover. Crust cover is predicted to affect seed burial and longevity of the seeds; in soil samples with a large portion of the crust constituents (i.e., lichens and mosses), seeds in our samples were often located among the microbiotic vegetation, hence relatively close to the soil surface. In general, seed survival nearer to the surface is reduced (Harper, 1977; Clements et al., 1996), but it is probable that seeds falling among microbiotic crust would be preserved longer than seeds occurring on the soil surface (Langhans et al., 2010). Seed bank studies in the southern Okanagan point to the need for more research in this area to enable ecosystem managers to fine-tune approaches to grazing levels, restoration plantings, and management of fire and other disturbances (Krannitz & Mottishaw, 2004; Clements et al., 2007).

Few studies have investigated the relationship of seed bank dynamics to microbiotic crusts experimentally, and many of the few studies that have been done under laboratory conditions, therefore not matching conditions of normal seed dispersal (Prasse & Bornkamm, 2000; Su et al., 2007). A study in the north-western Negev desert of Israel found that increased roughness of the surface due to the presence of crust reduced the probability of seeds coming to rest, therefore lowered seed emergence and survival (Prasse & Bornkamm, 2000). However, other have found increased seed bank emergence with increased crust presence (Su et al., 2007). A study in the Teng-ger Desert in northeastern China observed much greater vascular plant emergence with moss crusts than algae crusts, with seeds more likely to become lodged in the moss crusts, even under high wind conditions (Su et al., 2007). Water status was also a major factor in the Chinese study, as it was in a study of crust types in Idaho within the Great Basin of North America (Serpe et al., 2006), with more moisture producing greater seedling emergence. Serpe et al. (2006) also pointed out that seedling emergence responses can vary with specific crust structure; different seed and seedling morphotypes likewise have a major influence on the result. Thus while studies from other regions provide some basis for understanding relationships between soil crusts and recruitment of vascular plants from the seed bank, there is a need for experimental studies within specific regions in order to know how best to restore semi-arid steppe grasslands after crust degradation.

In our study broadcast seeding was more effective than hayseeding on undisturbed soils and seed rate did not affect establishment. Rates of greater than 1000 seeds per m² are thought to be high (Jacobs et al., 1999) and it appears applications that exceed that amount are unnecessary. The apparent repression by the hayseed of the most common grass species on the site, *S. cryptandrus,* may represent be an important management consideration if *S. cryptandrus* is considered a valuable component of the community, as a major constituent of the native seed bank. Strategically developing restoration management to encompass disturbance-adapted native species such as *S. cryptandrus* may be difficult.

In contrast, the hayseed cover did enhance other native grass species such as *P. spicata* and *H. comata* establishment, and likewise vascular arbuscular mycorrhizal inoculant was very helpful to the establishment of some of the native grasses as has been seen in other studies (Camill et al., 2004; Anderson, 2008). It would be interesting to explore the connections between health of the microbiotic crust and the vigour of vascular arbuscular mycorrhizae in this system.

In addition to its ineffectiveness in terms of destroying weed seeds, the result of solarization resembled that of applying broadcast herbicides because large expanses of bare soil were exposed after treatment. Manual and chemical control of *C. diffusa* did reduce weed biomass, but ultimately populations decreased with the removal of livestock removal coupled with the cumulative success of biological control. The populations of insects utilized in biological control of *C. diffusa* have increased enough to reduce the populations of *C. diffusa* to negligible levels at this site and many others in the region, even with the persistent seed bank of this species (Myers et al., 2009; Gayton, 2011).

Based on the preceding discussion and results, we can make the following recommendations for restoration of communities within these arid shrub-steppe ecosystems:

- Restoration measures should minimize soil disturbance
- Measures to restore a diverse and healthy native plant community must include the microbiotic crust in these ecosystems
- Livestock grazing must be monitored rigorously to limit unnecessary soil disturbance
- Restoration planting, particularly if involving warm-season grasses, may be enhanced by incorporation of vesicular arbuscular mycorrhizae
- Restoration planting and invasive species control should account for native species with seed banks such as *Sporobolus cryptandrous*
- Spatial dispersion and amount of shrub cover should be carefully managed to promote native plants and reduce influence of grazing and non-native plants
- Long-term planning should be developed to monitor ecosystem conditions, invasive species, and other important management factors such as biological control as well as the desired ecosystem trajectory

Fig. 7. Kiosk at the Osoyoos Desert Society site showing amidst the grassland ecosystem with scattered shrubs of antelope bitterbrush (*Purshia tridentata*); photo courtesy of the Osoyoos Desert Society

7. Acknowledgment

We would like to acknowledge the Habitat Conservation Trust Fund, Environment Canada, Human Resources Development Commission, the Osoyoos Desert Society, and Trinity Western University, and NSERC for funding the project. We are very grateful for the role of Geoff Scudder in the project in terms of both scientific guidance and ensuring funding through NSERC. Jordan Carbery was invaluable as field supervisor and primary data collector for the above-ground vegetation cover and the project would have been impossible without him. Likewise, TWU undergraduate thesis students Tonya Snidal who lead the data collection efforts for diffuse knapweed (*Centaurea diffusa*) germination and heatshocking studies and Holly Kunster who lead the data collection efforts for sand dropseed (*Sporobolus cryptandrus*) germination and heatshocking studies deserve tremendous thanks for the many hours they worked on the project. We are very grateful to Luke Bainard for the final tallying of seeds from the seed bank and organization of the data. We would also like to acknowledge Shannon Berch, Ministry of Forests Research Branch and Sharmin Gamiet for their assistance with the VAM experiment and members of the Osoyoos Desert Society Science Committee for their assistance in the early stages of the project. Committee members included Dr. P. Krannitz (Environment Canada), Judy Millar (Ministry of Water, Land, and Air Protection, and Lisa Scott (Eco-Matters Consulting). We thank Mirwais Mauj Qaderi and Luke Bainard for helpful comments on the manuscript.

8. References

Allen, P.S., Debaene-Gill, S.B. & Meyer, S.E. (1994). Regulation of germination timing in facultatively fall emerging grasses. Pages 215-219 *in* Monsen, Stephen B.; Kitchen, Stanley G., compilers. Proceedings--ecology and management of annual rangelands; 1992; Boise, ID. Gen. Tech. Rep. INT-GTR-313. Ogden, UT: U.S. Department of Agriculture, Forest Service, Intermountain Research Station.

Anderson, R.S. (2008). Growth and arbuscular mycorrhizal fungal colonization of two prairie grasses grown in soil from restorations of three ages. Restoration Ecology 16: 650-656.

Atwood, L.B. & Krannitz, P. (2000). Effect of the microbiotic crust of the antelope-brush (*Purshia tridentata*) shrub-steppe on soil moisture. Pages 809–812 *in* L. M. Darling, ed. Proc. Conf. Biology and Management of Species and Habitats at Risk, Kamloops, BC, 15-19 Feb. 1999. Vol. Two. B.C. Minist. Environ., Lands and Parks, Victoria, BC, and Univ. College of the Cariboo, Kamloops, BC. 520pp.

Atwood L.B. & Scudder G.G.E. (2003). Experimental ecological restoration in the SouthOkanagan shrub-steppe. Proceedings of Ecosystems at Risk: Antelope-Brush Restoration, March 28–30, 2003, Osoyoos, British Columbia, Canada, ed. R Seaton, pp. 1–12. New Westminster, BC: Society for Ecological Restoration Available from: http://www.ser.org/serbc/pdf/antelope_brush_restoration.pdf

Baker, H.G., (1974). The evolution of weeds. Annual Review of Ecology and Systematics 5: 1-24.

Bakker, J.P., Poschlod, P., Strykstra, R.J., Bekker, R.M. & Thompson, K. (1996). Seed banks and seed dispersal: important topics in restoration ecology. Acta Botanica Neerlandica 45: 461-490.

Braun, M., Koch, W., Mussa, H.H. & Stiefvater, M. (1988). Solarization for weed and pest control: possibilities and limitations. Pages 169-178 *in* R. Cavalloro and A. El Titi, eds. Weed Control in Vegetable Production. Balkema, Rotterdam, NL.

Bryan, A. (1996). Vaseux Lake habitat management plan. South Okanagan conservation strategy. B.C. Ministry of Environment, Lands and Parks, Penticton, BC. 93pp.

Camill, P., McKone, M.J., Sturges, S.T., Severud, W.J. Ellis, E., Limmer, J., Martin, W.J., Navratil, R.T., Purdie, A.J., Sandel, B.S., Talakder, S. & Trout, A. (2004). Community- and ecosystem-level changes in a species-rich tallgrass prairie restoration. Ecological Applications 14: 1680-1694.

Clapperton, J. & Ryan, M. (2001). Uncovering the real dirt on no-till. South Dakota No-till Association 2001 No-till Workshop Proceedings. Available from: http://www.sdnotill.com/proceedings.htm

Clements, D.R., Benoit, D.L., Murphy, S.D. & Swanton, C.J. (1996). Tillage effects on weed seed return and seedbank composition. Weed Science 44: 314-322.

Clements, D.R., Krannitz, P.G. & Gillespie, S.M. (2007). Seed banks of invasive and native plant species in a semi-desert shrub-steppe with varying grazing histories. Northwest Science 81: 37- 49.

Clements, D.R. & Scott, L. (2011). Weed invasion of the Montane Cordillera Ecozone. Pages 1-29 *in* Assessment of Species Diversity in the Montane Cordillera Ecozone. *Edited by* G.G.E. Scudder and I.M. Smith. Royal British Columbia Museum. Available from: http://www.royalbcmuseum.bc.ca/Content_Files/Files/mce/weed_invasion.pdf

Chilton, R.R.H. (1988). A summary of climatic regimes of British Columbia. Province of B.C.: Victoria, B.C.

Davis, M.A. (2009). Invasion Biology. Oxford University Press, Oxford.

Egley, G. H. (1983). Weed seed and seedling reductions by soil solarization with transparent polyethylene sheets. Weed Science 31: 404-409.

Egley, G.H. (1990). High-temperature effects on germination and survival of weed seeds in soil. Weed Science 38: 429-435.

Elliott, R.R. (1973). History of Nevada. University of Nebraska Press, Lincoln, Nevada. 118 p.

Erickson, W. (2003). Antelope Brush – Bluebunch Wheatgrass in the East Kootenay-Rocky Mountain Trench region of British Columbia. Proceedings of Ecosystems at Risk: Antelope-Brush Restoration, March 28–30, 2003, Osoyoos, British Columbia, Canada, ed. R Seaton, pp. 19–27. New Westminster, BC: Soc. Ecol. Restor. Available from: http://www.ser.org/serbc/pdf/antelope_brush_restoration.pdf

Gayton, D. (2011). Women, weeds and very small insects. Osoyoos Desert Society Newsletter April 2011 pp. 6-7. Available from: http://www.desert.org/pdfs/2011%20Spring%20Newsletter.pdf

Halvorson, G.A. & Lang, K.J. (1989). Revegetation of a salt water blowout site. Journal of Range Management 42: 61-65.

Harper, J.L., (1977). Population Biology of Plants. Academic Press, New York.

Hobbs, R.J. & Humphries, S.E. (1995). An integrated approach to the ecology and management of plant invasions. Conservation Biology 9: 761-770.

Holm, L.G., Plucknett, D.L., Pancho J.V. & Herberger, J.P. (1977). The world's worst weeds: Distribution and biology. University of Hawaii, Honolulu.

Horowitz, M., Regev, Y. & Herzlinger, G. (1983). Solarization for weed control. Weed Science 31: 170-179.

Jacobs, J.S. Carpinelli, M.F., and Sheley, R.L. (1999). Revegetating noxious weed infested rangeland. In Biology and management of noxious rangeland weeds. Pages 133-141 *in* R.L. Sheley and J.K. Petroff (eds.). Oregon State University Press: Corvallis.

Jacobsohn, R., Greenberger, A., Katan, J., Levi, M. & Alon, H. (1980). Control of Egyptian broomrape (*Orobanche aegyptiaca*) and other weeds by means of solar heating of the soil by polyethylene mulching. Weed Science 28: 312-316.

Krannitz, P.G. (2008). Response of antelope bitterbrush shrubsteppe to variation in livestock grazing. Western North American Naturalist 68: 138-152.

Krannitz, P. G. & Mottishaw, J. (2004). Fire effects and antelope-brush: fire not as detrimental as might be expected. Menziesia 9: 7-10 and 13.

Lakon, G. (1949). The topographical tetrazolium method for determining the germinatingcapacity of seeds. Plant Physiology 24: 289-394.

Larson, B.M.H. (2007). Who's invading what? Systems thinking about invasive species. Canadian Journal of Plant Science 87: 993-999.

Leck, M.A., Parker, V.T. & Simpson, R.L. (eds.). (1989). Ecology of soil seed banks. Academic Press, San Diego. 461 p.

Linke, K.-H. (1994). Effects of soil solarization on arable weeds under Mediterranean conditions: control, lack of response or stimulation. Crop Protection 13: 115-120.

Loope, W.L. & Gifford, G.F. (1972). Influence of a soil microfloral crust on select properties of soils under pinyon juniper in southeastern Utah Journal of Soil and Water Conservation 27: 164–167.

MacDougall, A.S. & Turkington, R. (2005). Are invasive species the drivers or passengers of change in degraded ecosystems? Ecology 86: 42-55.

Mack, R.N. (1981). The invasion of *Bromus tectorum* L. into western North America: an ecological chronicle. Agroecosystems 7: 145-165.

Mack, R.N. (1986). Alien plant invasion into the Intermountain West: a case history. Pages 191-213 *in* H.A. Mooney & J.A. Drake, eds., Ecology of Biological Invasions of North America and Hawaii. Springer-Verlag, New York.

Malone, C.R. (1967). A rapid method for enumeration of viable seeds in soil. Weeds 15: 381-382.

Mandle, L., Bufford, J.L., Schmidt, I.B. & Daehler, C.C. (2011). Woody exotic plant invasions and fire: reciprical impacts and consequences for native ecosystems. Biological Invasions 13: 1815-1827.

Mangrich, M.E. & Saltveit, M. (2000). Effect of chilling, heat shock, and vigor on the growth of cucumber (*Cucumis sativus*) radicals. Physiologia Plantarum 109: 137-142.

Meinig, D.W. (1968). The Great Columbia Plain. University of Washington Press, Seattle, WA.

Mosquin, T., P.G. Whiting & McAllister, D.E. (1995). Canada's biodiversity: the variety of life, its status, economic benefits, conservation costs and unmet needs. Canadian Museum of Nature, Ottawa, ON.

Myers J.H., Jackson C., Quinn H., White S. & Cory J.S. (2009). Successful biological control of diffuse knapweed, *Centaurea diffusa*, in British Columbia, Canada. Biological Control 50: 66–72

Nolan, D.G. & Upadhyaya, M.K. (1988). Primary seed dormancy in diffuse and spotted knapweed. Canadian Journal of Plant Science 68: 775-783.

O'Keefe, M.A. (1996). Rainfalls and prairie establishment. Restoration and Management Notes 4 (Summer): 26-29.

Prasse, R. & Bornkamm, R., (2000). Effect of microbiotic soil surface crusts on emergence of vascular plants. Plant Ecology 150: 65–75.

Roché, B.F., Jr. & Roché, C.T. (1999). Diffuse knapweed. Pages 217-230 *in* R.L. Sheley & J.K. Petroff (eds.). Biology and Management of Noxious Rangeland Weeds. Oregon State University Press, Corvalis, Oregon.

Rubin, B. & Benjamin, A. (1984). Solar heating of the soil: Involvement of environmental factors in the weed control process. Weed Science 32: 138-142.

Sabo, D.G., Johnson, G.V., Martin, W.C. & Aldon, E. F. (1979). Germination requirements of 19 species of arid land plants. Res. Pap. RM-210. Fort Collins, CO: U.S. Department of Agriculture, Forest Service, Rocky Mountain Forest and Range Experiment Station. 26 p.

Sauerborn, J., Linke, K.-H., Saxena, M.C. & Koch, W. (1989). Solarization, a physical control method for weeds and parasitic plants (*Orobanche* spp.) in Mediterranean agriculture. Weed Research 29: 391-397.

Serpe, M.D., Orm, J.M., Barkes, T. & Rosentreter, R. (2006). Germination and seed water status of four grasses on moss-dominated biological soil crusts from arid lands Plant Ecology 185: 163-178.

Simberloff D., Schmitz D.C. & T.C. Brown (eds.). (1997). Strangers in Paradise: Impact and Management of Nonindigneous Species in Florida. Island Press, Washington, D.C. 467 pp.

Standifer, L.C., Wilson, P.W. & Porche-Sorbet, R. (1984). Effects of solarization on soil weed seed populations. Weed Science 32: 569-573.

Stapleton, J.J. & DeVay, J.E. (1986). Soil solarization: a non-chemical approach for management of plant pathogens and pests. Crop Protection 5: 190-198.

Su, Y.-G., Li X.-R., Cheng Y.-W., Tan, H.-J., & Jia, R-L. (2007). Effects of biological soil crusts on emergence of desert vascular plants in North China. Plant Ecolgy 191: 11-19.

Trappe, J.M. (1981). Mycorrhizae and productivity of arid and semi-arid rangelands. Pages 581-599 *in* Manassah, J.T. & E.J. Briskey (eds.), Advancements in food producing systems for arid and semi-arid lands. Academic Press: New York.

Van Waes, J.M. & Debergh, P.C. (1986). Adaptation of the tetrazolium method for testing the seed viability, and scanning electron microscopy study of some Western European orchids. Physiologia Plantarum 66: 435-442.

Weaver, J.E. (1968). Prairie plants and their environment: A fifty-year study in the Midwest. Lincoln, NE: University of Nebraska Press. 276 p.

Wicklow, H. (1994). Mycorrhizal ecology of shrub-steppe habitat. Pages 207-210 *in* S.E. Monsen, S.E. & Kitchen, S.J. (eds). Proceedings: Ecology of Annual Rangelands. USDA INT-GTR-313.

18

Changes to Marine Trophic Networks Caused by Fishing

Andrés F. Navia[1,2], Enric Cortés[3], Ferenc Jordán[4],
Víctor H. Cruz-Escalona[2] and Paola A. Mejía-Falla[1]
*[1]Fundación Colombiana Para la Investigación y
Conservación de Tiburones y Rayas SQUALUS*
[2]Centro Interdisciplinario de Ciencias Marinas, Instituto Politécnico Nacional
[3]NOAA, National Marine Fisheries Service, Panama City Laboratory
[4]The Microsoft Research – University of Trento COSBI, Trento
[1]Colombia
[2]México
[3]USA
[4]Italy

1. Introduction

Multiple anthropogenic sources, such as contamination, habitat degradation, eutrophication, and, more recently, fishing, have steadily been impacting marine ecosystems for at least the past two centuries, generating probably irreversible structural and functional changes (Estes et al., 2011; Lotze & Milewski, 2004). In particular, increasing fishing pressure during the past 50 years and habitat degradation have had a wide range of impacts on ecosystems worldwide, which are reflected in changes in abundance, spatial distribution, productivity, and structure of exploited communities (Blaber et al., 2000; Hall, 1999; Jackson et al., 2001; Lotze et al., 2006; Myers & Worm, 2005). These impacts on community structure and function have been widely documented and quantified in many marine ecosystems (Haedrich & Barnes, 1997; Jennings & Kaiser, 1998; Pauly et al., 1998; Sala et al., 2004; Worm et al., 2006; Yemane et al., 2005).

Some authors have suggested that, although changes in species composition are an important indicator to identify perturbed ecosystems, a holistic knowledge allowing identification of structural and functional effects could emerge from the study of communities as networks interconnected by trophic interactions (Bascompte et al., 2005; Dunne et al., 2002). Owing to the relatively stable characteristics of trophic networks, these interactions can provide information on species relationships within a community and how human activities could be degrading ecosystems (Dell et al., 2005).

Recent publications that have assessed the relationship between fishing and possible alterations of direct and indirect trophic relationships within impacted ecosystems have detected strong ecological effects, such as trophic cascades and changes in ecosystem control equilibrium, either top-down or bottom-up (Barausse et al., 2009; Baum & Worm, 2009;

Ferretti et al., 2010). Thus, these publications have generated particular interest on top predators given their role as regulators of intermediate predator populations, or mesopredators, and further proposed that removal of top predators can result in changes in intraguild relationships that affect biodiversity and the equilibrium of the ecosystems under study (Baum & Worm, 2009; Ritchie & Johnson, 2009).

Extensive work has also been conducted on the relationship between species loss and secondary extinctions generated by predator-prey interactions (Dunne & Williams, 2009, Dunne et al., 2002, 2004), finding in many cases that the increase in the capacity of response of trophic networks is related to the increase in diversity (expressed as species richness) and the number of interactions among these species in the network (connectance).

Owing to the variety of approaches used to study trophic networks and the large number of topics studied in these systems, we conduct a review of the effects detected in trophic networks that have been proposed to originate in the fisheries impacting different ecosystems throughout the world's oceans. This review consists of three main themes: 1) effects of fishing on the structure of trophic networks, 2) effects of fishing on the function of trophic networks, and 3) effects of fishing on interspecific trophic relationships (direct and indirect effects). With the aim of exploring the effect of fishing on a trophic network, we conclude by presenting a topological analysis of a network by simulating fishery removals and assessing the effect of parameters considered to be important in network structure, but not addressed in sufficient detail in previous work.

2. Effects of fishing on trophic networks

Human-induced changes to marine ecosystems have been taking place for centuries, but have only reached global dimensions in the past few decades. Thus, there are three effects with major connotations generated by humans: 1) changes in nutrient cycles and climate which affect ecosystem structure from the bottom up, 2) fishing activity which could affect ecosystems, mainly from the top down, and 3) habitat alteration and contamination which affect ecosystems at all trophic levels.

This chapter focuses on the fact that human beings have used marine resources throughout history, from subsistence fishing activities to large-scale fisheries in almost all the oceans in the planet. With the often unrestrained increase in fishing activity, some have sounded the alarm on the possible effects of this practice on populations and the world's marine ecosystems (Jackson et al., 2001; Lotze & Worm, 2009). Since fisheries have both direct and indirect effects on the ecosystem, wherein commercial and non-commercial species establish feeding interactions, it is very likely that impacts from human activities, which are exerted on individuals, propagate to populations and finally emerge at the community level (Sandström et al., 2005).

Most studies have focused on the effects of fishing on the population dynamics of species, most often charismatic species or those with high commercial value (e.g. Lotze & Worm, 2009; Lotze et al., 2011). In terms of communities, many studies have tried to assess the effects of anthropogenic activities on functional groups or entire ecosystems, documenting various types of responses (e.g. Jackson et al., 2001; Pandolfi et al., 2003), but even then the consequences of these activities on the structure and functioning of ecosystems remain unclear (Lotze et al., 2011).

2.1 Effects of fishing on the structure of trophic networks

2.1.1 Structural attributes of trophic networks

Trophic networks display structural attributes that appear to be constant, or at least regular, throughout the planet's latitudinal range. These regularities have been linked to the stability of networks and their capacity to respond to different types of environmental stressors (Bascompte et al., 2005; Dunne et al., 2004; Solé & Montoya, 2001). These structural attributes are based mainly on the number of interactions between predators and prey, proportional abundance of predators, intermediate species, and basal species, and the number of species at different trophic levels.

Despite these being the most basic structural properties of trophic networks, very few studies on the effects of fishing have addressed them. In that respect, Lotze et al. (2011) found significant changes in the constant proportions that must exist among top predators (T), intermediate species (I) and basal species (B), known as "species scaling laws" (Briand & Cohen, 1984). Additionally, based on other structural indicators of trophic networks (link density, connectance, cannibalism), Lotze and collaborators concluded that the trophic network in the Adriatic Sea has been subjected to overfishing of high trophic levels, leading to its structural simplification, progressively becoming less connected and complex. This type of structural changes directly affects the capacity of the network to respond to species loss (robustness) and increases the likelihood of secondary extinctions, even with low values of species reductions, leading the network to structural collapse more easily (Dunne et al., 2004).

2.1.2 Structural simplification

In the study by Lotze et al. (2011), the trophic network was significantly simplified and the loss of slightly less than 50% of the original richness of the network would result in its total collapse. Although very few studies explicitly mention structural simplification of networks as a result of fishing (Coll et al., 2008a, 2009a, 2009b), or changes in "species scaling laws", they can be inferred in studies reporting changes in the distribution of biomass among different trophic levels through time. This phenomenon has been observed in a large number of studies that use mass balance trophic models as analytical tools (e.g. Albouy et al., 2010; Barausse et al., 2009; Coll et al., 2008a, 2008b, 2009a, 2009b; Hinke et al., 2004; Jones et al., 2009; Savenkoff et al., 2007a, 2007b).

Using this approach, Coll et al. (2008a, 2009a, 2009b) found significant differences in the composition of a trophic network subjected to different fishing levels through time. Their results showed a change in the biomass proportions (equivalent to species scaling laws) among trophic levels between the ecosystem without fishing pressure and the three scenarios with increasing fishing pressure. Coll and collaborators studied the Mediterranean Sea ecosystem, which has been subjected to increasing fishing pressure for over two millennia and displays, according to these authors, a high level of degradation in trophic network structures and thus a dangerous simplification. These authors also found that highly impacted environments, not only in terms of intensity but also of time, show stronger overexploitation effects at high trophic levels, larger network simplification, and high reductions in productivity and biomass. As a result of this drastic structural simplification, Mediterranean ecosystems showed less robustness to secondary extinctions than other less impacted ecosystems (Coll et al., 2009a, 2009b).

All these indicators of structural simplification of trophic networks have a very important implication, which is the possible reduction in functional redundancy, part of the biological insurance of ecosystems (Montoya et al., 2001). Thus, fishing pressure creates two highly dangerous scenarios for the stability of trophic networks: removing a functional level in the network, or transforming a species previously belonging to a feeding guild and sharing its ecological function with other species, into a key species, which upon being impacted would seriously challenge the stability of the whole ecosystem. Attempts to measure the effect of fishing on community structure based on indices that describe community attributes (equitability, richness, Hill indices, and others) have shown limited success (Piet & Jennings, 2005) because cause-effect relationships as well as the direction of the fishing effect on these indices are unclear (Bianchi et al., 2000; Rice, 2000). Other indicators have also been explored as evidence of fishing effects on populations, communities, and ecosystems (Friedlander & DeMartini, 2002; Fulton et al., 2005). These authors proposed that some community and ecosystem-based indicators could be useful in management actions, but that some, especially those based on network analysis and ecological models (e.g. ascendency), have low reliability because they depend on information that is difficult to collect, model formulation, and level of knowledge on the modeled systems.

2.1.3 Fishing down versus fishing through marine food webs

Pauly et al. (1998) proposed an important fishing effect related to the structure and composition of trophic networks called *fishing down marine food webs*. It postulates that selective catches of top predators have modified the composition of fishery landings and reduced their mean trophic level (MTL). Given the interpretation given to this phenomenon, whereby fishing has substantially modified trophic networks, from being dominated by large predators of high trophic level to small species of lower trophic levels, fishing down marine food webs was initially considered an effect of dire consequences. It was documented both at a global (Pauly et al., 1998) and regional (Pauly & Palomares, 2005; Pinnegar et al., 2002, 2003) scale, as well as at a local scale in countries such as Thailand (Christensen, 1998), Canada (Lotze & Milewski, 2004; Pauly et al., 2001), China (Pang & Pauly, 2001, as cited in Pauly, 2010), Portugal (Baeta et al., 2009), Iceland (Valtysson & Pauly, 2003), Namibia (Willemse & Pauly, 2004) Senegal (Laurans et al., 2004), USA (Steneck et al., 2004), Mexico (Sala et al., 2004), Spain (Sánchez & Olaso, 2004), Chile (Arancibia & Neira, 2005), Greece (Stergiou, 2005), Uruguay and Argentina (Jaureguizar & Milessi, 2008), India (Bhathal & Pauly, 2008), and Brazil (Frieire & Pauly, 2010).

However, several criticisms emerged to the general interpretation of this mechanism by Pauly and collaborators, the first one by Caddy et al. (1998). These authors argued that the degree of taxonomic resolution used for the analysis affected the trophic level assigned to species, that the trophic level of catches does not necessarily reflect the trophic level of the ecosystem, that the statistical data used (from FAO) were influenced by aquaculture production, and finally that eutrophication of coastal ecosystems has increased the abundance of lower trophic level organisms. Pauly (2010) countered each of these criticisms and we encourage the reader to judge how well they were addressed. Essington et al. (2006) in turn proposed that fishing down marine food webs is a phenomenon specific to North Atlantic fisheries caused by the sequential collapse and replacement of the fisheries in the region, and that the decline in mean trophic level of the catches in many other areas of the world is caused by increased harvesting of low trophic levels in marine networks (sequential addition of new fisheries), even when catches of high trophic level species remain constant or increase. This pattern of sequential addition to the fisheries of low trophic level species was termed *fishing through the food web*. The controversy thus arose of whether the drastic effect on the structure and mean trophic level of the landings was or not reflective of a real effect of fishing on marine trophic networks. Essington et al. (2006) argued that results based on catch or landing records do not necessarily reflect the composition and state of ecosystems because the indicators are biased by the interests of fisheries operating in each region.

In that regard, Litzow and Urban (2009) reported that the historical periods of decrease in the trophic levels of catches in Alaska obeyed to fishing through the food web and not fishing down the food web, adding as an argument that declines in the trophic level of catches are caused in many cases by temporary additions of fisheries targeting low trophic level species (e.g. crustaceans). Litzow and Urban (2009) concluded that it is clear that commercial exploitation has had profound effects on marine ecosystems in Alaska, but that due to the complexity of connections in marine trophic networks it is difficult to understand these effects. In terms of the ecological interpretation of fishing through the food web, Essington et al. (2006) noted that although they found increases in the catches of high trophic level species, this does not mean that these stocks are healthy and that their findings should not be used to make population inferences since they worked with species categories grouped by trophic level.

2.1.4 Mean Trophic Level (MTL) as an indicator of ecosystem health

Although the MTL of catches is the indicator most frequently used to assess the status of marine ecosystems, it has been widely questioned (Branch et al., 2010; Essington et al., 2006; Powers & Monk, 2010) because it is influenced by economic interests in the different fisheries. Branch et al. (2010) reported that the computation of catch MTL does not adequately correlate with ecosystem MTL and thus this index does not properly measure the magnitude of fishing effects or the rate at which ecosystems are being altered by fishing. Owing to the weaknesses of the catch MTL, it is unlikely that this indicator alone reliably shows any structural effects on trophic networks, let alone any effects of fishing on their complexity and stability. Another argument against this indicator is that in ecosystems where fisheries simultaneously harvest species at different trophic levels (multispecific fisheries), changes in the MTL become masked and the index remains more or less stable with time, potentially giving the impression of a sustainable fishery through time (Pérez-España et al., 2006).

An example of the above was reported for Colombia's Pacific Ocean coast where direct monitoring of shrimp fishery landings between 1995 and 2007 revealed that the MTL of elasmobranch fishes decreased from 3.60 to 3.55 (Mejía-Falla & Navia, 2010), suggesting that the fishery has not impacted these species considerably. However, using only the MTL value is not sufficient because the authors recorded the loss of shark species at trophic level 4 from the catches (*Carcharhinus* spp. and *Sphyrna* spp.) and an increase in the proportion of species at lower trophic levels. Thus, we suggest that indices based on the MTL of catches alone are insufficient to identify structural changes in trophic networks and to detect possible consequences of these changes on network function.

2.1.5 Topological analysis as a tool to detect the effect of fishing on trophic networks

Nearly all studies dealing with the structure of trophic networks mentioned thus far have focused on assessing the effects of fishing on specific characteristics of the network structure (i.e. proportion of species, trophic level, scaling laws), but almost none has attempted to evaluate the effect of fishing on the global structure of networks and how that structure responds to fishing pressure. Only Gaichas and Francis (2008) assessed the structural configuration of the Gulf of Alaska trophic web, finding that that network has small-world attributes and scale-free network properties, and concluding that fishery management actions should focus on highly connected species, which are those that maintain the structural integrity of the network. However, these authors did not carry out simulations supporting their choice of the most adequate management measures proposed for this trophic network.

2.2 Effects of fishing on the function of trophic networks

Fishing does not only affect network structure. Different levels of fishing pressure can generate multiple effects on the function of species and their interactions. These effects are much more difficult to detect and assess than structural effects and often cause the largest changes in ecosystems because they link the different types of ecosystem control spreading across trophic networks. These mechanisms are referred to as *top-down, bottom-up,* and *wasp-waist* (Cury et al., 2003; Pace et al., 1999).

Since fisheries have mostly targeted large species, which exert predatory functions within trophic networks, the most well-known effects to date are those based on the decrease in abundance of those species. A growing body of literature has reported a strong relationship between fishing and decreases in abundance of populations of top predators, with depletions reported to reach such critical levels as 90% of virgin. These reductions have been documented in coastal, benthic, demersal, and pelagic environments and are associated with different fisheries (Baum et al., 2003; Ferretti et al., 2008; Shepherd & Myers, 2005). The decrease in top predator abundance has allegedly led to community restructuring, with their composition (richness and abundance) now being dominated by medium-sized species with lower trophic levels (Ellis et al., 2005; Lotze et al., 2011; Myers et al., 2007). Estes et al. (2011) recently referred to the loss of top predators as "humankind's most pervasive influence on nature".

2.2.1 Trophic cascades and mesopredator release

The decrease in abundance of the large predators, and the associated reduction in top-down ecosystem control mediated through predation or "risk effect", can contribute to the increase in populations of intermediate predators (*mesopredators*)— marine mammals, sharks, rays, and turtles—, thus inducing the formation of trophic cascades (Ferretti et al., 2010; Heithaus et al., 2008). Most published studies have focused on assessing how the decrease in abundance of one species can affect relationships in the trophic network. Hence, the most widely cited and studied effect of fishing on trophic networks is indeed the trophic cascade (e.g. Baum & Worm, 2009; Essington, 2010; Estes et al., 2010; Sandin et al., 2010). This phenomenon has been documented in different marine environments (e.g. Albouy et al., 2010; Andersen & Pedersen, 2010; Casini et al., 2008; Daskalov, 2002; Daskalov et al.,

2007; Estes et al., 1998; Frank et al., 2005; Heithaus et al., 2008; Myers et al., 2007; Pace et al., 1999; Ritchie & Johnson, 2009; Scheffer et al., 2005), and in general all studies describe how a reduction of a large predator population and the ensuing increase in abundance of some of its prey (e.g. seabirds, turtles, reef sharks, and seals) lead to a rapid decline in abundance of species at lower trophic levels and even basal species.

These studies suggest that overfishing can initiate and maintain both structural and functional changes, whose indirect effects can result in a complete reorganization of the network. Frank et al. (2005) even suggested that the trophic cascade effect results from the virtual elimination of the structuring function of large predators in marine ecosystems. In that vein, Bascompte et al. (2005) proposed that trophic cascades reduce the percentage of omnivory and increase the vulnerability of trophic networks to different types of perturbations.

Most of these studies on trophic cascades gave rise to the concept of mesopredator, which has been basically used to refer to medium-sized predator species, which as a result of overfishing of top predators, are increasing in abundance in many marine environments around the world (Beentjes et al., 2002; Levin et al., 2006; Okey et al., 2004; Stevens et al., 2000) and even collapsing populations of their main prey (Myers et al., 2007). More specifically, mesopredator population increases have been mainly recorded in cold and temperate water and low diversity environments such as the western North Atlantic (Choi et al., 2004; Frank et al., 2005), eastern North Atlantic (Blanchard et al., 2005), North Sea (Daan et al., 2005), Baltic Sea (Österblom et al., 2007), and subtropical waters of the North Pacific (Polovina et al., 2009), and we are only aware of a few studies reporting this effect in tropical trophic networks, most of which were carried out in reef ecosystems (Dulvy et al., 2004a, 2004b; Heck et al., 2000; Huges, 1994; McClanahan, 1997, 2000; Ward & Myers, 2005). Sandin et al. (2010) concluded that results of research in tropical coastal ecosystems provide good evidence of "prey release", but only limited support for trophic cascades.

2.2.2 Functional redundancy and ecosystem control

In contrast to the ideas just exposed, Cox et al. (2002) reported that although North Pacific fisheries substantially decreased predator abundance, evidence for the onset of trophic cascades is very limited. Similarly, it has been documented that decreases in abundance of large predators, especially sharks, do not necessarily trigger a mesopredator effect, and that the results and magnitude of this phenomenon could be related to the ecological richness and redundancy of the ecosystem in question (Carlson, 2007; Kitchell et al., 2002; Navia et al., 2010).

Andersen and Pedersen (2010) proposed that fishing can potentially activate trophic cascades, normally buffered both upwards and downwards in trophic networks. They proposed that although the effects of fishing on large predators can be observed even at the plankton level, their intensity is low. They also suggested that when a fishery acts on the different trophic levels of a network, it eliminates the variability characteristic of trophic cascades. Frank et al. (2007) suggested that species diversity and temperature influence potential effects of trophic cascades because high-diversity, warm-water environments have high functional redundancy and if one species is reduced, another could occupy its niche and thus prevent or buffer the trophic cascade. These authors even proposed that while low-diversity, cold-water environments could succumb to top-down ecosystem control effects and their recovery would be very difficult (if at all possible), warmer water environments could oscillate between top-down and bottom-up ecosystem controls according to the level of fishing and shifting temperature regimes.

In general, the fact that most studies reporting top-down control effects and hence trophic cascades are based on cold-water, high-latitude environments (e.g. Estes et al., 2010; Frank et al., 2005, 2007) has to do with these ecosystems exhibiting several characteristics important for these phenomena to be observed in the first place: they are ecosystems of little complexity and low species richness, which translates into low levels of omnivory in the trophic network. These low-diversity marine ecosystems are generally strongly interconnected and highly dependent on trophic interactions that develop within their networks (e.g. Barents Sea between Norway and Russia), which makes them more vulnerable to fishing (Gislason, 2003).

In contrast, ecosystems in tropical latitudes seem to be a little more resistant to the effects of harvesting since time series studies on composition, diversity, and volume of catches show much weaker effects than those recorded in cold and temperate ecosystems (Harris & Poiner, 1991; Sainsbury, 1991; Sainsbury et al., 1997). For example, Hinke et al. (2004) modeled the effects of different oceanic fisheries on trophic networks in the Pacific Ocean finding that a population decline of scombrids of the genus *Auxis* led to increases in biomass of other species of similar trophic level and function. They attributed these population increases to the reduction of predation by tunas due to fishing, but ignored that the population reduction of *Auxis* spp. is precisely what allowed for increased prey availability for species in the same feeding guild, thus facilitating their increase in abundance.

Thus, owing to the importance in their capacity to respond, ecosystems must maintain functional redundancy and the fraction of omnivory to the extent possible since reductions in these characteristics are indicators of fragility and destabilization of the network (Bascompte et al., 2005). Ecosystems subjected to high levels of fishing pressure have already been found to show lower omnivory indices (Morissette et al., 2009). It has also been documented that, in addition to functional redundancy, the identity of predators could play an important role in regulating the lowest trophic levels in the food chain. This is because when the abundance of forage fish (i.e. engraulids and clupeids) declines as a result of fishing, populations of forage invertebrates would not be able to control the abundance of algae thus causing changes in the composition of the trophic network (O'Connor & Brunno, 2007).

Although wasp-waist ecosystem control has been proposed for ecosystems where species at an intermediate trophic level exert control on the flow of energy in the network (Cury et al., 2000; Micheli, 1999), very few studies have assessed the structural significance of this control. Because this control mechanism is based on a single or a few species in very high abundance but also commercially important, one can predict that these ecosystems may become even more vulnerable than those regulated by top-down and bottom-up mechanisms. Jordán et al. (2005) suggested that model ecosystems under wasp-waist are very sensitive to effects on key species because of two main reasons. First, because interactions between wasp-waist species (i.e. anchovies and sardines) are stronger than those between other species pairs because even if these two species do not have direct interactions between them, they share a large number of predators and prey. This allows for the change in abundance in one of them to spread indirect effects such as "apparent competition" or "exploitation competition" (Menge, 1995). Second, because wasp-waist species have higher population self-regulatory values than those of other species, which according to ecological theory could cause cyclical and chaotic dynamics (Hassell et al., 1976) and unpredictable oscillations in nature (Bakun & Broad, 2003).

2.2.3 Ecopath with Ecosim as a tool to detect the effect of fishing on trophic networks

Mass-balance models have been widely used to explore potential effects of fishing on the structure and function of trophic networks or important species or functional groups in those networks. This tool yields results on the energy and biomass balance of ecosystems as well as parameters or indicators needed to interpret the possible effects that fishing can generate on those ecosystems through time. Most analyses conducted with these models aimed at studying the effects of fishing on trophic networks reviewed in this chapter can be grouped into three categories that analyze different properties of the network: overexploitation of trophic levels, simplification of network structure, and imbalances in biomass and energy fluxes, the latter being the most widely studied.

Nearly all studies based on mass-balance models have shown fishing effects on the structure and function of trophic networks, with imbalances in the proportions of biomass among trophic levels being those most frequently found. More specifically, generalized effects are reductions in biomass of top predators and an increase in the proportion of species at intermediate trophic levels and basal species, suggesting that fishing is shifting ecosystem structure from large species with low abundance and slow developmental cycles to small species with high abundance and faster developmental cycles (Albouy et al., 2010; Arias-González et al., 2004; Barausse et al., 2009; Chen et al., 2008; Coll et al., 2007, 2008a, 2009a, 2009b, 2010; Duan et al., 2009; Lotze et al., 2011; Savenkoff et al., 2007a, 2007b). This generalized effect is reducing the naturally occurring competition among the original species in the network and facilitating the onset of indirect effects that generate competition among species that did not strongly interact before. Similar findings were obtained in southeast Australia with the Atlantis marine ecosystem model (Griffith et al., 2011).

Given the reduction in network complexity and abundance of high trophic level species, transfer efficiency of energy in ecosystems has increased through time. This has been identified as one of the main indicators of functional changes in trophic networks because this transfer indicates how efficient the flow of energy is from one trophic level to the next. Thus, an increase in this indicator suggests that energetic changes at low and intermediate levels can reach the upper portions of the network more quickly, making the ecosystem more vulnerable to the dynamics of basal species and thus more sensitive to environmental change. Many documents have reported an increase in the value of energy transfer of ecosystems subject to fishing (Chen et al., 2008; Coll et al., 2009a, 2009b; Duan et al., 2009; Lotze et al., 2011), and some of them suggested that when fishing stops the effect is reversed, that is, the upward energy transfer of the network decreases (Coll et al., 2009a). This happens because fisheries generally focus on high trophic level species with low levels of yield and biomass flow. Additionally, high connectivity values suggest that if an energy transfer pathway is altered, another will compensate for the loss so that total biomass changes at a given trophic level are minimal; thus, if the biomass of a particular prey declines, predators will shift to alternate prey (Link et al., 2009). Overholtz and Link (2009) proposed that if systems are dominated by processes from the medium and low trophic levels, they will not become affected by changes in energy fluxes at high trophic levels and will further be protected by the high connectivity of the network components.

Imbalances in energy fluxes caused by the effects of fishing on trophic networks have also been measured with other indices. For example, increasing fishing pressure through time has caused reductions in the "fishing-in-balance index", which is helpful to analyze energy transfer within ecosystems, suggesting that changes in biomass considerably alter energy transfer from the lowest to the highest trophic levels (Pauly & Watson, 2005). These changes in biomass also decrease fluxes between different network components, leading to a reduction in the total yield of the system with increasing fishing pressure (Duan et al., 2009).

Since ascendency is an indicator of ecosystem maturity and a higher value indicates higher resilience to anthropogenic effects on the network, one would expect pristine ecosystems to have high values of ascendency. This relationship between less perturbed trophic networks and a higher value of maturity has been reported for some ecosystems (Morissette et al., 2009), and it has even been documented that the ascendency values of an ecosystem varied during two different time scales, with higher values occurring when fishing pressure was lower (Duan et al., 2009).

Based on results of different mass-balance models, most of which modeled the effects of different levels of fishing pressure in historical or simulated scenarios, the general tendency is that ecosystems reduce their maturity and complexity in direct relation to time and fishing effort. Morissette et al. (2009) explored whether fishing intensity could lead to significant variations in the structure and dynamics of two contrasting ecosystems (one pristine and one exhausted) finding that pristine ecosystems have ecosystem indicators (e.g. system omnivory index, ascendency) that suggest higher resilience and capacity of recovery to potential modifications in the trophic network. Libralato et al. (2010) explored the differences between the trophic networks of a marine protected area and an exploited area and ratified that the environments devoid of fishing pressure show more complex trophic networks that maintain the proportions of species scaling laws and structural and functional properties, and are thus more resistant to different types of environmental or human pressure.

In addition to the dependency of mass-balance models on the quality of information available, Coll and Libralato (2011) highlighted another important limitation of this approach in terms of their capacity of prediction of ecosystem changes resulting from fishing effort: the absence of models describing truly pristine ecosystems to use as benchmarks for those that are highly impacted. This problem is particularly notorious in the Mediterranean Sea where many studies describing the effects of fishing on ecosystem structures have been carried out, yet not enough information is available on how Mediterranean trophic networks are structured in the absence or at low levels of fishing. This limitation does not only apply to models developed with mass-balance analyses, but also to all models built using information based on harvesting activities in the study areas.

However, as proposed by Essington (2007), ecosystem models in tandem with a reflective analysis incorporating uncertainty could serve as the starting point for management actions, and therefore it is important to incorporate this type of analysis to generate models in the future. Some studies have already made a first attempt at improving this deficiency (Ciavatta et al., 2009; Coll et al., 2008a).

2.3 Effects of fishing on interspecific trophic relationships

2.3.1 Direct and indirect effects

High species diversity has been linked to the stability of trophic networks through the complex interactions that arise among network components, which in turn create multiple spreading pathways of effects through alternate routes that buffer the magnitude of changes (spreading through indirect effects). However, the presence of an indirect effect does not always contribute to network stability.

Results from a large number of studies on fishing effects indicate that the changes in structure and biomass to which trophic networks are subjected to through time trigger indirect effects that can be "visualized" by the establishment of new interactions among network components. These new interactions can be mediated by direct relationships (e.g. predation) or by indirect relationships (e.g. competition) and form the basis of a "new organizational state of the network". If these effects are relatively strong, the network will slowly enter a new organizational state that will be very difficult to leave. These progressive changes have been called *phase shifts* (Scheffer, 2010) and can occur at different scales, from an inversion in the predator-prey relationship that does not affect other species to periodic species replacements to alternate ecosystem states. These changes have already been reported in some marine ecosystems with different degrees of intensity (Barkai & McQuaid, 1988; Cury & Shannon, 2004; Frank et al., 2005; Jackson et al., 2001; Österblom et al., 2007; Scheffer & Carpenter, 2003; Scheffer et al., 2001; Vasas et al., 2007).

Specifically, some authors have reported that in addition to trophic cascades, indirect effects such as *exploitation competition* (Menge, 1995) have been detected in some trophic networks as a result of fishing (Chen et al., 2008; Coll et al., 2007; Duan et al., 2009) and that they could be reflecting phase shifts in these ecosystems. Exploitation competition is an effect that can spread rapidly as a result of a reduction in the abundance of top predators and that has received little attention heretofore. For example, Barausse et al. (2009) suggested that intense fishing pressure in the northern Adriatic Sea reduced fish stocks to such an extent that many of them, without having gone extinct, do not seem to have an impact on the mortality rates of their prey. It is thus possible that different competitors take advantage of this trophic void to increase their feeding rates on new prey items. An example of this was reported by Worm et al. (2005), who proposed that the high degree of diet overlap between whales and adult pelagic fish enabled the onset of exploitation competition between these two groups after the decrease in whale populations between 1950 and 1970 caused by fishing, with a shift from an ecosystem dominated by marine mammals to one dominated by pelagic fish.

2.3.2 Structural changes and phase shifts

A frequently detected effect in this review was the structural change of trophic networks, generally shifting from the dominance of large piscivorous fish to that of small-bodied forage fish, or also leading to the replacement of top predators. However, these are not the only possible or documented changes in alternate stable states or phase shifts in an ecosystem (Scheffer, 2010). For example, Savenkoff et al. (2007a, 2007b) identified, in addition to a change in network dominance, a change in predator structure because the reduction in large pelagic fish abundance allowed for marine mammals to be the only top predators in the system. Along the same lines, a decrease in mean size of the catch has been found in different ecosystems as well as a reduction in mean weight of species and specimens caught (Zwanenburg, 2000). Levin et al. (2006) documented that between 1980 and 2001 catches of some fish species in the North Pacific increased in volume, but the mean weight of fish caught decreased between 56% and 67%, depending on the species. These authors attributed the changes in mean size of the catch and composition of the fish assemblage to fishing, noting that the species that now dominate the ecosystem have trophic levels and life history strategies very different from those of the species they replaced. Extreme examples of fishing-driven fluctuations in marine ecosystems are kelp forests in the Gulf of Maine, which have undergone three different stable states, from fish to urchins and from urchins to crabs (Bourque et al., 2007), and Nova Scotia coastal reefs, which have fluctuated between kelp forests and urchin barrens three times since 1965 (Steneck et al., 2002).

All these changes will be reflected in prey availability and trophic functions within an ecosystem, enabling the development of new interactions or modifying the strength of current interactions. The relationship between the presence or abundance of species and the strength of interactions has been identified as a force that facilitates the change of state at low trophic levels (Moréh et al., 2009). Some tropical ecosystems, especially coral reefs, have also undergone structural changes in composition, displaying phase shifts or alternate stable states (Hughes et al., 2003; Pandolfi et al., 2003; Sandin et al., 2008), switching from high coral cover, high rates of coral recruitment and low cover of competitive fleshy algae (McClanahan, 1997; Sandin et al., 2008) to ecosystems with low densities of fish biomass, high densities of echinoderms and high cover of fleshy algae (e.g. Dulvy et al., 2004a, 2004b; Pinnegar & Polunin, 2004).

Perhaps one of the most worrisome problems, if not the most important of all associated with the changes just described, is that, together with structural changes, the diet of some species has also been found to be altered, with a switch from diets dominated by fish to diets dominated by invertebrates. These changes cause the alteration of interaction forces between predators and prey and the establishment of a series of new predator-prey relationships and thus direct and indirect effects that could ultimately contribute, as proposed by Mangel and Levin (2005), to fishing modifying ecosystems so profoundly that it would lead them to alternate states where it would be virtually impossible for the species that have significantly decreased in abundance to recover. A possible example is the hypothesis by Springer et al. (2003) to explain the shift from a state dominated by sea otters to an urchin dominated phase in southwest Alaska. They proposed that the post-War World II whaling industry reduced prey availability for killer whales, leading to an expansion of their diet to include pinnipeds and sea otters, which in turn reduced the sea otter population and facilitated a population increase of sea urchins.

2.3.3 Trophic cycles

Population declines of some species could enhance the importance of the so-called trophic network cycles, which would play a central role in the pathways taken by networks under different levels of fishing pressure. Thus, phase shifts lead to the juvenile stages of top predators being more vulnerable to predation and competition than the adult stages of their prey (e.g. Köster & Möllmann, 2000). These changes between developmental stages and predation in fishes have already been well documented (de Ross & Persson, 2002; Worm & Myers, 2003). This phenomenon in turn leads to reduced recruitment rates of top predators, even to such low levels to impair stock recovery after fishing ends. Some authors have proposed that it was the cause for why cod stocks in the North Atlantic and Baltic Sea have not recovered more than 15 years after fishing ceased.

2.3.4 Keystone species and fisheries

Another way in which fisheries have caused changes in interspecific trophic relationships with various ecosystem effects is when they have acted directly or indirectly on the keystone species of an ecosystem resulting in the creation of new trophic or even spatial (habitat modification) organizations. The latter case has been reported in some reef ecosystems where fishing removed predators and competitors of echinoderms (i.e. sea urchins, crown of thorns starfish), spreading indirect effects of reduction in algal cover and reef erosion, allowing for new dominant species to become established and ultimately modifying the biotic structure of the reef (Dulvy et al., 2004a, 2004b; Hughes, 1994; McClanahan & Muthiga, 1988; McClanahan et al., 1996, 2002; Pinnegar & Polunin, 2004). McClanahan et al. (2007) reported that almost 40 years after fishing stopped, the biomass of reef fishes in Kenya has recovered and is close to virgin levels, which in turn has increased predation levels on sea urchins in marine protected areas, leading to the recovery of coral cover and benthos heterogeneity.

In the coasts of Alaska and Canada, sea otters and some fish species act as keystone species predating on sea urchins and regulating their effect on the abundance of kelp forests, which support a large number of species in these regions. Thus, strong fishery impacts on sea urchin predators intensified grazing and the deterioration of kelp forests, leading to marked changes in the fauna of that ecosystem (Duggins et al., 1989; Estes & Duggins, 1985; Reisewitz et al., 2005; Tegner & Dayton, 2000).

Below we present a map showing the geographical distribution of the different effects of fishing on trophic networks presented throughout this chapter and covering all literature cited (Figure 1). It is expected that the number of ecosystems affected by fishing will grow considerably as new studies on the effect of fishing on trophic webs are completed.

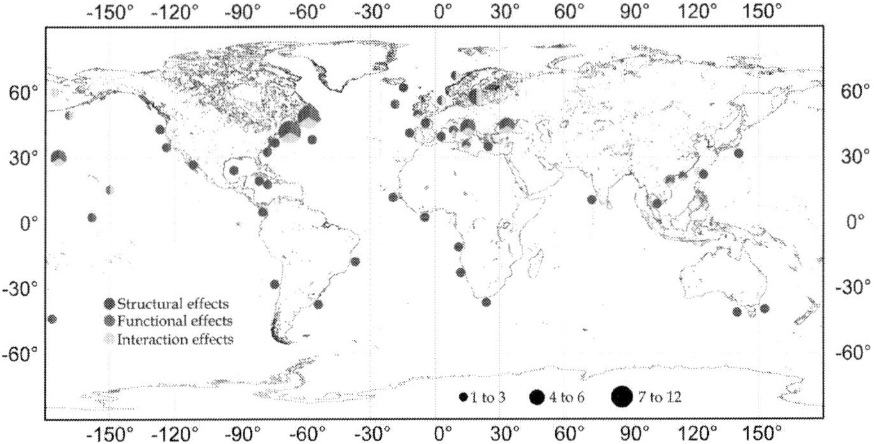

Fig. 1. Geographical distribution of studies reporting fishing effects on the structure, function, or interactions of trophic networks. The map is a simplification as these effects are often interrelated.

3. Case study

Although analyses of secondary extinction have been widely used as mechanisms to assess the resistance of trophic networks to species loss (Dunne et al., 2004; Solé & Montoya, 2001), we believe the assumption implicit in those analyses is a major simplification of the natural dynamics of trophic networks because a predator that loses one or more prey could adapt to use another available resource. In contrast, if the local extinction of a species does occur, it is very likely that the network becomes fragmented to some extent and, depending on the connectivity and topological importance of the species removed, the network could be led to a new organizational state.

Thus, assuming that the way in which species in a network organize and interact is important for network stability, it has been proposed that the consequences of *small-world* and *scale-free* structural patterns may be of great importance in recognizing the sensitivity to perturbations in biological networks (Montoya & Solé, 2002). It has also been demonstrated that a regular network can be transformed into a small-world network if a small proportion of nodes are reconnected to some randomly chosen nodes.

The small-world structural pattern, which is based on grouping of nodes, has shown to be useful to provide quick answers to different perturbations in some theoretical trophic networks, suggesting that this structural arrangement can be of benefit for network resilience. Solé and Montoya (2001) and Montoya and Solé (2002) determined that trophic networks with a small-world structural arrangement were more resistant to secondary species extinctions than those with random structure.

With the evidence presented throughout this chapter, we suggest that trophic networks are becoming simplified, some ecosystems are undergoing state changes, and in many others the proportions of species have been altered, all of which implies that large-scale structural patterns of trophic networks (e.g. scale-free, small-world) are being affected.

The Gulf of Tortugas in the Colombian Pacific Ocean has been subject to intense fishing since 1960. Although the target of these fisheries is shallow-water shrimp, a large number of fish and invertebrate species are also caught as bycatch. Several studies of the feeding habits of these species have been conducted in the area, facilitating the description and understanding of the community food web (Navia et al., 2010). However, the effect of fishing on that network has not yet been assessed. Thus, taking into account the structural consequences of fishing on networks, we designed an exercise to assess whether the trophic network of the Gulf of Tortugas displays a small-world structure, and if so, test the hypothesis that sustained fishing pressure can modify network structure, taking it from a small-world arrangement to a random one (Figure 2).

The first step was to assess whether the structure of the Gulf of Tortugas network meets the requirement of scale-free node distribution, and therefore can display a small-world arrangement. To that end we conducted two analyses. First, we computed a frequency distribution of the number of connections by node (node degree) of the original network (250 nodes and 579 interactions), and second, we calculated a frequency distribution (in log scale) of the node degree of a network with random structure, built with the same number of nodes (250) and interactions (579) as the original network. Results were as follows:

1. Frequency distribution of the node degree of the original network: this analysis showed a power distribution of connections by node, i.e., the network displays many nodes with very few connections and very few nodes with a large number of connections. This type of distribution is characteristic of networks structured in a small-world arrangement and is known as a scale-free distribution (Figure 3a).
2. Frequency distribution of the node degree of the random network: this analysis, which was generated with the computer program Pajek (http://pajek.imfm.si), showed a Poisson frequency distribution, which is expected in trophic networks that adhere to a random structure (Montoya & Solé, 2002) (Figure 3b).

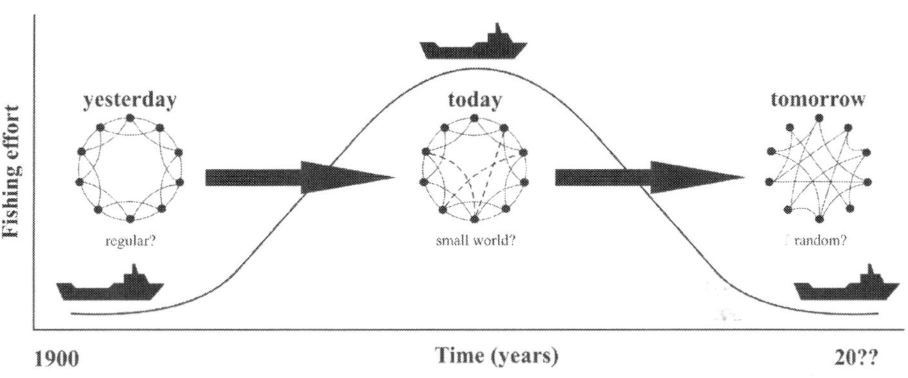

Fig. 2. Schematic representation of the hypothesis of the effect of fishing on the structure of trophic networks.

Based on these results, the fundamental principle for the trophic network under study to have a small-world structure is met, i.e., that the node degree follows a scale-free distribution. Next, to test our question of whether fishing effects can modify or at least induce changes in the structure of a trophic network, switching from a small-world to a random structure (or at least show a tendency), we chose two important structural features of networks: the *clustering coefficient* (CC) and the average *path length* (PL). The first index is helpful to determine the extent to which some groups of species are more connected internally than with other groups; the second index measures the average number of steps along the shortest paths to connect all possible node pairs in the network and is useful as a measure of the efficiency of information transport or mass transport in a network. Thus, a trophic network maintains its small-world structural arrangement if the CC of the observed network is greater than that of a random network of the same characteristics (nodes and interactions). Both small-world and random networks have low values of PL.

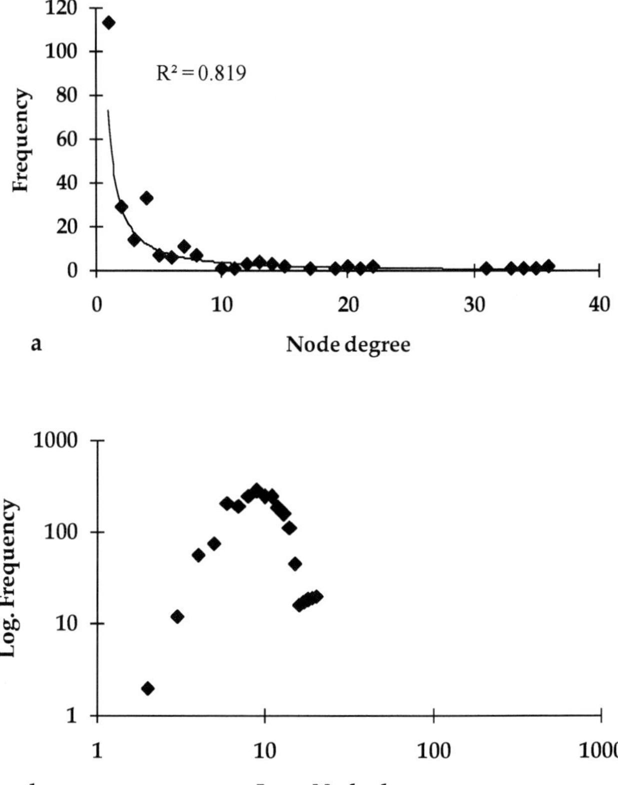

Fig. 3. Frequency distribution of the node degree of the trophic network in the Gulf of Tortugas. a) original network, b) random network generated with the same number of nodes and interactions as the original network.

These indices were selected because they have an ecological support that strengthens potential results. More specifically, CC is a measure that may indicate the degree of functional similarity of a trophic network, and therefore could be an indirect reflection of the functional redundancy in the network; trophic networks with high values of CC will be more interconnected and therefore greater functional redundancy is to be expected. On the other hand, PL is an indicator of the quantity of indirect interactions that can be established within a trophic network and therefore, of the buffering capacity of the network to a given effect. Changes in these indices, indicating the transition from a small-world to a random network, will imply that the new network will have a very short average path length and thus, since there is not much distance between species, an effect will rapidly spread throughout the network without finding many possible paths (indirect effects) for it to be buffered. Additionally, random networks do not have significant values of CC, which suggests that species with diet similarities have decreased and that functional redundancy perhaps is no longer significantly present to serve as biological insurance for the network.

To develop this exercise we decided to assess the effect of two criteria for selecting species to be removed from the network and simulated the removal of those species as a result of fishing. The first criterion was fishery importance (commercial value of species) in the study area, and the second criterion was topological importance of the species in the trophic network.

Species of commercial importance:

These species were selected based on the presence of fishing fleets that have exerted historical pressure on these resources and their trading prices because, since they are the species of higher commercial value, their fisheries have been more intense than those of other species in the network. The species selected were:

- Shrimps
- Snappers
- Clupeiforms, in particular *Cetengraulis mysticetus* and *Ophistonema libertate*, which make up the so-called "carduma", which is the object of a targeted fishery.

Species of topological importance:

Network analysis provides a number of tools that can support quantitative community ecology. In particular, there exist techniques for quantifying the positional importance of species (system components) in food webs. Species that are of high importance in a trophic network can be in either central (like hubs) or unique positions. The latter can be interpreted as species having non-redundant neighborhoods. As a result, their extinction (or overfishing) has profound effects on the ecosystem. Here we present and use the *TO* index for quantifying *topological overlap*. *TO* is a mesoscale network index, considering also non-local neighborhood, but weighted by distance (i.e. not considering the whole network equally important). This is quite sound in ecology, as a suite of field and theoretical results support the importance of indirect interactions.

Several mesoscale indices have already been suggested in network science, most of them considering distance between nodes (e.g. closeness and betweenness centrality [Wasserman & Faust, 1994]). Some of these indices have been applied to ecological problems (Estrada, 2007; Jordán et al., 2007). Others have been slightly modified and adapted to ecology (see net status [Harary, 1961] and keystone index [Jordán et al., 1999]) or simply developed by ecologists (measuring apparent competition [Godfray et al., 1999; Müller et al. 1999]). We use a sophisticated version of the latter index, as it is quite general and suitable for quantifying redundancy of neighborhoods (uniqueness and replaceability of species).

The *topological importance* (*TI*) index (Jordán et al., 2009) makes it possible to analyze indirect interactions of various lengths separately (up to a 3-step-length). It assumes a network with undirected links where interspecific effects may spread in any direction without bias (we are interested in interaction webs, in the broadest sense, but considering only indirect chain effects [Wootton, 1994]). The effect of species j on species i, when i may be reached from j in n steps, is defined as $a_{n,ij}$. When $n=1$ (*i.e.* the effect of j on i is direct): $a_{1,ij} = 1/D_i$, where D_i is the degree of node i (i.e. the number of its direct neighbors including both prey or predatory species). We assume that indirect chain effects are multiplicative and additive. When the effect of step n is considered, we define the effect received by species i from all N species in the same network (see Equation 1)

$$\Psi_{n,i} = \sum_{j=1}^{N} a_{n,ij} \tag{1}$$

which is equal to 1 (i.e. each species is affected by the same unit effect). Furthermore, we define the n-step effect originated from a species i (see Equation 2)

$$\sigma_{n,i} = \sum_{j=1}^{N} a_{n,ji} \tag{2}$$

which may vary among different species (i.e. effects originated from different species may be different). Here, we define the topological importance of species i, when effects up to n steps are considered (see Equation 3)

$$TI_i^n = \frac{\sum_{m=1}^{n} \sigma_{m,i}}{n} = \frac{\sum_{m=1}^{n} \sum_{j=1}^{N} a_{m,ji}}{n} \tag{3}$$

which is simply the sum of effects originated from species i up to n steps (one plus two plus three...up to n) averaged over by the maximum number of steps considered (n). With this index, it is possible to quantify the origins of effects influencing a particular species, i.e. the internal interaction structure of the network.

The $a_{n,ij}$-values for species j had been defined as its "trophic field" (Jordán, 2001). For long indirect effects, every species is connected to every other. It is reasonable to define a t threshold of $a_{n,ij}$-values separating strong interactive partners from weak interactors. Given a maximum length of indirect effects (n) and a threshold for interaction strength (t), every node may be characterized by its effective trophic range (Jordán et al., 2009). Since the sets of strong interactors of two, or more, nodes may overlap, it is important to quantify the positional uniqueness of graph nodes. The "trophic field overlap" (TO^n_{ij}) between nodes i and j is the number of strong interactors appearing in both i's and j's effective range. The sum of all TO-values between species i and others ($\Sigma TO^{n,t}_{ij}$ summed over all j with $i \neq j$) provides the summed trophic field overlap of species i ($TO^{n,t}_i$), and this may be normalized by dividing it with the maximum value ($TO^{n,t}_{max}$) for a given network ($relTO^{n,t}_i = TO^{n,t}_i / TO^{n,t}_{max}$). Note that all this is determined by t, n and the topology of the network. We define the "topological uniqueness" of species i as $TU^{n,t}_i = 1 - relTO^{n,t}_i$. Here, we used $n=3$ and $t=0.001$. This index may contribute to the problem of how to quantify species and role and redundancy in ecosystems (Bond, 1994; Luczkovich et al., 2003; Shannon & Cury, 2003).

Results of TU^3 showed that nodes 50 (*Carcharhinus leucas*), 218 (*Sphyrna lewini*), and 103 (*Galeocerdo cuvier*) had the greatest topological uniqueness (Figure 4) and were the most difficult to replace because there are no other species that can overlap their function.

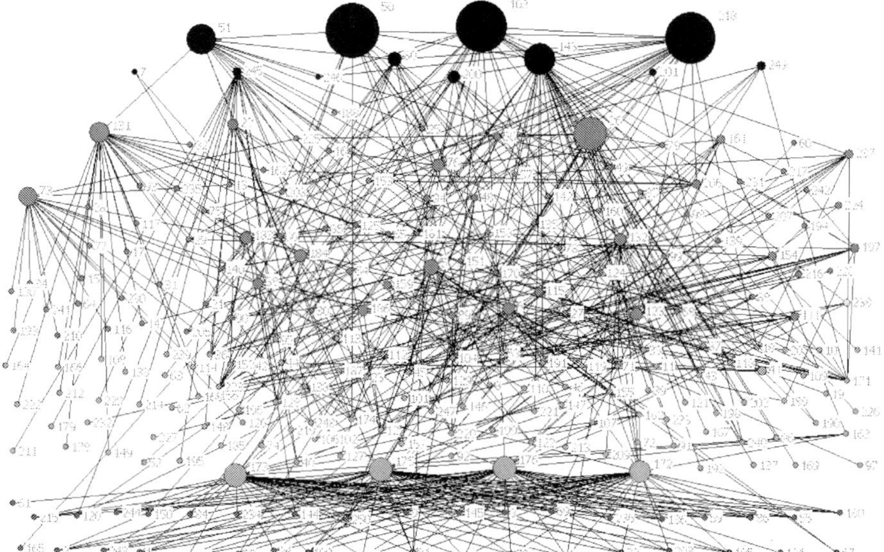

Fig. 4. Schematic image of the trophic network of the Gulf of Tortugas based on TU^3. The size of the nodes is directly proportional to topological uniqueness, indicating the low redundancy of their neighborhood. Black: elasmobranchs; red: teleosts; green: invertebrates; blue: zooplankton and phytoplankton.

To assess the effect of fishing on the structure of the trophic network of the Gulf of Tortugas, we compared values of CC and PL of the original network to those of the corresponding random network in different scenarios under the two criteria defined above (commercial importance and topological importance). The scenarios consisted of leaving the network unaltered and then sequentially removing the groups defined above to simulate the effect of fishing. The scenarios were as follows:

Species of commercial interest

Scenario 1. The trophic network was left unaltered (250 nodes, 579 interactions), without eliminating any nodes (initial network), and values of CC and PL calculated for this and the corresponding simulated network. Based on the analyses described at the beginning of this section (scale-free distribution) and the CC and PL results, we established that the Gulf of Tortugas network displays small-world structure.

Scenario 2. The nodes representing shrimps in the trophic networks were removed and values CC of PL calculated for this and the corresponding simulated network (232 nodes and 398 interactions).

Scenario 3. In addition to the shrimp nodes, nodes representing snappers were also removed and values of CC and PL calculated for this and the corresponding simulated network (229 nodes and 349 interactions).

Scenario 4. In addition to the previous removals, nodes representing the "carduma" category were also removed and values of CC and PL calculated for this and the corresponding simulated network (226 nodes and 329 interactions).

Species of topological importance (TU³)

Scenario 1. The trophic network was left unaltered (250 nodes, 579 interactions), without eliminating any nodes (initial network), and values of CC and PL calculated for this and the corresponding simulated network. Based on the analyses described at the beginning of this section (scale-free distribution) and the CC and PL results, we established that the Gulf of Tortugas network displays small-world structure.

Scenario 2. The node with the greatest topological importance in the trophic network was removed (50, *Carcharhinus leucas*) and values of CC and PL calculated for this and the corresponding simulated network (249 nodes and 565 interactions).

Scenario 3. In addition to node 50, the second-most important node topologically was removed (218, *Sphyrna lewini*) and values of CC and PL calculated for this and the corresponding simulated network (248 nodes and 543 interactions).

Scenario 4. In addition to the removal of the previous nodes, node 103 (*Galeocerdo cuvier*), which was the third-most important topologically, was also removed and values of CC and PL calculated for this and the corresponding simulated network (247 nodes and 520 interactions).

Fisheries and species of commercial interest

In terms of the loss of high-value commercial species, one can see that the relationship between the CC of the observed and the random networks is not significantly modified, with the value of the observed network always being higher (Figure 5a). These results suggest that the loss of these species does not cause any detectable effects on the compartmentalization of the network or on the interactions that take place in those compartments. Not even the loss of shrimps (one of the groups with the highest centrality within the network) results in any indication of alterations in the CC pattern of the network. Since the CC values of the observed networks are higher than those of the corresponding random networks, the small-world organization in the trophic network of the Gulf of Tortugas is maintained, even with the loss of species of high commercial value.

In all scenarios explored, the observed trophic networks had, on average, shorter paths between the most distant nodes than those of the random networks (< 2) (Figure 5b), which is characteristic of trophic networks with a small-world structure (Williams et al., 2002), and has been found in other work using this type of analysis (e.g. Gaichas & Francis, 2008). This feature is important in terms of spreading effects within the network because indirect effects with an average length of 2 (e.g. apparent competition or keystone predation) tend to dissipate when the average is greater than 3 steps, and therefore reduce the capacity of buffering the spread of effects within the network.

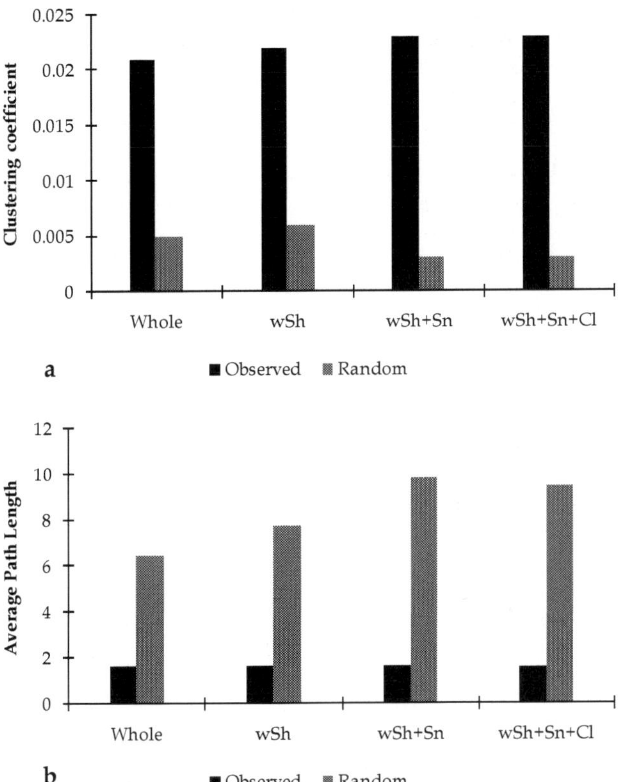

Fig. 5. Clustering coefficient (a) and average path length (b) values of the observed networks and their corresponding simulated networks in each scenario. These scenarios correspond to the removal of species with high commercial value. Whole: entire network; wSh: without shrimps; wSh+Sn: without shrimps or snappers; wSh+Sn+Cl: without shrimps, snappers or "carduma".

In terms of the relationship between the observed and random PL values, one can see that the loss of shrimps has the largest effect on the relationship because once snappers and clupeiforms are removed, PL values of the simulated networks increase relatively little. This could be due to the fact that shrimps are, among high-valued species, those with the highest centrality in the network and therefore their removal leads to a significant effect in the connectivity of the network.

Fisheries and species of topological importance

A rather different pattern from that previously described can be observed when removing species of high topological importance. The original trophic network also has higher CC values than the simulated network, but upon removal of nodes, CC values of the observed networks decrease whereas those of the simulated networks tend to increase. This tendency is rather strong because when reaching scenario 4 (removal of nodes 50, 218, and 103), the CC value of the observed network becomes lower than that of the simulated network (Figure 6a). These results suggest that the small-world organization of the trophic network of the Gulf of Tortugas has been significantly altered, perhaps to the extent of losing it.

In terms of the relationship between the observed and random PL values for species of topological importance, one can see that the loss of these nodes does not affect the network since the PL values of both the observed and simulated networks remain relatively constant in the four scenarios (Figure 6b).

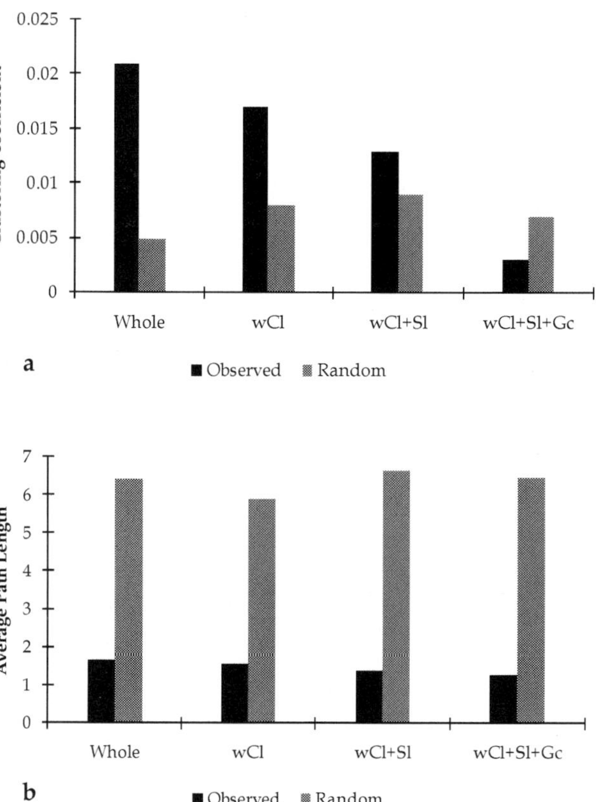

Fig. 6. Clustering coefficient (a) and average path length (b) values of the observed networks and their corresponding simulated networks in each scenario. These scenarios correspond to the removal of species with high topological importance. Whole: entire network; wCl: without *Carcharhinus leucas*; wCL+Sl: without *C. leucas* or *Sphyrna lewini*; wCL+Sl+Gc: without *C. leucas*, *S. lewini* or *Galeocerdo cuvier*.

Results of this exercise suggest that the trophic network of the Gulf of Tortugas in the Colombian Pacific Ocean display properties typical of a small-world structural arrangement and a scale-free interaction distribution. These results are thus relevant in terms of network stability and even fisheries management because it has been reported that these features are important for network stability. This seems consistent with the fact that removing the two groups with the highest connectivity in the network (shrimps and snappers) did not cause an effect indicative of a substantial alteration of the structural properties of this network. However, our results showed that trophic networks with a small-world organization are susceptible to the removal of species of high topological importance, especially those that have low positional redundancy within the network. In this case, three top predators are the species of highest topological importance in the network.

This exercise, which represents only a first attempt at assessing the effects of fishing on the structural features of a highly complex network, and the first at applying this type of approach to a purely tropical environment, must be supported with the exploration of additional scenarios to corroborate that the Gulf of Tortugas network is indeed highly resistant to the targeted removal of species with high connectivity and economic importance, but not to that of species of topological importance. Finally, based on these results, we highlight the importance of adopting fishery management measures involving not only species of high commercial value, but also those that play unique roles in the network by contributing disproportionately to the structure and stability of marine trophic networks.

4. Conclusions

Some have proposed that the solution to the current biomass reductions of some commercial species is a decrease in fishing mortality or even a complete cessation of fishing. As an example, Chen et al. (2008) proposed that adopting different strategies of reduction in fishing mortality in the Gulf of Beibu, China, would allow the biomass of most species in the network to increase by almost an order of magnitude, which seems like a simplistic proposal and ignores many important considerations of trophic network dynamics, which we present next.

Trophic networks are complex structures that establish high levels of interaction among their elements and therefore maintain dynamic processes that contribute to their stability. As reviewed throughout this chapter, fisheries can affect trophic networks from several perspectives: structural, functional, or in terms of the trophic interactions among species, and an effect generated by one of these aspects will likely affect the others.

Although we divided the effects in three themes for ease of understanding, they are all interrelated and become magnified as fishing pressure increases. For example, a trophic network that suffers a "simple" imbalance in the proportion between predators and prey could spread an indirect effect, which in turn could foster an interaction that was not previously significant. If the species involved are not adapted to adjust to this new dynamic, some of them may experience population declines that could spread another sequence of indirect effects like those mentioned, which could even modify some ecosystem functions. Thus, all these processes can add up and magnify to the point of producing a larger change in some of the species in the network and lead to the so-called phase shift. Alternatively, if the network is highly redundant, the effect may not be visible because it might dissipate throughout the network.

The problem with this hypothetical sequence of changes is that fishing pressure on trophic networks does not only affect the structural features of the network, but also promotes indirect effects that can force functional changes and lead to irreversible modifications. The magnitude of these changes is related to fishing intensity and the time it is exerted on the network.

The conclusion is therefore that under current levels of fishing pressure most trophic networks are headed towards experiencing phase shifts to a greater or lesser extent and in different time and intensity scales. For example, trophic networks in cold and temperate ecosystems and wasp-waist trophic networks have a higher likelihood of being impacted by fishing effects and cause phase shifts with little probability of return than those in tropical ecosystems. These changes have already been evidenced from trophic cascade effects reported in these ecosystems. Modifications of these trophic networks are so marked in some cases that, although fisheries have not operated for many years, predator populations have not returned to their initial states.

In the case of tropical ecosystems, the high diversity, connectance, and interactions among highly redundant species lead to fishing effects dissipating somewhat in the trophic web and thus being less apparent, yet still present. Another issue is that in tropical ecosystems, multispecific fisheries operate that extract species at all trophic levels, which could lead to imbalances in the structural and mass-balance properties not being easily observed, perhaps masking functional or interaction effects dangerous for network stability.

To know the real effects that fishing can generate on trophic networks it is first necessary to understand the forces that condition the interactions and dynamics among species, such as the capacity of species to switch to new prey types once the abundance of current prey is reduced, the effect of ontogenetic changes on the functional redundancy of species, and the importance of mutual predation on the population dynamics of species, among others.

5. Acknowledgments

AFN thanks IPN-CICIMAR and CONACYT, and PAMF thanks COLCIENCIAS for providing funding for their PhD degrees. VHCE thanks SNI, IPN-EDI, IPN-COFAA and SIP-IPN-20120949. This study was partially supported by COLCIENCIAS-SQUALUS (Grant RC No. 156-2010).

6. References

Albouy, C., Mouillot, D., Rocklin, D., Culioli, J. M. & Le Loc'h, F. (2010). Simulation of the Combined Effects of Artisanal and Recreational Fisheries on a Mediterranean MPA Ecosystem Using a Trophic Model. *Marine Ecology Progress Series*, Vol.412, (August 2010), pp. 207-221, ISSN 0171-8630

Andersen, K. H. & Pedersen, M. (2010). Damped Trophic Cascades Driven by Fishing in Model Marine Ecosystems. *Proceedings the Royal Society London B*, Vol.277, No. 1682, (November 2009), pp. 795-802, ISSN 1471-2954

Arancibia, H. & Neira, S. (2005). Long-Term Changes in the Mean Trophic Level of Central Chile Fishery Landings. *Scientia Marina*, Vol.69, No.2, (June 2005), pp. 295-300, ISSN 0214-8358

Arias-González, J. E., Nuñez-Lara, E., González-Salas, C. & Galzín, R. (2004). Trophic Models for Investigation of Fishing Effect on Coral Reef Ecosystems. *Ecological Modelling*, Vol.172, No.2-4, (March 2004), pp. 197-212, ISSN 0304-3800

Baeta, F., Costa, M. J. & Cabral, H. (2009). Changes in the Trophic Level of Portuguese Landings and Fish Market Price Variation in the Last Decades. *Fisheries Research*, Vol.97, No.3, (May 2009), pp. 216-22, ISSN 0165-7836

Bakun, A. & Broad, K. (2003). Environmental 'Loopholes' and Fish Population Dynamics: Comparative Pattern Recognition With Focus on El Nino Effects in the Pacific. *Fisheries Oceanography*, Vol.12, No.4-5, (September 2003), pp. 458– 473, ISSN 1365-2419

Barausse, A., Ducci, A., Mazzoldi, C., Artoli, Y. & Palmeri, L. (2009). Trophic Network Model of the Northern Adriatic Sea: Analysis of an Exploited and Eutrophic Ecosystem. *Estuarine Coastal Shelf Science*, Vol.83, No.4, (August 2009), pp. 577-590, ISSN 0272-7714

Barkai, A. & McQuaid, C. (1988). Predator-Prey Role Reversal in a Marine Benthic Ecosystem. *Science*, Vol.242, No.4875, (October 1988), pp. 62-64, ISSN 0036-8075

Bascompte, J., Melián, C. J. & Sala, E. (2005). Interaction Strength Combinations and the Overfishing of a Marine Food Web. *Proceedings of the National Academy of Sciences of the United States of America*, Vol.102, No.15, (March 2005), pp. 5443–5447, ISSN 1091-6490

Baum, J. K. & Worm, B. (2009). Cascading Top-Down Effects of Changing Oceanic Predator Abundances. *Journal of Animal Ecology*, Vol.78, No.4, (March 2009), pp. 699-714, ISSN 0021-8790

Baum, J. K., Myers, R., Keller, D. G., Worm, B., Harley S. J. & Doherty, P. A. (2003). Collapse and Conservation of Sharks Populations in the Northwest Atlantic. *Science*, Vol.299, No.5605, (January 2003), pp. 389-392, ISSN 0036-8075

Beentjes, M., Bull, B., Hurst, R. & Bagley, N. (2002). Demersal Fish Assemblages Along the Continental Shelf and Upper Slope of the East Coast of the South Island, New Zealand. *New Zealand Journal of Marine and Freshwater Research*, Vol.36, No.1, (March 2002), pp. 197–223. ISSN 0028-8330

Bhathal, B. & Pauly, D. (2008). Fishing Down Marine Food Webs and Spatial Expansion of Coastal Fisheries in India, 1950–2000. *Fisheries Research*, Vol.91, No.1, (May 2008), pp. 26–34, ISSN 0165-7836

Bianchi, G., Gislason, H., Graham, K., Hill, L., Jin, X., Koranteng, K., et al. (2000). Impact of Fishing on Size Composition and Diversity of Demersal Fish Communities. *ICES Journal of Marine Science*, Vol.57, No.3, (June 2000), pp. 558-571, ISSN 1095-9289

Blaber, S. J. M., Cyrus, D. P., Albaret, J. J., Ching, C. V., Day, J. W. & Elliot, M. (2000). Effects of Fishing on the Structure and Functioning of Estuarine and Nearshore Ecosystem. *ICES Journal of Marine Science*, Vol.57, No.3, (June 2000), pp. 590-602, ISSN 1095-9289

Blanchard, J. L., Dulvy, N. K., Jennings, S., Ellis, J. E., Pinnegar, J. K., Tidd, A. & Kell, L. T. (2005). Do Climate and Fishing Influence Size-Based Indicators of Celtic Sea Fish Community Structure? *ICES Journal of Marine Science*, Vol.62, No.3 (June 2005), pp. 405–411, ISSN 1095-9289

Bond, W. J. (1994). Keystone Species, In: *Biodiversity and Ecosystem Function*, Schulze, E. D. & Mooney, H. A (Eds.), 237-254, Springer, ISBN 3540581030, Berlin, Germany

Bourque, B. J., Johnson, B. & Steneck, R. S. (2007). Possible Prehistoric Hunter–Gatherer Impacts on Food Web Structure in the Gulf of Maine. In: *Human Impacts on Ancient*

Marine Environments, Erlandson J. & Torben, R. (Eds.), 165–187, University of California Press, ISBN 978-0-520253-43-8, Berkeley, USA

Branch, A. T., Watson, R., Fulton, E. A., Jennings, S., McGilliard, C. R., Pablico, G. T., et al. (2010). The Trophic Fingerprint of Marine Fisheries. *Nature*, Vol.468, (November 2010), pp. 431-435, ISSN 0028-0836

Briand, F. & Cohen, J. E. (1984). Community Food Webs Have Scale Invariant Structure. *Nature*, Vol.307, (January 1984), pp. 264–267, ISSN 0028-0836

Carlson, J. K. (2007). Modeling the Role of Sharks in the Trophic Dynamics of Apalachicola Bay, Florida. In: *Shark Nursery Grounds of the Gulf of Mexico and the East Coast Waters of the United States*, McCandless, C. T., Kohler, N. E. & Pratt, H. L. (Eds.), 281-300, American Fisheries Society, ISBN 978-1-888569-81-0, Bethesda, USA

Caddy, J. F., Csirke, J., García, S. J. & Grainier, R. J. R. (1998). How Pervasive is "Fishing Down Marine Food Webs"? *Science*, Vol. 282, No.5393, (November 1998), pp. 1383, ISSN 0036-8075

Casini, M., Lövgren, J., Hjelm, J., Cardinale, M., Molinero, J. C. & Kornilovs, G. (2008). Multi Level Trophic Cascades in a Heavily Exploited Open Marine Ecosystem. *Proceedings the Royal Society London B*, Vol.275, No. 1644, (August 2008), pp. 1793-1801, ISSN 1471-2954

Chen, Z., Qiu, Y., Jia, X. & Xu, S. (2008). Using an Ecosystem Modeling Approach to Explore Possible Ecosystem Impacts of Fishing in the Beibu Gulf, Northern South China Sea. *Ecosystems*, Vol.11, No.8, (December 2008), pp. 1318-1334, ISSN 1432-9840

Choi, J. S., Frank, K. T., Leggett, W. C. & Drinkwater, K. (2004). Transition to an Alternate State in a Continental Shelf Ecosystem. *Canadian Journal of Fisheries and Aquatic Sciences*, Vol.61, No.4, (April 2004), pp. 505–510, ISSN 1205-7533

Christensen, V. (1998). Fishery-Induced Changes in the Marine Ecosystem: Insights from Models of the Gulf of Thailand. *Journal of Fish Biology*, Vol.53, (December 1998), pp. 128–142, ISSN 1095-8649

Ciavatta, S., Lovato, T., Ratto, M. & Pastres, R. (2009) Global Uncertainty and Sensitivity Analysis of a Food-Web Bioaccumulation Model. *Environmental Toxicology and Chemistry*, Vol.28, No.4, (April 2009), pp. 718–732, ISSN 1552-8618

Coll, M., Santojanni, A., Palomera, I., Tudela, S. & Arneri E. (2007). An Ecological Model of the Northern and Central Adriatic Sea: Analysis of Ecosystem Structure and Fishing Impacts. *Journal of Marine Systems*, Vol.67, No.1-2, (August 2007), pp. 119-154, ISSN 0924-7963

Coll, M., Lotze H. K. & Romanuk, T. N. (2008a). Structural Degradation in Mediterranean Sea Food Webs: Testing Ecological Hypothesis Using Stochastic and Mass-Balance Modeling. *Ecosystems*, Vol.11, No.6, (September 2008), pp. 939-960, ISSN 1432-9840

Coll, M., Libralato, S., Tudela, S., Palomera, I. & Pranovi, F. (2008b.) Ecosystem Overfishing in the Ocean. *PLoS One*, Vol.3, No.12, (December 2008), e3881, ISSN 1932-6203

Coll, M., Palomera, I. & Tudela, S. (2009a). Decadal Changes in a NW Mediterranean Sea Food Web in Relation to Fishing Exploitation. *Ecological Modelling*, Vol.220, No.17 (September 2009), pp. 2088-2102, ISSN 0304-3800

Coll, M., Santojanni, A., Palomera, I. & Arneri, E. (2009b). Food Web Changes in the Adriatic Sea Over Last Three Decades. *Marine Ecology Progress Series*, Vol.381, (April 2009), pp. 17-37, ISSN 0171-8630

Coll, M., Santojanni, A., Palomera, I. & Arneri, E. (2010). Ecosystems Assessment of the North-Central Adriatic Sea: Towards a Multivariate Reference Framework. *Marine Ecology Progress Series*, Vol.417, (November 2010), pp. 193-210, ISSN 0171-8630

Coll, M. & Libralato, S. (2011). Contributions of Food Web Modelling to the Ecosystem Approach to Marine Resource Management in the Mediterranean Sea. *Fish and Fisheries*. doi: 10.1111/j.1467-2979.2011.00420.x, (May 2011), pp. 1-25, ISSN 1467-2979

Cox, S. E., Essington, T. E., Kitchell, J. F., Martell, S. J. D., Walters, C. J., Boggs, C. & Kaplan, I. (2002). Reconstructing Ecosystem Dynamics in the Central Pacific Ocean, 1952-1998. II. A preliminary Assessment of the Trophic Impacts of Fishing and Effects on Tuna Dynamics. *Canadian Journal of Fisheries and Aquatic Sciences*, Vol.59, No.11, (November 2002), pp. 1736-1747, ISSN 1205-7533

Cury, P. & Shannon, L. (2004). Regime Shifts in Upwelling Ecosystems: Observed Changes and Possible Mechanisms in the Northern and Southern Benguela. *Progress in Oceanography*, Vol. 60, No.2-4 (February 2004), pp. 223-243, ISSN 0079-6611

Cury, P., Bakun, A., Crawford, R. J. M., Jarre-Teichmann, A., Quinones, R., Shannon, L. J. & Verheye, H. M. (2000). Small Pelagics in Upwelling Systems: Patterns of Interaction and Structural Changes in "Wasp-Waist" Ecosystems. *ICES Journal of Marine Science*, Vol.57, No.3, (June 2000), pp. 603-618, ISSN 1095-9289

Cury, P., Shannon, L. & Shin, Y. J. (2003). The Functioning of Marine Ecosystems: a Fisheries Perspective. In: *Responsible Fisheries in the Marine Ecosystem*, Sinclair, M. & Valdimarsson, G (Eds.), 103-124, CAB International, ISBN 9251047677, Walllingford, UK.

Daan, N., Gislason, H., Pope J. G. & Rice, J. C. (2005). Changes in the North Sea Fish Community: Evidence of Indirect Effects of Fishing? *ICES Journal of Marine Science*, Vol.62, No.2, (March 2005), pp. 177-188, ISSN 1095-9289

Daskalov, G. M. (2002). Overfishing Drives a Trophic Cascade in the Black Sea. *Marine Ecology Progress Series*, Vol.225, (January 2002), pp. 53–63, ISSN 0171-8630

Daskalov, G. M., Grishin, A. N., Rodionov, S. & Mihneva, V. (2007). Trophic Cascades Triggered by Overfishing Reveal Possible Mechanisms of Ecosystem Regime Shifts. *Proceedings of the National Academy of Sciences of the United States of America*, Vol.104, No.25, (June 2007), pp. 10518–10523, ISSN 1091-6490

Dell, A. I., Kokkoris, G. D., Banasek-Richter, C., Bersier, L. F., Dunne, J. A., Kondoh, M., et al. (2005). How do Complex Food Webs Persist in Nature? In: *Dynamic Food Webs: Multispecies Assemblages, Ecosystem Development and Environmental Change*, de Ruiter, M. P. C., Wolters, V. & Moore, J. C. (Eds.), 425-436, Academic Press, ISBN 978-0-12-088458-2, San Diego, USA

De Ross, A. M. &. Persson, L. (2002). Size-Dependent Life History Traits Promote Catastrophic Collapses of Top Predators. *Proceedings of the National Academy of Sciences of the United States of America*, Vol.99, No.20, (October 2002), pp. 12907-12912, ISSN 1091-6490

Duan, L. J., Li, S. Y., Liu, Y., Moreau, J. & Christensen, V. (2009). Modeling Changes in the Coastal Ecosystem of the Pearl River Estuary from 1981 to 1998. *Ecological Modelling*, Vol.220, No.20, (October 2009), pp. 2802-2818, ISSN 0304-3800

Duggins, D. O., Simenstad, C. A. & Estes, J. A. (1989). Magnification of Secondary Production by Kelp Detritus in Coastal Marine Ecosystems. *Science*, Vol.245, No.4914, (July 1989), pp. 170–173, ISSN 0036-8075

Dulvy, N. K., Freckleton, R. P. & Polunin, N. V. C. (2004a). Coral Reef Cascades and the Indirect Effects of Predator Removal by Exploitation. *Ecology Letters*, Vol.7, No.5, (May 2004), pp. 410–416, ISSN 1461-0248

Dulvy, N. K., Polunin, N. V. C., Mill, A. C. &. Graham, N. A. J. (2004b). Size Structural Change in Lightly Exploited Coral Reef Fish Communities: Evidence for Weak Indirect Effects. *Canadian Journal of Fisheries and Aquatic Sciences*, Vol.61, No.3, (March 2004), pp. 466–475, ISSN 1205-7533

Dunne, J. A., Williams R. J. & Martinez, N. D. (2002). Network Structure and Biodiversity Loss in Food Webs: Robustness Increases with Connectance. *Ecology Letters*, Vol.5, No.4, (July 2002), pp. 558-567, ISSN 1461-0248

Dunne, J. A., Williams, R. J. &. Martinez N. D. (2004). Network Structure and Robustness of Marine Food Webs. *Marine Ecology Progress Series*, Vol.273, (June 2004), pp. 291-302, ISSN 0171-8630

Dunne, J. A. & Williams, R. J. (2009). Cascading Extinctions and Community Collapse in Model Food Webs. *Proceedings the Royal Society London B*, Vol.364, No.1524 (June 2009), pp. 711-1723, ISSN 1471-2954

Ellis, J. R., Dulvy, N. K., Jennings, S., Parker-Humphreys, M. & Rogers, S. I. (2005). Assessing the Status of Demersal Elasmobranchs in UK Waters: a Review. *Journal of the Marine Biological Association of the United Kingdom*, Vol.85, No.5, (October 2005), pp. 1025-1047, ISSN 0025-3154

Essington, T .E. (2007). Pelagic Ecosystem Response to a Century of Commercial Fishing and Whaling. In: *Whales, Whaling, and Ocean Ecosystems*, Estes, J. A., DeMaster, D. P., Doak, D. F., Williams, T. M. &. Brownell, R. L. Jr., (Eds.), 38–49, University of California Press, ISBN 9780520248847, California, USA

Essington, T. E. (2010). Trophic Cascades in Open Ocean Ecosystems. In: *Trophic Cascades Predators, Prey, and the Changing Dynamics of Nature*, Terborgh, J. & Estes, J. A. (Eds.), 91-105, Island Press, ISBN 9784-59726-486-0, Washington, USA

Essington, T. E., Beaudreau A. H. & Wiedenmann, J. (2006). Fishing Through Marine Food Webs. *Proceedings of the National Academy of Sciences of the United States of America*, Vol.103, No.9, (February 2006), pp. 3171–3175, ISSN 1091-6490

Estes, J. A. & Duggins, D. O. (1995). Sea Otters and Kelp Forests in Alaska: Generality and Variation in a Community Ecological Paradigm. *Ecological Monograph*, Vol.65, No.1, (February 1995), pp. 75-100, ISSN 0012-9615

Estes, J. A., Tinker, M. T., Williams, T. M. & Doak, D. F. (1998). Killer Whale Predation on Sea Otters Linking Oceanic and Nearshore Ecosystems. *Science*, Vol.282, No.5688, (October 1998), pp. 473-476, ISSN 0036-8075

Estes, J. A., Peterson, C. H. &. Steneck, R. S. (2010). Some Effects of Apex Predators in Higher–Latitude Coastal Oceans. In: *Trophic Cascades Predators, Prey, and the Changing Dynamics of Nature*, Terborgh, J. & Estes, J. A. (Eds.), 37-53, Island Press, ISBN 9784-59726-486-0, Washington, USA

Estes, J. A., Terborgh, J., Brashares, J. S., Power, M. E., Berger, J., Bond W. J., et al. (2011). Trophic Downgrading of Planet Earth. *Science*, Vol.333, No.6040, (July 2011), pp. 301-306, ISSN 0036-8075

Estrada, E. (2007). Characterization of Topological Keystone Species: Local, Global and "Meso-Scale" Centralities in Food Webs. *Ecological Complexity*, Vol.4, No.1-2, (March 2007), pp.48-57, ISSN 1476-945X

Ferretti, F., Myers, R. A., Serena, F. & Lotze, H. K. (2008). Loss of Large Predatory Sharks from the Mediterranean Sea. *Conservation Biology*, Vol.22, No.4, (August 2008), pp. 952-964, ISSN 0888-8892

Ferretti, F., Worm, B., Britten, G. L., Heithaus, M. R. & Lotze, H. K. (2010). Patterns and Ecosystem Consequences of the Shark Declines in the Ocean. *Ecology Letters*, Vol.13, No.8, (August 2010), pp. 1055-1071, ISSN 1461-0248

Frank, K. T., Petrie, B., Choi J. S. & Leggett, W. C. (2005). Trophic Cascades in a Formerly Cod Dominated Ecosystem. *Science*, Vol.308, No.5728, (June 2005), pp. 1621-1623, ISSN 0036-8075

Frank, K. T., Petrie, B. & Shackell, N. L. (2007). The Ups and Downs of Trophic Control in Continental Shelf Ecosystems. *Trends in Ecology and Evolution*, Vol.22, No.5 (May 2007), pp. 236-242, ISSN 0169-5347

Friedlander, A. M. & DeMartini, E. E. (2002). Contrasts in Density, Size, and Biomass of Reef Fishes Between the Northwestern and the Main Hawaiian Islands: the Effects of Fishing Down Apex Predators. *Marine Ecology Progress Series*, Vol.230, (April 2002), pp. 253–264, ISSN 0171-8630

Frieire, K. M. F. & Pauly, D. (2010). Fishing Down Brazilian Marine Food Webs, with Emphasis on the East Brazil Large Marine Ecosystem. *Fisheries Research*, Vol.105, No.1, (June 2010), pp. 57-62, ISSN 0165-7836

Fulton, E. A., Smith, A. D. M. & Punt, A. E. (2005). Which Ecological Indicators Can Robustly Detect Effects of Fishing? *ICES Journal of Marine Science*, Vol.62, No.3, (June 2005), pp. 540-551, ISSN 1095-9289

Gaichas, S. K. & Francis, R. C. (2008). Network Models for Ecosystem-Based Fishery Analysis: a Review of Concepts and Application to the Gulf of Alaska Marine Food Web. *Canadian Journal of Fisheries and Aquatic Sciences*, Vol.65, No.9, (September 2008), pp. 1965-1982, ISSN 1205-7533

Gislason, H. (2003). The Effects of Fishing on Non-Target Species and Ecosystem Structure and Function, In: *Responsible Fisheries in the Marine Ecosystem*, Sinclair, M. & Valdimarsson, G. (Eds.), 255-274, CAB International, ISBN 085199633, Roma, Italy

Godfray, H. C. J., Lewis, O. T. & Memmott, J. (1999). Studying Insect Diversity in the Tropics. *Proceedings the Royal Society London B*, Vol.354, No.1391, (November 1999), pp. 1811-1824, ISSN 1471-2954

Griffith, G., Fulton, E. A. & Richardson, A. J. (2011). Effects of Fishing and Acidification-Related Benthic Mortality on the Southeast Australian Marine Ecosystem. *Global Change Biology*, Vol.17, No.10, (October 2011), pp. 3058-3074, ISSN 1365-2486

Haedrich, R. L. & Barnes, S. M. (1997). Changes Over Time of the Size Structure in an Exploited Shelf Fish Community. *Fisheries Research*, Vol.31, No.3, (August 1997), pp. 229-239, ISSN 0165-7836

Hall, S. J. (1999). *The Effects of Fishing on Marine Ecosystems and Communities*. Blackwell Science, ISBN 978-0632041121, Oxford, UK

Harary, F. (1961). Who eats whom? *General Systems*, Vol.6, pp. 41-44, ISSN 0308-1079

Harris, A. N. & Poiner, I. R. (1991). Changes in Species Composition of Demersal Fish Fauna of Southeast Gulf of Carpentaria, Australia, after 20 Years of Fishing. *Marine Biology*, Vol.111, No.3, (June 1991), pp. 503-519, ISSN 0025-3162

Hassell, M. P., Lawton, J. H. & May, R. M. (1976). Patterns of Dynamical Behaviour in Single Species Populations. *Journal of Animal Ecology*, Vol.42, No.2, (June 1976), pp. 471-486, ISSN 0021-8790

Heck, K. L. Jr., Pennock, J. R., Valentine, J. F., Coen, L. D. & Sklenar, S. A. (2000). Effects of Nutrient Enrichment and Small Predator Density on Seagrass Ecosystems: An Experimental Assessment. *Limnology and Oceanography*, Vol.45, No.5, (July 2000), pp. 1041-1057, ISSN 0024-3590

Heithaus, M. R., Frid, A., Wirsing, A. J. & Worm, B. (2008). Predicting Ecological Consequences of Marine Top Predator Declines. *Trends in Ecology and Evolution*, Vol.23, No.4, (April 2008), pp. 202-210, ISSN 0169-5347

Hinke, J. T., Kaplan, I. C., Aydin, K., Watters, G. M., Olson R. J. & Kitchell, J. F. (2004). Visualizing the Food Web Effects of Fishing for Tunas in the Pacific Ocean. *Ecology and Society*, Vol.9, No.1, (August 2004), pp. 1-10, ISSN 1708-3087

Hughes, T. P. (1994). Catastrophes, Phase Shifts, and Large-Scale Degradation of a Caribbean Coral Reef. *Science*, Vol.265, No.5178, (September 1994), pp. 1547-1551, ISSN 0036-8075

Hughes, T. P., Baird, A. H., Bellwood, D. R., Card, M., Connolly, S. R., Folke, C., Grosberg, R., et al. (2003). Climate Change, Human Impacts, and the Resilience of Coral Reefs. *Science*, Vol.301, No.5635, (August 2003), pp. 929-933, ISSN 0036-8075

Jackson, J. B. C., Kirby, M. X., Berger, W. H., Bjorndal, K. A., Botsford, L. W., Bourque, B. J., et al. (2001). Historical Overfishing and the Recent Collapse of Coastal Ecosystems. *Science*, Vol.293, No.5530, (July 2001), pp. 629-638, ISSN 0036-8075

Jaureguizar, A. J. & Milessi, A. C. (2008). Assessing the Source of Fishing Down Marine Food Web Process in the Argentinean-Uruguayan Common Fishing Zone. *Scientia Marina*, Vol.72, No.1, (March 2008), pp. 25-36, ISSN 1886-8134

Jennings, S. & Kaiser, M. J. (1998). The Effects of Fishing on Marine Ecosystems. In: *Advances in Marine Biology*, D. W. Sims, (Ed.), 201-352, Academic Press, ISBN 978-0-12-374351-0, Plymouth, UK

Jones, E., Gray, T. & Umponstira, C. (2009). The Impact of Artisanal Fishing on Coral Reef Fish Health in Hat Thai Mueang, Phang-nga Province, Southern Thailand. *Marine Policy*, Vol.33, No.4, (July 2009), pp. 544-552, ISSN 0308-597X

Jordán, F., Takács-Sánta, A. & Molnár, I. (1999). A Reliability Theoretical Quest for Keystones. *Oikos*, Vol.86, No.3, (September 1999), pp. 453-462, ISSN 0030-1299

Jordán, F. (2001). Trophic Fields. *Community Ecology*, Vol.2, No.2, (December 2001), pp. 181-185, ISSN 1585-8553.

Jordán, F., Liu, W. C. &. Wyatt, T. (2005). Topological Constraints on the Dynamics of Wasp-Waist Ecosystems. *Journal of Marine System*, Vol.57, No.3-4, (September 2005), pp. 250-263, ISSN 0924-7963

Jordán, F., Benedek, Z. & Podani, J. (2007). Quantifying Positional Importance in Food Webs: a Comparison of Centrality Indices. *Ecological Modelling*, Vol. 205, No.1-2, (July 2007), pp. 270-275, ISSN 0304-3800

Jordán, F., Liu, W-C. & Mike, Á. (2009). Trophic Field Overlap: a New Approach to Quantify Keystone Species. *Ecological Modelling*, Vol.220, No.21, (November 2009), pp. 2899-2907, ISSN 0304-3800

Kitchell, J. F., Essington, T. E., Boggs, C. H., Schindler, D. E. & Walters, C. J. (2002). The Role of Sharks and Long-Line Fisheries in a Pelagic Ecosystem of Central Pacific. *Ecosystems*, Vol.5, No.2, (March 2002), pp. 202-216, ISSN 1432-9840

Köster, F. W. & Möllmann, C. (2000). Trophodynamic Control by Clupeid Predators on Recruitment Success in Baltic Cod? *ICES Journal of Marine Science*, Vol.57, No.2, (April 2000), pp. 310-323, ISSN 1095-9289

Laurans, M., Gascuel, D., Chassot, E. & Thiam D. (2004). Changes in the Trophic Structure of Fish Demersal Communities in West Africa in the Three Last Decades. *Aquatic Living Resources*, Vol.173, No.2, (November 2004), pp. 163-173, ISSN 0990-7440

Levin, P. S., Holmes, E. E., Piner, K. R. & Harvey, C. J. (2006). Shifts in a Pacific Ocean Fish Assemblage: the Potential Influence of Exploitation. *Conservation Biology*, Vol.20, No.4, (August 2006), pp. 1181-1190, ISSN 0888-8892

Libralato, S., Coll, M., Tempesta, M., Santojanni, A., Spoto, M., Palomera, I., et al. (2010). Food Webs Traits of Protected and Exploited Areas of the Adriatic Sea. *Biological Conservation*, Vol.143, No.9, (September 2010), pp. 2182-2194, ISSN 0006-3207

Link, J., Col, L., Guida, V., Dow, D., O'Reilly, J., Green, J., et al. (2009). Response of Balanced Network Models to Large-Scale Perturbation: Implications for Evaluating the Role of Small Pelagic in the Gulf of Maine. *Ecological Modelling*, Vol.220, No.3, (February 2009), pp. 351-369, ISSN 0304-3800

Litzow, M. A. & Urban, D. (2009). Fishing Through (and Up) Alaskan Food Webs. *Canadian Journal of Fisheries and Aquatic Sciences*, Vol.66, No.2, (February 2009), pp. 201-211, ISSN 1205-7533

Lotze, H. K. & Milewski, I. (2004). Two Centuries of Multiple Human Impacts and Successive Changes in a North Atlantic Food Web. *Ecological Applications*, Vol.14, No.5, (October 2004), pp. 1428-1447, ISSN 1051-0761

Lotze, H. K. & Worm, B. (2009). Historical Baselines for Large Marine Animals. *Trends in Ecology and Evolution*, Vol.24, No.5, (May 2009), pp. 254-262, ISSN 0169-5347

Lotze, H. K., Lenihan, H. S., Bourque, B. J., Bradbury, R. H., Cooke, R. G., Kay, M. C., et al. (2006). Depletion, Degradation and Recovery Potential of Estuaries and Coastal Seas. *Science*, Vol.312, No.5781, (June 2006), pp. 1806-1809, ISSN 1095-9289

Lotze, H. K., Coll, M. & Dunne, J. A. (2011). Historical Changes in Marine Resources, Food Web Structure and Ecosystem Functioning in the Adriatic Sea, Mediterranean. *Ecosystems*, Vol.14, No.2, (March 2011), pp. 198-222, ISSN 1432-9840

Luczkovich, J. J., Borgatti, S. P., Johnson, J. C. & Everett, M. G. (2003). Defining and Measuring Trophic Role Similarity in Food Webs Using Regular Equivalence. *Journal of Theoretical Biology*, Vol.220, No.3, (February 2003), pp. 303-321, ISSN 0022-5193

Mangel, M. &. Levin, P. S. (2005). Regime, Phase and Paradigm Shifts: Making Community Ecology the Basic Science for Fisheries. *Proceedings the Royal Society London B*, Vol.360, No.1453, (January 2005), pp. 95-105, ISSN 1471-2954

McClanahan, T. R. (1997). Primary Succession of Coral-Reef Algae: Differing Patterns on Fished Versus Unfished Reefs. *Journal of Experimental Marine Biology and Ecology*, Vol.218, No.1, (November 1997), pp. 77-102, ISSN 0022-0981

McClanahan, T. R. (2000). Recovery of a Coral Reef Keystone Predator, *Balistapus undulatus*, in East African Marine Parks. *Biological Conservation*, Vol.94, No.2, (July 2000), pp. 191–198, ISSN 0006-3207

McClanahan, T. R. &. Muthiga, N.A. (1988). Changes in Kenyan Coral Reef Community Structure and Function Due to Exploitation. *Hydrobiologia*, Vol.166, No.3, (September 1988), pp. 269-276, ISSN 0073-2087

McClanahan, T. R., Kamukuru, A. T., Muthiga, N. A., Gilagabher-Yebio, M. & Obura, D. (1996). Effect of Sea Urchin Reductions on Algae, Coral, and Fish Populations. *Conservation Biology*, Vol.10, No.1, (February 1996), pp. 136-154, ISSN 0888-8892

McClanahan, T. R., Polunin, N. & Done, T. (2002). Ecological States and the Resilience of Coral Reefs. *Conservation Ecology*, Vol.6, No.2, (December 2002), pp. 1-18, ISSN 1708-3087

McClanahan, T. R., Graham, N. A. J., Calnan, J. M. & MacNeil, M. A. (2007). Toward Pristine Biomass: Reef Fish Recovery in Coral Reef Marine Protected Areas in Kenya. *Ecological Applications*, Vol.17, No.4, (June 2007), pp. 1055–1067, ISSN 1051-0761

Mejía-Falla, P. A. & Navia A. F. (2010). Efectos de la pesca de arrastre sobre la estructura del ensamblaje de elasmobranquios costeros del Pacífico colombiano, *Memorias del II Encuentro Colombiano sobre Condrictios*, ISBN 978-958-44-2931-5, Cali, August 2010

Menge, B. (1995). Indirect Effects in Marine Rocky Intertidal Interaction Webs: Patterns and Importance. *Ecological Monograph*, Vol.65, No.1, (February 1995), pp. 21-74, ISSN 0012-9615

Micheli, F. (1999). Eutrophication, Fisheries and Consumer-Resource Dynamics in Marine Pelagic Ecosystems. *Science*, Vol.285, No.5432, (August 1999), pp. 1396-1398, ISSN 0036-8075

Montoya, J. M. & Solé, R. (2002). Small-World Patterns in Food Webs. *Journal of Theoretical Biology*, Vol.214, No.3, (February 2002), pp. 405-412, ISSN 0022-5193

Montoya, J. M., Solé, R. & Rodríguez, M. A. (2001). La Arquitectura de la Naturaleza: Complejidad y Fragilidad en Redes Ecológicas. *Ecosistemas, Revista de Ecología y Medio Ambiente*, Vol.10, No.2. (May 2001), pp. 1-14, ISSN 1697-2473

Moréh, A., Jordán, F., Slzilágyi, A. & Scheuring, I. (2009). Overfishing and Regime Shifts in Minimal Food Web Models. *Community Ecology*, Vol.10, No.2, (December 2009), pp. 236-243, ISSN 1585-8553

Morissette, L., Pedersen, T. & Nielsen, M. (2009). Comparing Pristine and Depleted Ecosystems: the Sofjord, Norway versus the Gulf of St. Lawrence, Canada. Effects of Intense Fisheries on Marine Ecosystems. *Progress in Oceanography*, Vol.81, No.1-4, (April 2009), pp. 174-187, ISSN 0079-6611

Müller, C. B., Adriaanse, I. C. T., Belshaw, R. & Godfray, H. C. J. (1999). The Structure of an Aphid-Parasitoid Community. *Journal of Animal Ecology*, Vol.68, No.2, (March 1999), pp. 346-370, ISSN 0021-8790

Myers, R. A. & Worm, B. (2005). Extinction, Survival or Recovery of Large Predatory Fishes. *Proceedings the Royal Society London B*, Vol.360, No.1453, (January 2009), pp. 13–20, ISSN 1471-2954

Myers, R., Baum, J. K., Shepered, T. D., Powers, S. & Peterson, C. H. (2007). Cascading Effects of the Loss of Apex Predatory Sharks from a Coastal Ocean. *Science*, Vol.315, No.5820, (March 2007), pp. 1846-1850, ISSN 0036-8075

Navia, A. F., Cortés, E. & Mejía-Falla, P. A. (2010). Topological Analysis of the Ecological Importance of Elasmobranch Fishes: a Food Web Study on the Gulf of Tortugas, Colombia. *Ecological Modelling*, Vol.221, No.24, (December 2010), pp. 2918-2926, ISSN 0304-3800

O'Connor, N. E. & Brunno J. F. (2007). Predatory Fish Loss Affects the Structure and Functioning of a Model Marine Food Web. *Oikos*, Vol.116, No.12, (December 2007), pp. 2017-2038, ISSN 0030-1299

Okey, T. A., Banks, S., Born, A. F., Bustamante, R. H., Calvopiña, M., Edgar, G. J., et al. (2004). A Trophic Model of a Galapagos Subtidal Rocky Reef for Evaluating Fisheries and Conservation Strategies. *Ecological Modelling*, Vol.172, No.2-4, (March 2004), pp. 383–401, ISSN 0304-3800

Österblom, H., Hansson, S., Larsson, U., Hjerne, O., Wulff, F., Elmgren R. & Folke, C. (2007). Human-Induced Trophic Cascades and Ecological Regime Shifts in the Baltic Sea. *Ecosystems*, Vol.10, No.6, (September 2007), pp. 877–888, ISSN 1432-9840

Overholtz, W. & Link, J. (2009). A Simulation Model to Explore the Response of the Gulf of Maine Food Web to Large-Scale Environmental and Ecological Changes. *Ecological Modelling*, Vol.220, No.19, (October 2009), pp. 2491-2502, ISSN 0304-3800

Pace, M. L., Cole, J. J., Carpenter S. R. & Kitchell J. F. (1999). Trophic Cascades Revealed in Diverse Ecosystems. *Trends in Ecology and Evolution*, Vol.14, No.12, (December 1999), pp. 483-488, ISSN 0169-5347

Pandolfi, J. M., Bradbury, R. H., Sala, E., Hughes, T. P., Bjorndal, K. A., Cooke, R. G., et al. (2003). Global Trajectories of the Long-Term Decline of Coral Reef Ecosystems. *Science*, Vol.301, No.5635, (August 2003), pp. 955–958, ISSN 0036-8075

Pauly, D. (2010). *5 Easy Pieces. The Impact of Fisheries on Marine Ecosystems*, Island Press, ISBN 9781597267199, Washington, USA

Pauly, D. & Palomares, M. L. (2005). Fishing Down Marine Food Webs: it is Far More Pervasive than we Thought. *Bulletin of Marine Science*, Vol.76, No.2, (April 2005), pp. 197-211, ISSN 0007-4977

Pauly, D. & Watson, R. (2005). Background and Interpretation of the 'Marine Trophic Index' as a Measure of Biodiversity. *Proceedings the Royal Society London B*, Vol.360, No.1454, (February 2005), pp. 415-423, ISSN 1471-2954

Pauly, D., Christensen, V., Dalsgaard, J., Froese, R. & Torres, F. (1998). Fishing Down Marine Food Webs. *Science*, Vol.279, No.5352, (February 1998), pp. 860-863, ISSN 0036-8075

Pauly, D., Palomares, M. L., Froese, R., Sa-a, P., Vakily, M., Preikshot D. & Wallace, S. (2001). Fishing Down Canadian Aquatic Food Webs. *Canadian Journal of Fisheries and Aquatic Sciences*, Vol.58, No.1, (November 2000), pp. 51-62, ISSN 1205-7533

Pérez-España, H., Abarca-Arenas L. G. & Jiménez-Badillo, M. L. (2006). Is Fishing Down Trophic Web a Generalized Phenomenon? The Case of Mexican Fisheries. *Fisheries Research*, Vol.79, No.3, (July 2006), pp. 349-352, ISSN 0165-7836

Piet, G. J. & Jennings, S. (2005). Response of Potential Fish Community Indicators to Fishing. *ICES Journal of Marine Science*, Vol.62, No.2, (March 2005), pp. 214-225, ISSN 1095-9289

Pinnegar, J. K. & Polunin, N. V. C. (2004). Predicting Indirect Effects of Fishing in Mediterranean Rocky Littoral Communities Using a Dynamic Simulation Model. *Ecological Modelling*, Vol.172, No.2-4, (March 2004), pp. 249-267, ISSN 0028-0836

Pinnegar, J. K., Jennings, S., O'Brien, C. M. & Polunin, N. V. C. (2002). Long-Term Changes in Trophic Level of the Celtic Sea Fish Community and Fish Market Price Distribution. *Journal of Applied Ecology*, Vol.39, No.3, (June 2002), pp. 377-390, ISSN 0021-8901

Pinnegar, J. K., Polunin, N. V. C. & Badalamenti, F. (2003). Long-Term Changes in the Trophic Level of Western Mediterranean Fishery and Aquaculture Landings. *Canadian Journal of Fisheries and Aquatic Sciences*, Vol.60, No.2, (February 2003), pp. 222-235, ISSN 1205-7533

Polovina, J. J., Abecassis, M., Howell, E. A. & Woodworth, P. (2009). Increases in the Relative Abundance of Mid-Trophic Level Fishes Concurrent with Declines in Apex Predators in the Subtropical North Pacific, 1996-2006. *Fishery Bulletin*, Vol.107, No.4, (December 2009), pp. 523-531, ISSN 0090-0656.

Powers, J. E. & Monk, M. H. (2010). Current and Future Use of Indicators for Ecosystem Based Fisheries Management. *Marine Policy*, Vol.34, No.3, (May 2010), pp. 723–727, ISSN 0308-597X

Reisewitz, S. E., Estes, J. A. & Simenstad, C. A. (2005). Indirect Food Web Interactions: Sea otters and Kelp Forest Fishes in the Aleutian Archipelago. *Oecologia*, Vol.146, No.4, (January 2006), pp. 623–631, ISSN 0029-8549

Rice, J. C. (2000). Evaluating Fishery Impacts Using Metrics of Community Structure. *ICES Journal of Marine Science*, Vol.57, No.3, (June 2000), pp. 682-688, ISSN 1095-9289

Ritchie, E. G. & Johnson, C. N. (2009). Predator Interactions, Mesopredator Release and Biodiversity Conservation. *Ecology Letters*, Vol.12, No.9, (July 2009), pp. 1-18, ISSN 1461-0248

Sainsbury, K. J. (1991). Application of an Experimental Management Approach to Management of a Tropical Multispecies Fishery with Highly Uncertain Dynamics. *ICES Marine Science Symposia*, 193, pp. 301-320.

Sainsbury, K. J., Campbell, R. A., Lindholm, R. & Whitelaw, A. W. (1997). Experimental Management of an Australian Multispecies Fishery: Examining the Possibility of Trawl-Induced Habitat Modification, In: *Global Trends: Fisheries Management*, Pikitch, K., Huppert D. D. & Sissenwine, M. P. (Eds.), 107-112, American Fisheries Society, ISBN 9781888569032, Maryland, USA

Sala, E., Aburto-Oropeza, O., Reza, M., Paredes G. & López-Lemus L. G. (2004). Fishing Down Coastal Food Webs in the Gulf of California, *Fisheries*, Vol.29, No.3, (March 2004), pp. 19-25, ISSN 03632415

Sánchez, F. & Olaso, I. (2004). Effects of Fisheries on the Cantabrian Sea Shelf Ecosystem. *Ecological Modelling*, Vol.172, No.2-4, (March 2004), pp. 151-174, ISSN 0304-3800

Sandin, S. A., Smith, J. E., DeMartini, E. E., Dinsdale, E. A., Donner, S. D., Friedlander, A. M., et al. (2008). Baselines and Degradation of Coral Reefs in the Northern Line Islands. *PLoS One*, Vol.3, No.2, (February 2008), e1548, ISSN 1932-6203

Sandin, S. A., Walsh S. M. & Jackson J. B. C. (2010). Prey Release, Trophic Cascades, and Phase Shifts in Tropical Nearshore Ecosystems. In: *Trophic Cascades Predators, Prey, and the Changing Dynamics of Nature*, Terborgh, J. & Estes, J. A. (Eds.), 71-90, Island Press, ISBN 9784-59726-486-0, Washington, USA

Savenkoff, C., Castonguay, M., Chabot, D., Hammill, M. O., Bourdages H. & Morissette L. (2007a). Changes in the Northern Gulf of St. Lawrence Ecosystem Estimated by Inverse Modeling: Evidence of a Fishery-Induced Regime Shift? *Estuarine, Coastal and Shelf Science*, Vol.73, No.3-4, (July 2007), pp. 711-724, ISSN 02727714

Savenkoff, C., Swain, D. P., Hanson, J. M., Castonguay, M., Hammill, M. O., Bourdages, H., et al. (2007b). Effects of Fishing and Predation in a Heavily Exploited Ecosystem: Comparing Periods Before and after the Collapse of Groundfish in the Southern Gulf of St. Lawrence (Canada). *Ecological Modelling*, Vol.204, No.1-2, (May 2007), pp. 115-128, ISSN 0304-3800

Shannon, L. J. & Cury, P. M. (2003). Indicators Quantifying Small Pelagic Fish Interactions: Application Using a Trophic Model of the Southern Benguela Ecosystem. *Ecological Indicators*, Vol.3, No.4, (January 2003), pp. 305-321, ISSN 1470-160X

Scheffer, M. (2010). Alternative States in Ecosystems. In: *Trophic Cascades Predators, Prey, and the Changing Dynamics of Nature,* Terborgh, J. & Estes, J. A. (Eds.), 287-298, Island Press, ISBN 9784-59726-486-0, Washington, USA

Scheffer, M. & Carpenter, S. R. (2003). Catastrophic Regime Shifts in Ecosystems: Linking Theory to Observation. *Trends in Ecology and Evolution*, Vol.18, No.12, (December 2003), pp. 648-656, ISSN 0169-5347

Scheffer, M., Carpenter, S. R., Foley, J. A., Folke, C. & Walker, B. (2001). Catastrophic Shifts in Ecosystems. *Nature*, Vol.413, (October 2011), pp. 591-596, ISSN 0028-0836

Scheffer, M., Carpenter S. R. & de Young, B. (2005). Cascading Effects of Overfishing Marine Systems. *Trends in Ecology and Evolution*, Vol.20, No.11, (November 2005), pp. 579-581, ISSN 0169-5347

Shepherd, T. D. & Myers R. A. (2005). Direct and Indirect Fishery Effects on Small Coastal Elasmobranchs in the Northern Gulf of Mexico. *Ecology Letters*, Vol.8, No.10, (October 2005), pp. 1095-1104, ISSN 1461-0248

Solé, R. & Montoya J. M. (2001). Complexity and Fragility in Ecological Networks. *Proceedings the Royal Society London B*, Vol.268, No.1480, (October 2001), pp. 2039-2045, ISSN 1471-2954

Springer, A. M., Estes, J. A., van Vliet, G. B., Williams, T. M., Doak, D. F., Danner, E. M., et al. (2003). Sequential Megafaunal Collapse in the North Pacific Ocean: an Ongoing Legacy of Industrial Whaling?. *Proceedings of the National Academy of Sciences of the United States of America*, Vol.100, No.21, (October 2003), pp. 12223–12228, ISSN 1091-6490

Sandström, O., Larsson, Å., Andersson, J., Appelberg, M., Bignert, A., Helene, E. K., et al. (2005). Three Decades of Swedish Experience Demonstrates the Need for Integrated Long-Term Monitoring of Fish in Marine Coastal Areas. *Water Quality Research Journal of Canada*, Vol.40, No.3, (July 2005), pp. 233-250, ISSN 1201-3080

Steneck, R. S., Graham, M. H., Bourque, B. J., Corbett, D., Erlandson, J. M., Estes, J. A. & Tegner, M. J. (2002). Kelp Forest Ecosystem: Biodiversity, Stability, Resilience and their Future. *Environmental Conservation*, Vol.29, No.4, (December 2002), pp. 436–459, ISSN 0376-8929.

Steneck, R. S., Vavrinec, J. &. Leland, A. V. (2004). Accelerating Trophic Level Dysfunction in Kelp Forest Ecosystems of the Western North Atlantic. *Ecosystems*, Vol.7, No.4, (June 2004), pp. 23–331. ISSN 1432-9840

Stergiou, K. I. (2005). Fisheries Impact on Trophic Levels: Long-Term Trends In Hellenic Waters, In: *State of the Hellenic Marine Environment*, Papathanassiou, E. & Zenetos, A. (Eds.), 326-329, Hellenic Centre for Marine Research, ISBN 9608665183, Athens, Greece

Stevens, J. D., Bonfil, R., Dulvy N. K. & Walker, P. A. (2000). The Effects of Fishing on Sharks, Rays, and Chimaeras (Chondrichthyans), and the Implications for Marine Ecosystem. *ICES Journal of Marine Science*, Vol.57, No.3, (June 2000), pp. 476–494, ISSN 1095-9289

Tegner, M. J. & Dayton, P. K. (2000). Ecosystem Effects of Fishing in Kelp Forests Communities. *ICES Journal of Marine Science*, Vol.57, No.3, (June 2000), pp. 579-589, ISSN 1095-9289

Valtysson, H. & Pauly, D. (2003). Fishing Down the Food Web: An Icelandic Case Study. In: *Competitiveness Within the Global Fisheries*, Guðmundsson E. & Valtysson H. (Eds.), pp. 12-24, University of Akureyri, ISBN 9979-834-26-9, Akureyri, Iceland

Vasas, V., Lancelot, C., Rousseau, V. & Jordán F. (2007). Eutrophication and Overfishing in Temperate Nearshore Pelagic Food Webs: A Network Perspective. *Marine Ecology Progress Series*, Vol.336, (April 2007), pp. 1-14, ISSN 0171-8630

Ward, P. & Myers R. A. (2005). Shifts in Open-Ocean Fish Communities Coinciding with the Commencement of Commercial Fishing. *Ecology*, Vol.86, No.4, (April 2005), pp. 835–847, ISSN 0012-9658

Wassermann, S. & Faust, K. (1994). *Social Network Analysis*, Cambridge University Press, ISBN 978-0521387071 Cambridge, UK

Williams, R. J., Berlow, E. L., Dunne, J. A., Barabasi, A. L. & Martínez, N. (2002). Two Degrees of Separation in Complex Food Webs. *Proceedings of the National Academy of Sciences of the United States of America*, Vol.99, No.20, (October 2002), pp. 12913-12916, ISSN 1091-6490

Willemse, N. & Pauly, D. (2004). Reconstruction and Interpretation of Marine Fisheries Catches from Namibian Waters, 1950 to 2000. In: *Namibia's Fisheries. Ecological, Economic and Social Aspects*, Sumaila, U. R., Boyer, D., Skogen, M. D. & Steinshamn, S. I. (Eds.), 99-112, The University of Chicago Press, ISBN 9789059720176, Chicago, USA

Wootton, J. T. (1994). The Nature and Consequences of Indirect Effects in Ecological Communities. *Annual Review of Ecology and Systematics*, Vol. 25, (November 1994), pp. 443-466, ISSN 0066-4162

Worm, B. & Myers R. A. (2003). Meta-Analysis of Cod-Shrimp Interactions Reveals Top-Down Control in Oceanic Food Webs. *Ecology*, Vol.84, No.1, (January 2003), pp. 162-173, ISSN 0012-9658

Worm, B., Lotze H. K. & Myers, R. A. (2005). Ecosystem Effects of Fishing and Whaling in the North Pacific and Atlantic Oceans, In: *Whales, Whaling and Ocean Ecosystems*, Estes, J. A., Brownell, R. L., de Master, D. P., Doak D. F. & Williams, T. M. (Eds.), 333-341, University of California Press, ISBN 9780520248847, Berkeley, USA.

Worm, B., Barbier, E. B., Beaumont, N., Duffy, J. E., Folke, C., Halpern, B. S., et al. (2006). Impacts on Biodiversity Loss on Ocean Ecosystem Services. *Science*, Vol.314, No.5800, (November 2006), pp. 787–790, ISSN 1095-9289

Yemane, D., Field, J., Leslie, G. & Rob, W. (2005). Exploring the Effects of Fishing on Fish Assemblages Using Abundance Biomass Comparison (ABC) Curves. *ICES Journal of Marine Science*, Vol.62, No.3, (June 2005), pp. 374-379, ISSN 1095-9289

Zwanenburg, K. C. T. (2000). The Effects of Fishing on Demersal Fish Communities of the Scotian Shelf. *ICES Journal of Marine Science*, Vol.57, No.3, (June 2000), pp. 503-509, ISSN 1095-9289

19

Diversity and Dynamics of Plant Communities in Niger River Valley (W Regional Park)

A. Mahamane[1,2], M. Zaman Allah[2], M. Saadou[1,2] and J. Lejoly[3]
[1]*Département de Biologie, Faculté des Sciences,*
Université Abdou Moumouni de Niamey, Niamey
[2]*Université de Maradi Maradi*
[3]*Laboratoire de Botanique Systématique et de Phytosociologie CP 169,*
Bruxelles, Université Libre de Bruxelles
[1,2]*Niger*
[3]*Belgium*

1. Introduction

The "W" Regional Park covers an area adjacent to the border of Benin, Burkina Faso and Niger (Fig. 1). The park hosts diverse flora and fauna. The large part of plant communities of this area remain poorly understood and only few reports on this subject are available (Garba, 1985; Boudouresque, 1995 and Couteron et al. 1992).

The objective of this work was to characterize plant communities along the Niger River bank during the flooding period and the dry season. These two periods play important roles in the ecosystem dynamics.

2. Materials and methods

The "W" Regional Park is located in the West African north-sudanian zone (Fig. 1; White, 1983) and covers an area of 1.024.280 ha. The average annual rainfall is 704.7 ± 180 mm with an average temperature of 37 ° C (Fig. 2). This park includes several Precambrian geological structures. In the river valley, the soil is of clayey gley/pseudo-gley type. The banks of this river host a "special vegetation" called "bourgou" by the local population in reference to "bourgoutiere" commonly used in the literature for this type of vegetation (Dulieu, 1989). The phytosociological investigation was conducted in several sites located in this area from 2002 to 2003 during the flooding period and the dry season. Each releve included the complete list of species with their abundance-dominance coefficients (Braun Blanquet, 1932).

Data analysis was performed with Canoco software (ter Braak and Smilauer. 1998). For the different plant communities that were studied, the specific diversity indices of Shannon and Weaver (1949 in Legendre and Legendre, 1998) and Pielou equitability index were calculated.

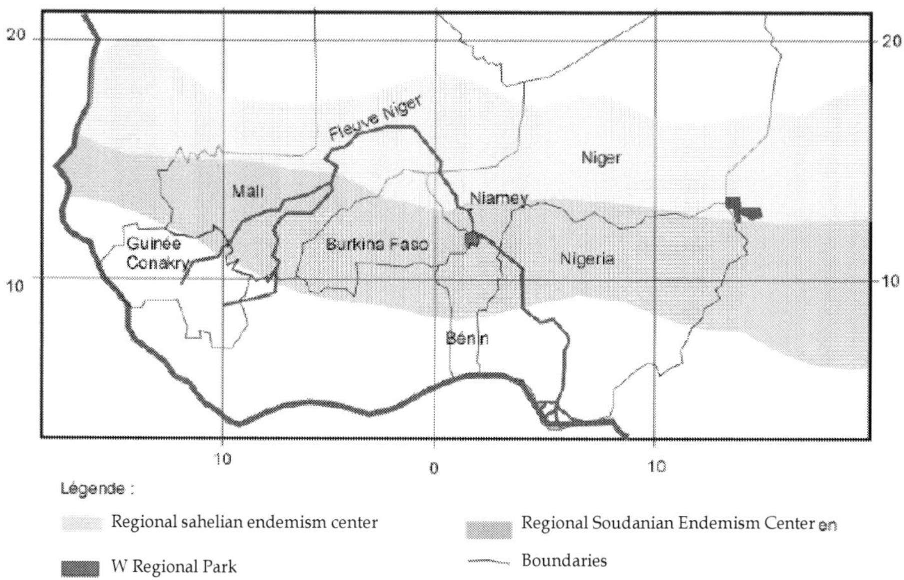

Fig. 1. "W" national park in West Africa

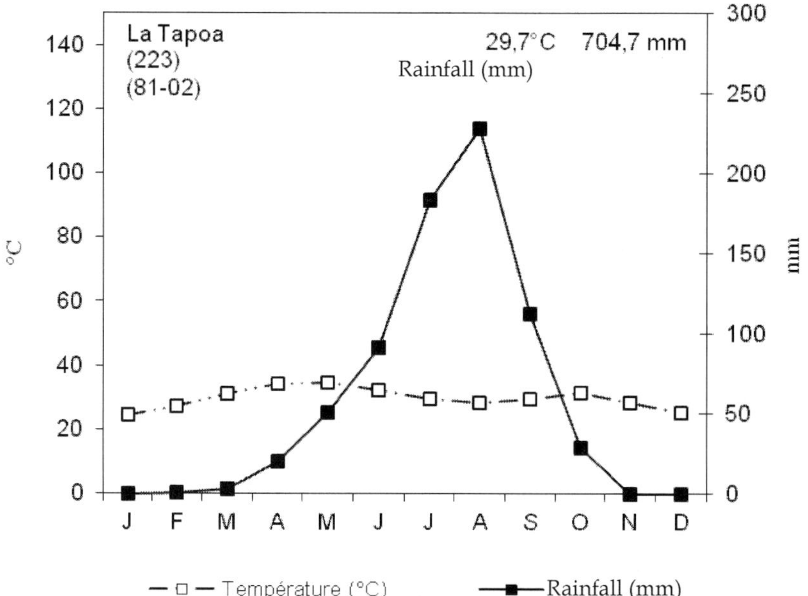

Fig. 2. Ombrothermic curve of Tapoa station (Niger)

The nomenclature after identification of species is referred to in Lebrun and Storck (1991-1997). The reported species were represented by *herbarium specimens* available in the herbaria of the UAM and ULB (BRLU).

3. Results

Detrended Correspondance Analysis was performed on a matrix of 42 relevés and 116 species. The first Axis reflected a gradient of water depth (Fig. 3). Near the origin of this axis were located the groups of plants adapted to deep water conditions occurring during flooding period while the dry season plant groups were positioned on its positive side.

The syntaxons described were:

- *Polygono senegalensis Echinochloetum colonae* ass. *nova*,
- *Eichhornietum crassipedis Vanderlyst* 1931,
- *Leptochloo coerulescentis Stachytarphetetum angustifoliae* ass. *nova*,
- *Cyperetum maculati* Mandango 1982.

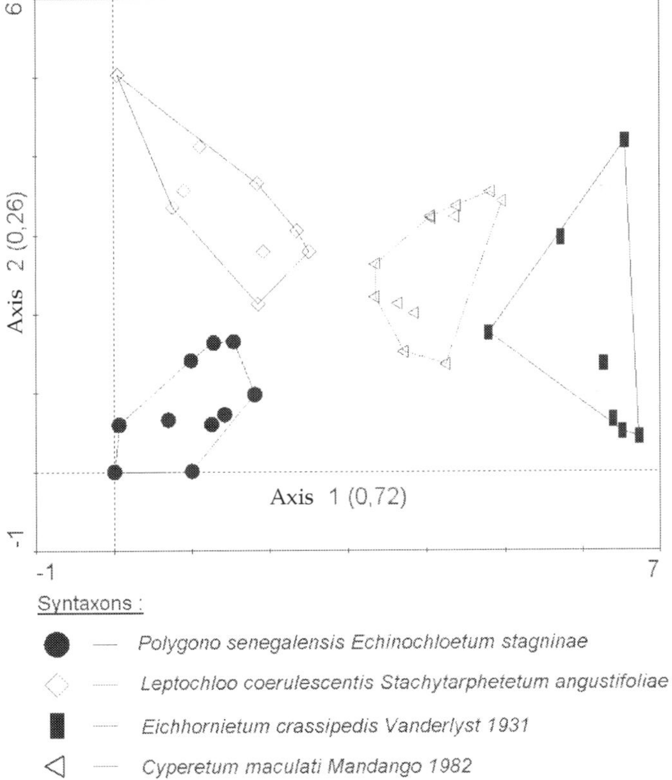

Fig. 3. Plant communities' classification in the banks of Niger River valley

3.1 *Polygono senegalensis Echinochloetum stagninae* ass. nova

Polygono Echinochloetum stagninae was defined by 13 releves and 22 species of which four were specific to this association: *Echinochloa stagnina, Polygonum senegalensis, Lemnapaucicostata* and *Azolla pinnata* (Table 1). Water depth may exceed 2 m. The pH was neutral and close to 7. The distribution of the biological types showed the predominance of therophytes (33.3%) followed by hydrophytes (28.6%) and phanerophytes (28.6%). Regarding the phytogeographical units, results showed the dominance of species with paleotropical distribution (33.33%) followed by pantropical species (28.57% and Sudan-Zambezian species (23.81%). The number of species per releve varied from 2 to 17 with an average of 4.19 ± 1.86. The Shannon diversity index was 2.69 and the maximum diversity index up to 4.64 while the Pielou equitability was 0.57.

A

	1	2	3	4	5	6	7	8	9	10	11	12	13	AL/C	CP
N°	1	2	3	4	5	6	7	8	9	10	11	12	13		
N° Author	577	474	562	563	565	566	567	390	392	26	27	28	29		
Surface (m²)	10	15	15	20	20	20	10	10	15	15	15	10	10		
Land cover (%)	15	24	78	42	7	77	68	86	44	53	9	77	74	AL/C	CP
Polygono Echinochloetum stagninae ass. nova.															
Polygonum senegalense	-	-	3	1	+	+	2	2	+	3	+	4	-	14,3	III
Echinochloa stagnina	+	+	3	3	1	4	3	4	3	2	+	+	4	29,3	V
Azolla pinnata	-	-	-	-	-	-	-	-	-	-	-	+	-	0,2	I
Lemna paucicostata	-	-	-	-	-	-	-	-	-	-	-	-	+	0,2	I
Oryza sativa L.	-	-	-	-	-	-	-	-	-	-	+	-	-	0,2	I
Eichhornietum crassipedis Vanderlyst 1931															
Eichhornia crassipes (Mart.) Solms Laub.	+	2	+	+	+	+	2	+	-	-	-	-	-	3,7	III
Oryza longistaminata	-	-	-	-	-	-	-	-	-	-	-	-	+	0,2	I
Cyperus dilatatus Schum. & Thonn.	-	+	-	-	-	-	-	-	-	-	-	-	-	0,2	I
Ceratopteris cornuta	-	-	-	-	-	-	-	-	-	-	-	-	-		
Saciolepis africana	-	-	-	-	-	-	-	-	-	-	-	-	-		
Echinochloa crus - pavonis	-	-	-	-	-	-	-	-	-	-	-	-	-		
Ludwidgia adscendens	-	-	-	-	-	-	-	-	-	+	-	-	-	0,2	I
Saciolepis ciliocincta	-	+	-	-	-	-	-	-	-	-	-	-	-	0,2	I
Echinochloa obtusifolia	-	-	-	-	-	-	-	-	-	-	-	-	-		
Nymphaea lotus	-	-	-	-	-	-	-	-	-	-	-	-	-		

B

	14	15	16	17	18	19	20	21	22	23	AL/C	CP
N°	14	15	16	17	18	19	20	21	22	23		
N° Author	479	557	476	558	30	31	571	439	576	551		
Surface (m²)	10	20	10	10	15	10	10	10	10	10		
Land cover (%)	30	10	63	59	57	54	36	30	47	23	AL/C	CP
Polygono Echinochloetum stagninae ass. nova.												
Polygonum senegalense	-	-	-	-	-	+	+	+	+	-	1,2	III
Echinochloa stagnina	-	-	-	-	-	2	-	-	-	-	1,5	I
Azolla pinnata	-	-	-	-	-	-	-	-	-	-		
Lemna paucicostata	-	-	-	-	-	-	-	-	-	-		
Oryza sativa L.	-	-	-	-	-	-	-	-	-	-		
Eichhornietum crassipedis Vanderlyst 1931												
Eichhornia crassipes (Mart.) Solms Laub.	+	+	3	+	1	2	2	+	+	2	9,9	V
Oryza longistaminata	-	-	-	-	3	-	2	+	-	-	5,6	II
Cyperus dilatatus Schum. & Thonn.	+	+	+	+	-	-	-	-	-	-	1,2	III
Ceratopteris cornuta	+	+	+	-	-	-	-	-	-	-	0,9	II
Saciolepis africana	-	-	-	-	+	+	-	-	-	-	0,6	II
Echinochloa crus - pavonis	-	-	-	-	-	-	+	-	+	-	0,6	II
Ludwidgia adscendens	-	-	-	-	-	+	-	-	-	-	0,3	I
Saciolepis ciliocincta	-	-	-	-	-	-	-	+	-	-	0,3	I
Echinochloa obtusifolia	2	-	-	-	-	-	-	-	-	-	1,5	I
Nymphaea lotus	-	-	-	-	-	2	-	-	-	-	1,5	I

Other species — A

Species	1	2	3	4	5	6	7	8	9	10	11	12	13	ALC	PC
Pterocarpus santalinoides DC.	-	-	-	-	-	-	-	-	-	-	-	-	-		
Acacacia ataxacantha	-	-	-	-	-	-	-	-	-	-	-	-	-		
Albizia zygia	-	-	-	-	-	-	-	-	-	-	-	-	-		
Sesbania sesban	-	-	-	-	-	-	-	-	-	-	-	+	-	0,2	I
Mitragyna inermis	-	-	-	-	-	-	-	-	-	-	-	-	-		
Dichrostachys cinerea	-	-	-	-	-	-	-	-	-	-	-	-	-		
Ipomomoea rubens	+	-	-	-	-	-	-	-	-	-	-	-	-	0,2	I
Mimosa pigra	+	-	-	-	-	-	-	-	-	-	-	-	-	0,2	I
Paspalum scrobicularum	-	-	-	-	-	-	-	-	-	-	-	-	-		
Phaseolus lunatus	-	-	-	-	-	-	-	-	-	-	-	-	-		
Luffa cylindrica	-	-	-	-	-	-	-	-	-	-	-	-	-		
Vetiveria nigritana	-	-	-	-	-	-	-	-	-	-	-	-	-		
Phyllanthus reticulathus	+	-	-	-	-	-	-	-	-	-	-	-	-	0,2	I
Taccazea apiculata	-	-	-	-	-	-	-	-	-	-	-	-	-		
Cynodon dactylon (L.) Pers.	-	-	-	-	-	-	-	-	-	-	-	-	-		
Merremia hederacea	-	-	-	-	-	-	-	-	-	-	-	-	-		
Ipomoea blepharophylla	-	-	-	-	-	-	-	-	-	-	-	-	-		
Dorstenia sp.	-	-	-	-	-	-	-	-	-	-	-	-	-		
Cerathophyllum demersum	-	-	-	-	-	-	-	-	-	-	-	-	-		
Utricularia stellaris L.f.	-	-	-	-	-	+	-	-	-	-	-	-	-	0,2	I
Ipomoea aquatica	-	-	-	-	-	-	-	-	-	-	-	-	-		

Other species — B

Species	1	2	3	4	5	6	7	8	9	10	ALC	PC
Pterocarpus santalinoides DC.	-	-	+	-	-	-	-	-	-	-	0,3	I
Acacacia ataxacantha	-	-	-	3	-	-	-	-	-	-	3,8	I
Albizia zygia	+	-	-	-	-	-	-	-	-	-	0,3	I
Sesbania sesban	-	-	-	-	-	-	-	-	-	-		
Mitragyna inermis	-	-	+	-	-	-	-	-	-	-	0,3	I
Dichrostachys cinerea	-	-	-	-	-	-	-	+	-	-	0,3	I
Ipomomoea rubens	-	1	+	+	+	-	-	+	-	1	1,4	IV
Mimosa pigra	-	-	1	-	-	-	-	+	3	+	4,5	III
Paspalum scrobicularum	-	-	+	-	-	-	-	-	-	-	0,3	I
Phaseolus lunatus	-	-	-	+	-	-	-	-	-	+	0,6	II
Luffa cylindrica	-	-	-	-	-	-	-	+	-	-	0,3	I
Vetiveria nigritana	-	-	-	-	+	-	-	-	-	-	0,3	I
Phyllanthus reticulathus	-	-	+	-	-	-	-	-	-	-	0,3	I
Taccazea apiculata	-	-	-	-	+	-	-	-	-	-	0,3	I
Cynodon dactylon (L.) Pers.	-	-	-	+	-	-	-	-	-	1	0,4	II
Merremia hederacea	+	-	-	-	-	-	+	-	-	-	0,6	II
Ipomoea blepharophylla	-	-	-	+	-	-	-	-	-	-	0,3	I
Dorstenia sp.	-	-	-	+	-	-	-	-	-	-	0,3	I
Cerathophyllum demersum	-	-	-	-	-	-	-	+	-	-	0,3	I
Utricularia stellaris L.f.	-	-	-	-	-	-	-	-	-	-		
Ipomoea aquatica	-	-	-	+	-	-	-	-	-	-	0,3	I

ALC: Average Land Cover, PC: Presence Coefficient

Table 1. *Polygono Echinochloetum stagninae* ass. nov. & *Eichhornietum crassipedis*.

3.2 *Eichhornietum crassipedis* Vanderlyst 1931

Eichhornietum crassipedis Vanderlyst 1931 was distributed along the linear fringe of the Niger River bank. Water depth was quite similar as for *Polygono senegalensis Echinochloetum stagninae*, since they colonized the same stations.

The distribution of biological types revealed the predominance of hydrophytes (40%), followed by therophytes (20%) and microphanerophytes (20%). The helophytes represented only 10%. Regarding the phytogeographical distribution results showed the dominance of cosmopolitan species. The *Eichhornietum crassipedis* association consisted of 10 releves and 17 species. The average number of species per releve was 4. The Shannon diversity index was 2.34 and Pielou equitability 0.57. These values indicated a very small number of dominant species within the plant community.

3.3 *Leptochloo Stachytarphetetum angustifoliae* ass. nova association

Leptochloo coerulescentis Stachytarphetetum angustifoliae developed in the late dry season and beginning of the rainy season on the banks of the river that were sufficiently dewatered to allow the development of an herbaceous layer. It corresponded to a more or less continuous linear strip along the banks that were battered by the waves.

This syntaxon was defined by 10 releves and 34 species of those three were specific to the association: *Stachytarpheta angustifolia, Leptochloa coerulescens, Cardiospermum halicacabum* (Table 2). The raw distribution of biological types showed that helophytes were dominant (58% -73%), followed by therophytes (25%) and hydrophytes (22%) in the pondered distribution. Species of Sudanian distribution represented only 8% of the spectrum. As for the weighted spectrum, it was largely dominated by the species of Sudanese-Zambezian distribution (59.4%), species of Sudanian (9%) and Afrotropical species distribution (8.79%). The number of species per survey varied from 3 to 13 with an average of 7.6 ± 3.2. The Shannon diversity index was 3.6 and the maximum diversity index 4.9. The equitability index of Pielou was 0.74. These results suggest an equal distribution and overlap between species.

3.4 *Cyperetum maculati* Mandango 1982 association

Cyperetum maculati grows in the dry season on the sandbanks of the river's main channel. *Cyperus maculatus* form clumps of variable size. It is a rhizomatous species which is completely submerged during flooding periods. During this period of prolonged immersion, the plants were represented by perennial rhizomes. The floristic association has many annuals germinating on wet sand (Table 2). The association was defined by 10 releves and 39 species of which seven were specific to this association: *Cyperus maculatus, Cleome viscosa, Glinus lotoides, Glinus oppositifolius, Cassia occidentalis, Trianthema portulacastrum and Bergia suffruticosa.*

Cyperetum maculati Mandago 1982 was represented by open herbaceous vegetation in dense clumps. The biological types distribution was dominated by therophytes and hydrophytes respectively 43.8% and 28.1%. The weighted distribution was represented by therophytes and phanerophytes with respectively 37.8% and 36.3% followed by hydrophytes (24.8%). The raw phytogeographical units distrbution was dominated by paleotropical species (30.3%), followed by species of Sudanese-Zambezian distribution (27.27%) and pantropical species (24.24%). Other types showed low phytogeographic values. For this group, the number of species per survey varied from 5 to 18 with an average of 7.3 ± 4.42. The Shannon diversity index was 3.14 with a maximum diversity index of 5.24. Pielou equitability index value was 0.64. These results support the conclusion that recovery is evenly distributed between species.

	C												D										
N°	24	25	26	27	28	29	30	31	32	33	ALC	CP	34	35	36	37	38	39	40	41	42	ALC	CP
N° auteurs	306	307	308	312	274	275	345	346	337	342			343	344	329	332	333	318	328	340	322		
Area (m²)	10	10	10	10	10	10	10	10	10	10			10	10	10	10	10	10	10	10	10		
Land cover (%)	27	89	56	41	100	77	65	24	83	25			100	27	100	100	100	100	19	59,5	42		
Cyperetum maculati Mandango 1982																							
Cyperus maculatus Böck.	-	+	+	3	4	3	-	-	-	+	15	IV	-	-	-	-	-	-	-	-	+	0,3	I
Glinus lotoides L.	+	+	-	-	+	-	+	+	-	1	2	IV	-	-	-	-	-	-	1	-	2	1,8	II
Mollugo nudicaulis	+	4	3	+	-	-	-	-	-	-	11	III	-	-	-	-	-	-	-	-	-		
Glinus oppositifolius (L.) A. DC.	+	-	-	-	-	-	+	+	-	+	1	III	-	-	-	-	-	-	-	-	+	0,3	I
Cleome viscosa L.	2	2	2	-	-	-	-	-	-	-	5	II	-	-	-	-	-	-	-	-	-		
Trianthema portulacastrum (L.) L.	-	-	-	-	+	-	-	-	+	-	1	II	-	-	-	-	-	-	-	-	-		
Bergia suffruticosa (Del.) Fenzl.	+	-	-	-	+	-	-	-	+	-	1	II	-	-	-	-	-	-	-	-	-		
Cassia occidentalis L.	-	-	-	-	+	-	-	-	4	-	7	II	-	-	-	-	-	-	-	-	-		
Heliotropium indicum L.	-	-	-	-	-	-	+	+	-	+	1	II	-	-	-	-	-	+	-	-	+	0,7	II
Leptochloo Stachytarphetetum angustifoliae ass. nova																							
Stachytarpheta angustifolia Mold.	-	-	-	-	-	-	-	-	-	-			3	+	3	3	3	2	-	1	-	19	IV
Leptochloa coerulescens Steud.	-	-	-	-	-	-	-	-	+	+	1	II	-	-	3	3	3	+	-	3	-	17	III
Albizia zygia (DC.) J. F. Mocer.	-	-	-	-	-	-	-	-	-	-			+	2	-	-	+	2	-	-	-	4	III
Cardiospermum halicacabum L.	-	-	-	-	-	-	-	-	-	-			-	-	-	2	-	3	-	-	-	5,8	II
Coldenia procumbens L.	-	-	-	-	+	-	-	-	-	-		I	3	-	-	-	-	-	-	-	-	4,2	I
Hyparrhenia involucrata Stapf. var involucrata	-	-	-	-	-	-	-	-	-	-			-	-	-	-	-	-	-	-	+	0,3	I
Eragrostis atrovirens (Desf.) Steud.	-	-	-	-	-	-	-	-	-	-			-	-	-	-	-	-	-	-	+	0,3	I

Other species	C1	C2	C3	C4	C5	C6	C7	C8	C9	C10	ALC	CP	D1	D2	D3	D4	D5	D6	D7	D8	D9	ALC	CP
Morelia senegalensis A. Rich.	-	-	-	-	-	-	-	-	-	-			-	+	3	3	3	2	-	-	-	15	III
Pterocarpus santalinoides DC.	-	-	-	-	-	-	-	-	-	-			-	-	-	3	-	-	-	-	-	4,2	I
Vitex chrysocarpa Planch. ex Benth.	-	-	-	-	-	-	+	-	-	-		I	-	-	+	+	-	-	-	-	-	0,7	II
Flueggea virosa (Rxb. ex. Willd.) Voigt	-	-	-	-	-	-	-	-	-	-			-	-	+	-	-	-	-	-	-	0,3	I
Cola laurifolia Mast.	-	-	-	-	-	-	+	-	-	-		I	-	-	+	-	-	-	-	-	-	0,3	I
Taccazea apiculata Oliv.	-	-	-	-	-	+	-	-	-	-		I	-	-	+	+	-	+	-	-	-	1	II
Acacia ataxacantha DC.	-	-	-	-	-	-	-	-	-	-			+	-	2	+	+	+	-	-	+	3,3	IV
Merremia hederacea Burm. f.	-	-	-	-	+	-	-	-	-	-		I	-	-	-	+	-	-	-	-	-	0,3	I
Diospyros mespiliformis Hochst. ex. A. DC.	-	-	-	-	-	-	-	-	-	-			+	+	-	-	-	+	-	-	-	1	II
Celtis toka (Forssk.) Hepper et Wood.	-	-	-	-	-	-	-	-	-	-			+	-	-	-	-	-	-	-	-	0,3	I
Hyptis spicigera Lam.	-	-	-	-	-	-	+	-	-	-		I	-	-	-	-	-	-	-	-	+	0,3	I
Corchorus tridens L.	-	-	-	-	+	-	-	-	+	-	1	II	-	-	-	-	-	-	-	-	-		
Tamarindus indica L	-	-	-	-	-	-	-	-	-	-			-	+	-	-	-	-	-	-	-	0,3	I
Mitragyna inermis	-	-	-	-	-	-	-	-	-	-			3	-	2	-	-	-	-	-	-	5,8	II
Ipomomoea rubens	-	-	-	-	-	-	+	-	-	-		I	-	-	-	-	-	-	-	+	-	0,3	I
Mimosa pigra	-	-	-	-	-	-	+	+	-	-	1	II	-	-	-	+	3	+	-	-	-	4,8	II
Cynodon dactylon (L.) Pers.	-	-	-	-	-	+	-	+	-	+	1	II	-	-	2	+	-	-	2	2	+	5,7	III
Polygonum senegalense	-	-	-	-	-	2	+	+	-	-	2	II	-	-	-	-	-	-	+	+	+	1	II
Eichhornia crassipes (Mart.) Solms Laub.	-	-	-	-	-	+	-	-	-	-		I	-	-	-	-	-	-	-	-	-		

Average Land Cover, CP: Coefficient de présence

Table 2. *Cyperetum maculati Mandango 1982 & Leptochloo- Stachytarphetetum angustifoliae ass. nov*

3.5 Synsystematique

class	Order	Alliance	Associations
Phragmitet ea Tüxen & Preising 1942,	*Papyretali a* Lebrun 1947	*Echinochloion crusis- pavonis* Léonard 1950	*Leptochloo- Stachytarphetetosu m angustifoliae* ass. nov.
		Jussieuion Léonard 1950	*Polygono senegalense Echinochloetum stagninea* ass. nov.
Potametea pectinati Tüxen & Preising 1942	*Nymphae etalia loti* Lebrun 1947	*Nymphaeion micranthae* E. Boud. 1995	*Eichhornietum crassipedis* Vanderlyst 1931
Ruderali- manihotete a (Léonard in Taton 1949) Schmitz 1988	*Amaranth o- Ecliptetali a* Schmitz 1971	*Ecliption albae* Lebrun 1947	*Cyperetum maculati* Mandango 1982

4. Discussion

The initial phase of the bourgoutiere included Leptochloa coerulescens, Echinochloa stagnina, Echinochloa pyramidalis, Cyperus cylindrostachyus, Saciolepis africana and Stachytarpheta angustifolia. These plant species were progressively established in the late dry season and early rainy season. Leptochloa coerulescens bear fruits during this period. This phase corresponded to the ecological amplitude of Leptochloo-Stachytarphetetum angustifoliae. With the increase in water level, Polygonum senegalense and Echinochloa stagnina actively developed by vegetative propagation.

The optimal phase or aquatic prairie of bourgoutiere corresponded to Polygono *Echinochloetum stagninae*. The group covered a broad variable on the shores of the river. During the dry season, *Echinochloa stagnina* as *Polygonum senegalense* fall on the dewatered river banks. At that time, both species have spread their seeds. In the riverbed, Cypretum maculata Mandango 1982 was subject to strong variations in relation to alternating periods of flood and dry period. During the dry period characterized with a low water flow in the river, the species happened to complete its cycle. The start of this cycle, as and when the water recedes, is characterized by buds on the stolons. The flooding period was characterized by a progressive invasion by water hyacinth and resulted in the formation of *Eichhornietum crassipedis*. This determined syntaxon pollution of aquatic stressed environments (Brendonck et al. (2003)).

These stations were characterized by variability of plant communities determined by the river water regime (Duvigneaud, 1946). Indeed, this group was not identified by Atta and DanJimo (2003) in the same river valley during the dry season. Gradually, as the water level dropped, hydrophytes population declined and progressively replaced by therophytes. This resulted in a decrease of the number of species.

5. Conclusion

Bougoutières vegetation plays an important role in the ecosystem of the W regional park. Apart from *Eichhornietum crassipedis*, the different syntaxons provide forage and habitat for wildlife. These syntaxons were characterized by low diversity indices in relation to dominance effects between species and disturbance (fire bank, harvested biomass) related to human activities.

6. Acknowledgments

The authors thank the Department of Wildlife Fisheries and Fish the Parc du W du Niger. They also thank for facilitating access to the Free University of Brussels, the University of Niamey, the Project ECOPAS and the African Academy for Science for their permanent support during the different phases of this work. They also thank Mr L. Pauwels the Botanical Garden of Meise, Prof. B. Sinsin Abomey Calavi University of Benin and Prof.. van der Maesen of the National Herbarium of Wageningen who have made valuable contributions in the identification of species.

7. List of species

Species	Family
Abrus precatorius L.	Fabaceae
Acacia ataxacantha DC.	Mimosaceae
Albizia zygia (DC.) J.F. Macbr.	Mimosaceae
Azolla pinnata R. Brown var pinnata	Azollaceae
Bergia suffruticosa (Del.) Fenzl	Elatinaceae
Caperonia fistulosa Beille	Euphorbiaceae
Cardiospermum halicacabum L.	Sapindaceae
Cassia occidentalis L.	Caesalpiniaceae
Celtis toka	(Forssk.) Hepper & Wood	Ulmaceae
Ceratophyllum demersum L.	Ceratophyllaceae
Ceratopteris cornuta	Adianthaceae
Cleome viscosa L.	Capparaceae
Cola laurifolia Mast.	Sterculiaceae
Coldenia procumbens L.	Boraginaceae
Commelina benghalensis L.	Commelinaceae
Corchorus tridens L.	Tiliaceae
Cynodon dactylon (L.) Pers.	Poaceae
Cyperus dilatatus Schum. & Thonn.	Cyperaceae
Cyperus maculatus Boeck.	Cyperaceae

Dactyloctenium aegyptium (L.) P. Beauv.	Poaceae
Dichrostachys cinerea (L.) Wight & Arn.	Mimosaceae
Diospyros mespiliformis Hochst. ex A. DC.	Ebenaceae
Eichhornia crassipes (Mart.) Solms	
Echinochloa crus-pavonis (Kunth) Schult.	Poaceae
Echinochloa obtusiflora Stapf	Poaceae
Echinochloa stagnina (Retz.) P. Beauv.	Poaceae
Eichhornia natans (P. Beauv.) Solms-Laub.	Pontederiaceae
Eleusine indica (L.) Gaertn.	Poaceae
Eragrostis atrovirens (Desf.) Trin. ex Steud.	Poaceae
Eragrostis tremula Hochst. ex Steud.	Poaceae
Fluegga virosa (Roxb. ex Willd.) Voigt	Euphorbiaceae
Glinus lotoides L.	Aizoaceae
Glinus oppositifolius (L.) DC.	Aizoaceae
Heliotropium indicum L.	Boraginaceae
Hyparrhenia involucrata Stapf	Poaceae
Hyptis spicigera Lam.	Lamiaceae
Indigofera hirsuta L.	Fabaceae
Ipomoea aquatica Forssk.	Convolvulaceae
Ipomoea blepharophylla Hall. f.	Convolvulaceae
Ipomoea rubens Choisy	Convolvulaceae
Lemna paucicostata Hegelm. ex Engelm.	Lemnaceae
Leptochloa caerulescens Steud.	Poaceae
Cleome viscosa L.	Capparaceae
Ludwigia octovalvis (Jacq.) Raven	Onagraceae
Luffa cylindrica (L.) M.J. Roem.	Cucurbitaceae
Merremia hederacea (Burm. f.) Hallier f.	Convolvulaceae
Mimosa pigra L.	Mimosaceae
Mitragyna inermis (Willd.) O. Ktze.	Rubiaceae
Mollugo nudicaulis Lam.	Molluginaceae
Morelia senegalensis A. Rich. ex DC.	Rubiaceae
Nymphaea lotus L.	Nympheaceae
Oryza longistaminata A. Chev. & Roehr.	Poaceae
Oryza sativa L.	Poaceae
Paspalum scrobiculatum L.	Poaceae
Phaseolus lunatus L	Fabaceae
Phyllanthus reticulatus Poir.	Euphorbiaceae
Polygonum senegalense Meisn.	Polygonaceae
Pterocarpus santalinoides DC.	Fabaceae
Sacciolepis africana C.E. Hubbard & Snowden	Poaceae
Sacciolepis ciliocincta (Pilger.) Stapf.	Poaceae
Sesbania leptocarpa DC.	Fabaceae
Sesbania sesban (L.) Merr.	Fabaceae

Stachytarpheta angustifolia (Mill.) Vahl	Verbenaceae
Tacazzea apiculata Oliv.	Asclepiadaceae
Tamarindus indica L.	Caesalpiniaceae
Trianthema portulacastrum L.	Aizoaceae
Utricularia stellaris L.f.	Lentibulariaceae
Vetiveria nigritana (Benth.) Stapf	Poaceae

8. References

Atta S. and Danjimo B., 2002. - Study of the vegetation of the river Niger, in the zone of influence of the W National Park of Niger. Regional Park Programme - W (ECOPAS) 7 ACP RPR 742. Research mission: 28 + appendices

Braun-Blanquet J., 1932. Plant sociology. The study of plant Communities. Ed McGray Hill, New York, London.: 439 p.

Brendonck L, Maes J, Rommens W, Dekeza N, T Nhiwatiwa, Barson M, Callebaut V, Phiri C, Moreau K, Gratwick B, Stevens M, N Alyn, Holsters E, F Ollevier, Marshall B, 2003. - The impact of water hyacinth (Eichhornia crassipes) in a eutrophic subtropical impoundment (Lake Chivero, Zimbabwe). II. Species diversity. Hydrobiology 158 (3): 389-405.

Dulieu D., 1989. - The bourgoutières Niger River: constraints, opportunities. Republic of Niger, thematic study. Republic of Niger Ministry of Animal Resources and Water / Republic French Ministry of Cooperation and Development. Institute of Animal Husbandry and Veterinary Medicine in Tropical Countries, Department of CIRAD, 77P.

Duvigneaud P., 1946. - The variability of plant associations. Bull. Soc. Roy. Bot. Belg., 78: 107-134.

Garba M., 1984. - Contribution to the study of flora and vegetation of aquatic and hydromorphic soils of western Niger Republic, longitude Dogondoutchi of the Niger River. PhD Thesis 3rd cycle, University of Niamey and University of Bordeaux II, 149p.

J. Lebrun P. A. Stock and L., 1991-1999. - Listing of flowering plants of tropical Africa. Editions of Conser. and Jard. Bot. Geneva, 4 volumes.

Legendre P. & L. Legendre, Numerical ecology. - 1998: Developments in Environmental Modelling 20. Elsevier: 235-245.

Mandango M. A., 1982. - Flora and vegetation of the islands of the Zaire River in the sub region Tshopo (Haut-Zaire). University of Kisangani, Faculty of Science, 429 p.

Schmitz A., 1988. - Revision of plant described from Zaire to Rwanda and Burundi. Annals Economics, Royal Museum for Central Africa Tervuren Belgium. Vol 17: 315P.

ter Braak, C. J. F. and P. Smilauer. 1998. Canoco Reference Manual and User's Guide to Canoco for Windows: Software for Canonical Community Ordination (version 4). Microcomputer Power (Ithaca, NY United States), 352 p.

White F., 1983. - The vegetation map of Africa. A description memoir, Unesco, Natural Resources Research 20 January -1356.

20

Ecological Flexibility of the Top Predator in an Island Ecosystem – Food Habit of the Iriomote Cat

Shinichi Watanabe
[1]Graduate School of Engineering and Science, University of the Ryukyus
[2]Faculty of Life Science and Biotechnology, Fukuyama University
Japan

1. Introduction

The responses of predators to changes in prey availability, i.e. density and distribution of prey in an environment, are of great interest, especially when they involve both temporal and spatial variations (Lodé, 2000). Variation of prey availability leads to various predator responses (Holling, 1959). For instance, predators specialized in catching particular prey often produce numerical responses to prey availability so that density of the predators mutually fluctuates with the prey density (Krebs & Myers, 1974; Hansson & Henttonen, 1985, 1988). In contrast, non-specialized predators often produce functional responses to prey availability, allowing these predators to switch prey types in relation to availability of alternative resources (Krebs, 1996).

The family Felidae is known as the most successfully evolved and developed predators specialized in feeding on mammalian prey (Kleiman & Eisenberg, 1973; Kruuk, 1982). Hence, they often show numerical response to density of a particular prey (e.g., Andersson & Erlinge, 1977; Krebs & Myers, 1974).

Members of the Felidae, being found at the top of the trophic hierarchy in an ecosystem, usually require extremely large habitat ranges. Thus, most cat species are found only on continents or large islands. The leopard cat *Prionailurus bengalensis*, the most widespread species of East Asian cat, is an exception to this rule, occurring on several small islands as well as larger islands and the Asian continent (Watanabe, 2009). The leopard cat chiefly preys on rodents but occasionally also feeds on other types of animals depending on region. On small islands with a small number of carnivore species, the cat frequently feeds on non-mammalian prey. As an extreme example of this, the Iriomote cat *Prionailurus bengalensis iriomotensis* lives on Iriomote Island, this being the smallest island (284 km^2) known to be inhabited by this predator (Watanabe, 2009; Fig. 1).

The felid population on Iriomote Island remained unknown to science until its discovery in 1965 due to the inaccessibility of the dense forest which the cat occupies and to the remoteness of the island (Imaizumi, 1967). The cat had been long considered a separate species due to morphological differences among other small felids of south-east Asia (Imaizumi, 1967; Leyhausen & Pfleiderer, 1999). This species was listed as endangered in

International Union for Conservation of Nature and Natural Resources [IUCN] (2000) Red List because of its restricted habitat and small population size, estimated at about 100 individuals (Izawa et al., 2000). The population has declined during the last decade due to habitat loss from development and mortality from traffic accidents (Watanabe et al., 2002). Recently, their taxonomical specific distinction has been questioned by molecular methods (Johnson et al., 1999; Masuda et al., 1994).

Fig. 1. An Iriomote cat taken by photo-trap (Mammal Ecology Laboratory, University of the Ryukyus).

The insular fauna of Iriomote Island is unique; there are no autochthonous terrestrial small mammals such as rodents that generally form the principal prey of wild felids. On the other hand, the herpetofauna is relatively abundant. In particular, the density of anurans on the forest floor is extremely high compared with those reported in other tropical forests (Watanabe et al., 2005). In addition, the avifauna is also abundant including many resident and migratory species (see the results). Thus, it is likely that there are unique characteristics of the ecology of the cat as the top predator in the ecosystem of Iriomote Island.

The Iriomote cat preys upon a variety of animals such as birds, reptiles, amphibians, and insects, besides mammals (Sakaguchi & Ono, 1994; Watanabe et al., 2003, Watanabe, 2009). Thus, the Iriomote cat may show functional responses to availability of various alternative prey. Indeed, Sakaguchi & Ono (1994) found a functional response of the Iriomote cat to seasonal variation in prey availability. The cat frequently preys upon *Eumeces* skinks, one of the dominant prey items of the cat, while skinks are abundant during spring and summer; however the cat changes its principal prey during periods of low skink availability.

The purpose of this study is to understand the mechanism that causes temporal variations of diet of the Iriomote cat. I investigate both prey availability and the cat diet in various types of habitats through all seasons. I then consider how the cat responds to variations of prey availability, in terms of preferences and seasonal patterns of prey eaten. Moreover, I will discuss the role of such ecological flexibility of the cat in order to adapt to the island ecosystem.

2. Materials and methods

2.1 Study area

Field survey was conducted on Iriomote Island (24°20′N, 123°49′E) in the Ryukyu Archipelago, southern Japan (Fig. 2). The island largely consists of highly folded mountains with the highest peak (Mt. Komi) being 469 m above sea level. Its vegetation is mostly natural subtropical evergreen broadleaved forest (83% of the island in area), which is largely composed of *Castanopsis sieboldii* and *Quercus miyagii* in the tree layer of 8-12 m height. The understorey flora is more diverse, consisting of species such as *Daphniphyllum glaucescens*, *Elaeocarpus japonicus*, *Rapanea neriifolia* and *Ardisia quinquegona* (Miyawaki, 1989; Numata, 1974).

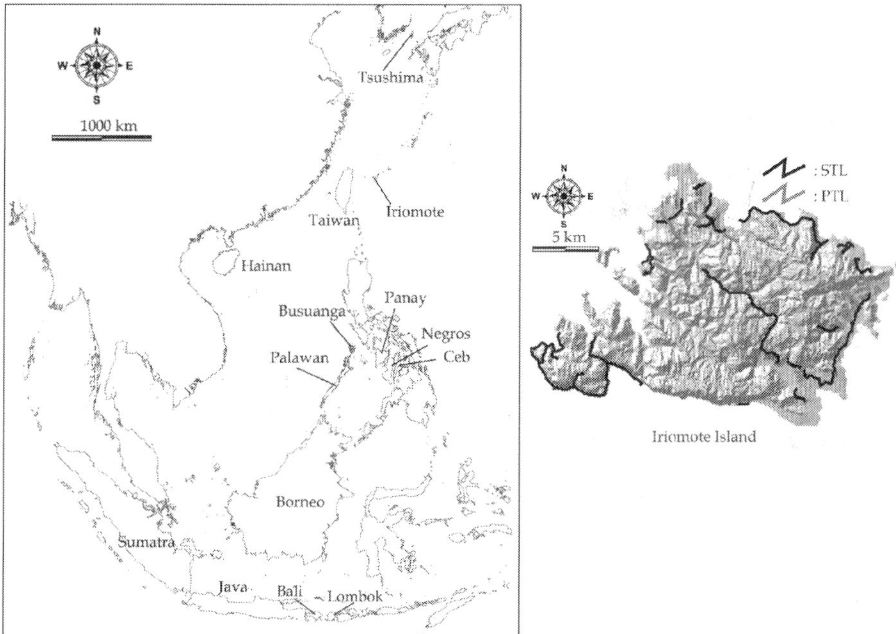

Fig. 2. Maps of East Asia showing islands with wild felid populations as cited in Watanabe (2009), and Iriomote Island (Ryukyu Archipelago, Japan) showing the locations of line transects along which scats of the Iriomote cat were collected (STL) and prey availability was estimated (PTL).

The island, which is located in the humid subtropical zone, includes the long chain of small islands (the Ryukyu Archipelago) which lies to the south of the main island of Japan between the latitudes 24° and 30°N. However, most of the original, natural vegetation has been lost in this region, as a result of widespread land cultivation, and also due to heavy damage during World War II. Only Iriomote Island, which is now protected as one of the Japanese National Parks, contains good examples of the natural subtropical forests (Numata, 1974).

The climate of Iriomote Island, like that of other Ryukyu Islands, is seasonal with spring (April to June; average monthly temperature is 22.5-27.4°C), summer (July to September; 27.3-28.9°C), autumn (October to December; 19.9-25.1°C), and winter (January to March; 18.3-19.9), and which is largely affected by the south or south-west monsoon during the summer and the north or north-west monsoon during the winter (Ayoade, 1983). The monthly rainfall and relative humidity are 142-274 mm and 73-82%, respectively; the annual rainfall is 2300 mm (Iriomote Meteorological Station).

2.2 Prey census

Prey distribution and abundance, i.e. prey availability, of the cat was estimated by line transect sampling carried out along routes set throughout the island (Fig. 2). The cat is considered as an opportunistic mobile predator which moves around an area seeking out ground-living animals visually (Sakaguchi & Ono, 1994). I thus evaluated potential prey availability for cats by using a line transect method. This method has been effectively used to determine relative abundance of animals in different habitat types (Buckland et al., 2001), which is presumably similar to the feeding pattern of the cat. I estimated relative abundance of various prey in different habitat types, and its distribution and seasonal change were assessed.

Eighteen transects (PTL: prey transect line) varying in length between 190 and 1510 m, with a total transect length of 11.5 km were laid along unpaved trails in several types of forests, along rice fields, crop lands, and pastures (Fig. 2). The trails are 3-4 m in width, and are covered with sparse grass on their ground surface. The PTLs were monitored over 14 periods between August 2001 and April 2003; total transect length in each period varied and resulted in 254 km of transect work. Study periods were divided into four seasons depending on climatic change of the island, particularly depending on the temperature and rainfall (see the study area).

Field surveys were conducted twice a day, daytime (8:00-12:00) and night time (20:00-24:00) depending upon subject animals; diurnal birds, lizards, and skinks were sighted in daytime, nocturnal mammals (flying foxes and bats), nocturnal birds (owls), snakes, frogs, and some insects in night time. Wild boars and crickets were sighted in both time periods. The transect work was carried out under dry weather conditions, by walking at about 2 km h-1 by the author throughout the study. Animals observed or singing in the transect lines and its vicinity (birds within 10 m in radius, lizards, skinks, and frogs within 2 m in radius, and insects were only counted on the transect lines) were located using a handheld Global Positioning System (GPS; model Garmin GPS II plus).

2.3 Scat analysis

Scat analysis was used to estimate the diet of the Iriomote cat, since this method is non-destructive and cost and time effective (Sakaguchi & Ono, 1994; Watanabe et al., 2003; Watanabe & Izawa, 2005). The scats of the Iriomote cat have been collected by trail censuses as a part of endangered species conservation projects administrated by the Forest Agency and Ministry of the Environment, Japan since 1993. Observers walked along a total length of 91.0 km of transect (STL: scat transect line, Fig. 2) including most parts of the PTLs, along mountain trails, farm roads, paved traffic roads, and coastal lines. Locations of scats found were recorded on the route divided at 250 m intervals (Environment Agency & Forestry Agency, 1998). Although domestic cats (*Felis catus*) also inhabit Iriomote Island, they live mainly in residential areas. In addition, domestic cat scats and those of the Iriomote cat are easily distinguished by specific odours and grooming hairs contained within them (Watanabe et al., 2003).

Collected scats were washed on 1-mm mesh sieve, and dried at a laboratory. Remains such as hair, bone fragments, teeth, feathers, scales, and arthropod chitin of the prey consumed were separated for species identification (Sakaguchi & Ono, 1994; Watanabe et al., 2003). Scats which contained no prey items were excluded from the analysis. I tried to identify prey found in the scats to the level of species using my own reference collection of potential prey species collected on Iriomote Island, e.g. as cited in Watanabe & Izawa (2005) for amphibians. The identification was achieved depending on the type and the quality of the scat sample.

I calculated the frequency of occurrence of each prey item in the scats, i.e. the percentage of the total number of scats that contained a particular prey type for each prey item.

2.4 Prey availability and predation by the cat

2.4.1 Prey preference

To examine food preference in the cat diet, I calculated Jacobs' (1974) modification (*D*) of Ivlev's Electivity Index for quantification of the diet selection.

$$D = (r - p) / (r + p - 2\,r\,p) \tag{1}$$

Where *r* is the proportion of a given prey item in the total identified items in the scat contents, and *p* is the proportion of the same prey item in the total number of items found in the prey transects. *D* varied from -1 to 0 for negative selection and from 0 to +1 for positive selection.

I only evaluated prey preference of each bird and amphibian species, respectively, which are two of the major prey groups in the cat diet (see the results). I derived *D*s for only prey items which were frequently observed within the transect survey, because these indices are vulnerable to sampling errors for prey that are rare in the environment (Jacobs, 1974). From the derived electivity index for each prey item, I considered the feeding pattern of the cat in terms of principal habitats of the major prey species.

2.4.2 Seasonal patterns

To evaluate seasonal patterns of prey availability, frequency of occurrence of each prey species per kilometre on the PTL was calculated in each study period and each season. I used a likelihood ratio test (*G*-test: Sokal & Rohlf, 1995) to examine the goodness of fit into expected values derived from proportion of transect distance in each season.

Likewise, seasonality of predation of each prey type found in scats was also evaluated. I calculated the frequency of occurrence of prey items in the scats for each season. I used a *G*-

test to examine the goodness of fit into expected values derived from proportion of analysed scats in each season.

I also performed Principle Component Analysis (*PCA*: Sokal & Rohlf, 1995) to classify the various prey species in terms of their relative similarity of seasonal variation, availability and predation. This *PCA* was run after standard *VARIMAX* rotation of data, the frequency occurrences in PTL and in a scat of each major prey item for each season.

All statistical analyses were carried out using SPSS for Windows version 11.5 (SPSS, Inc.; Chicago, Illinois, USA). Statistical significance was set at the 0.05 level.

3. Results

3.1 Prey availability

From the total of 254 km of transects sampled, I found 8096 individual animal belonging to 72 different prey items including animals which could not be identified to the level of species but were distinguishable from other groups. Most of the prey items are birds (61.6% of the total number of animals found in PTLs), while mammals, reptiles, amphibians, and insects occupied 4.1%, 15.1%, 9.59%, 9.59%, respectively. Insects occupied a low percentage proportion because only three major prey items of insects were recorded. In spite of the species composition, the number of individuals found was dominated by amphibians (62.3% of total number of individuals found in PTLs); particularly, the Indian rice frog *Fejervarya limnocharis* occupied 41.8% of all animals found. In contrast, mammals, birds, reptiles, and insects occupied 1.05%, 24.4%, 7.31%, and 4.92%, respectively.

In the case of mammals, I encountered 85 individuals belonging to three different groups, most of which were the Yaeyama flying fox *Pteropus dasymallus yaeyamae* (65.9% of all encountered mammals). Small-sized chiropterans (24.7%) were the second most frequently encountered animals but not all of them could be identified because I only observed them briefly as they flew across the transect.

In birds, I encountered 1973 individuals belonging to 50 different bird groups in 25 families. The relative abundance of each species was the highest in the Brown-eared bulbul *Hypsipetes amaurotis stejnegeri* (19.6% of all encountered birds), followed by the Oriental turtledove *Streptopelia orientalis* (12.2%), the Jungle crow *Corvus macrorhynchos osai* (9.28%), thrushes in the genus *Turdus* (mostly *T. pallidus* or *T. chrysolaus* although they were difficult to distinguish from each other during the censuses: 7.65%), and *Parus* spp., mostly *P. major* or *P. varius* (7.60%). Thirty-six percent of the identified bird species are arboreal foragers which are sighted mainly on trees or in shrubs, and 52% of the bird species are ground-living terrestrial foragers which are sighted mainly on the ground; the rest of the bird species (12%) are sighted in both habitats, according to our observations during the censuses.

In the case of reptiles, I encountered 572 individuals belonging to seven different groups, most of which were the Sakishima tree lizard *Japalura polygonata ishigakiensis* (41.4% of all encountered reptiles) and *Eumeces* skinks (*Eumeces kishinouyei* or *E. stimpsonii*: 42.1%). I observed four species of snakes, of which the Sakishima habu *Trimeresurus elegans* (10.1%) which is the only venomous snake on the island, was the most frequently observed during censuses.

In amphibians, only anurans were found, of which I encountered 3386 individuals belonging to seven species in three families. The relative abundance of each species was the highest in the Indian rice frog *F. limnocharis* (67.1% of all encountered frogs), followed by the Ryukyu kajika frog *Buergeria japonica* (13.3%), the Ornata narrow-mouthed toad *Microhyla ornata* (10.4%), and the Eiffinger's tree frog *Chirixalus eiffingeri* (4.8%). Most frog species are terrestrial ground-living foragers except for *C. eiffingeri* that inhabits mostly the tree canopy and the Owston's green tree frog *Rhacophorus owstoni* which inhabits mainly tree canopies but comes down to the forest floor during their breeding season (Maeda & Matsui, 1999).

3.2 Predation by the cat

I examined the contents of 947 scats collected in the study area from December 1993 to November 2002. I identified 76 different prey items (four items of mammals, 21 items of birds, seven items of reptiles, five items of amphibians, one item of fish, 26 items of insects, two items of crustaceans, three items of centipedes, six items of arachnids, and one item of gastropod) which include prey items that could not be identified to the level of species but were distinguishable from other prey items. Species identification of mammals and reptiles was accomplished with ease due to the small number of potential species in each taxon and the presence of clearly defined hairs, scales, teeth or mandibles in the scat content. However, birds and insects found in the scat contents were difficult to identify due to a large number of potential prey species, and unclearly characterized body parts in the scat content. Amphibians were also difficult to identify due to only a small number of bone fragments. Thus, these prey taxa include unidentified animal remains at relatively high frequencies. Meanwhile, prey species were diversified in birds (27.6% of total number of identified prey items) and in insects (34.2%).

Only one prey item was found to be contained in 31.5% of total examined scats, while 31.6%, 22.3%, 10.2%, 3.17%, 0.74%, 0.42%, and 0.11% of the scats contained two, three, and four to eight prey items, respectively. Scats containing larger prey items (the estimated liveweight is more than 200 g: Imaizumi et al., 1977) contained significantly fewer prey items (1.92±1.15, ± a standard deviation of number of prey items found in the scats) than scats with only smaller prey (< 200 g, 2.37±1.19: Mann-Whitney U-test, $U = 60202$, $P < 0.001$).

Principle prey species which were frequently found in the scats belonged to a variety of taxonomical groups ranging from mammals to crustaceans. The major prey species are described as follows: the Yaeyama flying fox *Pteropus dasymallus yaeyamae* and the black rat *Rattus rattus* in mammals, egrets *Egretta* spp., the banded crake *Rallina eurizonoides*, the white breathed water hen *Amaurornis phoenicurus*, and thrushes *Turdus* spp. (presumably most of them were the pale thrush *T. pallidus* but difficult to distinguish birds in the genus), in birds, the Sakishima tree lizard *J. p. ishigakiensis*, the Kishinoueno's giant skink *E. kishinouyei*, and the Sakishima habu *T. elegans*, in reptiles, the Indian rice frog *F. limnocharis*, the Owston's green tree frog *Rh. owstoni*, and the Ornata narrow-mouthed toad *M. ornata*, in amphibians, the spotted cockroaches *Rhabdoblatta* spp., camel crickets, *Diestrammena* spp., and the spotted cricket *Cardiodactylus novaeguineae*, in insects, and freshwater prawns *Macrobrachium* spp. in crustaceans. Among these prey items, black rats, flying foxes, and rice frogs were the most frequently present species in the cat diet.

The overall frequency of occurrence in each taxon was calculated as follows: 30.9% in mammals, 47.7% in birds, 33.1% in reptiles, 34.3% in amphibians, 2.96% in fish, 24.3% in insects, 4.22% in crustaceans, and 3.27% in others.

3.3 Prey preference

I calculated Jacobs' electivity indices for 18 bird groups in which taxonomically similar species with a small number of samples were combined, and for seven amphibian species (Table 1). The result show that the cat preferred preying on some bird items; egrets, rails (*R. eurizonoides* and *A. phoenicurus*), *Halcyon coromanda*, and wagtails in the genus *Motacilla*, all of which were terrestrial foragers. However, *Hirundo tahitica*, *Pericrocotus divaricatus*, *H. amaurotis*, *Cettia diphone*, *Terpsiphone atrocaudata*, *Parus* spp., *Zosterops japonica*, and *Passer montanus*, all of which were arboreal foragers, were not preferred or were avoided by the cat.

Among amphibians, the cat preferred preying on *R. supranarina* and *Rh. owstoni* while *R. psaltes*, *C. eiffingeri*, *B. japonica*, *F. limnocharis*, and *M. ornata* were not preferred or was avoided for predation by the cat (Table 1).

	Prey item	Habitat[*1]	Migration[*2]	Liveweight (g)[*3]	Available N (%)	Consumed N (%)	Electivity Index (*D*)	Preference
Bird;								
Ardeidae	Egrets	T	R,W	>500	176 (9.2)	22 (12.7)	0.18	+
Anatidae	Ducks	T	R,W	>500	131 (6.8)	15 (8.7)	0.13	+
Rallidae	Rails	T	R	200	49 (2.6)	48 (27.7)	0.87	+
Scolopacidae	Snipes	T	W	200	21 (1.1)	2 (1.2)	0.03	+
Columbidae	Pigeon	A,T	R	200	260 (13.6)	21 (12.1)	-0.07	-
Strigidae	Owls	A	R	120	89 (4.7)	6 (3.5)	-0.15	-
Alcedinidae	*Halcyon coromanda*	T	S	100	18 (0.9)	6 (3.5)	0.58	+
Hirundinidae	*Hirundo tahitica*	A	W	20	22 (1.2)			-
Motacillidae	*Motacilla* spp.	T	W	30	34 (1.8)	7 (4.0)	0.40	+
Campephagidae	*Pericrocotus divaricatus*	A	R	20	26 (1.4)			-
Pycnonotidae	*Hypsipetes amaurotis*	A	R	30	386 (20.2)	2 (1.2)	-0.91	-
Turdidae	*Turdus* spp.	T	W	80	159 (8.3)	27 (15.6)	0.34	+
Sylviidae	*Cettia diphone*	A	R	20	61 (3.2)			-
Monarchidae	*Terpsiphone atrocaudata*	A	R	20	23 (1.2)			-
Paridae	*Parus* spp.	A	R	20	150 (7.8)	2 (1.2)	-0.76	-
Zosteropidae	*Zosterops japonica*	A	R	20	87 (4.6)	2 (1.2)	-0.61	-
Ploceidae	*Passer montanus*	A	R	20	36 (1.9)			-
Corvidae	*Corvus macrorhynchos*	A,T	R	300	183 (9.6)	13 (7.5)	-0.13	-
Frog;								
Ranidae	*Rana supranarina*	T		60	27 (0.5)	7 (2.8)	0.69	+
	Rana psaltes	T		10	128 (2.5)			-
	Fejervarya limnocharis	T		20	3,386 (67.1)	142 (57.0)	-0.20	-
Rhacophoridae	*Rhacophorus owstoni*	A,T		20	68 (1.3)	55 (22.3)	0.91	+
	Chirixalus eiffingeri	A		4	242 (4.8)			-
	Buergeria japonica	T		4	669 (13.3)	14 (5.7)	-0.44	-
Microhylidae	*Microhyla ornata*	T		4	526 (10.4)	29 (11.7)	0.07	+

Table 1. Prey preferences of 25 prey items by the Iriomote cat, indicated by Jacobs (1974) electivity index (*D*). [*1]Principal habitat of each species: arboreal (A), terrestrial (T), and both (A, T), depending on where species were often observed during the censuses. [*2]Migration of each species; residents (R), summer migrators (S) and winter migrators (W), according to Takano (1982) and Committee for Check-List of Japanese Birds (2000). [*3]Estimated liveweight from weighing my reference collection or Imaizumi et al. (1977).

3.4 Seasonal variation

3.4.1 Prey availability

The frequency of occurrence of most species changed seasonally. Within an identified 72 species found in the PTLs, 66 species were absent in some periods or present in limited periods of the study years. Although six species, *S. orientalis*, *H. amaurotis*, and *C. macrorhynchos*, of birds, and *B. japonica* and *M. ornata* of amphibians, and *Diestrammena* spp. of insects, were found in all the study periods, these frequencies also largely varied depending on season.

In the bird species, 44 percent of species are residents on the island, while the rest are seasonal migrators: winter birds migrating from north to the island during autumn and winter for overwintering (40%), summer birds migrating from south to the island during spring and summer for breeding (2.0%), and transient migrators stopping off briefly on the island during migration from north to further south of the island (4.0%: Takano, 1982; Committee for Check-List of Japanese Birds, 2000). The relative abundances of principal bird species observed in prey censuses during each season were derived. During all seasons, four species of resident birds, *H. amaurotis*, *S. orientalis*, *C. macrorhynchos*, and *Parus* spp., were relatively abundant. Meanwhile, during autumn and winter, two groups of migrators, *Turdus* spp. and ducks, were also abundant, which occupied respectively high proportions of the avifauna.

Among the amphibians, in spring to autumn, *F. limnocharis* was the most abundant, followed by *M. ornata*, *B. japonica*, and *C. eiffingeri*. Meanwhile, *M. ornata* was the most abundant during winter, followed by *F. limnocharis*, *R. supranarina*, *C. eiffingeri*, and *Rh. owstoni*.

To assess seasonal variation of more general compositions of prey availability, I calculated seasonal variations of frequency occurrences of 18 major prey groups in prey censuses which were also frequently eaten by the cat. The results are summarized in Table 2. Some prey groups including taxonomically similar prey species (ducks in the family Anatidae, the genera of egrets *Egretta* spp., thrushes *Turdus* spp., skinks *Eumeces* spp., cockroaches *Rhabdoblatta* spp., and camel crickets *Diestrammena* spp.) were combined within the taxonomical group and the data were pooled.

Frequency occurrence of 14 prey groups showed significant seasonal variations (*G*-test, *P* < 0.05). Although other groups, water hens, *S. orientalis*, *T. elegans*, and *R. supranarina*, thrushes, and *J. polygonata*, did not show significant seasonal variations (*G*-test, *P* > 0.05), these frequency occurrences also differed among seasons in varying degrees.

A more detailed picture of seasonality of prey availability of the cat emerges from applying a *PCA* model to the data matrix in Table 2. The first and second principal components of *PCA* explained 61.2% and 25.0% of the total variability, respectively, and the scores of the first and second factors are presented in Table 3. In the score plot of each prey species on factors, the main variables ordering the various species along factor 1 were "winter-summer", and those ordering the species along factor 2 were "spring-autumn", which were sufficient to illustrate the main structure of prey availability among seasons (Fig. 3).

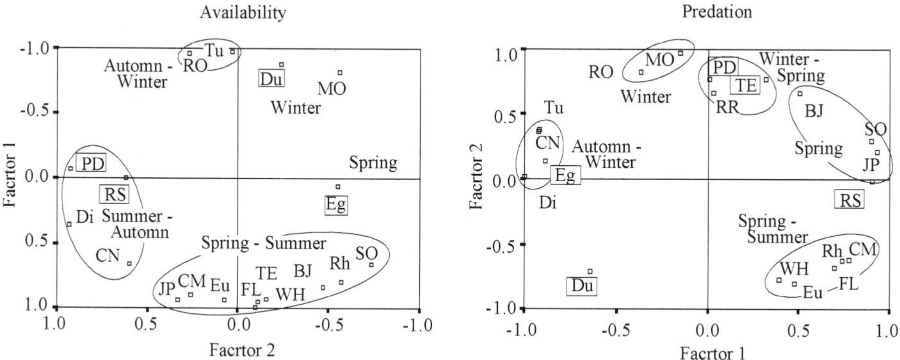

Fig. 3. Seasonal variation of potential prey availability and diet of the Iriomote cat. Two dimensional-plots of scores for each prey items on first two factors using *VARIMAX* rotation model *PCAs* showing mainly four season prey groups. PD: *Pteropus dasymallus*, RR: *Rattus rattus*, CM: *Corvus macrorhynchos*, Du: ducks, Eg: egrets, SO: *Streptopelia orientalis*, Tu: *Turdus* spp., WH: water hens, Eu: *Eumeces* spp., JP: *Japalura polygonata*, TE: *Trimeresurus elegans*, BJ: *Buergeria japonica*, MO: *Microhyla ornata*, FL: *Fejervarya limnocharis*, RS: *Rana supranarina*, RO: *Rhacophorus owstonii*, Di: *Diestrammena* spp., Rh: *Rhabdoblatta* spp., CN: *Cardiodactylus novaeguineae*.

Prey items	Frequency of occurrence (%) in scats							Frequency of occurrence (individuals km^{-1}) in PTLs						
	Spring	Summer	Autumn	Winter	Overall	G	P	Spring	Summer	Autumn	Winter	Overall	G	P
Mammals														
Pteropus dasymallus (Yaeyama flying fox)	12.9	5.3	10.4	11.2	10.5	8.43	0.038 *	0.29	0.46	0.69	0.31	0.42	14.9	0.002 *
Rattus rattus (black rat)	17.7	19.0	13.4	26.2	20.2	11.8	0.008 *							
Birds														
Egrets	1.29	0.53	5.22	3.19	2.32	12.2	0.007 *	2.45	0.31	1.41	0.79	1.26	31.0	<0.001 **
Ducks	0.32	2.65	5.22	0.64	1.58	15.3	0.002 *	0.60	0.31	1.29	4.45	1.08	52.0	<0.001 **
Water hens	3.22	6.35	2.99	2.56	3.59	6.14	0.105	0.47	0.42	0.32	0.22	0.39	0.63	0.890
Streptopelia orientalis (Oriental turtle dove)	1.93	1.06	0.75	0.96	1.27	1.61	0.657	2.17	2.01	1.65	1.87	1.97	1.39	0.708
Turdus spp. (thrushes)	1.29	1.06	4.48	4.47	2.75	5.85	0.119	0.18	0.00	3.02	5.53	1.30	155	<0.001 **
Corvus macrorhynchos (jungle crow)	1.93	2.65	0.75	0.32	1.37	6.59	0.086	1.23	2.39	0.85	0.57	1.50	16.8	<0.001 **
Reptiles														
Japalura polygonata (Sakishima tree lizard)	6.75	4.23	2.99	3.51	4.65	4.76	0.190	1.64	3.22	1.21	0.00	1.95	43.7	<0.001 **
Eumeces spp. (skinks)	30.5	34.4	27.6	3.83	22.1	112	<0.001 **	1.80	3.68	0.24	0.07	1.98	93.1	<0.001 **
Trimeresurus elegans (viper)	2.25	1.59	0.00	3.51	2.22	8.46	0.037 *	0.64	0.59	0.34	0.12	0.44	0.97	0.810
Amphibians														
Rana supranarina (large tip-nosed frog)	3.54	1.59	0.75	0.32	1.69	8.85	0.031 *	0.20	0.15	0.50	0.03	0.20	3.73	0.292
Fejervarya limnocharis (Indian rice frog)	17.4	24.3	11.2	8.63	15.0	25.7	<0.001 **	36.5	45.1	9.95	3.30	25.7	383	<0.001 **
Rhacophorus owstoni (Owston's green tree frog)	1.93	0.00	1.49	15.0	5.81	75.0	<0.001 **	0.06	0.05	1.03	1.14	0.52	60.9	<0.001 **
Buergeria japonica (Ryukyu kajika frog)	2.89	0.00	0.75	1.28	1.48	9.67	0.022 *	10.53	6.42	1.68	0.37	5.07	36.7	<0.001 **
Microhyla ornata (Ornata narrow-mouthed toad)	2.89	0.53	1.49	5.43	3.06	12.4	0.006 *	4.28	1.87	3.60	6.54	3.99	102	<0.001 **
Insects														
Rhabdoblatta spp. (spotted cockroaches)	8.04	13.2	2.24	0.96	5.91	41.1	<0.001 **	0.20	0.18	0.04	0.09	0.14	9.19	0.027 *
Diestrammena spp. (camel crickets)	0.64	3.70	10.4	8.31	5.17	32.1	<0.001 **	1.40	2.48	2.30	1.48	1.92	29.0	<0.001 **
Cardiodactylus novaeguineae (spotted cricket)	1.29	1.06	13.4	13.7	6.97	61.5	<0.001 **	0.00	1.07	0.26	0.00	0.41	93.7	<0.001 **
Number of examined scats	311	189	134	313	947			Distance (km) Day	38.32	44.78	24.83	13.92	122	
								Night	34.38	39.07	26.14	32.40	132	
								Total	72.70	83.85	50.97	46.32	254	

Table 2. Seasonal variations of availability and predation of principal prey items observed in prey censuses and found in scats of the Iriomote cat. * $P < 0.05$; ** $P < 0.001$.

Prey items	Availability		Predation	
	Factor 1	Factor 2	Factor 1	Factor 2
Mammals				
Pteropus dasymallus (Yaeyama flying fox)	-0.072	0.920	0.004	0.773
Rattus rattus (black rat)	—	—	0.027	0.665
Birds				
Egrets	0.059	-0.553	-0.888	0.138
Ducks	-0.882	-0.239	-0.639	-0.709
Water hens	0.929	-0.161	0.393	-0.771
Streptopelia orientalis (Oriental turtle dove)	0.659	-0.737	0.896	0.301
Turdus spp. (thrushes)	-0.978	0.025	-0.920	0.379
Corvus macrorhynchos (jungle crow)	0.899	0.256	0.774	-0.617
Reptiles				
Japalura polygonata (Sakishima tree lizard)	0.940	0.324	0.930	0.211
Eumeces spp. (skinks)	0.938	0.068	0.476	-0.802
Trimeresurus elegans (viper)	0.954	-0.116	0.317	0.771
Amphibians				
Rana supranarina (large tip-nosed frog)	-0.003	0.614	0.902	-0.016
Fejervarya limnocharis (Indian rice frog)	0.991	-0.102	0.695	-0.676
Rhacophorus owstoni (Owston's green tree frog)	-0.964	0.266	-0.366	0.826
Buergeria japonica (Ryukyu kajika frog)	0.837	-0.476	0.505	0.663
Microhyla ornata (Ornata narrow-mouthed toad)	-0.824	-0.563	-0.156	0.973
Insects				
Rhabdoblatta spp. (spotted cockroaches)	0.797	-0.576	0.736	-0.624
Diestrammena spp. (camel crickets)	0.359	0.927	-1.000	0.012
Cardiodactylus novaeguineae (spotted cricket)	0.657	0.592	-0.927	0.369
Eigenvalue	11.01	4.51	10.07	5.93
% Explained variance	61.17	25.05	53.02	31.20

Table 3. Results of principle component analysis for seasonal variations of availability and predation of principal prey items found in prey censuses and scats of the Iriomote cat.

Ducks and *M. ornata* were the most abundant in winter. Egrets were the most abundant in spring. *Streptopelia orientalis, Rhabdoblatta* spp., *B. japonica,* water hens, *T. elegans, F. limnocharis, Eumeces* spp., *C. macrorhynchos,* and *J. polygonata,* were more abundant in spring and summer. *Cardiodactylus novaeguineae, Diestrammena* spp., *R. supranarina, P. dasymallus,* were more abundant in summer and autumn. *Rhacophorus owstoni* and *Turdus* spp. were more abundant in autumn and winter.

3.4.2 Predation by the cat

The frequency occurrence of most prey species in scat contents changed monthly or seasonally. Within an identified 76 prey items in the scat content, 72 prey items were absent in some months or present in limited months of the years. Although four species, black rats, flying foxes, rice frogs, and freshwater prawns, were found at all months, these frequencies also greatly varied on a monthly basis.

The frequency occurrences for each season of 19 major prey groups which were frequently found in the scats were calculated, and the seasonal variations were compared among seasons (Table 2). Fifteen groups of the prey items showed significant seasonality (G-test, $P < 0.05$). Four prey species, water hens, *S. orientalis*, thrushes, and *J. polygonata* did not show statistically significant seasonality (G-test, $P > 0.05$), but these frequency occurrences also varied among seasons.

A more detailed picture of seasonality of the diet composition emerges from applying a *PCA* model to the data matrix in Table 2. The first and second principal components of *PCA* explained 53.0% and 31.2% of the total variability, respectively, and the scores of the first and second factors are presented in Table 3. In the score plot of each prey species on factors, the main variables ordering the various species along factor 1 were "autumn-spring", and those ordering the species along factor 2 were "summer-winter", which were sufficient to illustrate the main structure of diet composition among seasons (Fig. 3).

Rhacophorus owstoni and *M. ornata* were most frequently eaten in winter. *P. dasymallus, Rattus ratuus*, and *T. elegans* were more frequently eaten in winter and spring. *Buergeria japonica, S. orientalis, J. polygonata*, and *R. supranarina* were most frequently eaten in spring. *Corvus macrorhynchosi, Rhabdoblatta* spp., *F. limnocharis*, water hens, and *Eumeces* spp., were more frequently eaten in spring and summer. Ducks were eaten most frequently in autumn. *Diestrammena* spp., egrets, *C. novaeguineae*, and *Turdus* spp., were eaten more frequently in autumn and winter.

4. Discussion

The results of this scat analysis show that the Iriomote cat fed on various types of prey, as observed in earlier studies (Sakaguchi & Ono, 1994; Watanabe et al., 2003). I observed 76 prey items of the cat demonstrating considerable diversity among the dietary studies. In other studies, scats analysed were collected from parts of the island during several years while scats analysed in the present study were collected from the whole island over a long-time period. Thus, a large number of scat contents were analysed, and many differences such as seasonal variation and localities of a variety of prey items could be estimated. Therefore, I believe that the results show general representation of the cat diet in the present study.

The Iriomote cat fed considerably more on avian prey than other animals in the present study although birds are almost always not principal prey among other wild felids (Kruuk, 1982; M. Sunquist & F. Sunquist, 2002). Some other small cats such as *Felis margarita, F. silvestris, Herpailurus yaguarondi, Leptailurus serval*, and *Oncifelis geoffroyi*, occasionally hunt bird prey, but the frequency of these occurrences are much lower than that of the Iriomote cat (M. Sunquist & F. Sunquist, 2002). It is therefore likely that the hunting pattern of the Iriomote cat is the most developed in focusing on bird prey among the cat family.

Of eighteen avian groups which were frequently observed in the prey censuses, thirteen groups were eaten by the cat; others are all arboreal foragers. Sakaguchi & Ono (1994) suggested that the cat may hunt mainly on the ground because all nine bird species found in their analysis were mainly foraging on the ground. I found that the cat occasionally preyed on arboreal species such as *H. amaurotis* and *Parus* spp., suggesting the Iriomote cats arboreal activities. However, prey preferences of the cat for arboreal birds were much lower than those for terrestrial foraging bird species. Thus, I support the prediction of Sakaguchi & Ono (1994) that the cat is a species foraging mainly on the ground, as opposed to in trees.

Species compositions of avifauna as potential prey resources of the cat drastically changed in relation to migrations of birds. This island with such abundant small vertebrates is known as a way station or breeding place for many migrant birds; 191 species belonging to 15 orders, 43 families, and 106 genera are recorded on this island, 83% of which are seasonal migrators while resident bird species occupy at least 17% of the total avifauna on the island (Okinawa Prefectural Education, 1985). In the cat diet, these avian prey items were present at high frequency. Thus, I presumed that such drastic seasonal changes of potential food resources of the cat in relation to migrations of large numbers of birds would strongly influence the feeding habits of the Iriomote cat.

Consequently, bird migrations largely affected the cat diet, particularly during winter by some migrators such as *Turdus* spp., ducks, *Motacilla* spp., and egrets. Most species of resident birds on the island breed during spring and summer (Takano, 1985); many juvenile birds were observed from June to August during the prey censuses. Particularly, *A. phoenicurus* and *C. macrorhynchos* were frequently preyed upon during spring and summer. It is likely that more vulnerable juveniles were frequently eaten during these periods. Although reptiles are principal prey items during spring and summer, the availability of this prey decrease during winter due to their low physiological activity under low temperature (Sakaguchi & Ono, 1994). It is, therefore, likely that the migrations of winter birds largely contribute to supply a deficiency of food resource during winter.

In dietary studies of other *P. bengalensis* populations, the cat diet is monopolized by mammalian prey in a continental habitat (Thailand), bird prey were present at less than 2% of their scats (Rabinowitz, 1990). On an island site, Tsushima Islands, Japan, although mammalian prey are also the most important food in the diet, the cat also preys heavily on birds (41.7% for the frequency of occurrence in the scats: Inoue, 1972). Unfortunately, there is little available information for diets of other wild felids on island sites. However, I could find those of feral domestic cat, *Felis catus*, populations which are widespread all over the world, and they had been introduced to many islands (Fitzgerald & Tuner, 2000). Dietary studies of domestic cats on continents show that mammals are usually the main prey eaten by cats, while birds are much less important than mammals. However, birds are much more important in the diet of domestic cats on islands than on continents. Particularly, on islands where seabirds are recorded in the diet, birds are present on average at 60% of frequency of occurrence in their scats (Fitzgerald & Tuner, 2000). Although small mammals such as rodents or lagomorphs are available on many islands, feral cat populations can persist on islands that lack mammalian prey but are abundant in birds (Kirkpatrick & Rauzon, 1986). Therefore, I suggest that the Iriomote cat can endure on its small island despite a poor mammalian fauna by means of shifting principal prey items to seasonal migrators, particularly during winter. It is during this winter period, that the mating season of the cat takes place(Okamura et al., 2000); home range size and activity of male cats increase (Nakanishi et al., 2005; Schmidt et al., 2003, 2009). Thus, the food requirement of the cat should increase in that time. Therefore, it is likely that winter migrators supply an increment of food requirement for reproduction of the cat.

In addition, the number of migrant birds on this island largely fluctuated year by year, particularly *Turdus* spp.; the species were rarely encountered in some years while they have been abundant in most years during this decade (Iriomote Wildlife Conservation Center pers. comm.). The incidence of winter migrators may influence survival rate and breeding success of the Iriomote cat.

In the case of the other main prey items of the Iriomote cat, amphibians were the second most frequent prey eaten in the cat diet. The Iriomote cat is also unique among the cat family in feeding on amphibians. For other carnivores, however, numerous mustelids are found to prey upon anurans. Although otters *Lutra lutra*, American mink *Mustela vision*, and badger *Meles meles* sometimes consume a significant quantity of frogs and toads, the occurrences of amphibians in the diet of carnivores remain generally low (Chanin & Linn, 1980; Erlinge, 1969; Webb, 1975; Wise et al., 1981). Exceptionally, polecats *Mustela putorius* feed heavily on anurans. Anurans occupy about 46% of the total number of prey which is almost the same as that of mammalian prey. However, the predation of anurans is limited only during spring while the availability of anurans is particularly high in relation to their breeding season (Lodé, 1996). In the present study, the Iriomote cat frequently feeds on anurans throughout the year. The abundance of amphibian fauna on the island is the highest so far reported in Old and New world tropics; that is relatively high throughout the year due to different breeding periods of several species (Watanabe et al., 2005). The availability of the frog species most frequently eaten, *F. limnocharis*, is high during spring and summer due to their breeding season (Watanabe et al., 2005). The frogs were mainly eaten during these periods. From autumn to winter, the availability of the frogs drastically decreased and they were not frequently eaten by the cat. During winter, the availability of two other species, *Rh. owstoni* and *M. ornata*, were high and they were mainly eaten during this period. Therefore anurans on the island are available food for the Iriomote cat throughout the year.

There are eight species of the amphibian fauna of Iriomote Island, of which five species were found in the scat contents of the cat. Of the other three species, *R. utsunomiyaorum*, lives chiefly near streams in montane forests (Maeda & Matsui, 1999); this species is rarely found and was not found during these censuses. Although *C. eiffingeri* was frequently observed during the censuses, I found this frog by mainly identifying their calls when made by individuals located on trees, whilst these frogs were rarely found on the ground. This species was not observed on the ground during the prey census in this study. Thus, the result also supports the prediction that the cat mainly hunts animals on the ground. *Rana psaltes* was a ground-dwelling species but the species was not found in the scat samples. It is not clear why the species was absent in the cat diet but it is possible that the density of this species is particularly low compared with those of other anurans in the habitats of the cat, or the identification of this species was not successfully achieved in the scat analysis.

The frogs eaten by the cat, *R. supranarina* and *Rh. Owstoni,* were only selectively preyed upon. For estimation of prey value for each species, a species with large mass is probably higher than a species with a small mass. It is because, from the results estimating body mass of *F. limnocharis* in scat contents of the Iriomote cat, the cat selectively preys on large sized frogs and avoids immature small frogs (Watanabe & Izawa, 2005). *Rana supranarina* is the largest frog species of the amphibian fauna. The size of adult *Rh. owstoni* is almost same as *F. limnocharis* but only breeding adults *Rh. owstoni* were observed during winter. *Buergeria japonica* and *M. ornata* are both small species, thus the values for the cat diet are presumably lower than those of larger frogs. Therefore the selectivity of frog species by the cat resulted from the differences of prey values, environments inhabited, and breeding cycles of each species.

From spring to summer, the availability of many reptiles, frogs, and resident birds are high; thus, the cat feeds on various prey types during the period. Meanwhile, availability of these prey species decreases during winter. Thus, the cat changes principal prey during winter. Additionally, the cat frequently preys on small-sized prey items such as frogs and insects, *M. ornata*, *B. japonica*, and *Diestrammena* spp., during winter.

Even if prey resources seasonally decreased or occasionally disappeared in habitats of open environment such as on continents, predators would move to search for other habitats encompassing plenty of food. Indeed, home ranges of leopard cats on the continent largely shift depending on wet and dry seasons and its sizes are much larger (Rabinowitz, 1990; Grassman, 2000, 2005) than those of Iriomote cats (Nakanishi et al., 2005; Schmidt et al., 2003). In restricted habitats such as small islands, however, predators have to remain in the same habitats during periods of prey resource scarcity.

The feeding pattern of leopard cats is presumed to be one of randomly foraging within a wide area with a relatively high abundance of their prey, and these cats prey on encountered animals; that is indicating an opportunistic feeding pattern (Fitzgerald & Tuner, 2000). The results in this study support that Iriomote cats are also opportunistic predators due to their non-selective varied diet representing prey availability with seasonal variations. The Iriomote cat is more likely to opportunistically forage in areas of abundant prey resources, feeding on various prey. However, it is unlikely that cats travel within their habitat and chase encountered volant birds. It is more effective that cats ambush and then attack a bird coming within close reach of the cat; that is using an ambushing feeding pattern. Therefore, the cats may shift their feeding tactics when switching their principal prey items. The movement pattern of the Iriomote cat varies in relation to seasons and local differences of habitats (Nakanishi et al., 2005; Sakaguchi, 1991; Schmidt et al., 2003, 2009). It may also vary in relation to shifting feeding patterns of the cat depending on seasonal difference of principal prey items. These patterns will be more clarified by comparison of movement patterns and prey availability of the cat in the further study.

From the above results, I considered that the broad range of the food niche of the Iriomote cat probably resulted from making the best possible use of fauna on the small subtropical island. Furthermore, the cat adapts to the islandwide environment to change principal prey items and feeding patterns in relation to temporal variations of food availability. I suggest that such flexibility of food habits of the Iriomote cat plays an important role for its existence on the small isolated habitat. Most small felids may potentially have such flexible properties in their ecology. However, I believe that it is revealed only in the peculiar fauna of the insular environment of Iriomote Island.

5. Conclusion

The leopard cat, *Prionailurus bengalensis*, is one of most widespread felids and is distributed throughout Asia. The Iriomote cat *Prionailurus bengalensis iriomotensis*, a subspecies of the leopard cat, lives only on Iriomote Island of the Ryukyu Archipelago in southern Japan. Although there are thousands of islands of various sizes within the range of distribution of the species, the Iriomote cat is the only population on such a small island (284 km²). Moreover, on the island, there are no autochthonous terrestrial small mammals such as rodents that are generally the principal prey of wild felids. Thus, it is likely that there are unique characteristics of the ecology of the cat as the top predator in the ecosystem. In the present study, I aim to clarify characteristics of the cat population inhabiting the most limited environment for wild felids.

The cat diet was examined regarding prey preference and seasonal pattern of each prey item by analysing their 947 scat contents collected from various environments throughout the island. I also investigated potential prey availability for the cat using a total length of 242 km of transects conducted over two years.

In the cat diet, 76 prey items were found, this being the most diversified diet in the cat family. Seasonal patterns of 19 principal prey items were examined, which was compared with those of prey availability. Both prey availability and predation varied between seasons. The cat seasonally shifted principal prey items in relation to prey availability. The seasonal patterns of prey availability were chiefly influenced by seasonal migration of birds, by temperature variation for reptiles, by reproductive cycles for insects and by temperature variation and reproductive cycles for amphibians, respectively.

From the above results, I considered that the broad range of the food niche of the Iriomote cat probably resulted from making the best possible use of fauna on the small subtropical island. Furthermore, the cat adapts to the islandwide environment to change principal prey items and feeding patterns in relation to temporal variations of prey availability.

6. Acknowledgment

This study could not have been carried out without the support from numerous people and institutions. I would especially like to acknowledge Profs. M. Izawa, M. Tsuchiya, A. Hagiwara, and Dr. N. Nakanishi, at University of the Ryukyus, and Prof. T. Doi at Nagasaki University, Dr. M. Okamura and Dr. N. Sakaguchi at Ministry of the Environment, Dr. A. Takahashi at the National Institute of Polar Research, for their valuable comments relating to this study. I also thank Mr. M. Dunn at the British Antarctic Survey for the language correction of the manuscript. I am also grateful to the Kumamoto Regional Forestry Office of Forestry Agency and the Iriomote Wildlife Center of Japan Ministry of the Environment (IWCC) for supporting this research. The scat censuses of the Iriomote cat were conducted as a part of endangered species conservation projects administered by the Kumamoto Regional Forestry Office. Field work was carried out using the facility of IWCC. I also thank the members of the Ecology laboratory, Faculty of Science, University of the Ryukyus. I also appreciate the kindness of people at IWCC, and many inhabitants of Iriomote Island, Ms. C. Matsumoto and Mr. M. Matsumoto in particular, during the field work. This study was partially funded by the Inui Memorial Trust for Research on Animal Science, Fujiwara Natural History Foundation, Pro Natura Fundation-Japan, and the 21st Century COE program of University of the Ryukyus.

7. References

Andersson, M. & Erlinge, S. (1977). Influence of predation on rodent populations. *Oikos*, Vol.29, No.3, pp.12-16, ISSN 0030-1299

Ayoade, J.O. (1983). *Introduction to climatology for the tropics*, John Wiley and Sons Ltd., ISBN 978-047-1104-07-0, Chichester, UK

Buckland, S.T.; Anderson, D.; Burnham, K; Laake, J.; Borchers, D.L. & Thomas, L. (2001). *Introduction to distance sampling: estimating abundance of biological populations*, Oxford University Press, ISBN 978-019-8509-27-1, New York, USA

Chanin, P.R.F. & Linn, I. (1980). The diet of the feral mink in southwest Britain. *Journal of Zoology*, Vol.192, No.2, (October 1980), pp.205-223, ISSN 0952-8369

Committee for Check-List of Japanese Birds (2000). *Check-list of Japanese birds 6th ed.*, The Ornithological Society of Japan, Hokkaido, Japan

Environment Agency & Forestry Agency (1998). *The report of patrol of the Iriomote cat, conservation project for rare species in 1996*, Okinawa District Forest Office, Okinawa, Japan (in Japanese)

Erlinge, S. (1969). Food habits of the otter *Lutra lutra* L. and the mink *Mustela vison* Schreber, in a trout water in Southern Sweden. *Oikos*, Vol.20, No.1, pp.1-7, ISSN 0030-1299

Fitzgerald, B.M. & Tuner, D.C. (2000). IV Predatory behaviour, In: *The domestic cat: the biology of its behaviour 2nd ed.*, D.C. Turner & P. Bateson (Ed.), 149–175, Cambridge University Press, ISBN 978-052-1636-48-3, Cambridge, UK

Grassman, L.I. (2000). Movements and diet of the leopard cat *Prionailurus bengalensis* in a seasonal evergreen forest in south-central Thailand. *Acta Theriologica*, Vol.45, pp.421-426, ISSN 0001-7051

Grassman, L.I.Jr.; Tewes, M.E.; Silvy, N.J.; & Kreetiyutanont, K. (2005). Spatial organization and diet of the leopard cat *Prionailurus bengalensis* in north-central Thailand. *Journal of Zoology*, Vol.266, No.1, pp.45–54, ISSN 0952-8369

Hansson, L. & Henttonen, H. (1985). Gradients in density variations of small rodents: the importance of latitude and snow cover. *Oecologia*, Vol.67, pp.394–402, ISSN 1432-1939

Hansson, L. & Henttonen, H. (1988). Rodent dynamics as community processes. *Trends in Ecology and Evolution*, Vol.3, No.8, (August 1988) pp.195–200, ISSN 0968-0012

Holling, C.S. (1959). Some characteristics of simple types of predation and parasitism. *Canadian Entomologist*, Vol.91, No.7, (July 1959), pp.385-398, ISSN 000-8347

Imaizumi, Y. (1967). A new genus and species of cat from Iriomote, Ryukyu Islands. *Journal of the Mammalian Society of Japan*, Vol.3, No.4, pp.74-107, ISSN 0914-1855

Imaizumi, Y.; Imaizumi, T. & Tyabata, T. (1977). *Study on the ecology and conservation of the Iriomote cat, 3rd report*. Environment Agency, Tokyo, Japan (in Japanese)

Inoue, T. (1972). The food habit of Tsushima leopard cat, *Felis bengalensis* sp., analyzed from their scats. *Journal of the Mammalian Society of Japan*, Vol.5, pp.155-169, ISSN 0914-1855 (in Japanese with English abstract)

IUCN. (2000) *2000 IUCN Red list of Threatened Species*. IUCN Publication Service Unit, Cambridge, UK

Izawa, M.; Sakaguchi, N. & Doi, T. (2000). The recent conservation program for the Iriomote cat, *Felis iriomotensis*. *Tropics*, Vol.10, pp.79-86, ISSN 0917-415

Jacobs, J. (1974). Quantitative measurement of food selection a modification of the forage ratio and Ivlev's electivity index. *Oecologia*, Vol.14, pp.413–417, ISSN 1432-1939

Johnson, W.E; Shinyashiki, F.; Raymond M.M.; Driscoll, C.; Leh, C.; Sunquist, M.; Johnston, L.; Bush, M.; Wildt, D.; Yuhki, N. & O'Brien, S.J. (1999). Molecular genetic characterization of two insular Asian cat species, Bornean bay cat and Iriomote cat, In: *Evolutionary theory and processes: modern perspectives*, S.P. Wasser, (Ed.), 223–248, Klewar Academic Publishers, ISBN 978-079-2355-18-2, Dordrecht, Netherland

Kirkpatrick, R. D. & Rauzon, M.J. (1986). Foods of feral cats, *Felis catus*, on Jarvis and Howland Islands, Central Pacific Ocean. *Biotropica,* Vol.18, No.1, (March 1986), pp.72-75, ISSN 1744-7429

Kleiman, D.C. & Eisenberg, J.F. (1975). Comparisons of canid and felids social systems from an evolutionary perspective. *Animal Behaviour,* Vol.21, No.4, (November 1973), pp.637-659, ISSN 0003-3472

Krebs, C.J. & Myers, J.H. (1974). Population cycles in small mammals. *Advances in Ecological Research,* Vol.8, pp.267-399, ISSN 0065-2504

Krebs, C.J. (1996). Population cycles revisited. . *Journal of Mammalogy,* Vol.77, No.1, (February 1996), pp.8-24, ISSN 0022-2372

Kruuk, H. (1982). Interactions between Felidae and their prey species: a review. In: *Cats of the world: Biology, conservation, and management,* S.D. Miller & D.D. Everett, (Ed.), 353–374, National Wildlife Federation, ISBN 978-0912-186-79-5,Washington D.C., USA

Leyhausen, P. & Pfleiderer, M. (1999). The systematic status of the Iriomote cat (*Prionailurus iriomotensis* Imaizumi 1967) and the subspecies of the leopard cat (*Prionailurus bengalensis* Kerr 1792). *Journal of Zoological Systematics and Evolutionary Research,* Vol.37, No.3, (September 1999), pp.121-131, ISSN 1439-0469

Lodé, T. (1996). Polecat predation on frogs and toads at breeding sites in western France. *Ethology Ecology and Evolution l,* Vol.8, pp.115-124, ISSN 0394-9370

Lodé, T. (2000). Functional response and area-restricted search of a predator: seasonal exploitation of anurans by European polecat *Mustela putorius. Austral Ecology,* Vol.25, pp.223-231, ISSN 1442-9985

Maeda, N. & Matsui, M. (1999). *Frogs and toads of Japan, revised edition,* Bun-Ichi Sogo Shuppan Co. Ltd, ISBN 978-482-9921-30-2, Tokyo, Japan

Masuda, R.; Yoshida, M.C.; Shinyashiki, F. & Bando, G. 1994. Molecular phylogenetic status of the Iriomote cat *Felis iriomotensis*, inferred from mitochondrial DNA sequence analysis. *Zoological Science,* Vol.11, pp.597-604, ISSN 0289-0003

Miyawaki, A. (1989). *Vegetation of Japan vol.10 Okinawa and Ogasawara,* Shibundo Co. Ltd., ISBN 478-4300-40-6, Tokyo, Japan (in Japanese).

Nakanishi, N.; Okamura, M.; Watanabe, S.; Izawa, M. & Doi, T. (2005). The effect of habitat on home range size in the Iriomote cat *Prionailurus bengalensis iriomotensis. Mammal Study,* Vol.30, No.1, pp.1-10, ISSN 1343-4152

Numata, M. (1974). *The flora and vegetation of Japan.* Kodansha Elsevier Scientific Pub. Co., ISBN 978-044-4998-80-4, Tokyo, Japan

Okamura, M.; Doi, T.; Sakaguchi, N. & Izawa, M. (2000). Annual reproductive cycle of the Iriomote cat *Felis iriomotensis. Mammal Study,* Vo.25, pp.75–85, ISSN 1343-4152

Okinawa Prefectural Education. (1985). *The report on Natural treasures of Iriomote Island Iriomote Island 3 (Animal).* Okinawa Prefectural Education, Okinawa, Japan (in Japanese)

Rabinowitz, A. (1990). Notes on the behavior and movements of leopard cats, *Felis bengalensis*, in a dry tropical forest mosaic in Thailand. *Biotropica*, Vol.22, No.4, (December 1990), pp.397-403, ISSN 1744-7429

Sakaguchi, N. (1991). Habitat of the Iriomote cat and changes in its response to prey availability, *Proceedings of the International Symposium on Wildlife Conservation in Tsukuba and Yokohama*, pp. 137-140, ISBN 491-5959-05-8, Tokyo, Japan, August 21-25, 1990

Sakaguchi, N. & Ono, Y. (1994). Seasonal change in the food habits of the Iriomote cat *Felis iriomotensis*. *Ecological Research*, Vol.9, No.2, pp.167-174, ISSN 0912-3814

Schmidt, K.; Nakanishi, N.; Okamura, M.; Doi, T. & Izawa, M. 2003. Movements and use of home range in the Iriomote cat (*Prionailurus bengalensis iriomotensis*). *Journal of Zoology*, Vol.261, No.3, (November 2003), pp.273-283, ISSN 0952-8369

Schmidt, K.; Nakanishi, N.; Izawa, M.; Okamura, M.; Watanabe, S.; Tanaka, S. & Doi, T. (2009). The reproductive tactics and activity patterns of solitary carnivores: the Iriomote cat. *Journal of Ethology*, Vol.27, No.1, (January 2009), pp.165-174, ISSN 0289-0771

Sokal, R.R. & Rohlf, F.J. (1995). *Biometry : the principles and practice of statistics in biological research 3rd ed.*, Freeman, ISBN 071-6724-11-1, New York, USA

Sunquist, M. & Sunquist, F. (2002). *Wild cats of the World*. The University of Chicago Press, ISBN 978-022-6779-99-7, Chicago and London.

Takano, S. (1982). *A field guide to the birds of Japan*. Wild Bird Society of Japan, ISBN 978-477-0012-46-3, Tokyo, Japan (in Japanese)

Watanabe, S. (2009). Factors affecting the distribution of the leopard cat *Prionailurus bengalensis* on East Asian islands. *Mammal Study*, Vol.34, No.4, (December 2009), pp.201-207, ISSN 1343-4152

Watanabe, S.; Nakanishi, N. & Izawa, M. (2003). Habitat and prey resource overlap between the Iriomote cat *Prionailurus iriomotensis* and introduced feral cat *Felis catus* based on assessment of scat content and distribution. *Mammal Study*, Vol.28, No.1, pp.47-56, ISSN 1343-4152

Watanabe, S. & Izawa, M. (2005). Species composition and size structure of frogs preyed by the Iriomote cat *Prionailurus bengalensis iriomotensis*. Mammal Study, Vol. 30, No.2, pp.151-155, ISSN 0266-4674

Watanabe, S.; Nakanishi, N.; Sakagichi, N.; Doi, T. & Izawa, M. (2002). Temporal changes of land covers analyzed by satellite remote sensing and habitat states of the Iriomote cat *Prionailurus iriomotensis*. *Bulletin of the International Association for Landscape Ecology-Japan*, Vol.7, No.2, (December 2002), pp.25-34, ISSN 1880-0092 (in Japanese with English abstract).

Watanabe, S.; Nakanishi N. & Izawa, M. (2005). Seasonal abundance of the floor-dwelling frog fauna on Iriomote Island of the Ryukyu Archipelago, Japan. *Journal of Tropical Ecology*, Vol.21, No.1, pp.85-91, ISSN 0266-4674

Webb, J.B. (1975). Food of the otter (*Lutra lutra* L.) on the Somerset Levels. *Journal of Zoology*, Vol.177, No.4, (December 1975), pp.486-491, ISSN 0952-8369

Wise, M.H.; Linn, I. & Kennedy, C.R. (1981). A comparison of the feeding biology of mink, *Mustela vison*, and otter, *Lutra lutra*. *Journal of Zoology*, Vol.195, No.2, (October 1981), pp.181-213, ISSN 0952-8369